PEARSON BACCALAUREATE

Environmental Systems and Societies

2nd Edition

ANDREW DAVIS • GARRETT NAGLE

Supporting every learner across the IB continuum

Published by Pearson Education Limited, 80 Strand, London, WC2R 0RL.

www.pearsonglobalschools.com

Edited by Penelope Lyons
Proofread by Judith Shaw and Alison Nick
Designed by Astwood Design
Typeset by Phoenix Photosetting, Chatham, Kent

Illustrated by Phoenix Photosetting
Cover design by Pearson Education Limited

First published 2015

19 18 17 16 15
IMP 10 9 8 7 6 5 4 3 2 1

British Library Cataloguing in Publication Data
A catalogue record for this book is available from the British Library

ISBN 978 1 447 99042 0
eBook only ISBN 978 1 447 99043 7

Printed in Slovakia by Neografia

Dedications
Andrew Davis – For my mother, and in memory of my father, who is remembered here.

Garrett Nagle – For Angela, Rosie, Patrick and Bethany.

Acknowledgements
The authors would like to thank Harriet Power for coordinating production of this book and keeping it on schedule. Penelope Lyons edited both 1st and 2nd editions with tremendous attention to detail – her contribution is greatly valued by us both. Our especial thanks to Stephen Marshall, who reviewed the second edition – his perceptive and constructive comments have been invaluable and materially improved the text.

The publisher would like to thank the following for their kind permission to reproduce their photographs:
(Key: b-bottom; c-centre; l-left; r-right; t-top)

123RF.com: Alexander Petrenko 119br, Alexey Kokoulin 51, andesign101 458t, Cathy Kovarik 319tr, dirk ercken 60, Fernando Grergory 173tr, Laurin Rinder 458b, Mauro Rodrigues 64t, Philip Bird 107t, Sean Pavone 220, Sergii Koval 244tl, Steven Prorak 319cr; **Alamy Images:** Bill Attwell 209, Chris Hellier 189, Design Pics Inc 66t, FLPA 132bl, Jenny Matthews 228, Martin Bond 6c, Martin Shields 452br, Matthew Hart 179t, Mint Images Limited 153, Nando Machado 158b, Peter Adams Photography Ltd 398, Photos12 7bl, Photoshot Holdings Ltd 178, RIA Novosti 5bl, Richard Ellis 338b, Robert Gilhooly 238, Victor de Schwanberg 52; **Andrew J Davis:** 105b, 113b, 115, 129b, 141t, 141b, 154, 159t, 162tr, 173c, 175, 197, 198bl, 198br, 206, 447r, 448tl, 448bl, 448br; **Bridgeman Art Library Ltd:** Canis Antarcticus (b/w photo), English School / Private Collection 181, Three flint tools (stone), Palaeolithic / Musee des Antiquites Nationales, St. Germain-en-Laye, France 420; **Corbis:** Aas, Erlend / epa 12b, Any Rain / epa 156r, Bettmann 6bl, Charles & Josette Lenars 190, DLILLC 160c, Elmer Frederick Fischer 185, Enzo @ Paolo Ragazzani 452tr, Frans Lanting 179c, G. Bowater 174, Galen Rowell 165b, Hubert Stadler 178r, Hupton-Deutsch Collection 156l, Image Source 183, 184b, James Marshall 245, Kevin Schafer 203t, Made Nagi / epa 7tr, Marcelo Del Pozo / Reuters 182, Matthias Breiter / National Geographic Creative 113t, Michael & Patricia Fogden 184t, Michael S. Yamashita 162tl, Natural Selection David Ponton / Design Pics 160b, Reuters 5br, Richard Hamilton Smith 37t, Rungroj Yongrit / epa 12t, Stephanie Maze 122tr, Tom Bean 37b, Tui De Roy / Minden Pictures 158tl, Visuals Unlimited 179b; **Creatas:** 454tl; **Digital Stock:** 108b; **Digital Vision:** 293, 453t, 453c, 454tc, 455; **Dreamstime.com:** 447tr, DVmsimages 447bl; **Fisheries and Oceans Canada:** 258t, E. Debruyn 258b; **FLPA Images of Nature:** Desmond Dugan 122tl, Mark Moffett / Minden Pictures 135, Wayne Hutchinson 122c; **Fotolia.com:** 7activestudio 67cr, 180t, Alphacandy 85, Alta Oosthuizen 70, alvaher 67tl, avkost 126 (eagle), birdiegal 203b, borisoff 76b, Christian Delbert 358, claffra 353r, davemhuntphoto 148, designua 126 (amoeba), dinar12 126 (fruit tree), dmussman 63t, Don Perucho 77, drimafilm 227t, EcoView 200, 216r, emeraldphoto 152, Eric Isselee 126 (shark), finecki 376, Helen Hotson 172, hramovnick 76t, Kitch Bain 62tl, Lars Johansson 126 (pine tree), Les Cunliffe 149, Mopic 47, Morenovel 2, nicolasprimola 180c, 180b, nito 310, norrie39 216l, Oleg Znamenskiy 75, Perseomedusa 151, peshkova 224bl, phb.me 268, Rawpixel 126 (car), Richard Griffin 126 (flower), Sebastien Burel 186, smp928s 68br, Tamara Kulikova 301, thatreec 67cl, TristanBM 126 (horse), Vasily Merkushev 126 (stone), Vera Kuttelvaserova 126 (beetle), Vitalii Hulai 126 (mouse), volff 126 (spoon); **Garrett E Nagle:** 50, 68bl, 119bl, 130t, 131, 136, 215, 218, 230b, 250, 255l, 255r, 261, 272, 280, 287, 300, 304, 337, 345, 346, 359, 366, 389, 403, 422, 434, 435, 436t, 436b, 438t, 457t, 460; **Getty Images:** Chlaus Lotscher 454r, Christer Fredriksson 204, Deborah Harrison 444, Dimitar Dilkoff 210, dragen 281, Emory Kristof 246, jian wan 446tl, Jim Xu 424, Joel Sartore 453b, Kelvin Yam 199, Ken Lucas 452l, Lintao Zhang 337tr, Mike Kemp 459, Rob Broek 350, View Pictures 367; **Illustrated London News Picture Library:** Ingram Publishing / Alamy 360; **Magnum Photos Ltd:** Chris Steele-Perkins 5tr; **PaleoScene:** Glen J Cuban 450br, Glen J Kuban 450bl; **Pearson Education Ltd:** Jules Selmes 456t; **PhotoDisc:** 454bl, Edmund van Hoorick 109, Joseph Green / Life File 107b, Photolink 288; **Photos.com:** Jupiterimages 43; **Photoshot Holdings Limited:** Imago 158tr; **Science Photo Library Ltd:** Alex Hyde 119t, British Antarctic Survey 33, David M. Schleser / Nature's Images 21, David Parker 129t, David Taylor 46, Digital Globe 6tl, Dr Jeremy Burgess 98bl, Georgette Douwma 82t, Kjell B. Sandved 67bl, NASA 4, NSIL / Dick Roberts, Visuals Unlimited 270, P Rona / OAR / National Undersea Research Program / NOAA 111tr, P.G.Adam, Publiphoto Diffusion 23, Patrick Landmann 132br, Paul & Paveena McKenzie, Visuals Limited 264, Sinclair Stammers 20, 67tr, US Geological Survey 144, Woods Hole Oceanographic Institution, Visuals Unlimited 79b; **Shutterstock.com:** Paolo Bona 286, Philip Lange 225, Stoonn 229; **Thames21:** 423br; **The Kobal Collection:** Lawrence Bender Prods. 379; **UNOG Library, League of Nations Archives:** Michos Tzovaras 8; **www.CartoonStock.com:** 444bl

Cover images: *Front:* **Corbis:** Marc Dozier

We are grateful to the following for permission to reproduce copyright material:

Figures

Figure 1.2 "The pattern of environmentalist ideologies", adapted from *Environmentalism, 2nd edition* by Timothy O'Riordan, Figure 10.1, page 376, copyright © 1981, Pion Ltd, London, www.pion.co.uk and www.envplan.com; Figure 4.19 MSC Ecolabel and text, www.msc.org. Reproduced with permission of Marine Stewardship Council; Figure 6.15 from *The Montreal Protocol on Substances that Deplete the Ozone Layer, The Achievements in Stratospheric Ozone Protection, Progress Report 1987-2012,* pages 20-21, http://ozone.unep.org/new_site/en/Information/Information_Kit/UNEP-MP_Achievements_in_Stratospheric_Oz.pdf, United Nations Environment Programme; and Figure 8.16 'The Great Pacific Garbage patch, copyright 5W Infographics, http://www.5wgraphics.com, Reproduced with permission.

Image

Book cover on p.380, *The Skeptical Environmentalist* by Bjorn Lomborg, Cambridge University Press, 2001. Reproduced with permission from Cambridge University Press.

Tables

Tables 7.4, 7.5 'Causes of Impacts, vulnerable areas and impacted sectors' and ' Intensity of impacts on different sectors due to Climate Change', pp.34,35, National Adaptation Programme of Action (NAPA), Updated Version, June 2009, http://unfccc.int/resource/docs/napa/ban02.pdf, Source: United Nations.

Text

Quote on p.445 by George E.P. Box from "Science and Statistics", *Journal of the American Statistical Association,* Vol 71, No.356, pp.791–79, December 1976, Taylor & Francis, US; Quote on p.446 by Richard Dawkins from *The Extended Phenotype: The Gene as the Selection,* Oxford University Press, 1999. p.113. Reproduced with permission of the author; Quote on p.447 by Rachel Carson from *The Sense of Wonder,* HaperCollins, 1998, p.56, copyright © 1956 by Rachel Carson. Reproduced with permission from Pollinger Limited (www.pollingerltd.com) on behalf of the Estate of Rachel Carson and Francis Collin, Trustee; Quote on p.449 by Buckminster Fuller from *Operating Manual for Spaceship Earth,* Lars Muller Publishers, 2008, p.60. Reproduced by permission of the Estate of R. Buckminster Fuller; Quote on p.454 by Rachel Carson from *Guarding Our Wildlife Refuges, Conservation in Action* #5. US Fish and Wildlife Service, 1948. http://www.fws.gov/CNO/docs/ConservationTransition_final.pdf, p.4. Reproduced by permission of US Fish & Wildlife Service; and Extract on p.461 from *Agricultural issues in the UK* by Garret Nagel, GeoActive, 160, Nelson Thornes, 1997. Reproduced by permission of the publishers Oxford University Press.

Every effort has been made to contact copyright holders of material reproduced in this book. In some instances we have been unable to trace the owners of copyright material, and we would appreciate any information that would enable us to do so. Any omissions will be rectified in subsequent printings if notice is given to the publishers.

The following material has been reproduced from IB documents and past examination papers: the Significant Ideas, Knowledge and Understanding, selected Applications and Skills, Guidance, International-Mindedness, and Theory of Knowledge from the sub-topics of the IB ESS Guide; selected past exam questions and corresponding mark schemes; the Assessment Objectives, Aims, Internal Assessment criteria, Big Questions, Command Terms, Paper 2 markbands, Topic and Sub-topic headers, and selected figures and tables from the IB ESS Guide.

Our thanks go to the International Baccalaureate for permission to reproduce its intellectual copyright.

There are links to relevant websites in this book. In order to ensure that the links remain up to date we have made them available on our website at www.pearsonhotlinks.co.uk. Search for this title or the ISBN 9781447990420.

Contents

Introduction

Welcome to Environmental systems and societies (ESS). We hope you enjoy the course. This book is designed to be a comprehensive coursebook, covering all aspects of the syllabus. It will help you prepare for your examinations in a thorough and methodical way as it follows the syllabus outline section by section, explaining and expanding on the material in the course guide. Each chapter deals with one topic from the syllabus and, within each chapter, each subtopic is named and numbered following the ESS guide. This makes the book readily accessible for use and reference throughout the course. There are also short chapters offering advice on completing the Internal Assessment (IA), writing the Extended Essay (EE), and developing examination strategies. In addition, there is an appendix in the ebook covering basic statistics and data analysis. Links between different parts of the syllabus are emphasized, and key facts essential to your understanding are highlighted throughout. At the end of each section, you will find practice questions to test your knowledge and understanding of that part of the course. You can self-assess your answers using the mark-schemes that can be found in the ebook.

The nature of Environmental systems and societies

The course recognizes that to understand the environmental issues of the 21st century, and suggest suitable management solutions, both the human and environmental aspects must be studied. The issues you will cover are complex, and include the actions required for the fair and sustainable use of shared resources.

The ESS course is the first fully transdisciplinary course within the IB. It covers group 3 (individuals and societies) and group 4 (experimental sciences). It is a multifaceted course that will require you to develop a diverse set of skills, enabling you to explore the cultural, economic, ethical, political, and social interactions of societies with the environment. As a group 4 subject, it demands the scientific rigour expected of an experimental science, and has a large practical component. The group 3 approach balances a scientific approach with a human-centred perspective which examines environmental issues from a social and cultural viewpoint. Throughout the book you will look at the environment from the perspective of human societies and assess their response in light of the scientific framework used in environmental sciences. As a result of studying this course you will become equipped with the ability to recognize and evaluate the impact of societies on the natural world. The book therefore looks at environmental issues from economic, historical, cultural, and socio-political viewpoints as well as a scientific one, to provide a holistic perspective.

The aims of the course and this book are to enable you to:

- acquire the knowledge and understanding of environmental systems at a variety of scales
- apply the knowledge, methodologies and skills to analyse environmental systems and issues at a variety of scales
- appreciate the dynamic interconnectedness between environmental systems and societies
- value the combination of personal, local and global perspectives in making informed decisions and taking responsible actions on environmental issues
- be critically aware that resources are finite, and that these could be inequitably distributed and exploited, and that management of these inequities is the key to sustainability
- develop awareness of the diversity of environmental value systems
- develop critical awareness that environmental problems are caused and solved by decisions made by individuals and societies that are based on different areas of knowledge
- engage with the controversies that surround a variety of environmental issues
- create innovative solutions to environmental issues by engaging actively in local and global contexts.

Approaches to learning

Approaches to teaching and learning (ATL) reflects the IB learner profile attributes, and is designed to enhance your learning and assist preparation for IAs and examinations. ATL runs throughout the IB Middle Years Programme (MYP) and Diploma Programme (DP), and encourages you to think of common skills that are necessary to all subjects. The variety of skills covered will equip you to continue to be actively engaged in learning after you leave your school or college.

There are five categories of ATL skills: thinking skills, communication skills, social skills, self-management skills and research skills. These skills encompass the key values that underpin an IB education.

ATL is addressed in the challenge yourself boxes and worksheets available online (see below). Exercises that test knowledge at the end of each subsection specifically address ATL thinking skills.

Information boxes

Throughout the book you will see a number of coloured boxes interspersed throughout each chapter, as well as information boxes at the start of each subtopic. Boxes may be in the margins or in the main text area. Each type provides different information and stimulus, as you can see in the following examples.

Significant ideas are listed at the beginning of each section: these are the overarching principles that define and encapsulate the learning within each subtopic.

Significant ideas

Historical events, among other influences, affect the development of environmental values systems and environmental movements.

There is a wide spectrum of environmental value systems each with their own premises and implications.

Big questions are designed to help you appreciate the overarching principles that are essential to the course, and to encourage you to revisit these central ideas in different contexts.

Big questions

As you read this section, consider the following big questions:

- What strengths and weaknesses of the systems approach and the use of models have been revealed through this topic?
- How are the issues addressed in this topic of relevance to sustainability or sustainable development?

Knowledge and understanding lists the main ideas for the section you are about to read, and sets out the content and aspects of learning to be covered.

Knowledge and understanding

- Biodiversity is a broad concept encompassing total diversity which includes diversity of species, habitat diversity and genetic diversity.
- Species diversity in communities is a product of two variables, the number of species (richness) and their relative proportions (evenness).
- Communities can be described and compared by the use of diversity indices. When comparing communities that are similar, then low diversity could be evidence of pollution, eutrophication or recent colonization of a site. The number of species present in an area is often used to indicate general patterns of biodiversity.

The green **key fact** boxes highlight the key information in the section you are reading. This makes them easily identifiable for quick reference. The boxes also enable you to identify the core learning points within a section.

Ocean circulation systems are driven by differences in temperature and salinity that affect water density. The resulting difference in water density drives the ocean conveyor belt which distributes heat around the world, so affecting climate.

Hotlink boxes direct you to the publisher's website, which in turn will take you to the relevant website(s). On the web pages there, you will find additional information to support the topic (e.g. video simulations, background reading and the like).

To learn more about the importance of pollinators, go to www. pearsonhotlinks.co.uk, enter the book title or ISBN, and click on weblink 5.3.

Hints for success provide insights into what you need to know or how to answer a question in order to achieve the highest marks in an examination. They also identify common pitfalls when answering such questions and suggest approaches that examiners like to see. These boxes highlight the applications and skills you are expected to have covered, so you know what you need to revise for exams.

You should be able to discuss the role of the albedo effect from clouds in regulating global average temperature.

In addition to the Theory of Knowledge chapter, there are **ToK** boxes like this throughout the book. These boxes are there to stimulate thought and consideration of any ToK issues as they arise and in context. Often, they will just contain a question to stimulate your own thoughts and discussion.

The poor are more vulnerable to global warming than the rich. However, on average, people in rich countries produce a larger amount of greenhouse gases per person than people in poor countries. Is this morally just?

The **interesting fact** boxes contain interesting information which will add to your wider knowledge but which does not fit within the main body of the text.

Up to 60 per cent of household waste in the USA is recyclable or compostable. But Americans compost only 8 per cent of their waste. Surveys suggest that the main reason Americans don't compost is because they think it is a complicated process. In contrast, the Zabbaleen who are responsible for much of the waste collection in Cairo, Egypt, recycle as much as 80 per cent of the waste collected.

International-mindedness is important to the International Baccalaureate. These boxes indicate examples of internationalism within the area of study. The information given offers you the chance to think about how Environmental systems and societies fits into the global landscape.

Popular books such as *Silent Spring*, and films such as Al Gore's *An Inconvenient Truth*, can provide knowledge about environmental issues on a global scale. People who previously had limited understanding of the environment are enabled to make up their own minds about global issues. But do they have enough information to see all sides of the argument? A good education would certainly put these arguments in a wider context. Is it a problem that many people received only one side of the argument?

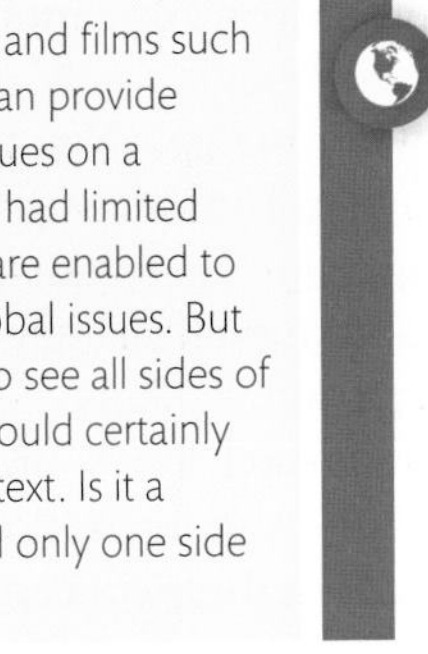

Case studies are self-contained examples that you can use to answer questions on specific points. They are usually longer than this example and often contain photographs or other illustrations.

Case study

Deforestation in Borneo

Deforestation in Borneo has progressed rapidly in recent years (Figure 2.53). It affects people, animals and the environment. A recent assessment by the United Nations Environment Program (UNEP) predicts that the Bornean orang-utan (endemic to the island) will be extinct in the wild by 2025 if current trends continue. Rapid forest loss and degradation threaten many other species, including the Sumatran rhinoceros and clouded leopard. The main cause of forest loss in Borneo is logging and the clearance of land for oil palm plantations.

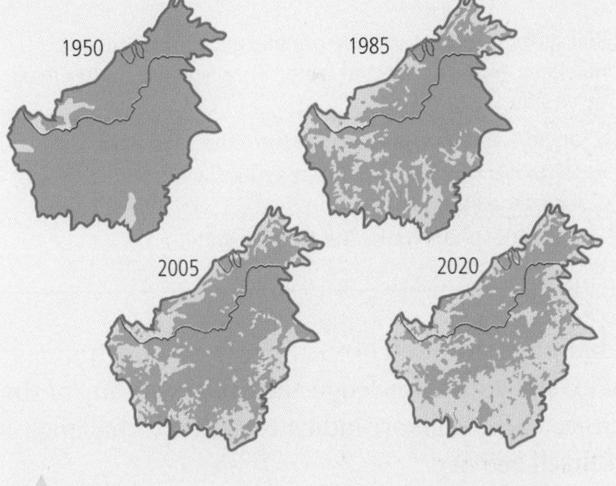

Figure 2.53 Loss of primary forest between 1950 and 2005, and a projection for forest cover in 2020 based on current trends

The **central concepts** of the ESS course include sustainability, equilibrium, strategy, biodiversity, and environmental value systems (EVSs – discussed in detail in Topic 1, pages 11–17). These concepts are highlighted in colour-coded textboxes at appropriate places in each chapter, to put these central ideas into context. Many of the issues encountered in the course, such as resource management, conservation, pollution, globalization, and energy security, are linked to these concepts.

CONCEPTS: Biodiversity

Mass extinctions have led to initial massive reductions in the Earth's biodiversity. These extinction events have resulted in new directions in evolution and therefore increased biodiversity in the long term.

The **systems approach** (see page 19) is central to the course: it helps you to understand complex and dynamic ecosystems, allows you to make connections with other subjects, and enables you to integrate new ideas into what you know already. The systems approach boxes alert you to key systems ideas at appropriate parts of the book.

SYSTEMS APPROACH

Soil system storages include organic matter, organisms, nutrients, minerals, air, and water. Thus, soil has matter in all three states:

- organic and inorganic matter form the solid state
- soil water (from precipitation, groundwater, and seepage) forms the liquid state
- soil atmosphere forms the gaseous state.

Challenge yourself boxes offer you opportunities to extend your knowledge and understanding of the subject. ATL skills are indicated for each challenge yourself activity.

CHALLENGE YOURSELF

Thinking skills 

To what extent can weather and climate be considered a natural system if human activities are affecting it so much?

Exercise questions are found at the end of each subtopic. They allow you to apply your knowledge and test your understanding of what you have just been reading. All of the answers to these questions can be found by reading the preceding text.

Exercises

1. Compare the areas that are predicted to have water stress in 2025 (Figure 4.8) with those that experienced water stress in 1999.
2. Explain why poor people often pay more for their water than rich people.
3. Evaluate the use of drip irrigation.

Practice questions are found at the end of each topic. They are mostly taken from previous years' IB examination papers. The answers to these questions can be found in the eBook.

Practice questions

1 a Discuss the potential ecological services and goods provided by a named ecosystem. [6]
  b Outline how culture, economics and technology have influenced the value of a named resource in different regions or in different historical periods. [6]

2 Evaluate landfill and incineration as strategies for the disposal of solid domestic waste. [6]

3 Compare and contrast the ecological footprint of the MEDCs with that of LEDCs. [6]

The ebook

In the ebook you will find the following features.

- Worksheets to accompany each chapter. These might contain extension exercises, suggestions for Internal Assessment (IA), revision, or other sources of information. Each worksheet has a specific ATL focus.
- Answers to all of the practice questions in the book.
- Interactive quizzes, which can be used to help test your understanding and practise answering exam-style essay questions.
- Animations that bring to life some of the more complex processes in the book.

- Larger versions of some of the photos and figures in the book, so the images can be studied in more detail.
- A glossary that includes definitions of all of the words printed in bold in the book.

Big questions

The big questions, listed at the start of each subtopic in this book, provide a focus for exploring the central ESS concepts in a variety of ways as the course progresses. At the end of each subtopic, big questions are used to review and highlight the central principles covered in the chapter. As well as being used to introduce and review topics, big questions can also be used as the basis for classroom discussions and assignments, and be used for revision exercises at the end of the course.

The six big questions are listed below. Their purpose is to get you to think in a holistic way about the course, and to consider the relationship between human societies and natural systems. They have been designed to help you appreciate the overarching principles that are essential to the course, and to encourage you to revisit these central ideas in different contexts.

A Which strengths and weaknesses of the systems approach and of the use of models have been revealed through this topic?

B To what extent have the solutions emerging from this topic been directed at preventing environmental impacts, limiting the extent of the environmental impacts or restoring systems in which environmental impacts have already occurred?

C What value systems are at play in the causes and approaches to resolving the issues addressed in this topic?

D How does your personal value system compare with the others you have encountered in the context of issues raised in this topic?

E How are the issues addressed in this topic relevant to sustainability or sustainable development?

F In which ways might the solutions explored in this topic alter your predictions for the state of human societies and the biosphere decades from now?

The big questions are not part of the required syllabus content, although they do identify an approach that will be reflected in more open-ended questions in examinations. The following table indicates which big questions have particular relevance to each chapter in this book.

Big question	Possible relevant chapters/topics
A	1, 2, 4, 5, 7, 8
B	3, 4, 5, 6, 7, 8
C	1, 3, 7, 8
D	1, 3, 7, 8
E	1, 2, 3, 4, 5, 6, 7, 8
F	3, 4, 5, 6, 7, 8

Approaches to Environmental systems and societies

Systems approach

The course requires a systems approach to environmental understanding and problem solving, and so the idea of a systems approach is a concept that is central to the course. The approach is explained in detail in Topic 1, and is used throughout the book. Science often uses a reductionist approach to examine phenomena, breaking a system down into its components and studying them separately. Environmental science cannot work in this way, as understanding the functioning of the whole topic (e.g. an ecosystem) is essential (i.e. a holistic approach is needed). The traditional reductionist approach of science inevitably tends to overlook or understate this important holistic quality. Furthermore, the systems approach is common to many disciplines (e.g. economics, geography, politics, and ecology). It emphasizes the similarities between all these disciplines in the ways in which matter, energy, and information flow, allowing common terminology to be used when discussing different systems and disciplines. This approach, therefore, integrates the perspectives of different subjects. Throughout this book, the integrated nature of this subject is stressed by examining the links between different areas of the syllabus and between different disciplines.

Sustainability

Sustainability refers to the use of natural resources in ways that do not reduce or degrade them so that they are available for future generations. This is central to an understanding of the nature of interactions between environmental systems and societies. Throughout the book, we look at resource management issues and show that these are essentially ones of sustainability.

Local and global approaches

Inevitably, appreciation of your local environment will enable you to appreciate these issues from a local perspective, through carrying out field-work in nearby ecosystems and research on local issues. Certain issues such as resource and pollution management require a national or regional perspective, and others an international perspective (e.g. global warming). This book explores all these perspectives in detail. On a broader scale, the course naturally leads us to an appreciation of the nature of the international dimension, since the resolution of the major environmental issues rests heavily on international relationships and agreements – case studies and key facts are used to illustrate these points throughout the book.

Holistic evaluation and human impact

The systems approach and the interaction between environmental systems and societies, encourages a holistic appreciation of the complexities of environmental issues. This course requires you to consider the costs and the benefits of human activities, both to the environment and to societies, over the short and long term, and on a local and global scale. In doing so, you will arrive at informed personal viewpoints. This book explains how you can justify your own position, and appreciate the views of others, along a continuum of EVSs.

Now you are ready to start. Good luck with your studies.

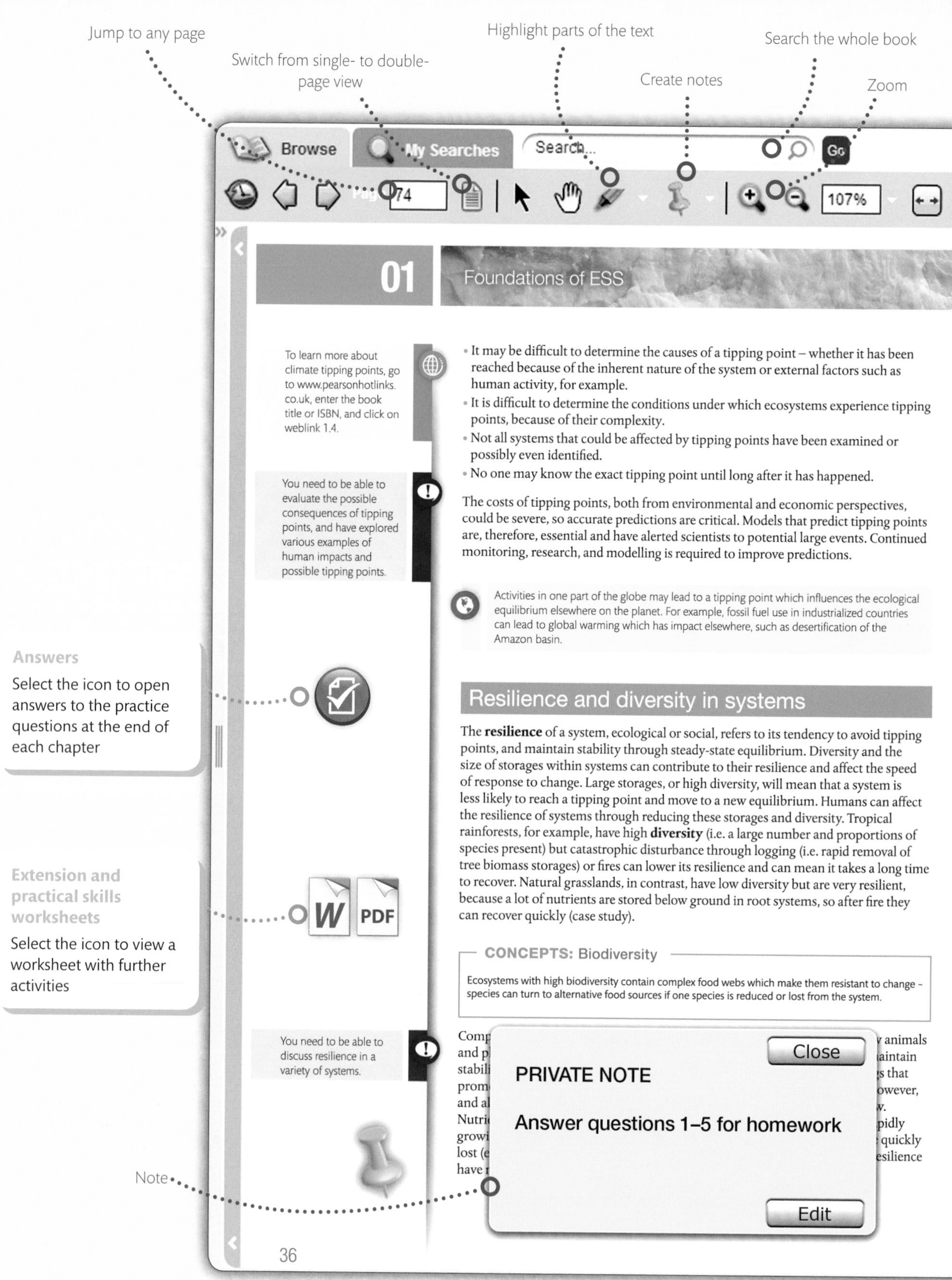
Jump to any page
Switch from single- to double-page view
Highlight parts of the text
Create notes
Search the whole book
Zoom
Browse
My Searches
Search...
Go
74
107%
01
Foundations of ESS
To learn more about climate tipping points, go to www.pearsonhotlinks.co.uk, enter the book title or ISBN, and click on weblink 1.4.
• It may be difficult to determine the causes of a tipping point – whether it has been reached because of the inherent nature of the system or external factors such as human activity, for example.
• It is difficult to determine the conditions under which ecosystems experience tipping points, because of their complexity.
• Not all systems that could be affected by tipping points have been examined or possibly even identified.
• No one may know the exact tipping point until long after it has happened.
You need to be able to evaluate the possible consequences of tipping points, and have explored various examples of human impacts and possible tipping points.
The costs of tipping points, both from environmental and economic perspectives, could be severe, so accurate predictions are critical. Models that predict tipping points are, therefore, essential and have alerted scientists to potential large events. Continued monitoring, research, and modelling is required to improve predictions.
Activities in one part of the globe may lead to a tipping point which influences the ecological equilibrium elsewhere on the planet. For example, fossil fuel use in industrialized countries can lead to global warming which has impact elsewhere, such as desertification of the Amazon basin.
Answers
Select the icon to open answers to the practice questions at the end of each chapter
Resilience and diversity in systems
The resilience of a system, ecological or social, refers to its tendency to avoid tipping points, and maintain stability through steady-state equilibrium. Diversity and the size of storages within systems can contribute to their resilience and affect the speed of response to change. Large storages, or high diversity, will mean that a system is less likely to reach a tipping point and move to a new equilibrium. Humans can affect the resilience of systems through reducing these storages and diversity. Tropical rainforests, for example, have high diversity (i.e. a large number and proportions of species present) but catastrophic disturbance through logging (i.e. rapid removal of tree biomass storages) or fires can lower its resilience and can mean it takes a long time to recover. Natural grasslands, in contrast, have low diversity but are very resilient, because a lot of nutrients are stored below ground in root systems, so after fire they can recover quickly (case study).
Extension and practical skills worksheets
Select the icon to view a worksheet with further activities
W
PDF
CONCEPTS: Biodiversity
Ecosystems with high biodiversity contain complex food webs which make them resistant to change - species can turn to alternative food sources if one species is reduced or lost from the system.
You need to be able to discuss resilience in a variety of systems.
PRIVATE NOTE
Close
Answer questions 1–5 for homework
Edit
Note
36

 This is an approximation of what your eBook will look like and not an exact reproduction

See the definitions of key terms in the glossary

ate a bookmark

Switch to whiteboard view

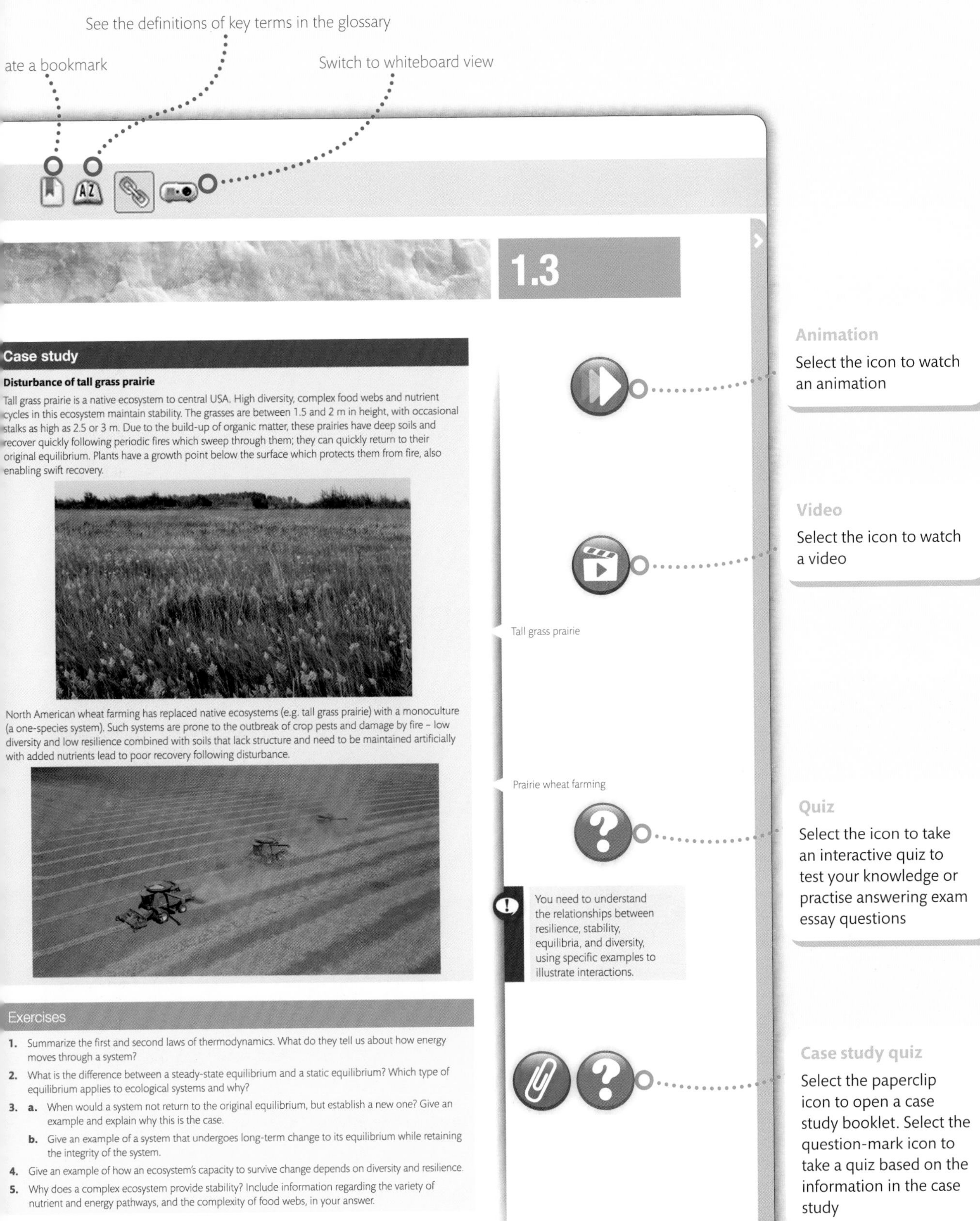
1.3
Case study
Disturbance of tall grass prairie
Tall grass prairie is a native ecosystem to central USA. High diversity, complex food webs and nutrient cycles in this ecosystem maintain stability. The grasses are between 1.5 and 2 m in height, with occasional stalks as high as 2.5 or 3 m. Due to the build-up of organic matter, these prairies have deep soils and recover quickly following periodic fires which sweep through them; they can quickly return to their original equilibrium. Plants have a growth point below the surface which protects them from fire, also enabling swift recovery.
Tall grass prairie
North American wheat farming has replaced native ecosystems (e.g. tall grass prairie) with a monoculture (a one-species system). Such systems are prone to the outbreak of crop pests and damage by fire – low diversity and low resilience combined with soils that lack structure and need to be maintained artificially with added nutrients lead to poor recovery following disturbance.
Prairie wheat farming
You need to understand the relationships between resilience, stability, equilibria, and diversity, using specific examples to illustrate interactions.
Exercises
1. Summarize the first and second laws of thermodynamics. What do they tell us about how energy moves through a system?
2. What is the difference between a steady-state equilibrium and a static equilibrium? Which type of equilibrium applies to ecological systems and why?
3. a. When would a system not return to the original equilibrium, but establish a new one? Give an example and explain why this is the case.
b. Give an example of a system that undergoes long-term change to its equilibrium while retaining the integrity of the system.
4. Give an example of how an ecosystem's capacity to survive change depends on diversity and resilience.
5. Why does a complex ecosystem provide stability? Include information regarding the variety of nutrient and energy pathways, and the complexity of food webs, in your answer.
37

Animation
Select the icon to watch an animation

Video
Select the icon to watch a video

Quiz
Select the icon to take an interactive quiz to test your knowledge or practise answering exam essay questions

Case study quiz
Select the paperclip icon to open a case study booklet. Select the question-mark icon to take a quiz based on the information in the case study

01 Foundations of ESS

1.1 Environmental value systems

Opposite: Argentina's Perito Moreno glacier, in the Patagonian province of Santa Cruz. Glaciers have historically been in equilibrium with their environment, although the increased melt of certain ice fields suggests that the planet may be heading for a tipping point where higher temperatures are the norm.

Significant ideas

Historical events, among other influences, affect the development of environmental values systems and environmental movements.

There is a wide spectrum of environmental value systems each with their own premises and implications.

Big questions

As you read this section, consider the following big questions:

- What value systems can you identify at play in the causes and approaches to resolving the issues addressed in this topic?
- How does your own value system compare with others you have encountered in the context of issues raised in this topic?

Knowledge and understanding

- Significant historical influences on the development of the environmental movement have come from literature, the media, major environmental disasters, international agreements, and technological developments.
- An environmental value system (EVS) is a world view or paradigm that shapes the way an individual or group of people perceive and evaluate environmental issues. This will be influenced by cultural, religious, economic, and socio-political context.
- An EVS might be considered as a 'system' in the sense that it may be influenced by education, experience, culture, and media (inputs) and involves a set of interrelated premises, values, and arguments that can generate consistent decisions and evaluations (outputs).
- There is a spectrum of EVSs from ecocentric through anthropocentric to technocentric value systems.
- An ecocentric viewpoint integrates social, spiritual, and environmental dimensions into a holistic ideal. It puts ecology and nature as central to humanity, and emphasizes a less materialistic approach to life with greater self-sufficiency of societies. An ecocentric viewpoint prioritizes biorights, emphasizes the importance of education, and encourages self-restraint in human behaviour.
- An anthropocentric viewpoint believes humans must sustainably manage the global system. This might be through the use of taxes, environmental regulation, and legislation. Debate is encouraged to reach a consensual, pragmatic approach to solving environmental problems.
- A technocentric viewpoint believes that technological developments can provide solutions to environmental problems. This is a consequence of a largely optimistic view of the role humans can play in improving the lot of humanity. Scientific research is encouraged in order to form policies and understand how systems can be controlled, manipulated, or changed to solve resource depletion. A pro-growth agenda is deemed necessary for society's improvement.
- There are extremes at either end of this spectrum – for example, deep ecologists (ecocentric) and cornucopian (technocentric). However, in practice, EVSs vary greatly with culture and time, and rarely fit simply or perfectly into any classification.
- Different EVSs ascribe different intrinsic values to components of the biosphere.

What is ESS?

The title of this course has three components – 'environment', 'systems', and 'societies'. The **environment** of an animal or plant can be defined as the external surroundings that act on it and affect its survival – our environment extends from our immediate surroundings to ultimately, at its greatest extent, the whole Earth. A **system** is something that is made from separate parts that are linked together and affect each other. A **society** is a group of individuals who share some common characteristic such as geographical location, cultural background, historical timeframe, religious perspective, or value system. ESS can best be appreciated when each of these components are viewed holistically, that is to say, as a whole.

The development of the environmental movement

Few photos can have had a greater impact than the one taken by NASA's Apollo 8 mission on 24 December 1968. Before this image, the Earth had seemed vast with almost limitless resources. But once people saw Earth suspended in space, with the moon much larger in the foreground, they gained an appreciation of the vulnerability of the planet and its uniqueness in the Solar System and the universe beyond. Some think that this photo was the beginning of the **environmental movement** – the worldwide campaign to raise awareness, and coordinate action, to tackle the negative effects that humans are having on the planet. But although the image was pivotal in helping to highlight environmental issues, the environmental movement existed before this milestone photograph.

Earthrise from the Moon. This photograph was taken during the Apollo 8 mission of 21–27 December 1968.

Significant historical influences on the development of the environmental movement have come from major environmental disasters, literature, the media, international agreements, and technological developments.

The environmental movement advocates sustainable development through changes in public policy and individual behaviour. The modern movement owes much to developments in the latter part of the 20th century, although its history stretches back for as long as humans have been faced with environmental issues. Some significant moments in the environmental movement are outlined below.

Environmental disasters

- In 1956, a new disease was discovered in Minamata City in Japan. It was named Minamata disease and was found to be linked to the release of methyl mercury into the waste water produced by the Chisso Corporation's chemical factory. The mercury accumulated in shellfish and fish along the coast. The contaminated fish and shellfish were eaten by the local population and caused mercury poisoning. The symptoms were neurological – numbness of the hands, damage to hearing, speech, and vision, and muscle weakness. In extreme cases, Minamata disease led to insanity, paralysis, and death. The pollution also led to birth defects in new-born children.

- At midnight on 3 December 1984, the Union Carbide **pesticide** plant in the Indian city of Bhopal released 42 tonnes of toxic methyl isocyanate gas. This happened because one of the tanks involved with processing the gas had overheated and burst. Some 500 000 people were exposed to the gas. It has been estimated that between 8000 and 10 000 people died within the first 72 hours following the exposure, and that up to 25 000 have died since from gas-related disease.
- On 26 April 1986, early in the morning, reactor number four at the Chernobyl plant in the Ukraine (then part of the Soviet Union) exploded. A plume of highly radioactive dust (fallout) was sent into the atmosphere and fell over an extensive area. Large areas of the Ukraine, Belarus, and Russia were badly contaminated. The disaster resulted in the evacuation and resettlement of over 336 000 people. The fallout caused increased incidence of cancers in the most exposed areas. An area immediately surrounding the plant, covering approximately 2600 km^2, still remains under exclusion due to radiation. The incident raised issues concerning the safety of Soviet nuclear power stations in particular, but also the general safety of nuclear power. These worries remain to this day.
- For many years, the Chernobyl disaster was the only major nuclear incident. That changed on 11 March 2011 when an earthquake in northern Japan caused a tsunami that hit the coastal Fukushima nuclear power plant, causing a meltdown in three of the six nuclear reactors. The damage resulted in radioactive material

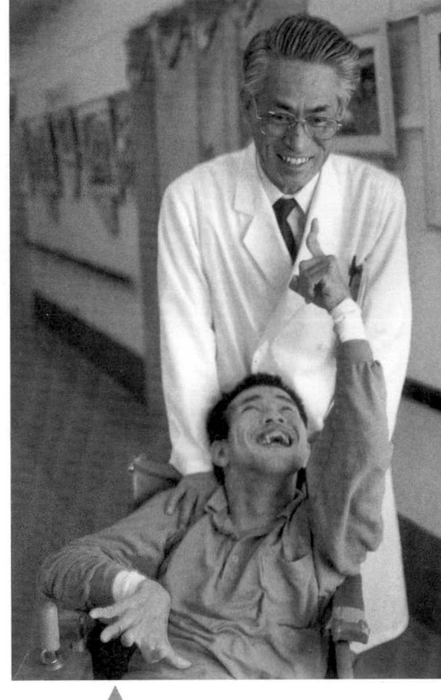

Minamata disease caused severe birth defects, ranging from malformed limbs to complete paralysis.

The Bhopal disaster made headlines around the world. Despite protests, little has been done for families of the victims.

Damage to the Chernobyl nuclear reactor

Satellite image of the Fukushima Dai-ichi nuclear power plant in Okuma, Japan, taken after the 2011 earthquake and tsunami.

escaping into the sea. Following the incident, all 48 of the country's reactors were closed so that new safety checks could be done, leading to an increased dependence on fossil fuels: before Fukushima, nuclear had provided 30 per cent of Japan's energy needs. The move away from nuclear power was replicated around the world. Germany, in particular, backtracked on its nuclear ambitions, even though the disaster at Fukushima was caused by specific local issues (the coastal location of the plant, and the inadequacy of defences for extreme tidal events such as tsunamis).

Literature

A selection of books on environmental issues, including some that have influenced the Green movement

- In 1962, American biologist Rachel Carson's influential book *Silent Spring* was published. It remains one of the most influential books of the environmental movement. The case against chemical pollution was strongly made as Carson documented the harmful effects of pesticides along **food chains** to top predators. The book led to widespread concerns about the use of pesticides and the pollution of the environment.
- Many other significant publications have contributed to the environmental movement. In 1972, the Club of Rome – a global think tank of academics, civil servants, diplomats, and industrialists that first met in Rome – published *The Limits to Growth*. This report examined the consequences of a rapidly growing world population on finite natural resources. It has sold 30 million copies in more than 30 translations and has become the best-selling environmental book in history.
- James Lovelock's book *Gaia* (1979) proposed the hypothesis that the Earth is a living organism, with self-regulatory mechanisms that maintain climatic and biological conditions. He saw the actions of humanity upsetting this balance with potentially catastrophic outcomes. Subsequent books, up to the present day, have developed these ideas.

Rachel Carson, a well-known biologist, wrote many popular natural history magazine articles and books.

Media

Protests about environmental disasters and concern about the unsustainable use of the Earth's resources have led to the formation of pressure groups, both local and international. All these groups have at their centre the concept of stewardship. This is the belief that every person has a responsibility to look after the planet, for themselves and for future generations, through wise management of natural resources. Such groups have resulted in increased media coverage that has raised public awareness about these issues. One of the most influential of these groups is Greenpeace, founded in the early 1970s, and which made its name in 1975 by mounting an anti-whaling campaign. The campaign actively confronted Soviet whalers in the Pacific Ocean off the Californian coast, and eventually developed into the 'Save the Whale' campaign, which made news headlines around the world and became the blueprint for future environmental campaigns. In the 1980s, Greenpeace made even bigger headlines with its anti-nuclear testing campaign.

The sinking of Greenpeace's flagship *Rainbow Warrior* in the port of Auckland, New Zealand, in July 1985, raised an international protest. The sinking was coordinated by French intelligence services to prevent the ship interfering with nuclear tests in the Polynesian island of Moruroa. For France, it was a public relations disaster that did much to promote Greenpeace's environmental campaign against nuclear testing.

- In 2006, the film *An Inconvenient Truth* examined the issues surrounding climate change, and increased awareness of environmental concerns. The publicity surrounding the film meant that more people than ever before heard about global warming, and its message was spread widely and rapidly through modern media, such as the internet. The film made the arguments about global warming very accessible to a wider audience, and raised the profile of the environmental movement worldwide. The film was supported by a book that recorded hard scientific evidence to support its claims.

An Inconvenient Truth, a documentary of Al Gore (former US Vice President) giving a lecture on climate change, marked a significant change in public opinion in the USA. It was the first time a mainstream political figure had championed environmental issues.

- Earth Day is marked each year on 22 April, coordinated globally via the internet and other media. It was founded in 1970 by a US Senator from Wisconsin, Gaylord Nelson, after he had seen the effects of a massive oil spill in Santa Barbara, California, in 1969. By creating a day that celebrated the Earth, he saw a way of moving environmental protection more centrally onto the national political agenda. Earth Day is celebrated simultaneously around the world, encouraging people to participate in environmental campaigns both local and global.

To learn more about Earth Day, go to www.pearsonhotlinks.co.uk, enter the book title or ISBN, and click on weblink 1.1.

International agreements

- In 1972, the United Nations held its first major conference on international environmental issues in Stockholm, Sweden – the UN Conference on the Human Environment, also known as the Stockholm Conference. It examined how human activity was affecting the global environment. Countries needed to think about how

United Nations Conference on Environment and Development, Rio de Janeiro, Brazil, 3–14 June 1992.

they could improve the living standards of their people without adding to pollution, habitat destruction and species extinction. The conference led to the Stockholm Declaration, which played a pivotal role in setting targets and shaping action at both an international and local level. These early initiatives ultimately led to the Rio Earth Summit in 1992, coordinated by the United Nations, which produced **Agenda 21** and the Rio Declaration. The Stockholm Declaration and subsequent global summits have played a leading role in shaping attitudes to sustainability.

- In 1987, a report by the UN World Commission on Environment and Development (WCED) was published, intended as a follow-up to the Stockholm Conference. The report was called *Our Common Future*; it took the ideas from Stockholm and developed them further. It linked environmental concerns to development and sought to promote sustainable development through international collaboration. It also placed environmental issues firmly on the political agenda. *Our Common Future* is also known as the Brundtland Report after the Chair of the WCED, former Norwegian Prime Minister, Gro Harlem Brundtland.
- The publication of *Our Common Future* and the work of the WCED provided the groundwork for the UN's Earth Summit in Rio in 1992. The conference was unprecedentedly large for a UN conference. It was attended by 172 nations: the wide uptake and international focus meant that its impact was likely to be felt across the world. The summit's radical message was that nothing less than a transformation of our attitudes and behaviour towards environmental issues would bring about the necessary changes. The conference led to the adoption of Agenda 21: a blueprint for action to achieve sustainable development worldwide (21 refers to the 21st century). Agenda 21 is a comprehensive plan of action to be taken globally, nationally and locally by organizations of the UN, governments, and environmental groups in every area in which humans affect the environment. It was adopted by more than 178 governments at the Earth Summit.

The Earth Summit changed attitudes to sustainability on a global scale, and changed the way in which people perceived economic growth (i.e. that sometimes this is at the expense of the environment and not necessarily a good thing). It encouraged people to think of the indirect values of ecosystems (e.g. ecosystem services, pages 42–43) rather than the purely economic ones. It also was important for emphasizing the relationships between human rights, population, social development, women, and human settlements, and the need for environmentally sustainable development. Its emphasis was on change in attitude affecting all economic activities, ensuring that its impact could be extensive. The conference meant that environmental issues came to be seen as mainstream rather than the preserve of a few environmental activists. Particular achievements were steps towards preserving the world's **biodiversity** (through the Convention on Biological Diversity, CBD) and steps to address the **enhanced greenhouse effect** (via the UN Framework Convention on Climate Change, UNFCCC), which in turn led to the Kyoto Protocol.

Acronyms are formed from the first letter or first few letters of each word in a phrase or title (e.g. CBD, UNFCCC, COP). Using such shortened forms can speed up communication. International conventions widely use acronyms.

Both the CBD and UNFCCC are legally binding conventions, and both are governed by the Conference of the Parties (COP) which meet either annually or biennially to assess the success and future directions of the Convention. For example, COP 11 of the CBD took place in Pyeongchang, Republic of Korea, in October 2014; COP 15 of the UNFCCC took place in Copenhagen, Denmark, in December 2009; and COP 20 of the UNFCCC took place in Lima, Peru, in December 2014. The Copenhagen Accord was a document produced at COP 15 of the UNFCCC, in which attending parties were asked to 'take note' of the concerns raised at the meeting about climate change – the document was not legally binding.

CONCEPTS: Biodiversity

Biodiversity is a broad concept encompassing the total variety of living systems.

Some national and state governments have legislated or advised that local authorities take steps to implement Agenda 21. Known as 'Local Agenda 21' (LA21), these strategies apply the philosophy of the Earth Summit at the local level. Each country is urged to develop an LA21 policy, with the agenda set by the community itself rather than by central or local government, as ownership and involvement of any initiatives by society at large is most likely to be successful.

- The 1992 Earth Summit was followed up 10 years later by the Johannesburg World Summit on Sustainable Development (Figure 1.1). The Johannesburg meeting looked mainly at social issues, and targets were set to reduce poverty and increase people's access to safe drinking water and sanitation (problems that cause death and disease in many **less economically developed countries (LEDCs)**).
- In 2012, the UN Conference on Sustainable Development (UN CSD, or Rio+20) took place to commemorate the 20th anniversary of the Earth Summit. The meeting had three main objectives:
  - to secure political commitment from nations to sustainable development
  - to assess progress towards internationally agreed commitments (e.g. CO_2 reductions)
  - to examine new and emerging challenges.

  Issues focused on two themes:
  - economic development in the context of sustainable development (i.e. the encouragement of 'green economies') with an emphasis on poverty eradication
  - how institutional frameworks can be developed on a more sustainable basis.

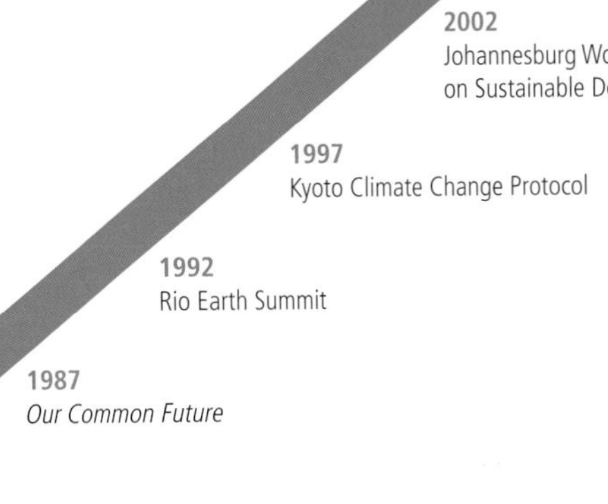

Figure 1.1 Important milestones in the environmental movement

Major landmarks in the modern environmental movement include: Minamata, Rachel Carson's *Silent Spring*, the Save the Whale campaign, Bhopal, and the Chernobyl disaster. These led to:

- environmental pressure groups, both local and global
- the concept of stewardship
- increased media coverage raising public awareness.

Rio+20 again brought governments, international institutions, and major groups together to agree on a range of measures to reduce poverty while promoting good jobs, clean energy, and a more sustainable and fair use of resources.

- The effect of climate change, both in terms of sustainable development and its effect on the planet in general, was discussed at a UN conference in Kyoto in 1997. Agreements were made to reduce emissions of **greenhouse gases**, and gave participant **more economically developed countries (MEDCs)** legally binding targets for cuts in emissions from the 1990 level. The Kyoto Protocol stipulated that these targets should be reached by the year 2012. The meetings that followed the UNFCCC meeting at Copenhagen in 2009 worked towards finding a successor to the Kyoto Protocol.
- At the 2011 Durban conference (South Africa), the debate about a legally binding global agreement was reopened: countries were given until 2015 to decide how far and how fast to cut their carbon emissions. Before the Durban conference, most countries were going to follow national targets for carbon emissions after 2012, which would be voluntary and not legally binding. The Durban Agreement differs from the Kyoto Protocol in that it includes both MEDCs and LEDCs rather than just MEDCs, and also differs from other summits in that it is working towards a legally binding treaty.

It is true that countries can break these agreements and there is little the international community can do about this. Moreover, summits may not achieve their initial goals, but they do act as important catalysts in changing the attitudes of governments, organizations and individuals.

Technological innovation

You need to cover a variety of significant historical influences that affected the environmental movement, and be able to recall a minimum of three in-depth examples in exams. It is a good idea to select a range of historical influences that includes both local and global examples.

The **Green Revolution** refers to a time between the 1940s and the late 1960s when developments in scientific research and technology in farming led to increased agricultural productivity worldwide. The Club of Rome (page 6) claimed in their *The Limits to Growth* report that, within a century, a mixture of human-made pollution and resource depletion would cause widespread population decline. But the intervention of the Green Revolution meant that by 2000, world population had reached 6 billion, and is predicted to rise to nearly 9 billion by 2050. The intensification of agriculture raised many questions for the environmental movement (pages 280–282), as has the increase in human population (pages 404–405).

Other technological innovations have created alternatives to fossil fuel use (e.g. solar panels and wind turbines) which make the arguments proposed by environmentalists (i.e. a switch to more sustainable sources of energy) a real possibility, and drive the environmental movement forward still further.

CONCEPTS: Environmental value systems

Environmental disasters have affected the way people view human impacts on the planet. Realization of the negative influences people have had has led to the development of the environmental movement, which in turn has affected the views of people around the globe.

Popular books such as *Silent Spring*, and films such as Al Gore's *An Inconvenient Truth*, can provide knowledge about environmental issues on a global scale. People who previously had limited understanding of the environment are enabled to make their own minds up about global issues. But do they have enough information to see all sides of the argument? A good education would certainly put these arguments in a wider context. Is it a problem that many people receive only one side of the argument?

Environmental value systems

An **environmental value system (EVS)** is a particular worldview or set of **paradigms** that shapes the way an individual, or group of people, perceive and evaluate environmental issues. A person's or group's EVS is shaped and influenced by cultural factors (including religion), economic (e.g. whether from a LEDC or MEDC), and socio-political context (e.g. democratic or authoritarian society).

An environmental value system (EVS) is a worldview or paradigm that shapes the way an individual or group of people perceive and evaluate environmental issues. It is influenced by the cultural, religious, economic and socio-political context.

SYSTEMS APPROACH

An EVS might be considered as a system in the sense that it may be influenced by education, experience, culture, and media (inputs), and involves a set of interrelated assumptions, values, and arguments that can generate consistent decisions and evaluations (outputs).

The systems approach is explained in detail on pages 19–22. EVSs, like all systems, are assemblages of parts and the relationships between them, which together constitute a whole. Systems have inputs, outputs (which are determined by the processing of inputs), and storages. The outputs generate consistent decisions and evaluations.

EVS inputs are:

- education
- cultural influences
- economic factors
- socio-political factors (the interaction of social and political factors; for example, communism, capitalism)
- religious texts and doctrine
- the media.

EVS outputs are:

- perspectives
- decisions on how to act regarding environmental issues
- courses of action.

Flows of information into individuals within societies are processed or transformed into changed perceptions of the environment and altered decisions about how best to act on environmental matters. At their strongest, such information flows cause people to take direct action to alleviate environmental concerns. It is possible that inputs transfer through the individual or group without processing, but it is unlikely that an input has no effect at all.

A society is a group of individuals who share some common characteristics, such as geographical location, cultural background, historical timeframe, religious perspective, value system, and so on.

EVSs act within social systems. Social systems are more general than **ecosystems**. There are lots of different types of social system: class-based; democratic or authoritarian; patriarchal (male dominance) or matriarchal (female dominance); religion-based; industrial (technology-based) or agrarian (agriculture-based); capitalist or communist. Rather than the flows of energy and matter we see in ecosystems (Chapter 2, pages 87–100), social systems have flows of information, ideas and people. Both ecosystems and social systems exist at different scales, and have common features such as feedback and equilibrium (pages 30–32). Trophic levels exist in ecosystems while in social systems there are social levels within society, and both contain consumers and producers. Producers in social systems are responsible for new input (e.g. ideas, films, books, documentaries) and consumers absorb and process this information.

The development of environmental value systems is influenced by differences in culture and society. Buddhist societies, for example, see the human being as an intrinsic part of nature. A society's EVS influences the actions taken by its citizens in response to environmental issues. Buddhist monks in Thailand, for example, are part of a growing environmental movement. They are involved in ecological conservation projects, and teach ecologically sound practices among Thai farmers. Unsustainable development based on rapid economic development is seen to be one of the primary causes of Thailand's environmental crisis. The respect in which Buddhist monks are held means that their views are listened to and can have a profound effect on the population.

Buddhist monks are frequently active in a range of campaigns including forest conservation in Thailand.

The range of EVSs

There is a range of EVSs, from ecocentric to technocentric.

EVSs can broadly be divided into **technocentric** and **ecocentric** with **anthropocentric** between the two (Figure 1.2). Technocentrists believe that technology will keep pace with and provide solutions to environmental problems. Ecocentrists are nature-centred and distrust modern large-scale technology; they prefer to work with natural environmental systems to solve problems, and to do this before problems get out of control. The anthropocentrists include both technocentric and ecocentric viewpoints. An anthropocentrist believes humans must sustainably manage the global system: this might be through taxes, environmental regulation, and legislation. Debate is encouraged so that a consensual, pragmatic approach to solving environmental problems can be reached.

The technocentrist approach is sometimes termed a **cornucopian** view: a belief in the unending resourcefulness of humans and their ability to control their environment. This leads to an optimistic view about the state of the world. Ecocentrists, in contrast, see themselves as subject to nature rather than in control of it. Ecocentrists see a world with limited resources where growth needs to be to be controlled so that only beneficial growth occurs. At one end of the ecocentrist worldview are the self-reliance soft ecologists – those who reject materialism and have a conservative view regarding environmental problem-solving. At the other end are the **deep ecologists** – those who put more value on nature than humanity.

Norwegian philosopher Arne Næss pioneered and first named the ecocentrist EVS known as deep ecology.

Although there are extremes at either end of this range (i.e. deep ecologists at the ecocentric end of the spectrum and cornucopians at the technocentric end), in practice, EVSs vary greatly with culture and time and rarely fit simply or perfectly into any classification.

Environmental Value System

Ecocentrism (nature centred)

An ecocentric viewpoint integrates social, spiritual, and environmental dimensions into a holistic ideal. It puts ecology and nature as central to humanity, and emphasizes a less materialistic approach to life with greater self-sufficiency of societies. An ecocentric viewpoint prioritizes biorights, emphasizes the importance of education and encourages self-restraint in human behaviour.

Anthropocentrism (people centred)

An anthropocentric viewpoint believes humans must sustainably manage the global system. This might be through the use of taxes, environmental regulation, and legislation. Debate would be encouraged to reach a consensual, pragmatic approach to solving environmental problems.

Technocentrism (technology centred)

A technocentric viewpoint believes that technological developments can provide solutions to environmental problems. This is a consequence of a largely optimistic view of the role humans can play in improving the lot of humanity. Scientific research is encouraged in order to form policies and understand how systems can be controlled, manipulated or changed to solve resource depletion. A pro-growth agenda is deemed necessary for society's improvement.

Deep ecologists

1 Intrinsic importance of nature for the humanity of man.

2 Ecological (and other natural) laws dictate human morality.

3 Biorights – the right of endangered species or unique landscapes to remain unmolested.

Self-reliance soft ecologists

1 Emphasis on smallness of scale and hence community identity in settlement, work, and leisure.

2 Integration of concepts of work and leisure through a process of personal and communal improvement.

3 Importance of participation in community affairs, and of guarantees of the rights of minority interests. Participation seen as both a continuing education and a political function.

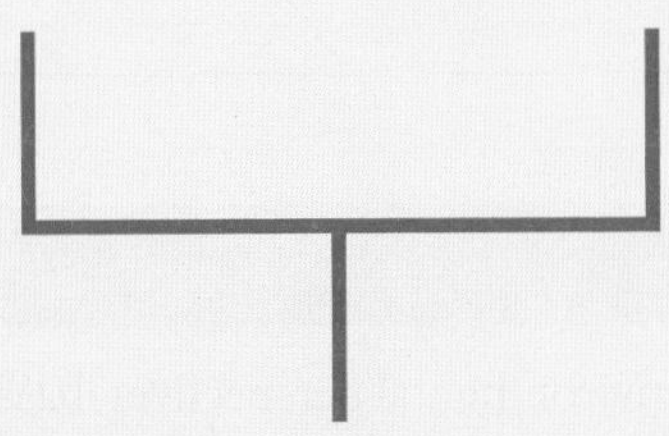

4 Lack of faith in modern large-scale technology and its associated demands on elitist expertise, central state authority, and inherently anti-democratic institutions.

5 Implication that materalism for its own sake is wrong and that economic growth can be geared to providing for the basic needs of those below subsistence levels.

Environmental managers

1 Belief that economic growth and resource exploitation can continue assuming:
  a suitable economic adjustments to taxes, fees, etc.
  b improvements in the legal rights to a minimum level of environmental quality
  c compensation arrangements satisfactory to those who experience adverse environmental and/or social effects.

2 Acceptance of new project appraisal techniques and decision review arrangements to allow for wider discussion or genuine search for consensus among representative groups of interested parties.

Cornucopians

1 Belief that people can always find a way out of any difficulties, whether political, scientific, or technological.

2 Acceptance that pro-growth goals define the rationality of project appraisal and policy formulation.

3 Optimism about the ability of humans to improve the lot of the world's people.

4 Faith that scientific and technological expertise provides the basic foundation for advice on matters pertaining to economic growth, and public health and safety.

5 Suspicion of attempts to widen basis for participation and lengthy discussion in project appraisal and policy review.

6 Belief that all impediments can be overcome given a will, ingenuity, and sufficient resources arising out of growth.

'The pattern of environmentalist ideologies', adapted from *Environmentalism, 2nd edition* by Timothy O'Riordan, Figure 10.1, page 376, copyright © 1981, Pion Ltd

Figure 1.2 The range of environmental value systems

Case study

A technocentrist approach to reducing carbon dioxide emissions

Energy and gasoline companies have been developing technological solutions to carbon dioxide emissions in order to alleviate global warming. Carbon-capture-and-storage (CCS) techniques involve taking the carbon dioxide produced from industrial processes and storing it in various ways (Figure 1.3). This means it is not released into the atmosphere and does not contribute to global warming. A BP project at In Salah in Algeria aims to store 17 million tonnes of carbon dioxide - an emission reduction equivalent to removing 4 million cars from the road. Such projects have yet to be made available on a large-scale commercial basis because of the costs involved.

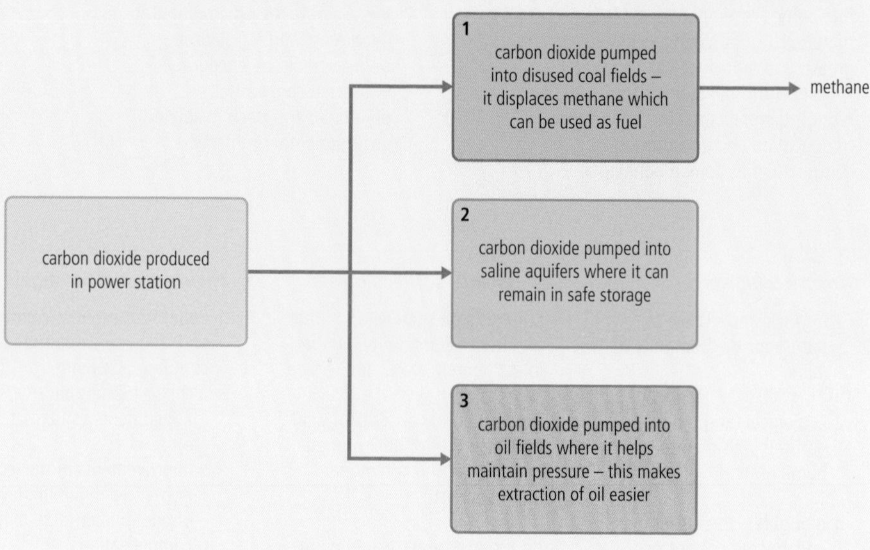

Figure 1.3 Options for carbon capture and storage

Intrinsic value means that something has value in its own right, i.e. inbuilt/inherent worth.

CONCEPTS: Environmental value systems

Discuss with your neighbour in class how the environment can have its own intrinsic value. Think of some specific examples and talk about these. Do you have the same view of what these intrinsic values are? Feedback to the rest of the class and discover how many different viewpoints there are.

Ecosystems may often cross national boundaries and conflict may arise from the clash of different value systems about exploitation of resources (for example, migration of wildlife across borders in southern Africa). This is discussed in Chapter 2.

Contrasting EVSs

Different types of society have different environmental perspectives, based on their individual EVSs. Two case studies examine two pairs of contrasting societies.

Case study

Judaeo-Christian and Buddhist societies

The view of the environment in Judaeo-Christian religions is one of stewardship, where humans have a role of responsibility towards the Earth. The Genesis story suggests that God gave the planet to humans as a gift. Other biblical stories indicate that humanity should make the most of this gift as stewards.

An example of such a story is the parable of the talents told by Jesus. A rich employer sets off on a journey. He leaves his money in the care of his three employees. On his return, he calls his employees together to give an account of their activities in his absence.

- The first employee invested the money, and increased it 10 times.
- The second also invested the money, and managed to increase it five times.
- The third, fearing his employer's reaction if he lost the money, buried it.

The employer fires the third man, and praises the other two for being good stewards and making something of the monies they were responsible for.

This contrasts with the Buddhist approach to the environment, which sees the human being as an intrinsic part of nature, rather than a steward. Buddhism is sometimes seen as an ecological philosophy (because of its worldview rather than anything that appears in its writings, which are not explicitly environmental). Buddhism emphasizes human interrelationships with all other parts of nature, and supports the belief that to think of ourselves as isolated from the rest of nature is unrealistic. The Buddhist approach can be summarized as:

- compassion is the basis for a balanced view of the whole world and of the environment
- a 'save and not waste' approach means that nothing in nature is spoiled or wasted; wanton destruction upsets the vital balance of life
- ecology is rebuilt through the philosophy 'uplift of all', which is based on people acting compassionately and working together **altruistically**.

Vegetarianism is part of the Buddhist tradition; it is a reflection of Buddhist respect for all life. Reincarnation, the belief that human consciousness (or spirit) is immortal and can be reborn after death in either human or animal form, also emphasizes humanity's interconnectedness with nature. Buddhists believe that nothing has a fixed and independent existence; all things are without self-existence and are impermanent. From this perspective, humans are intimately related to their environment and cannot exist separately from the rest of the world. Recognizing this principle of interdependence inspires an attitude of humility and responsibility towards the environment.

TOK

Different societies have different environmental perspectives, based on individual EVSs. Individual and societal understanding and interpretation of data regarding environmental issues is influenced by these perspectives. Can there be such a thing as an unbiased view of the environment? Can we ever expect to establish a balanced view of global environmental issues?

Case study

Native Americans and European pioneers

Prior to the colonization of North America by Europeans from the late 16th century onward, the country was occupied solely by native American Indian tribes. Native Americans, in general, saw their environment as communal, and had a subsistence economy based on barter. Their low-impact technologies meant that they lived in harmony with the environment – something supported by their animistic religion where all things have a soul – animals, plants, rocks, mountains, rivers, and stars.

The incoming European pioneers operated frontier economics, which involved the exploitation of what they saw as seemingly unlimited resources. This inevitably led to environmental degradation through over-population, lack of connectivity with the environment, heavy and technologically advanced industry, and unchecked exploitation of natural resources.

Decision-making and EVSs

EVSs influence our decision-making processes. Let's consider the contrasting perspectives of ecocentrism and technocentrism in relation to three specific cases.

You need to be able to evaluate the implications of two contrasting environmental value systems in the context of given environmental issues, and to be able to justify the implications using evidence and examples to make the justification clear.

Environmental challenges posed by the extensive use of fossil fuels

Fossil fuels have problems associated with their use (i.e. **global warming**). The cornucopian belief in the resourcefulness of humans and their ability to control their environment would lead to a technocentric solution, where science is used to find a useful alternative (e.g. hydrogen fuel cells). As technocentrists, cornucopians would see this as a good example of resource replacement: an environmentally damaging industry can be replaced by an alternative one. Rather than seeing it as necessary to change their lifestyles to reduce the use of fuel, cornucopians would look to develop technology to reduce the output of carbon dioxide from fuel use. A cornucopian would say that economic systems have a vested interest in being efficient so the existing problems will self-correct given enough time, and that development (which requires energy) will increase standards of living thus increasing demands for a healthy environment. Scientific efforts should be devoted to removing carbon dioxide from the atmosphere, and reducing its release, rather than curtailing economic growth.

A technocentrist would predict that market pressure would eventually result in the lowering of carbon dioxide emission levels.

An ecocentrist approach to the same problem would call for the reduction of greenhouse gases through curtailing existing gas-emitting industry, even if this restricts economic growth.

Approaches of resource managers to increasing demand for water resources

The technocentric manager would suggest that future needs can be met by technology, innovation and the ability to use untapped reserves. They would support such measures as removal of fresh water from seawater (desalination) if they were near an ocean, iceberg capture and transport, wastewater purification, synthetic water production (water made through chemical reactions, or hydrogen fuel cell technology), cloud seeding (Figure 1.4), and extracting water from deep aquifers. They would also look at innovative ways to reduce the use of water, both in industry and at a domestic level.

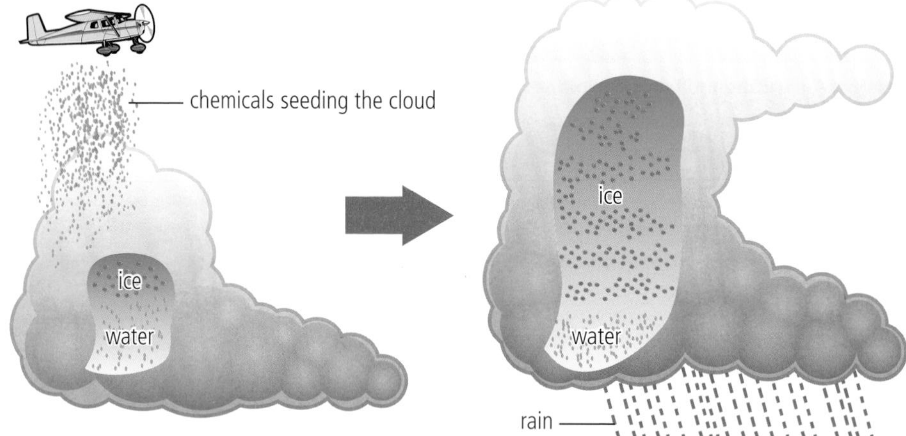

Figure 1.4 Chemicals such as silver iodide or frozen carbon dioxide are released into clouds. They offer surfaces around which water and ice crystals form. When they are large enough, they fall out of the cloud and become rain.

The ecocentric manager would highlight the overuse and misuse of water. They would encourage the conservation of water and greater recycling, and say that water use should be within sustainable levels. Monitoring would be recommended to ensure that water use remained within sustainable limits. An ecocentrist would encourage water use that had few detrimental impacts on habitat, wildlife, and the environment.

Methods for reducing acid rain

Acid rain is produced when sulfur dioxide, produced by burning fossil fuels such as coal, dissolves in atmospheric water, ultimately falling as rain (see page 341). The ecocentrist would argue for a change in lifestyle that reduces the need for either the energy produced by coal, or the products that are made with that energy. For example, a reduction of heat in the home could be achieved by dressing more warmly instead of raising the indoor temperature. Changes in transport use would reduce reliance on fossil fuels, and could be achieved by walking or bicycling to work or when doing the shopping. Reducing the use of cars would reduce the release of acid deposition precursors. Ecocentrists would also encourage the 'reuse, reduce, recycle' philosophy (Figure 1.5; Chapter 8, pages 435–436). Central to their worldview would be the idea that life should cherish spiritual well-being, rather than the satisfaction of material desires. This would reduce desires for continuously purchasing consumer goods.

Figure 1.5 The 'Reduce, Reuse, Recycle' campaign encourages people to care for goods (making them last as opposed to frequently replacing them with new ones), reduce consumption, and recycle waste.

Technocentrists would again argue for use of alternative technology and encouraging continued economic growth irrespective of the effect of greenhouse gas emissions because they see humanity as able to control the problem as and when necessary.

Intrinsic value

Different EVSs view the different components of the **biosphere** (the living part of the Earth) in very different ways, and attribute to them different values. For example, indigenous farmers using shifting cultivation in the Amazonian rainforest in Brazil would see the rainforest as a natural resource that should be used in a way that minimizes human impact on the environment (i.e. an ecocentrist EVS), whereas city-dwellers in Brasilia (federal capital of Brazil) are more likely to see the rainforest as a resource to be exploited for economic gain, and underestimate the true value of pristine rainforest (i.e. a technocentrist EVS). Intrinsic values may also vary between different EVSs (case studies, pages 14–15). An intrinsic value is one that has an inherent worth, irrespective of economic considerations (Chapter 8, pages 418–419), such as the belief that all life on Earth has a right to exist. For example, a visitor to a friend's garden in the summer may value the abundance of insect life not seen in their city home, whereas the owner appreciates the services provided by the insects in sustaining the garden, such as woodlice that recycle fallen leaves and bees that pollinate the flowers. Intrinsic values include values based on cultural, aesthetic, and bequest significance (i.e. of value to children and grandchildren).

CONCEPTS: Environmental value systems

EVSs determine the decision-making processes regarding environmental issues, such as choice of energy usage, reaction to limited natural resources such as water, responses to pollution, and attitudes towards ecological deficit. The ESS course helps you to appreciate and evaluate your own EVS. Such an understanding will enable you to appreciate how worldviews influence the way in which you perceive and act regarding environmental issues. During this course, you will be encouraged to develop your own EVS and justify your decisions on environmental issues based on this EVS. This is a personal thing: EVSs are individual and there are no 'wrong' EVSs, but you should be able to justify your viewpoint.

Exercises

1. Draw a timeline from the 1950s to the present day to summarize development of the modern environmental movement.
2. What is meant by an environmental value system? List three inputs and three outputs of these systems.
3. Environmental value systems range between *ecocentric* and *technocentric* perspectives. What do these terms mean?
4. Summarize the differences between ecocentric and technocentric philosophies with regard to the following issues:
   a. environmental challenges posed by fossil fuels
   b. the response of resource managers to increasing demands for water
   c. methods for reducing acid rain.

Big questions

Having read this section, you can now discuss the following big questions:

- What value systems can you identify at play in the causes and approaches to resolving the issues addressed in this topic?
- How does your own value system compare with others you have encountered in the context of issues raised in this topic?

CHALLENGE YOURSELF

Thinking skills 

Create a table that summarizes the ecocentric and technocentric approaches to each of the three case studies discussed above. The table will help you compare and evaluate the response of managers with contrasting EVSs to different environmental issues.

Different EVSs ascribe different intrinsic values to components of the biosphere.

You need to be able to discuss the view that the environment can have its own intrinsic value.

TOK

There are assumptions, values and beliefs, and worldviews that affect the way in which we view the world. These are influenced by the way we are raised by our parents, by education, by our friends and by the society we live in.

Points you may want to consider in your discussions:

- What have you learned about EVSs and how different people can view the environment in different ways?
- How do EVSs affect how people respond to environmental issues?
- How have historical events and the development of the environmental movement affected peoples' EVSs around the world?
- What have you learned about your own EVS from this chapter?

1.2 Systems and models

Significant ideas

A systems approach can help in the study of complex environmental issues.

The use of models of systems simplifies interactions but may provide a more holistic view than reducing issues to single processes.

Big questions

As you read this section, consider the following big question:

- What strengths and weaknesses of the systems approach and the use of models have been revealed through this topic?

Knowledge and understanding

- A systems approach is a way of visualizing a complex set of interactions which may be ecological or societal.
- These interactions produce the emergent properties of the system.
- The concept of a system can be applied to a range of scales.
- A system is comprised of storages and flows.
- The flows provide inputs and outputs of energy and matter.
- The flows are processes that may be either transfers (a change in location) or transformations (a change in the chemical nature, a change in state, or a change in energy).
- In systems diagrams, storages are usually represented as rectangular boxes, and flows as arrows with the arrow indicating the direction of the flow. The size of the box and the arrow may represent the size/magnitude of the storage or flow.
- An open system exchanges both energy and matter across its boundary while a closed system only exchanges energy across its boundary.
- An isolated system is a hypothetical concept in which neither energy nor matter is exchanged across the boundary.
- Ecosystems are open systems. Closed systems only exist experimentally although the global geochemical cycles approximate to closed systems.
- A model is a simplified version of reality, and can be used to understand how a system works and predict how it will respond to change.
- A model inevitably involves some approximation and loss of accuracy.

What are systems?

A system is an assemblage of parts and the relationships between them, which together constitute an entity or whole.

A systems approach is a way of visualizing a complex set of interactions which may be ecological or societal.

There are different ways of studying **systems**. A *reductionist* approach divides systems into parts, or components, and each part is studied separately. This is the way of traditional scientific investigations. But a system can also be studied as a whole, with patterns and processes described for the whole system. This is the holistic approach, and is usually used in modern ecological investigations. The advantage of using the systems method is that it can show how components within the whole system relate to one another. A systems approach is a way of visualizing a complex set of interactions, which can be applied across a wide range of different disciplines. This course focuses on systems as they relate to ecosystems and society, although the systems approach can equally be applied to other subjects such as economics or politics.

Diagrams are used to represent systems. Using the systems approach, a tree can be summarized as shown in Figure 1.6.

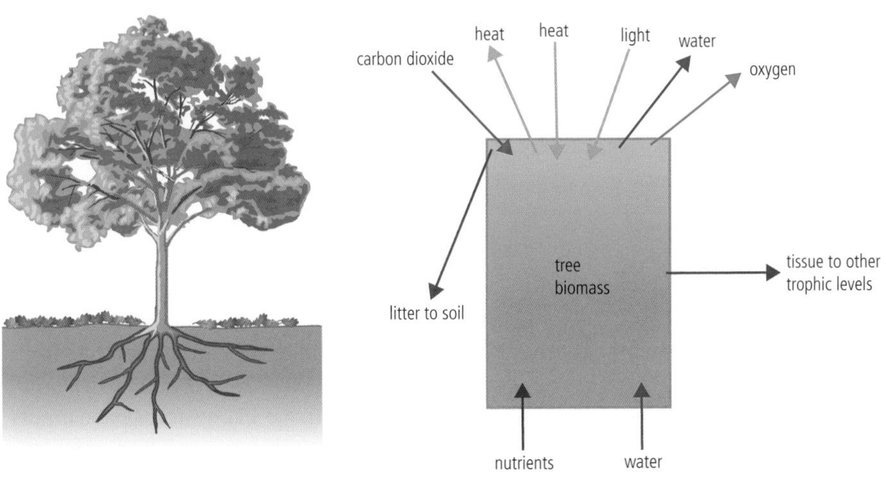

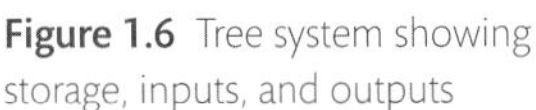

Figure 1.6 Tree system showing storage, inputs, and outputs

The arrows into and out of the tree systems diagram indicate inputs and outputs. In addition, the diagram could be labelled with processes on each arrow. Processes in this example would include:

- **photosynthesis** – transforming carbon dioxide (CO_2), water (H_2O), and light into **biomass** and oxygen (O_2)
- **respiration** – transforming biomass into carbon dioxide and water
- diffusion – allowing the movement of nutrients and water into the tree
- consumption – transferring tissue (i.e. biomass) from one **trophic level** to another.

The interdependent components of systems are connected through the transfer of energy and matter, with all parts linked together and affecting each other. Examples of systems, with increasing levels of complexity, include particles, atoms, molecules, cells, organs, organ systems, communities, ecosystems, biomes, the Earth, the Solar System, galaxies, and the universe.

The systems approach emphasizes similarities in the ways in which matter, energy and information link together in a variety of different disciplines. This approach, therefore, allows different subjects to be looked at in the same way, and for links to be made between them. Although the individual parts of a complex system can be looked at using the reductionist approach, this ignores the way in which such a system operates as a whole. A holistic approach is necessary to fully understand the way in which the parts of a complex system operate together. These interactions produce the emergent properties of the system.

SYSTEMS APPROACH

The concept of systems has been used in science for many years, especially in biology where the functioning of the whole organism can be understood in terms of the interactions between various systems, such as the breathing and circulatory system. The reductionist approach often used in traditional science tends to look at the individual parts of a system, rather than the whole, so that the 'big picture' is missed (TOK Chapter, page 445). The nature of the environment and how we relate to it demands a holistic treatment. A systems approach emphasizes the ways in which matter, energy and information flow, and can be used to integrate the perspectives of different disciplines to better represent the complex nature of the environment.

The systems concept can be applied across a range of scales, from global-scale biomes to the small scale of life contained within a bromeliad in the rainforest canopy.

The systems concept on a range of scales

Bromeliads are flowering plants found in abundance in the canopy of the rainforest. They capture rainwater, which enables a small ecosystem to exist containing tree frogs, snails, flatworms, tiny crabs and salamanders. Animals within the bromeliad may spend their entire lives inside the plant.

An ecosystem is a community of interdependent organisms and the physical environment they inhabit. Different ecosystems exist where different species and physical or climatic environments are found. An ecosystem may, therefore, be of any size up to global.

For example, a tropical rainforest contains lots of small-scale ecosystems, such as the complex web of life that exists within a single bromeliad in the canopy (Interesting fact box, page 21). As you have just learned (page 11), systems have inputs, outputs, and storages. The rainforest can be viewed as an ecosystem with particular inputs (e.g. sunlight energy, nutrients, and water), outputs (e.g. oxygen, soil litter, and water), and storages (e.g. biomass within trees and plants; nutrients within soil). Such ecosystems can be viewed on the local scale (i.e. within one country) or more widely (in many different countries where the same climatic conditions apply). On the global scale, ecosystems with similar climatic conditions in different parts of the world are called **biomes**. Examples of different biomes include tundra, tropical rainforest, and desert (Chapter 2, pages 102–103).

Biosphere refers to the part of the Earth inhabited by organisms that extends from upper parts of the atmosphere to deep in the Earth's crust.

The holistic approach and the reductionist approach used by conventional science use almost identical methodologies: the difference between them may, therefore, be only one of perspective.

At the largest scale, our entire planet can be seen as an ecosystem, with specific energy inputs from the Sun and with particular physical characteristics. The Gaia hypothesis, formulated by scientist James Lovelock in the mid-1960s (page 6), proposes that our planet functions as a single living organism. The hypothesis says that the Earth is a global control system of surface temperature, atmospheric composition, and ocean salinity. It proposes that Earth's elements (water, soil, rock, atmosphere, and the living component called the biosphere) are closely integrated in a complex interacting system that maintains the climatic and biogeochemical conditions on Earth in a preferred homeostasis (i.e. in the balance that best provides the conditions for life on Earth).

Dendrobates pumilio, the strawberry poison dart frog, is common in the Atlantic lowland tropical forests of Central America, especially Costa Rica. The female typically lays 3–5 eggs on the forest floor in a jelly-like mass that keeps them moist. Once the eggs are ready to hatch, one of the parents steps into the jelly surrounding the eggs: the tadpoles respond to the movement and climb onto the parent's back, where they stick to a secretion of mucus. The parent carries the tadpoles up to the canopy where they are deposited in water caught by the upturned leaves of a bromeliad. Each tadpole is put in a separate pool to increase the likelihood that some offspring will survive predators. The bromeliad ecosystem is a vital part of the frog's life-history.

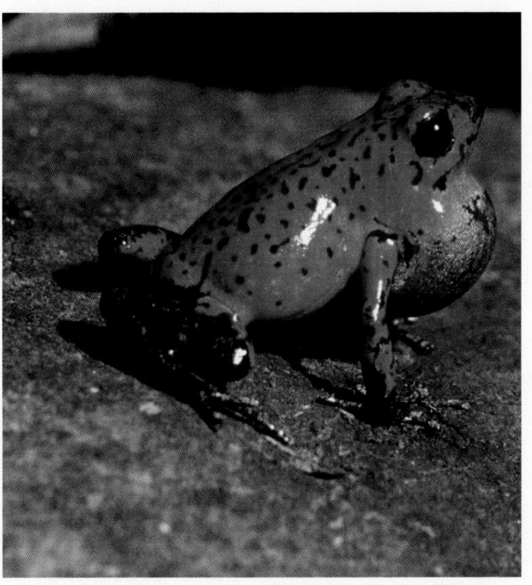

Strawberry poison-dart frog

The characteristics of systems

A system consists of storages and flows. **Storages** are places where matter or energy is kept in a system, and **flows** provide inputs and outputs of energy and matter. The flows are processes that may be either **transfers** (a change in location) or **transformations** (a change in the chemical nature, a change in state or a change in energy).

Systems can be represented as diagrams (page 19). In these **systems diagrams**, storages are usually represented as rectangular boxes, and flows as arrows with the arrow indicating the direction of the flow. Figure 1.7 shows flows and storage for the social system discussed on page 11.

TOK

The systems approach gives a holistic view of the issues, whereas the reductionist approach of conventional science is to break the system down into its components and to understand the interrelations between them. The former describes patterns and models of the whole system, whereas the latter aims at explaining cause-and-effect relationships within it.

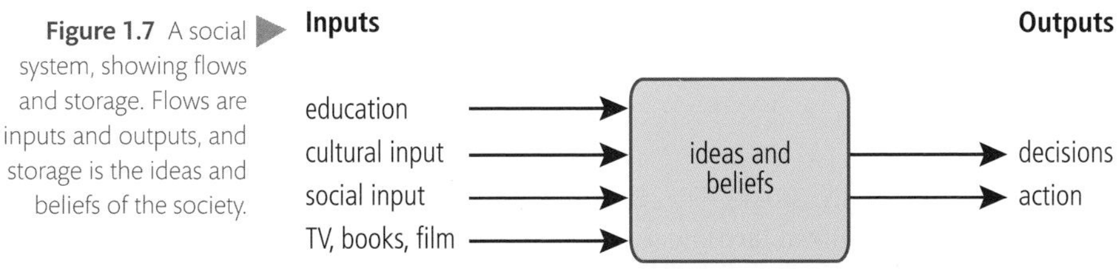

Figure 1.7 A social system, showing flows and storage. Flows are inputs and outputs, and storage is the ideas and beliefs of the society.

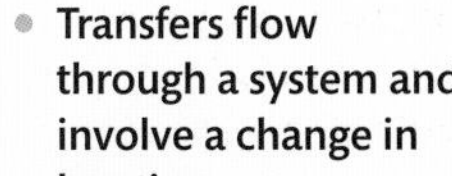

- **Transfers flow through a system and involve a change in location.**
- **Transformations lead to an interaction within a system in the formation of a new end product, or involve a change of state.**

A diagram can show several storages and the flows between them, and show the relationship between the parts of a complex system. For example, Figure 1.8 shows a systems diagram for the movement of energy and matter in a forest ecosystem.

Boxes in Figure 1.8 show storages, such the atmosphere and biomass (i.e. biological matter). Arrows show flows – inputs to and outputs from storages. The arrows are labelled with different processes, either transfers or transformations.

Transfers include:

- harvesting of forest products
- the fall of leaves and wood to the ground.

Figure 1.8 A forest ecosystem shown as a systems diagram

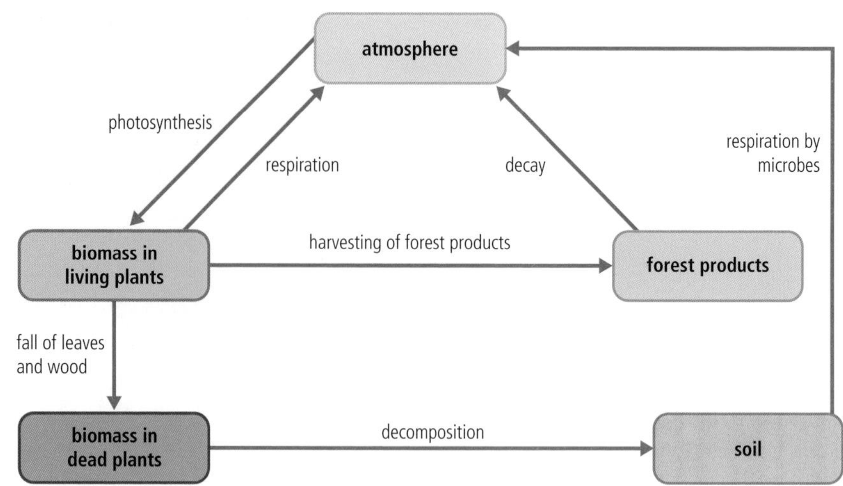

Transformations include:

- photosynthesis – transforming carbon dioxide (CO_2), water (H_2O), and light into biomass and oxygen (O_2)
- respiration – transforming biomass into carbon dioxide and water.

Figure 1.9 Nutrient cycles for (a) a temperate deciduous woodland and (b) an area nearby where the woodland has been cleared for mixed farming. Biomass is all the living material in the ecosystem. Arrows are proportional to the amount of energy present (i.e. larger arrows show greater energy flow than smaller ones).

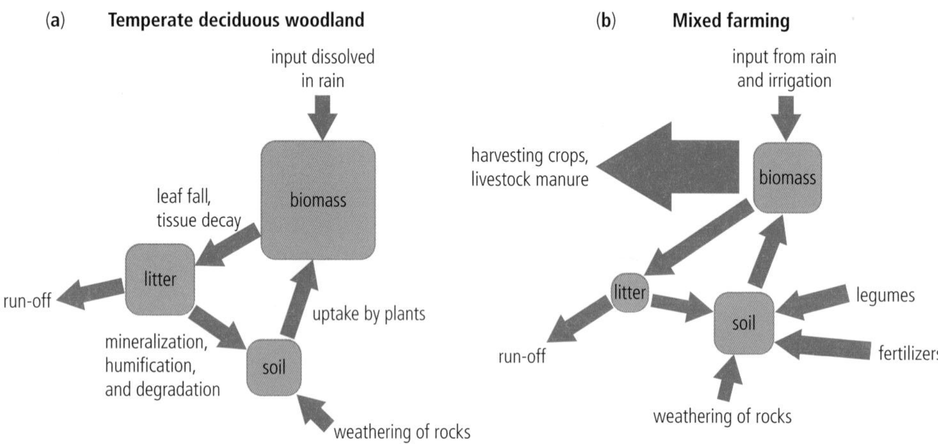

The size of boxes and arrows in systems diagrams can be drawn to represent the size (i.e. magnitude) of the storage or flow. The system diagrams in Figure 1.9 offer information about the different systems by drawing flows and stores proportionally (e.g. biomass store is larger in the woodland; litter store is larger in the woodland; and there is a large output in mixed farming due to the harvested crops and livestock). The diagrams also show that legumes and fertilizers are additional inputs in mixed farming. Extra value can be given to systems diagrams, even simple ones, by showing data quantitatively.

You are expected to be able to apply a systems approach to all the topics covered in this course. You should be able to interpret system diagrams and use data to produce your own for a variety of examples (such as carbon cycling, food production, and soil systems). These ideas are explored in subsequent chapters.

Open, closed, and isolated systems

Systems can be divided into three types, depending on the flow of energy and matter between the system and the surrounding environment.

- **Open systems** – Both matter and energy are exchanged across the boundaries of the system (Figure 1.10a). Open systems are organic (i.e. living) and so must interact with their environment to take in energy and new matter, and to remove wastes (e.g. an ecosystem). People are also open systems in that they must interact with their environment in order to take in food, water, and obtain shelter, and produce waste products.

- **Closed systems** – Energy but not matter is exchanged across the boundaries of the system (Figure 1.10b). Examples are atoms and molecules, and mechanical systems. The Earth can be seen as a closed system: input = solar radiation (Sun's energy or light), output = heat energy. Matter is recycled within the system. Although space ships and meteorites can be seen as moving a small amount of matter in and out of the Earth system, they are generally discounted. Strictly, closed systems do not occur naturally on Earth, but all the global cycles of matter (e.g. the water and nitrogen cycles) approximate to closed systems. Closed systems can also exist experimentally (e.g. Biosphere II).
- **Isolated systems** – Neither energy nor matter is exchanged across the boundary of the system (Figure 1.10c). These systems do not exist naturally, although it is possible to think of the entire universe as an isolated system.

An open system exchanges both energy and matter with its surroundings, a closed system exchanges energy but not matter, and an isolated system does not exchange anything with its surroundings.

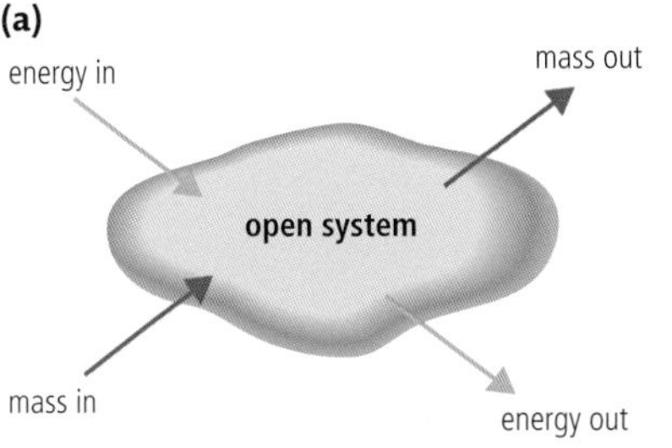

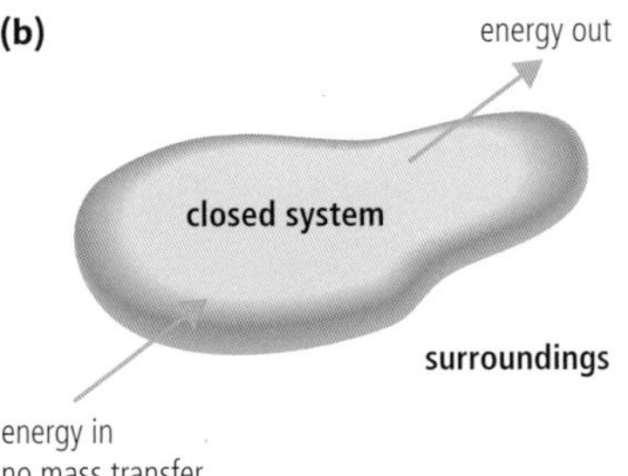

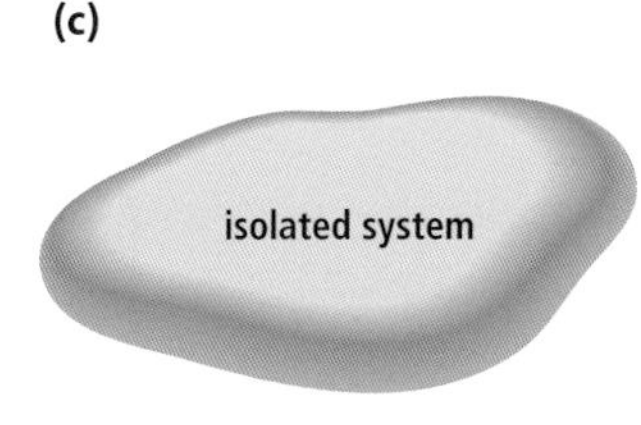

Figure 1.10 The exchange of matter (mass) and energy across the boundary of different systems. Open systems (a) exchange both; closed systems (b) exchange only energy, and isolated systems (c) exchange neither.

Biosphere II encloses an area equivalent to 2.5 football pitches, and contains five different biomes (ocean with coral reef, mangrove, rainforest, savannah, and desert). Further areas explore agricultural systems and human impact on natural systems.

Biosphere II is an experiment to model the Earth as a closed system. It was constructed in Arizona between 1987 and 1991 and enables scientists to study the complex interactions of natural systems (e.g. the constantly changing chemistry of the air, water, and soil within the greenhouses), and the possible use of closed biospheres in space colonization. It allows the study and manipulation of a biosphere without harming the Earth. The project is still running and has resulted in numerous scientific papers showing that small, closed ecosystems are complex and vulnerable to unplanned events, such as fluctuations in CO_2 levels experienced during the experiment, a drop in oxygen levels due to soils over-rich in soil microbes, and an over-abundance of insect pests that affected food supply.

TOK

An isolated system does not exchange matter or energy with its surroundings, and therefore cannot be observed. Is this a useful concept?

TOK

Systems are hierarchical, and what may be seen as the whole system in one investigation may be seen as only part of another system in a different study (e.g. a human can be seen as a whole system, with inputs of food and water and outputs of waste, or as part of a larger system such as an ecosystem or social system). Difficulties may arise as to where the boundaries are placed, and how this choice is made.

Models

A model is a simplified description designed to show the structure or workings of an object, system or concept.

A **model** is a simplified version of reality. Models can be used to understand how systems work and predict how they will respond to change. Computer models use current and past data to generate future predictions. For example, they are used to predict how global surface temperatures will change during the 21st century (Chapter 7). All models have strengths and limitations and inevitably involve some simplification and loss of accuracy.

You are expected to be able to construct a system diagram or a model from a given set of information.

Some models are complex, such as the computer models that predict the effect of climate change. Other models, even of complex systems, are simpler (Figure 1.11).

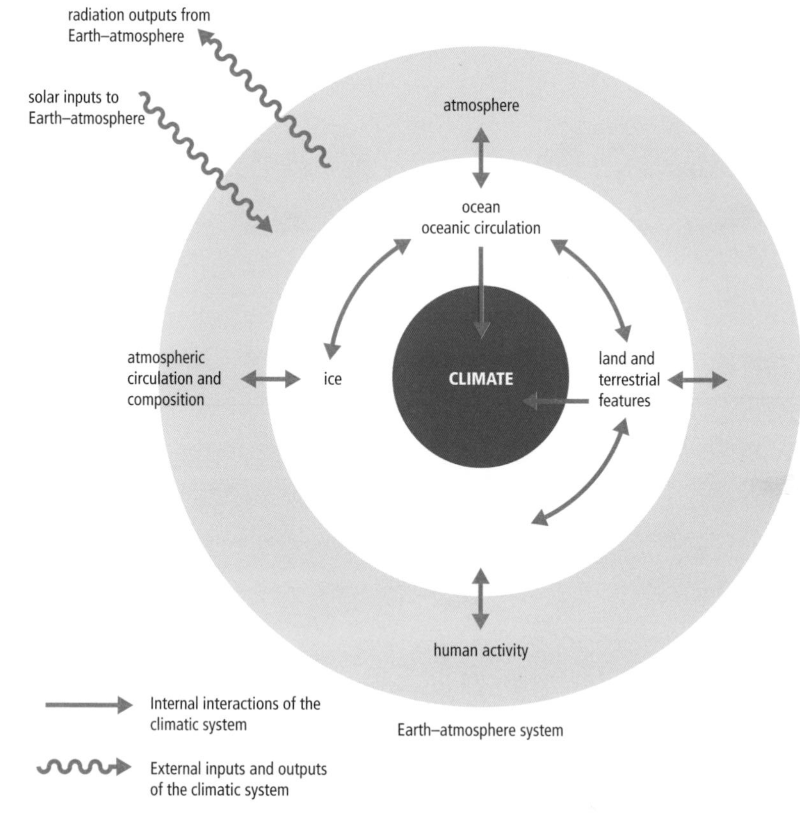

Figure 1.11 A model of the climatic system

Figure 1.12 Models are simplified versions of reality. They can show much about the main processes in an ecosystem and show key linkages. This is a model of a forest ecosystem.

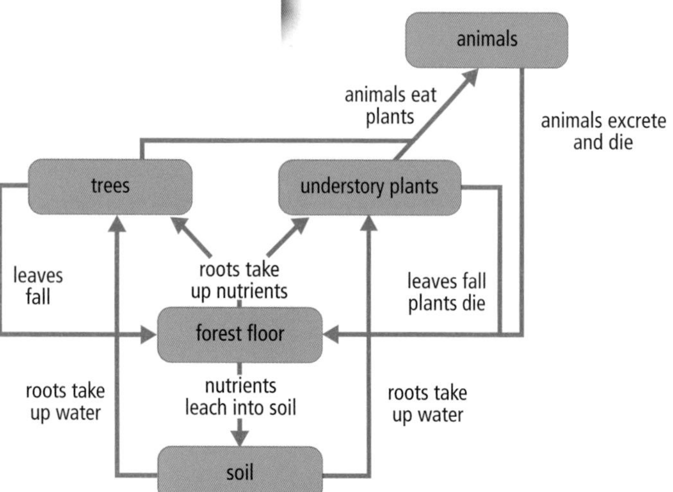

Models can be used to show the flows, storages, and linkages within systems, using the diagrammatic approach (pages 21–22). For example, Figure 1.12 shows a model of an ecosystem. While unable to show much of the complexity of the real system, Figure 1.12 still helps us to understand basic ecosystem function.

Evaluating the use of models

Strengths of models

- Models allow scientists to simplify complex systems and use them to predict what will happen if there are changes to inputs, outputs, or storages.
- Models allow inputs to be changed and outcomes examined without having to wait a long time, as we would have to if studying real events.
- Models allow results to be shown to other scientists and to the public, and are easier to understand than detailed information about the whole system.

Limitations of models

- Environmental factors are very complex with many interrelated components, and it may be impossible to take all variables into account.
- Different models may show different effects using the same data. For example, models used to predict the effect of climate change can give very different results.
- Models themselves may be very complex and when they are oversimplified they may become less accurate. For example, there are many complex factors involved in atmospheric systems.
- Because many assumptions have to be made about these complex factors, models such as climate models may not be accurate.
- The complexity and oversimplification of climate models, for example, has led some people to criticize these models.
- Different models use slightly different data to calculate predictions.
- Any model is only as good as the data used. The data put into the model may not be reliable.
- Models rely on the expertise of the people making them and this can lead to impartiality.
- As models predict further into the future, they become more uncertain.
- Different people may interpret models in different ways and so come to different conclusions. People who would gain from the results of the models may use them to their advantage.

You need to be able to evaluate the use of models as a tool in a given situation, for example climate change predictions.

The need for models to summarize complex systems requires approximation techniques to be used: these can lead to loss of information and oversimplification. A model inevitably involves some approximation and therefore loss of accuracy. The advantage of models is that they can clearly illustrate links between parts of the system, and give a clear overview of complex interrelationships.

Exercises

1. How does the holistic approach to systems differ from the reductionist approach of conventional science? What are the advantages of the holistic approach compared to the conventional approach?
2. Draw a table comparing and contrasting open, closed, and isolated systems. Comparisons should be made in terms of the exchange of matter and energy with their surroundings. Give examples for each.
3. What is meant by transfer within a system? How does this differ from transformation processes?
4. Draw a systems diagram showing the inputs, outputs, and storages of a forest ecosystem.
5. Draw a table listing the strengths and weaknesses of models. Your table could summarize the issues regarding one particular model (e.g. climate change) or be more generally applicable.

Big questions

Having read this section, you can now discuss the following big questions:

- What strengths and weaknesses of the systems approach and the use of models have been revealed through this topic?

Points you may want to consider in your discussions:

- How does a systems approach facilitate a holistic approach to understanding?
- What are the strengths and weaknesses of the systems you have examined in this section?
- What have you learned about models and how they can be used, for example, to predict climate change? Do their benefits outweigh their limitations?

1.3 Energy and equilibria

Significant ideas

The laws of thermodynamics govern the flow of energy in a system and the ability to do work.

Systems can exist in alternative stable states or as equilibria between which there are tipping points.

Destabilizing, positive feedback mechanisms drive systems towards these tipping points, whereas stabilizing, negative feedback mechanisms resist such changes.

Big questions

As you read this section, consider the following big questions:

- What strengths and weaknesses of the systems approach and the use of models have been revealed through this topic?
- How are the issues addressed in this topic of relevance to sustainability or sustainable development?

Knowledge and understanding

- The first law of thermodynamics is the principle of conservation of energy, which states that energy in an isolated system can be transformed but cannot be created or destroyed.
- The principle of conservation of energy can be modelled by the energy transformations along food chains and energy production systems.
- The second law of thermodynamics states that the entropy of a system increases over time. Entropy is a measure of the amount of disorder in a system. An increase in entropy arising from energy transformations reduces the energy available to do work.
- The second law of thermodynamics explains the inefficiency and decrease in available energy along a food chain and energy generation systems.
- As an open system, an ecosystem will normally exist in a stable equilibrium, either a steady-state or one developing over time (e.g. succession), and maintained by stabilizing negative feedback loops.
- Negative feedback loops (stabilizing) occur when the output of a process inhibits or reverses the operation of the same process in such a way to reduce change – it counteracts deviation.
- Positive feedback loops (destabilizing) will tend to amplify changes and drive the system towards a tipping point where a new equilibrium is adopted.
- The resilience of a system, ecological or social, refers to its tendency to avoid such tipping points, and maintain stability.
- Diversity and the size of storages within systems can contribute to their resilience and affect the speed of response to change (time lags).

- Humans can affect the resilience of systems through reducing these storages and diversity.
- The delays involved in feedback loops make it difficult to predict tipping points and add to the complexity of modelling systems.

Laws of thermodynamics and environmental systems

Energy exists in a variety of forms (light, heat, chemical, electrical, and kinetic). It can be changed from one form into another but cannot be created or destroyed. Any form of energy can be converted to any other form, but heat can be converted to other forms only when there is a temperature difference. The behaviour of energy in systems is defined by the laws of thermodynamics. There are two laws, which relate to how energy moves through systems.

First law of thermodynamics

The **first law of thermodynamics** states that energy can neither be created nor destroyed: it can only change form. This means that the total energy in any system, including the entire universe, is constant and all that can happen is change in the form the energy takes. This law is known as the law of conservation of energy. In ecosystems, energy enters the system in the form of sunlight, is converted into biomass via photosynthesis, passes along food chains as biomass, is consumed, and ultimately leaves the ecosystem in the form of heat. No new energy has been created – it has simply been transformed and passed from one form to another (Figure 1.13). Heat is released because of the inefficient transfer of energy (as in all other systems).

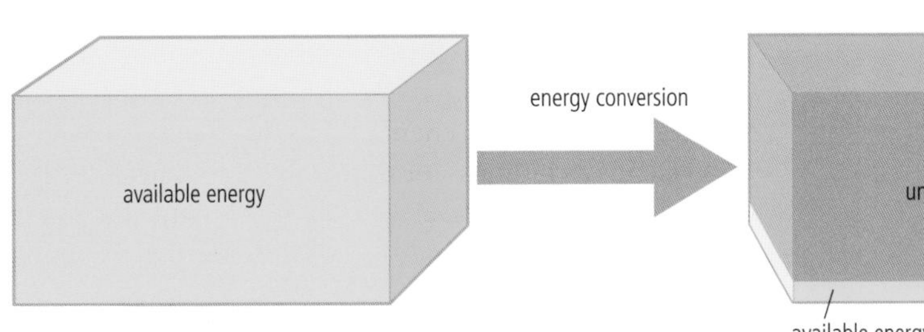

Figure 1.13 Energy cannot be created or destroyed: it can only be changed from one form into another. The total energy in any system is constant, only the form can change.

Available energy is used to do work such as growth, movement, and the assembly of complex molecules. Although the total amount of energy in a system does not change, the amount of available energy does (Figure 1.13).

The available energy in a system is reduced through inefficient energy conversions. The total amount of energy remains the same, but less is available for work. An increasing quantity of unusable energy is lost from the system as heat (which cannot be recycled into useable energy).

The first law of thermodynamics concerns the conservation of energy (i.e. energy can be neither created nor destroyed); whereas the second law explains that energy is lost from systems when work is done, bringing about disorder (entropy).

Second law of thermodynamics

The transformation and transfer of energy is not 100 per cent efficient: in any energy conversion there is less usable energy at the end of the process than at the beginning (Figure 1.15). This means there is a dissipation of energy which is then not available for work. The **second law of thermodynamics** states that energy goes from a concentrated form (e.g. the Sun) into a dispersed form (ultimately heat): the availability of energy to do work therefore decreases and the system becomes increasingly disordered.

The first law of thermodynamics explains how some of the energy entering an ecosystem is lost as heat energy, because energy entering must equal energy remaining in the system plus energy leaving the system. The second law of thermodynamics explains how energy transformations in living systems can lead to loss of energy from the system. The order in living systems is only maintained by constant input of new energy from the Sun.

Entropy is a measure of the amount of disorder in a system. An increase in entropy arising from energy transformations reduces the energy available to do work.

TOK

The laws of thermodynamics are examples of scientific laws. In what ways do scientific laws differ from the laws of human science subjects, such as economics?

You need to be able to explain the implications of the laws of thermodynamics to ecological systems.

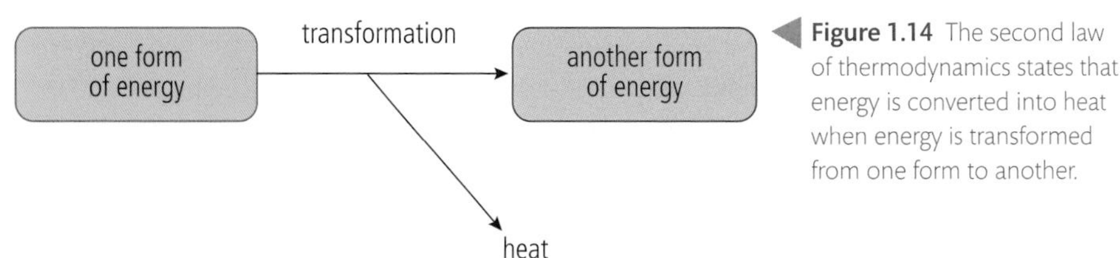

Figure 1.14 The second law of thermodynamics states that energy is converted into heat when energy is transformed from one form to another.

Energy is needed to create order (e.g. to hold complex molecules together). The second law states that the disorder in a system increases over time. Disorder in a system is called **entropy**. An increase in entropy arising from energy transformations reduces the energy available to do work. Therefore, as less energy becomes available, disorder (entropy) increases. In any isolated system, where there is no new input of energy, entropy tends to increase spontaneously. The universe can be seen as an isolated system in which entropy is steadily increasing so eventually, in billions of years' time, no available energy will be present.

The laws of thermodynamics and environmental systems

Natural systems can never actually be isolated because there must always be an input of energy for work (to replace energy that is dissipated). The maintenance of order in living systems requires a constant input of energy to replace available energy lost through inefficient transfers. Although matter can be recycled, energy cannot, and once available energy has been lost from a system in the form of heat energy it cannot be made available again.

One way energy enters an ecosystem is as sunlight energy. This sunlight energy is then changed into biomass by photosynthesis: this process captures sunlight energy and transforms it into chemical energy. Chemical energy in producers is passed along food chains as biomass, or transformed into heat during respiration. Available energy is used to do work such as growth, movement, and making complex molecules. As we know from the second law of thermodynamics, the transfer and transformation of energy is inefficient with all energy ultimately being lost into the environment as heat. This is why food chains tend to be short.

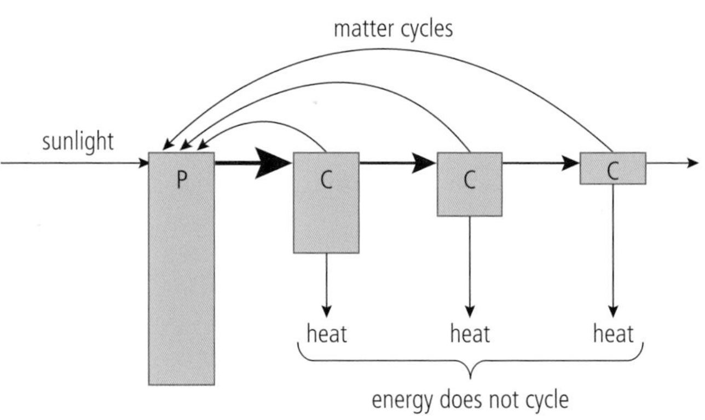

Figure 1.15 Energy flow through a food chain; P = producers, C = consumers. The boxes show energy available to do work at each feeding level. Energy decreases through the food chain as some is converted to heat. The '10 per cent rule' indicates that on average only around 10 per cent of the available energy is passed on to the next trophic level.

The nature of equilibria

Open systems tend to have a state of balance among the components of a system – they are in a state of **equilibrium**. This means that although there may be slight fluctuations in the system, there are no sudden changes and the fluctuations tend to be between closely defined limits. Equilibrium allows systems to return to an original state following disturbance. Two different types of equilibrium are discussed below.

Steady-state equilibrium

A **steady-state equilibrium** is the common property of most open systems in nature. Despite constant inputs and outputs of energy and matter, the overall stability of the system remains. In steady-state equilibrium there are no changes over the longer term, but there may be oscillations in the very short term. Fluctuations in the system are around a fixed level and deviation above or below results in a return towards this average state (Figure 1.16).

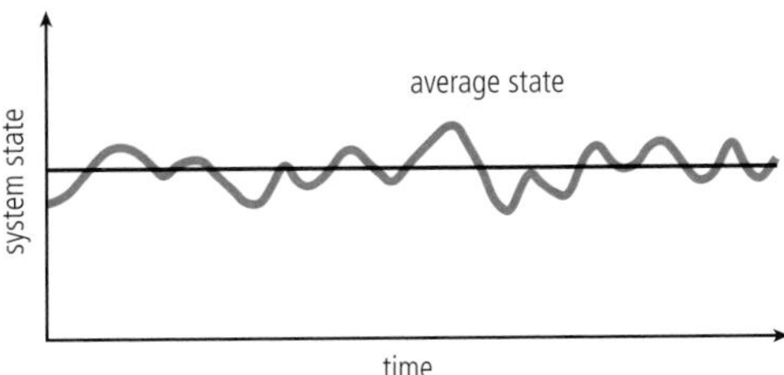

Figure 1.16 The conditions of an open system fluctuate around an average state in steady-state equilibrium.

There is a tendency in natural systems for the equilibrium to return after disturbance, but some systems (e.g. succession) may undergo long-term changes to their equilibrium until reaching a steady-state equilibrium with the climax community (Chapter 2, pages 118–119).

The stability of steady-state equilibrium means that the system can return to the steady state following disturbance. For example, the death of a canopy tree in the rainforest leaves a gap in the canopy, which eventually closes again through the process of succession (page 32 and Chapter 2, pages 114–115). Homeostatic mechanisms in animals maintain body conditions at a steady state – a move away from the steady state results in a return to the equilibrium (for example, temperature control in humans – see page 31). (You may come across the term 'dynamic equilibrium' to describe this phenomenon, but it is not used in this course.)

A steady-state equilibrium is the condition of an open system in which there are no changes over the longer term, but in which there may be oscillations in the very short term.

Static equilibrium

In **static equilibrium**, there are no inputs or outputs of matter or energy and no change in the system over time (Figure 1.17).

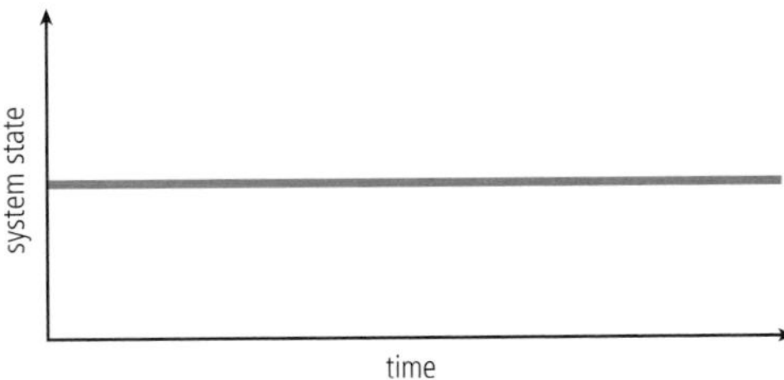

Figure 1.17 Static equilibrium

Inanimate objects such as a chair or table are in static equilibrium. No natural systems are in static equilibrium because all have inputs and outputs of energy and matter.

Most open systems have steady-state equilibrium, where any change to a stable system results in a return to the original equilibrium after the disturbance. Negative feedback (page 31) mechanisms return the system to the original state. This is because there are inputs and outputs of energy and matter to the system that allow this to happen. Static equilibrium is when there is no input or output from the system, and no change occurs; this does not apply to any natural system.

Stable and unstable equilibrium

A stable equilibrium is the condition of a system in which there is a tendency for it to return to the previous equilibrium following disturbance.

If a system returns to the original equilibrium after a disturbance, it is a **stable equilibrium** (Figure 1.18a and b). A system that does not return to the same equilibrium but forms a new equilibrium is an **unstable equilibrium** (Figure 1.19a and b). Positive feedback mechanisms (see below) can lead to a system moving away from its original equilibrium.

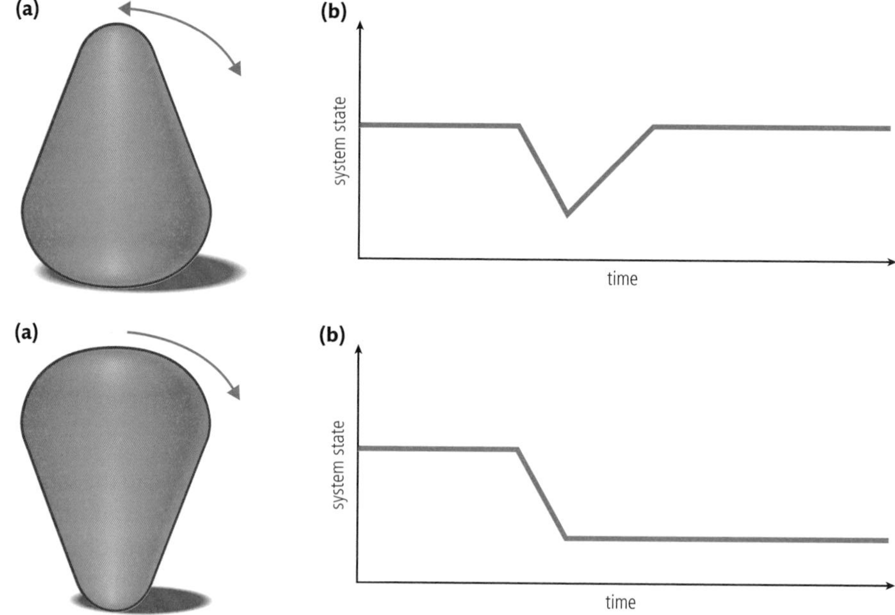

Figure 1.18 (a) Disturbance to the system results in it returning to its original equilibrium. (b) Immediately following disturbance, conditions may be very different in the system, but eventually return to the original equilibrium.

Figure 1.19 (a) Disturbance results in a new equilibrium very different from the first (in this case the object lying horizontally rather than standing vertically). (b) Scientists believe that the Earth's climate may reach a new equilibrium following the effects of global warming, with conditions on the planet dramatically altered.

Positive and negative feedback

Homeostatic systems in animals require **feedback** mechanisms to return them to their original steady state. This is also true of all other systems. Such mechanisms allow systems to self-regulate (Figure 1.20). Feedback loops can be positive or negative.

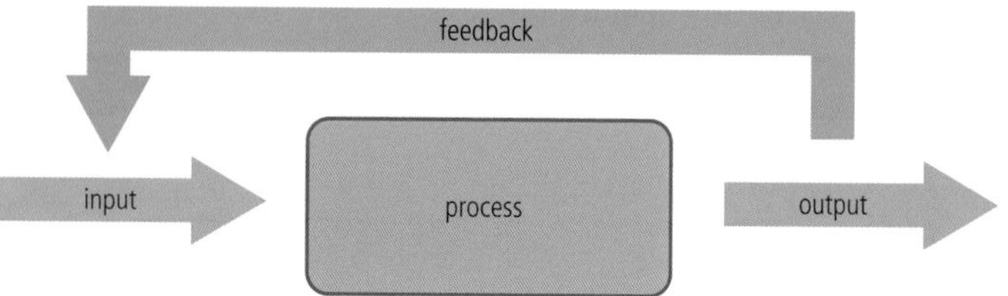

Figure 1.20 Changes to the processes in a system lead to changes in the level of output. This feeds back to affect the level of input.

Positive feedback

Positive feedback occurs when a change in the state of a system leads to additional and increased change. Thus, an increase in the size of one or more of the system's outputs feeds back into the system and results in self-sustained change that alters the state of a system away from its original equilibrium towards instability (Figure 1.21). For example, increased temperature through global warming melts more of the ice in the polar ice caps and glaciers, leading to a decrease in the Earth's **albedo** (reflection from the Earth's surface) – the Earth absorbs more of the Sun's energy which makes the temperature increase even more, melting more ice (Chapter 7, pages 377–378). Exponential population growth is also an example of positive feedback.

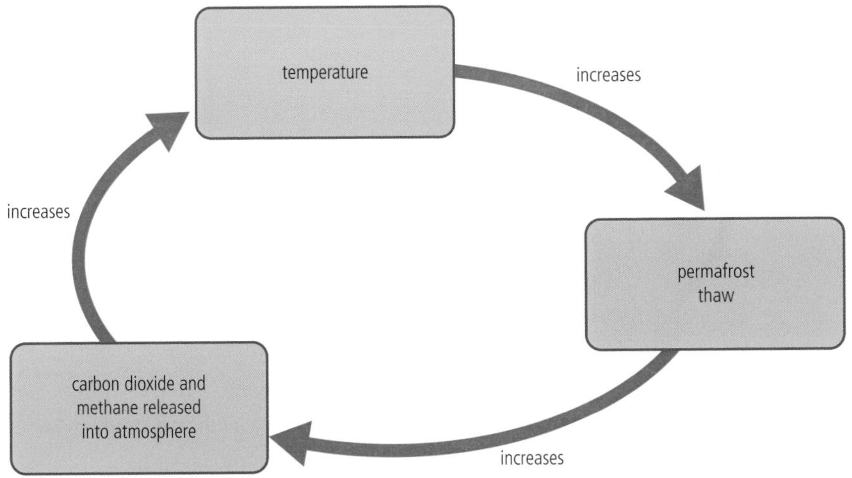

Figure 1.21 A positive feedback mechanism enhancing climate change. Such mechanisms are often linked to tipping points, when the system becomes unstable and forms a new equilibrium.

Case study

Humans, resources, and space

Human population is growing at an ever-increasing rate – more people on the planet produce more children (positive feedback) and the rate will continue to increase as long as there are sufficient resources available to support the population. Human population is growing exponentially, which means that growth rate is proportional to its present size.

Some 2000 years ago, the Earth's population was about 300 million people. In 2015, it was 7.3 billion. It took the human population thousands of years to reach 1 billion, which it did in 1804. However, it took only 123 years to double to 2 billion in 1927. The population doubled again to 4 billion in 1974 (after only 47 years), and if it continues at the current rate it will reach 8 billion in 2028. Doubling from the 2015 figure of 7.3 billion to 14.6 billion will have a much greater impact than any previous doubling because of the increased gap between the potential food supply (arithmetic growth) and population size (geometric growth).

Negative feedback

Negative feedback can be defined as feedback that counteracts any change away from equilibrium, contributing to stability. Negative feedback is a method of control that regulates itself. An ecosystem, for example, normally exists in a stable equilibrium, either a steady-state equilibrium or one developing over time (e.g. succession, page 114), because it is maintained by stabilizing negative feedback loops. Steady-state equilibrium in the human body is also maintained by negative feedback. For example, in temperature control, an increase in the temperature of the body results in increased sweat release and vasodilation, thus increasing evaporation of sweat from the skin, cooling the body and returning it to its original equilibrium. On a larger scale, increased release of carbon dioxide through the burning of fossil fuels leads to enhanced plant growth through increased photosynthesis. This reduces atmospheric levels of carbon dioxide. Negative feedback mechanisms are stabilizing forces within systems. They counteract deviation. Consider Figure 1.22: if high winds blow down a tree in the rainforest, a gap is left in the canopy and more light is let in to the forest floor. This encourages new growth: rates of growth are rapid as light levels are high, so new saplings compete to take the place of the old tree in the canopy and equilibrium is restored. In this way, negative feedback and succession (page 114) have closed the gap.

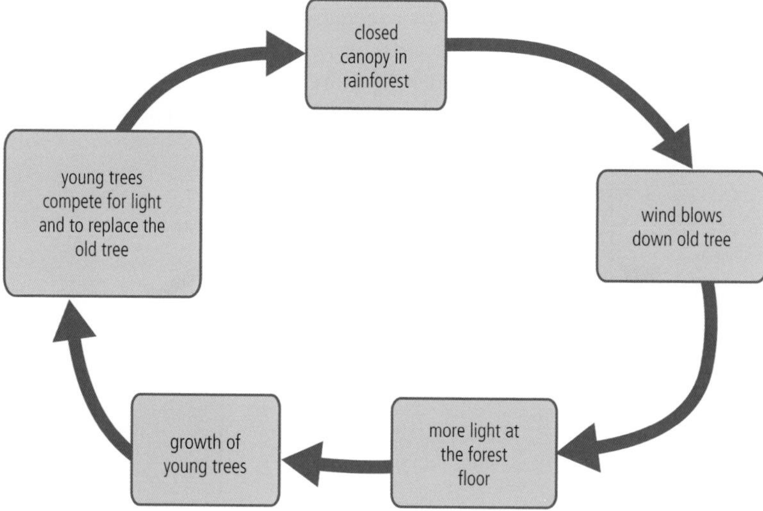

Figure 1.22 Negative feedback can lead to steady-state equilibrium in a rainforest. Gaps in the forest canopy are closed when young trees compete for light and replace the old tree.

Predator–prey relationships are another example of negative feedback (page 66).

Feedback refers to the return of part of the output from a system as input, so as to affect succeeding outputs. There are two type of feedback.

- **Negative feedback tends to reduce, neutralize, or counteract any deviation from an equilibrium, and promotes stability.**
- **Positive feedback amplifies or increases change; it leads to exponential deviation away from an equilibrium.**

A system may contain both negative and positive feedback loops resulting in different effects within the system (Figure 1.23).

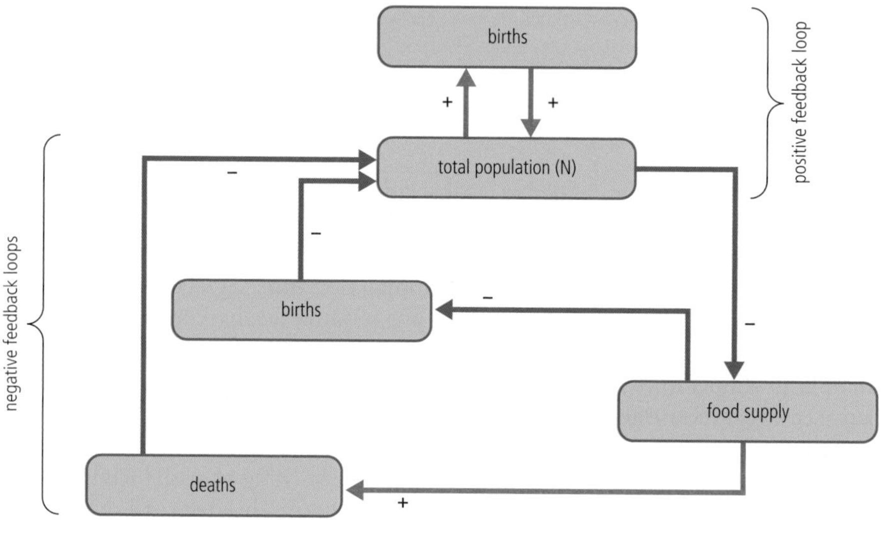

Figure 1.23 Population control in animal populations contains both negative and positive feedback loops.

Tipping points

A **tipping point** is a critical threshold when even a small change can have dramatic effects and cause a disproportionately large response in the overall system. Positive feedback loops are destabilizing and tend to amplify changes and drive the system towards a tipping point where a new equilibrium is adopted (Figure 1.24; Figure 1.19, page 30). Most projected tipping points are linked to climate change (Chapter 7), and represent points beyond which irreversible change or damage occurs. Increases in CO_2 levels above a certain value (450 ppm) would lead to increased global mean temperature, causing melting of the ice sheets and permafrost (Chapter 7). Reaching such a tipping would, for example, cause long-term damage to societies, the melting of Himalayan mountain glaciers, and a lack of fresh water in many Asian societies.

Positive feedback loops (destabilizing) will tend to amplify changes and drive the system towards a tipping point where a new equilibrium is adopted.

A tipping point is the minimum amount of change within a system that will destabilize it, causing it to reach a new equilibrium or stable state.

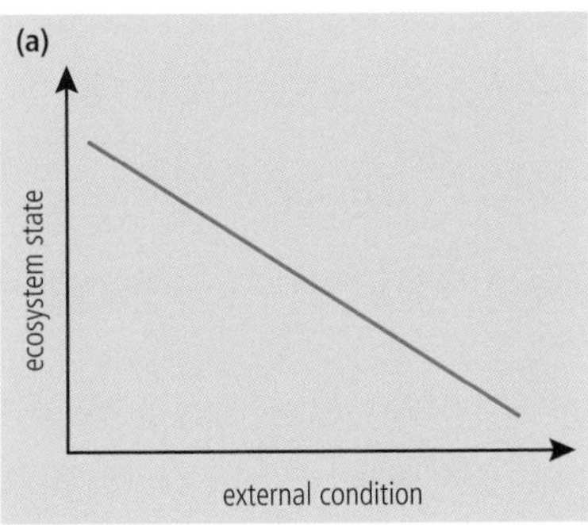

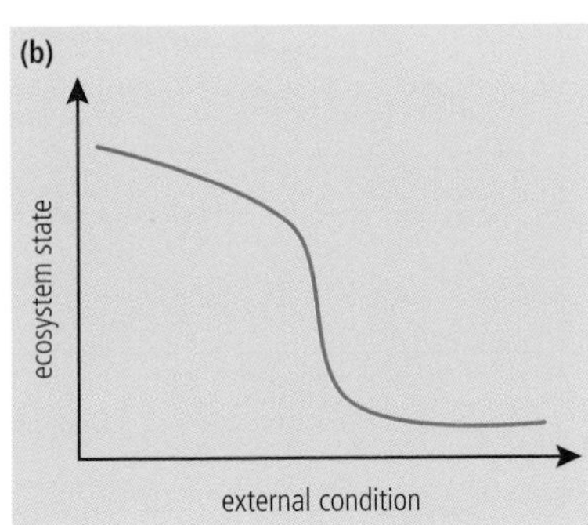

Figure 1.24 How different types of ecosystem may respond to changing external conditions

If external conditions in the environment, such as nutrient input or temperature, change gradually, then ecosystem state may respond gradually (Figure 1.24a), in which case there are no tipping points involved. In other cases, there may be little response below a certain threshold, but fast changes in the system can occur once the threshold is reached (Figure 1.24b), even though a small change in environmental conditions has occurred – in such cases, a tipping point has been reached.

Case study

Krill harvesting in the southern ocean

Krill is a small shrimp-like crustacean that is a food source for seals, whales, penguins, and other seabirds. Krill is harvested to produce food for farmed fish and nutritional supplements for people. Research into the effects of Antarctic krill in the seas near South Georgia have indicated the level of fishing that is sustainable, beyond which a tipping point would be reached leading to rapid change in the southern ocean ecosystem. Krill form the base of the food chain, and so significant reduction in their population density severely affects other animals (e.g. gentoo and macaroni penguins, and Antarctic fur seal). The study showed that animals that feed on the krill begin to suffer when the krill population declined below a critical level of 20 g m^{-2}, which is approximately one-third of the maximum measured amount of krill available (Figure 1.25). This critical level is also shown in seabird species around the world, from the Arctic to the Antarctic, and from the Pacific to the Atlantic.

Antarctic krill (*Euphausia superba*). Krill live in huge swarms which can be kilometres across and reach densities of 10 000 individuals per cubic metre. Each individual is at most 6 cm long. They feed on algae. Krill are a major food species for a wide array of oceanic creatures, ranging from small fish, such as sardines and herring, to blue whales.

continued

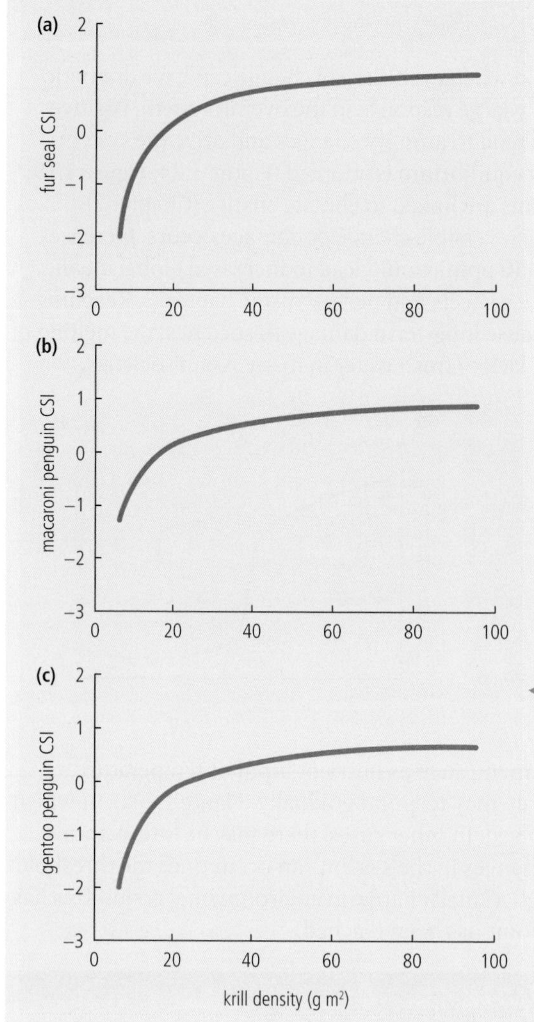

Figure 1.25 Graphs showing the effect of changes in krill density on upper trophic level predators (fur seals, macaroni penguins, and gentoo penguins). The combined standardized index (CSI) uses a range of variables to assess the health of predator populations (e.g. population size, breeding performance, offspring growth rate, foraging behaviour and diet). Data show that a tipping point is reached at 20 g m^{-2} of krill.

To learn more about conservation of the Antarctic ocean ecosystem, go to www.pearsonhotlinks.co.uk, enter the book title or ISBN, and click on weblink 1.2.

Systems at threat from tipping points include:

- Antarctic sea ecosystems (case study)
- Arctic sea-ice
- Greenland ice sheet
- West Antarctic ice sheet
- El Niño Southern Oscillation (ENSO)
- West African monsoon
- Amazon rainforest
- boreal forest.
- thermohaline circulation (THC) (Chapter 7).

Some of these are discussed below.

El Niño Southern Oscillation

To learn more about climate patterns, go to www.pearsonhotlinks.co.uk, enter the book title or ISBN, and click on weblink 1.3.

El Niño Southern Oscillation (ENSO) refers to fluctuation in sea surface temperatures across the Pacific Ocean, with oscillations occurring every 3 to 7 years. Warming and cooling of the tropical eastern Pacific Ocean (i.e. off the west coast of South America) are known as El Niño and La Niña, respectively. Because ocean circulation has a global extent (Chapter 4, pages 221–222), ENSO can have large-scale effects on the global climate system, and cause extreme weather such as droughts and floods. El Niño events, for example, can lead to warm and very wet weather in the months April to October with flooding along the western coast of South America (in countries such as

Peru and Ecuador). At the same time, drought occurs in Australia, Malaysia, Indonesia, and the Philippines; warmer than normal winters occur in northern USA and Canada, with greater rainfall in south-west USA, and droughts in Africa and India. Developing countries bordering the Pacific Ocean (on both its eastern and western extremes) are particularly affected by ENSO events.

West African monsoon

The heavy rains that occur in West Africa are affected by sea surface temperature. A change in global mean temperature of 3–5 °C could lead to a collapse of the West African monsoon. With reduced rainfall in western Africa, more moisture would reach areas such as the Sahara, which could lead to increased rainfall and a 'greening' as more vegetation grows.

Amazon rainforest

Increased temperatures due to climate change, and the effects of deforestation through logging and land clearance, could lead to a tipping point in the Amazon. Rainforest creates its own weather patterns, with high levels of **transpiration** (evaporation of water from leaves) leading to localized rainfall. Drier condition would lead to increased likelihood of forest fires, and reduced forest extent through forest dieback: loss of trees would result in less transpiration, with more water ending up in rivers and ultimately the sea rather than in the forest. The ultimate decrease in water circulating locally would result in a tipping point being reached, leading to the desertification of the Amazon basin.

Boreal forest

Boreal forest, or Taiga (Chapter 2, page 76), is characterized by coniferous trees such as pines. It is the Earth's most extensive biome and is found throughout the northern hemisphere. Research suggests that a 3 °C increase in mean global temperature may be the threshold for loss of the boreal forest, caused by increased water stress, decreased tree reproduction rates, increased vulnerability to disease, and fire.

Figure 1.26 Likelihood and possible and impacts of tipping points resulting from climate change

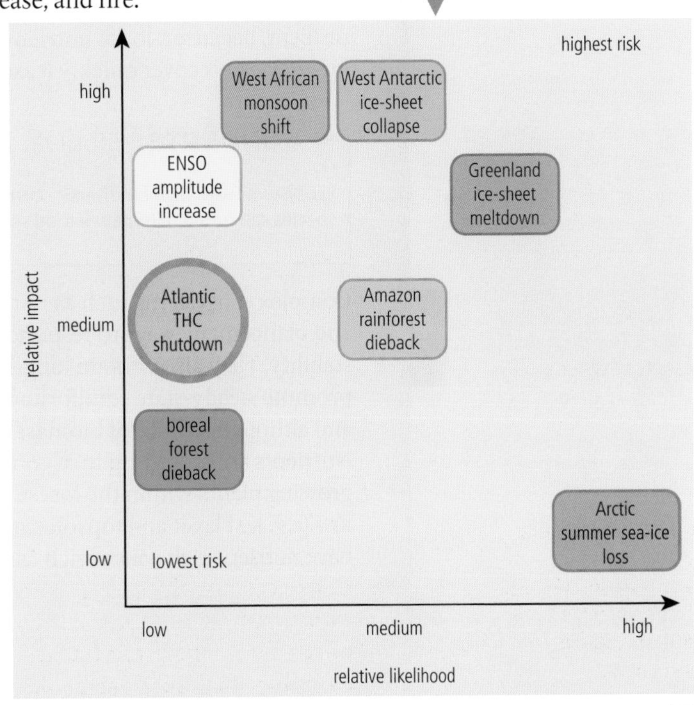

The likelihood and possible impacts of these tipping points are shown in Figure 1.26.

Models are used to predict tipping points and, as you have already seen, such models have strengths and limitations (page 25). The delays involved in feedback loops make it difficult to predict tipping points and add to the difficulty of modelling systems. Other problems with predicting tipping points include:

- There is no globally accepted definition of what is meant by the term *tipping point*: how different do two system states need to be to say that a tipping point has been reached?
- Not all properties of a system will change abruptly at one time, and so it may be difficult to say when a tipping point has been reached.
- The exact size of the impacts resulting from tipping points have not been fully identified for all systems.

To learn more about climate tipping points, go to www.pearsonhotlinks.co.uk, enter the book title or ISBN, and click on weblink 1.4.

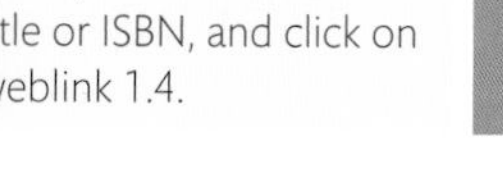

You need to be able to evaluate the possible consequences of tipping points, and have explored various examples of human impacts and possible tipping points.

- It may be difficult to determine the causes of a tipping point – whether it has been reached because of the inherent nature of the system or external factors such as human activity, for example.
- It is difficult to determine the conditions under which ecosystems experience tipping points, because of their complexity.
- Not all systems that could be affected by tipping points have been examined or possibly even identified.
- No one may know the exact tipping point until long after it has happened.

The costs of tipping points, both from environmental and economic perspectives, could be severe, so accurate predictions are critical. Models that predict tipping points are, therefore, essential and have alerted scientists to potential large events. Continued monitoring, research, and modelling is required to improve predictions.

Activities in one part of the globe may lead to a tipping point which influences the ecological equilibrium elsewhere on the planet. For example, fossil fuel use in industrialized countries can lead to global warming which has impact elsewhere, such as desertification of the Amazon basin.

Resilience and diversity in systems

The **resilience** of a system, ecological or social, refers to its tendency to avoid tipping points, and maintain stability through steady-state equilibrium. Diversity and the size of storages within systems can contribute to their resilience and affect the speed of response to change. Large storages, or high diversity, will mean that a system is less likely to reach a tipping point and move to a new equilibrium. Humans can affect the resilience of systems through reducing these storages and diversity. Tropical rainforests, for example, have high **diversity** (i.e. a large number and proportions of species present – see page 138) but catastrophic disturbance through logging (i.e. rapid removal of tree biomass storages) or fires can lower its resilience and can mean it takes a long time to recover. Natural grasslands, in contrast, have low diversity but are very resilient, because a lot of nutrients are stored below ground in root systems, so after fire they can recover quickly (case study).

CONCEPTS: Biodiversity

Ecosystems with high biodiversity contain complex food webs which make them resistant to change – species can turn to alternative food sources if one species is reduced or lost from the system.

You need to be able to discuss resilience in a variety of systems.

Complex ecosystems such as rainforests have complex food webs which allow animals and plants many ways to respond to disturbance of the ecosystem and thus maintain stability. They also contain long-lived species and dormant seeds and seedlings that promote steady-state equilibrium. Rainforests have thin, low-nutrient soils, however, and although storage of biomass in trees is high, nutrient storage in soils is low. Nutrients are locked-up in decomposing plant matter on the surface and in rapidly growing plants within the forest, so when the forest is disturbed, nutrients are quickly lost (e.g. leaf layer and topsoil can be washed away). Ecosystems with higher resilience have nutrient-rich soils which can promote new growth.

Case study

Disturbance of tall grass prairie

Tall grass prairie is a native ecosystem to central USA. High diversity, complex food webs and nutrient cycles in this ecosystem maintain stability. The grasses are between 1.5 and 2 m in height, with occasional stalks as high as 2.5 or 3 m. Due to the build-up of organic matter, these prairies have deep soils and recover quickly following periodic fires which sweep through them; they can quickly return to their original equilibrium. Plants have a growth point below the surface which protects them from fire, also enabling swift recovery.

Tall grass prairie

North American wheat farming has replaced native ecosystems (e.g. tall grass prairie) with a monoculture (a one-species system). Such systems are prone to the outbreak of crop pests and damage by fire – low diversity and low resilience combined with soils that lack structure and need to be maintained artificially with added nutrients lead to poor recovery following disturbance.

Prairie wheat farming

You need to understand the relationships between resilience, stability, equilibria, and diversity, using specific examples to illustrate interactions.

Exercises

1. Summarize the first and second laws of thermodynamics. What do they tell us about how energy moves through a system?
2. What is the difference between a steady-state equilibrium and a static equilibrium? Which type of equilibrium applies to ecological systems and why?
3. **a.** When would a system not return to the original equilibrium, but establish a new one? Give an example and explain why this is the case.

   b. Give an example of a system that undergoes long-term change to its equilibrium while retaining the integrity of the system.
4. Give an example of how an ecosystem's capacity to survive change depends on diversity and resilience.
5. Why does a complex ecosystem provide stability? Include information regarding the variety of nutrient and energy pathways, and the complexity of food webs, in your answer.

Big questions

Having read this section, you can now discuss the following big questions:

- What strengths and weaknesses of the systems approach and the use of models have been revealed through this topic?
- How are the issues addressed in this topic of relevance to sustainability or sustainable development?

You may want to consider the following points in your discussions:

- The principle of conservation of energy can be modelled by the energy transformations along food chains and energy production systems: what are the strengths and limitations of such models?
- How do the delays involved in feedback loops make it difficult to predict tipping points and add to the complexity of modelling systems?
- Do the benefits of the models used to predict tipping points outweigh their limitations?
- How does sustainability reduce the chance that tipping points will be reached?

1.4 Sustainability

Significant ideas

All systems can be viewed through the lens of sustainability.

Sustainable development meets the needs of the present without compromising the ability of future generations to meet their own needs.

Environmental indicators and ecological footprints can be used to assess sustainability.

Environmental Impact Assessments (EIAs) play an important role in sustainable development.

Big questions

As you read this section, consider the following big questions:

- What strengths and weaknesses of the systems approach and the use of models have been revealed through this topic?
- How are the issues addressed in this topic of relevance to sustainability or sustainable development?

Knowledge and understanding

- Sustainability is the use and management of resources that allow full natural replacement of the resources exploited and full recovery of the ecosystems affected by their extraction and use.
- Natural capital is a term used for natural resources that can produce a sustainable natural income of goods or services.
- Natural income is the yield obtained from natural resources
- Ecosystems may provide life-supporting services such as water replenishment, flood and erosion protection, and goods such as timber, fisheries, and agricultural crops.

- Factors such as biodiversity, pollution, population, or climate may be used quantitatively as environmental indicators of sustainability. These factors can be applied on a range of scales from local to global. The Millennium Ecosystem Assessment gave a scientific appraisal of the condition and trends in the world's ecosystems and the services they provide using environmental indicators, as well as the scientific basis for action to conserve and use them sustainably.
- Environmental Impact Assessments (EIAs) incorporate baseline studies before a development project is undertaken. They assess the environmental, social, and economic impacts of the project, predicting and evaluating possible impacts and suggesting mitigation strategies for the project. They are usually followed by an audit and continued monitoring. Each country or region has different guidance on the use of EIAs.
- EIAs provide decision-makers with information in order to consider the environmental impact of a project. There is not necessarily a requirement to implement an EIA's proposals and many socio-economic factors may influence the decisions made.
- Criticisms of EIAs include the lack of a standard practice or training for practitioners, the lack of a clear definition of system boundaries and the lack of inclusion of indirect impacts.
- An ecological footprint (EF) is the area of land and water required to sustainably provide all resources at the rate at which they are being consumed by a given population. Where the EF is greater than the area available to the population, this is an indication of unsustainability.

What is sustainability?

CONCEPTS: Sustainability

Sustainability is the use of natural resources in ways that do not reduce or degrade the resources, so that they are available for future generations. The concept of sustainability is central to an understanding of the nature of interactions between environmental systems and societies. Resource management issues are essentially issues of sustainability.

Sustainability is the use and management of resources that allows full natural replacement of the resources exploited and full recovery of the ecosystems affected by their extraction and use.

Sustainability means using global resources at a rate that allows natural regeneration and minimizes damage to the environment. If continued human well-being is dependent on the goods and services provided by certain forms of **natural capital**, then long-term harvest (and pollution) rates should not exceed rates of capital renewal. For example, a system harvesting renewable resources at a rate that enables replacement by natural growth shows sustainability. Sustainability is living within the means of nature (i.e. on the 'interest' or sustainable income generated by natural capital) and ensuring resources are not degraded (i.e. natural capital is not used up) so that future generations can continue to use the resource. The concept can be applied in our everyday lives. Deforestation can be used to illustrate the concept of sustainability and unsustainability: if the rate of forest removal is less than the annual growth of the forest (i.e. the **natural income**), then the forest removal is sustainable. If the rate of forest removal is greater than the annual growth of the forest, then the forest removal is unsustainable.

When processing a natural resource to create income, sustainability needs to be applied at every level of the supply chain (Figure 1.27).

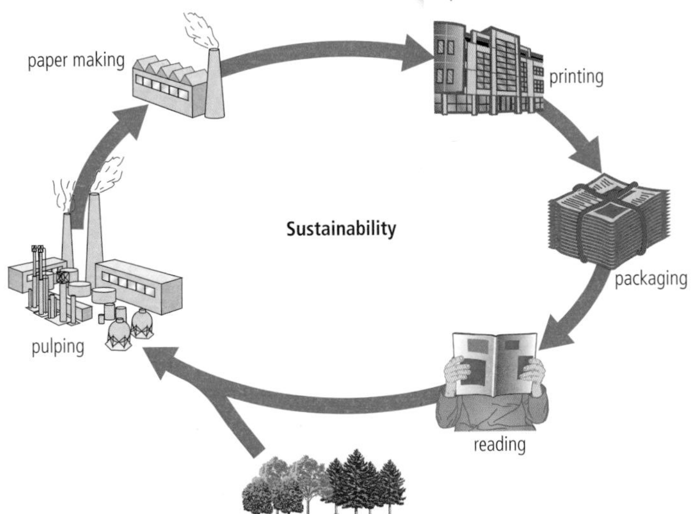

Figure 1.27 Sustainability applies to harvesting natural capital, to the generation of energy to process the product, and to how the product is packaged and marketed.

The term *sustainability* has a precise meaning in this course. It means the use of global resources at a rate that allows natural regeneration and minimizes damage to the environment. For example, a system of harvesting resources at a rate that allows replacement by natural growth demonstrates sustainability.

Sustainability can be encouraged though careful application of:

- ecological land use to maintain habitat quality and connectivity for all species
- sustainable material cycles (e.g. carbon, nitrogen, and water cycles) to prevent the contamination of living systems
- social systems that contribute to a culture of sufficiency that eases the consumption pressures on natural capital.

Humans often use resources beyond sustainable limits through over-population (unrealistic demand for limited resources), financial motives (exploitation of resources for short-term financial gain), or ignorance (lack of knowledge of the resource's sustainable level). For example, unsustainable practice with regard to soils includes:

- overgrazing (trampling and feeding by livestock lead to loss of vegetation and exposure of the underlying soil)
- overcultivation (loss of soil fertility and structure leave topsoil vulnerable to erosion by wind and water).

Local or global?

A global perspective for managing resources sustainably is desirable because many problems have worldwide impact (e.g. global warming, Chapter 7, page 373). Such a perspective allows for understanding the knock-on effects of environmental problems beyond national boundaries and helps governments to be more responsible. Ecosystems are affected by global processes, so sustainability needs to be understood as a global issue (e.g. the atmospheric system with regard to climate change). A global perspective also helps us to understand that our actions have an impact on others, which is useful for getting societies to think about impacts on different generations, as well as different countries. A worldview stresses the interrelationships between systems so knock-on effects are reduced. But because ecosystems exist on many scales, a more local perspective is sometimes appropriate. Human actions are often culturally specific (e.g. traditional farming methods) and so global solutions may not be locally applicable.

Often local methods have evolved to be more sustainable and appropriate for the local environment. It is also often the case that individual and small-scale community action can be very effective for managing resources sustainably (e.g. local recycling schemes). Sometimes a global approach is not appropriate because environmental problems are local in nature as, for example, point-source pollution (page 50).

Sustainable development

Sustainable development means 'meeting the needs of the present without compromising the ability of future generations to meet their own needs'. Sustainable development consists of three pillars: economic development, social development and environmental protection (Figure 1.28). Sustainable development was first clearly defined in 1987 in the Brundtland Report, *Our Common Future,* produced by the United Nations World Commission on Environment and Development (page 8).

However, the definition of sustainable development varies depending on viewpoint, which makes it a problematic term. For example, some economists view sustainable development in purely commercial terms as a stable annual return on investment regardless of the environmental impact, whereas some environmentalists may view it as a stable return without environmental degradation. In the minds of

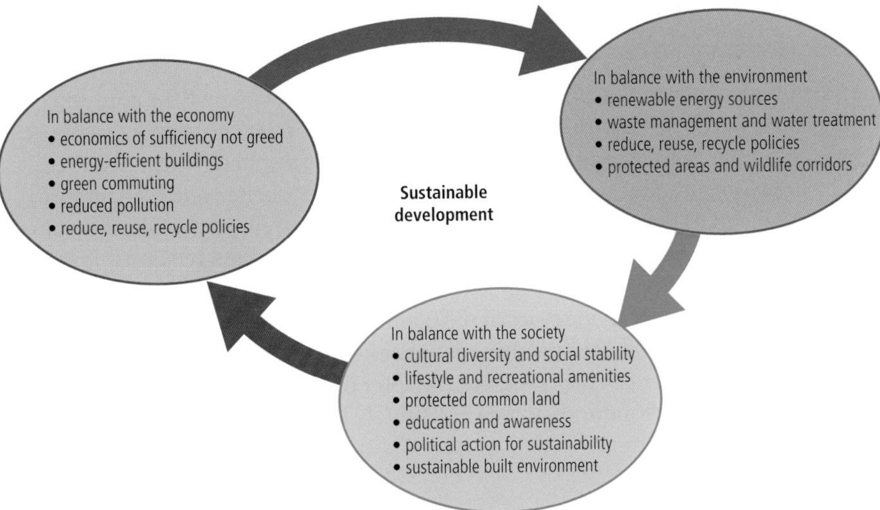

Figure 1.28 Sustainable development focuses on the quality of environmental, economic, and social and cultural development. The concept encompasses ideas and values that inspire individuals and organizations to become better stewards of the environment and promote positive economic growth and social objectives.

many economists, development and sustainability are contradictory positions. Environmentalists hold the concept of sustainable development to be the best way forward for society and the planet. Some people believe that development (particularly development designed to allow LEDCs to compete with MEDCs) can never be sustainable within a free market as the relationship is unequal. The value of this approach is, therefore, a matter of considerable debate.

Is sustainable development possible in the long term? You might think not, because we have finite resources which may not be enough for everyone to use as they want. If people are not prepared to reduce their standards of living, this may well be true. In less developed countries, people are using increasing amounts of resources, and as these countries contain 80 per cent of all people, sustainable development will be difficult. If we cannot find new technologies fast enough to replace fossil fuels, and do not increase our use of renewable resources, non-renewable resources will run out. Population growth is a key factor in sustainable development. If we prove incapable of stopping this growth, or at least slowing it down, there will be more and more pressure on natural resources and increased likelihood of many being used unsustainably.

On the other hand, you may see sustainable development as being possible, given certain precautions. We should be able to develop the technology to use renewable resources for all our needs – micro-generation using wind turbines and solar power, for example (page 358). Renewable resources could provide energy for domestic homes and factories. Transport could use hydrogen-powered engines replacing the need for fossil fuels. We could use less energy in general by insulating our homes and places of work. Personal choices backed up by legislation could make us reuse and recycle more. Technological developments in crop growing could mean more production. Given all these factors, it is possible and certainly desirable for sustainable development to be possible in the long term.

CONCEPTS: Environmental value systems

Ultimately, the choices people make depend on their environmental value system. People with a technocentric worldview see the technological possibilities as central to solving environmental problems. An ecocentric worldview leads to greater caution and a drive to use Earth's natural resources in a sustainable way rather than rely on technology to solve the problems.

Natural capital (resources) and natural income

Natural capital is a term used for resources that can produce a sustainable natural income of goods or services. Natural income is the yield obtained from resources.

The Earth contains many resources that support its natural systems: the core and crust of the planet; the biosphere (the living part) containing forests, grassland, deserts, tundra, and other biomes, and the upper layers of the atmosphere. These resources are all extensively used by humans to provide food, water, shelter, and life-support systems. We tend to have an anthropocentric (human-centred) view of these resources and their use. Resources are discussed in terms of their use by and relationship to human populations. Ecologically minded economists describe resources as natural capital. This is equivalent to the store of the planet (or stock) – the present accumulated quantity of natural capital. **Renewable resources** can be used over and over again. If properly managed, renewable resources (Chapter 8, pages 414–415) are forms of wealth that can produce natural income indefinitely in the form of valuable goods and services (Figure 1.29).

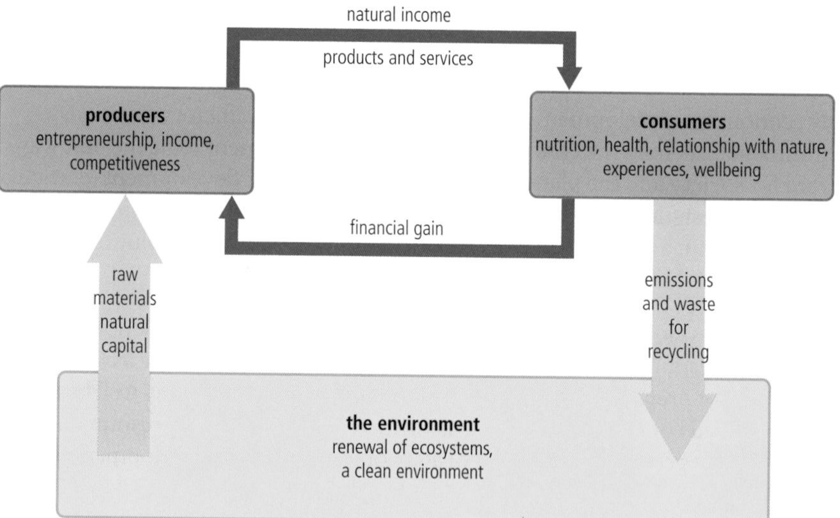

Figure 1.29 Natural capital and natural income. Raw materials from the environment (natural capital) are harvested and used by producers to generate products and services (natural income) that are then used by consumers. A renewable resource is a natural resource that the environment continues to supply or replace as it is used, and whose emissions and waste are recycled in a sustainable way.

In order to provide income indefinitely, the products and services used should not reduce the original resource (or capital). For example, if a forest is to provide ongoing income in the form of timber, the amount of original capital (the forest) must remain the same while income is generated from new growth. This is the same idea as living on the interest from a bank account – the original money is not used and only the interest is removed and spent.

Ecosystems may provide life-supporting services such as water replenishment, flood and erosion protection, and goods such as timber, fisheries, and agricultural crops. The income from natural capital may be in the form of goods or services:

- goods are marketable commodities such as timber and grain
- ecological services might be flood and erosion protection, climate stabilization, maintenance of soil fertility (Table 1.1, and Chapter 8, pages 417–418).

Other resources may not be replenished or renewed following removal of natural capital. These **non-renewable resources** (Chapters 5 and 7) will eventually run out if they are not replaced. Using economic terms, these resources can be considered as equivalent to those forms of economic capital that cannot generate wealth (i.e. income) without liquidation of the estate. In other words, the capital in the bank account is spent. Predictions about how long many of Earth's minerals and metals will last before they run out are usually basic. They may not take into account any

Table 1.1 Ecosystem types and the services they provide

Ecosystem	Service provided										
	fresh water	food	timber and fibre	new products	regulate diversity	cycle nutrients	quality air and climate	human health	detox	regulate hazards	cultural
cultivated		✔	✔	✔	✔	✔	✔				✔
dry land		✔		✔	✔	✔	✔	✔	✔		✔
forest	✔	✔	✔	✔	✔	✔	✔	✔	✔	✔	✔
urban		✔			✔		✔	✔	✔		✔
lakes and rivers	✔	✔		✔	✔	✔	✔	✔	✔	✔	✔
coastal	✔	✔	✔		✔	✔	✔	✔	✔	✔	✔
marine		✔		✔	✔	✔	✔		✔		✔
polar	✔	✔			✔		✔				✔
mountain	✔	✔			✔		✔			✔	✔
island		✔			✔		✔				✔

increase in demand due to new technologies, and they may assume that production equals consumption. Accurate estimates of global reserves and precise figures for consumption are needed for more exact predictions. However, it is clear that key non-renewable natural resources are limited and that there is a need to minimize waste, recycle, reuse and, where possible, replace rare elements with more abundant ones.

International summits and conferences aim to produce international tools (bodies, treaties, agreements, and so on) that address environmental issues and direct countries towards a sustainable future. The Kyoto Protocol is an example (pages 390–391).

This oil rig is Norwegian. Oil is a non-renewable resource the use of which has been a key feature of the 20th century, and continues to be into the 21st.

CONCEPTS: Sustainability

A sustainability lens should be used throughout the course where appropriate to explore how sustainability is being applied to a range of resource uses, such as energy provision, soil utilization in farming, and water consumption.

You need to be able to explain the relationship between natural capital, natural income and sustainability, and discuss the value of ecosystem services to a society.

The Millennium Ecosystem Assessment

Factors such as biodiversity, pollution, population, or climate may be used quantitatively (i.e. measuring by giving values) as environmental indicators of sustainability. These factors can be applied on a range of scales from local to global.

The Millennium Ecosystem Assessment (MA) gave a scientific appraisal of the condition and trends in the world's ecosystems and the services they provide using environmental indicators, as well as the scientific basis for action to conserve and use them sustainably. The United Nations Secretary-General at the time, Kofi Annan, called for the MA in 2000, and it was initiated in 2001, with the aim of assessing the consequences of ecosystem change for human well-being. The MA also looked at how the conservation and sustainable use of those systems could be implemented, and their contribution to human well-being improved. The reports produced by the MA provide an up-to-date review of the conditions of the world's ecosystems and the services they provide, and the options to restore, conserve or enhance the sustainable use of ecosystems. The main findings of the MA are as follows.

- Humans have changed ecosystems more rapidly in the past 50 years than in any previous period in history, resulting in a substantial and largely irreversible loss in the diversity of life on Earth.
- Changes that have been made to ecosystems have contributed to substantial overall gains in human well-being and economic development, but at the cost of many ecosystems and the services they provide.
- Changes have increased the poverty of some groups of people.
- The problems caused by ecosystem degradation will, unless addressed, substantially reduce the benefits that future generations obtain from ecosystems.
- Restoring ecosystems while meeting increasing demands for services can be achieved, but will involve significant changes in policies and practices.
- Overall, human actions are depleting Earth's natural capital at a faster rate than it is being restored, which is putting such strain on the environment that the ability of the planet's ecosystems to sustain future generations can no longer be taken for granted. However, the MA indicates that it may be possible to reverse changes as long as appropriate actions are taken quickly.

To learn more about the different MA reports, go to www.pearsonhotlinks.co.uk, enter the book title or ISBN, and click on weblink 1.5.

You need to be able to discuss how environmental indicators (such as Millennium Ecosystem Assessment) can be used to evaluate the progress of a project to increase sustainability.

CONCEPTS: Biodiversity

Measurements of biodiversity may be used quantitatively as environmental indicators of sustainability.

Environmental Impact Assessments

EIAs provide decision-makers with information in order to consider the environmental impact of a project. There is not necessarily a requirement to implement an EIA's proposals and many socio-economic factors may influence the decisions made.

Demand for resources, for new housing, for new energy supplies, and for new transport links are inevitable, but before any development project gets permission to begin, an **Environmental Impact Assessment (EIA)** must be carried out. The purpose of an EIA is to establish the impact of the project on the environment. It predicts possible impacts on habitats, species and ecosystems, and helps decision-makers decide if the development should go ahead. An EIA also addresses the mitigation of potential environmental impacts associated with the development. The report should provide a non-technical summary at the conclusion so that lay-people and the media can understand the implications of the study.

Some countries incorporate EIAs within their legal framework, with penalties and measures that can be taken if the conditions of the EIA are broken. Other countries may simply use the assessment to inform policy decisions. In some countries, the information and suggestions of the EIA are often ignored, or take second place to economic concerns.

The first stage of an EIA is to carry out a **baseline study**. This study is undertaken because it is important to know what the physical and biological environment is like before the project starts so that it can be monitored during and after the development. Variables measured as part of a baseline study should include:

- habitat type and abundance – record total area of each habitat type
- species list – record number of species (faunal and flora) present
- species diversity – estimate the abundance of each species and calculate diversity of the community
- list of endangered species
- land use – assess land use type and use coverage
- hydrology – assess hydrological conditions in terms of volume, discharge, flows, and water quality
- human population – assess present population
- soil – quality, fertility, and pH.

CONCEPTS: Biodiversity

A baseline study establishes the biodiversity of an area to be developed so that potential negative effects can be prevented.

EIAs incorporate a baseline study and assess the environmental, social, and economic impacts of the project, predicting and evaluating possible impacts and suggesting mitigation strategies. They are usually followed by an audit and continued monitoring. Different countries have different guidance on the use of EIAs.

CONCEPTS: Environmental value systems

EIAs offer advice to governments, but whether or not they are adopted depends on the EVS of the government involved. In China, for example, the EIA for the Three Gorges Dam showed the damage that would be done to the environment, but the government chose to focus on the benefits to the country.

It is often difficult to put together a complete baseline study due to lack of data, and sometimes not all of the impacts are identified. An EIA may be limited by the quality of the baseline study. The value of EIAs in the environmental decision-making process can be compromised in other ways. Environmental impact prediction is speculative because of the complexity of natural systems and the uncertainty of feedback mechanisms. The predictions of an EIA may, therefore, prove to be inaccurate in the long term. On the other hand, at their best, EIAs can lead to changes in the development plans, avoiding negative environmental impact. It could be argued that any improvement to a development outweighs any negative aspects.

You need to be able to evaluate the use of EIAs. Criticisms of EIAs include the lack of a standard practice or training for practitioners, the lack of a clear definition of system boundaries, and the lack of inclusion of indirect impacts.

Case study

Three Gorges Dam, Yangtze River

The Three Gorges Dam is 12.3 km long and 185 m tall – five times larger than the Hoover Dam in the USA. On completion, it created a reservoir nearly 660 km long, flooded to a depth of 175 m above sea level.

The Three Gorges Dam is the largest hydroelectric dam development in the world. Located on the Yangtze River in the People's Republic of China (Figure 1.30), construction began in 1993 and the dam was fully operational by the end of 2009. Project engineers estimated that the dam could generate an eighth of the country's electricity. This energy would be produced without the release of harmful greenhouse gases. The Chinese government cites other improvements that the development produces: reduced seasonal flooding and increased economic development along the edges of the new reservoir.

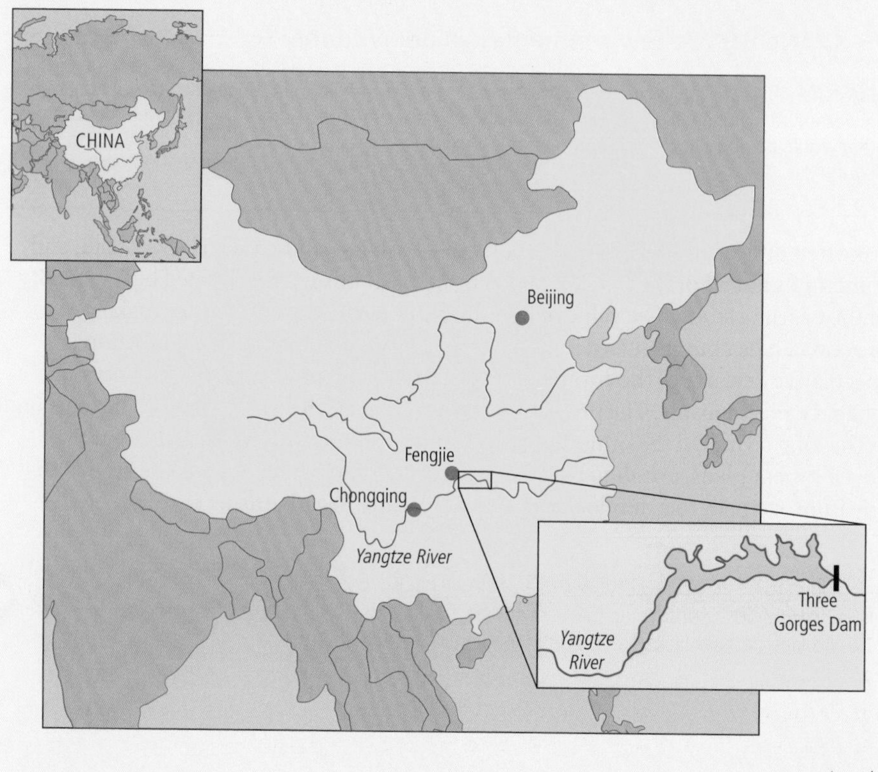

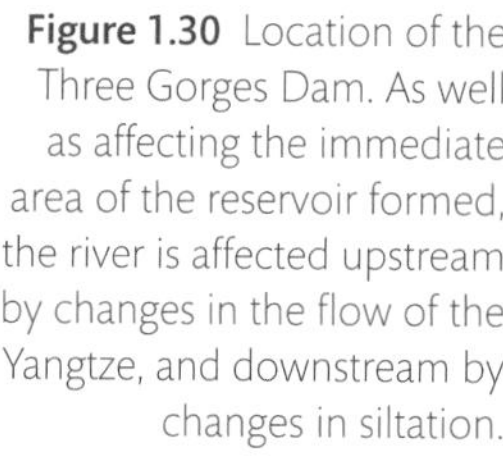

Figure 1.30 Location of the Three Gorges Dam. As well as affecting the immediate area of the reservoir formed, the river is affected upstream by changes in the flow of the Yangtze, and downstream by changes in siltation.

To learn more about the Three Gorges Dam, go to www.pearsonhotlinks.co.uk, enter the book title or ISBN, and click on weblink 1.6.

continued

An EIA was required to look at potential ecosystem disruption; relocation of people in areas to be flooded, and the social consequences of resettlement; the effects of sedimentation in areas behind the dam which would reduce water speed; the effects of landslides from the increase in geological pressure from rising water; and earthquake potential.

The EIA determined that there are 47 endangered species in the Three Gorges Dam area, including the Chinese river dolphin and the Chinese sturgeon. The report identified economic problems as well as the environmental problems that disruption of the ecosystem would cause. For example, the physical barrier of the dam would interfere with fish spawning and, in combination with pollution, which would have a serious impact on the fishing economy of the Yangtze River. In terms of social costs, the dam would flood 13 cities, 140 towns, 1352 villages, and 100 000 acres of China's most fertile land. Two million people would have to be resettled by 2012, and 4 million by 2020. Geological problems included the growing risk of new landslides and increased chance of earthquakes (due to the mass of water in a reservoir altering the pressure in the rock below), and a reduction in sediment reaching the East China Sea (reducing the fertility of the land in this area).

The overall view of the people responsible for the development was that the environmental and social problems did not reduce the feasibility of the project, and that the positive impact on the environment and national economy outweighed any negative impact.

Ecological footprints

An **ecological footprint** (**EF**) focuses on a given population and its current rate of resource consumption, and estimates the area of environment necessary to sustainably support that particular population. How great this area is, compared to the area actually available to the population, gives an indication of whether or not the population is living sustainably. If the EF is greater than the area available to the population, the population is living unsustainably.

The concept of ecological footprint: how great is your impact on the planet?

This issue of EFs is examined in Chapter 8 (pages 437–441). The concept is introduced here to give you a sense of your own impact on the planet at the start of the course, and as something for you to think about as the course progresses: what can you do to reduce your EF, or that of your school or college, based on information you are given?

Exercises

1. Explain the concept of resources in terms of natural income.
2. Explain the concept of sustainability in terms of natural capital and natural income.
3. Discuss the concept of sustainable development.
4. Outline the difference between sustainability and sustainable development.
5. Describe the stages of an environmental impact assessment.
6. Describe and evaluate the use of environmental impact assessments.

CHALLENGE YOURSELF

Research skills 

Find your own example of an EIA. What were the conclusions of the study? Were the recommendations followed?

TOK

EIAs incorporate baseline studies before a development project is undertaken. To what extent should environmental concerns limit our pursuit of knowledge?

Ecological footprints (EF) represent the hypothetical area of land required by a society, group, or individual to fulfil all their resource needs and assimilation of wastes.

CHALLENGE YOURSELF

Research skills 

It is also possible to calculate an individual's EF. There are many websites available to help you do this. Work out your ecological footprint using the hotlink on the next page.

How many planets would we need if everyone lived the same lifestyle as you? What issues are taken into account when calculating EF? How can you reduce your EF? What steps can you take today to start this process?

To calculate your ecological footprint, go to www.pearsonhotlinks.co.uk, enter the book title or ISBN, and click on weblink 1.7.

You need to be able to explain the relationship between ecological footprint and sustainability.

Big questions

Having read this section, you can now discuss the following big questions:

- What strengths and weaknesses of the systems approach and the use of models have been revealed through this topic?
- What have you learned about sustainability and sustainable development in this chapter?

You may want to consider the following points in your discussions:

- What are the differences between sustainability and sustainable development?
- Ecological Footprint is a model used to estimate the demands that human populations place on the environment: what are the limitations and benefits of these models.
- How do EIAs help ensure that development is sustainable?

1.5 Humans and pollution

Significant ideas

Pollution is a highly diverse phenomenon of human disturbance in ecosystems.

Pollution management strategies can be applied at different levels.

Big questions

As you read this section, consider the following big questions:

- What strengths and weaknesses of the systems approach and the use of models have been revealed through this topic?
- What value systems can you identify at play in the causes and approaches to resolving the issues addressed in this topic?

Knowledge and understanding

- Pollution is the addition of a substance or an agent to an environment by human activity, at a rate greater than that at which it can be rendered harmless by the environment, and which has an appreciable effect on the organisms within it.
- Pollutants may be in the form of organic/inorganic substances, light, sound, or heat energy, biological agents, or invasive species, and derive from a wide range of human activities including the combustion of fossil fuels.
- Pollution may be non-point or point source, persistent or biodegradable, acute or chronic.
- Pollutants may be primary (active on emission) or secondary (arising from primary pollutants undergoing physical or chemical change).
- DDT exemplifies a conflict between the utility of a pollutant and its effect on the environment.

What is pollution?

Pollution is contamination of the Earth and atmosphere to such an extent that normal environmental processes are adversely affected. Polluted elements are disagreeable, toxic, and harmful.

- Pollution can be natural, such as from volcanic eruptions, as well as human in origin (TOK chapter, page 457).
- It can be deliberate or it may be accidental.
- It includes the release of substances which harm the sustainable quality of air, water, and soil, and which reduces human quality of life.

Pollution is the addition of a substance or an agent to an environment by human activity, at a rate greater than that at which it can be rendered harmless by the environment, and which has an appreciable effect on the organisms within it.

Pollutants come in various forms, such as:

- organic or inorganic substances (such as pesticides and plastics)
- light, sound, or heat energy
- biological agents (i.e. organisms introduced to control agricultural pests but which may become pests themselves, such as the cane toad in Australia)
- invasive species (i.e. species that are not native to a country but which have been introduced, such as Japanese knotweed in the UK)
- derived from a wide range of human activities including the combustion of fossil fuels (Figure 1.31).

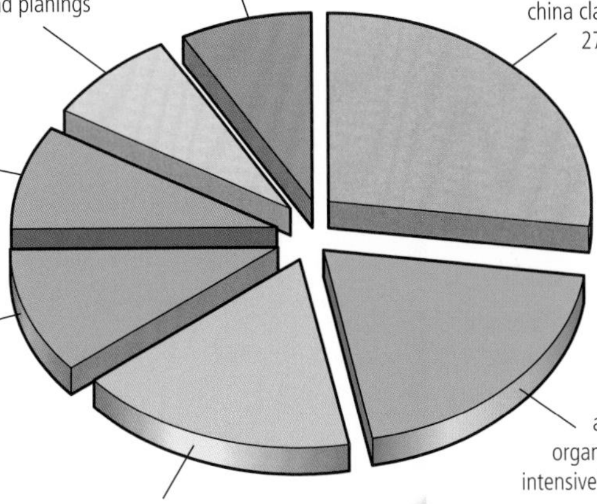

Figure 1.31 The major sources of pollution include the combustion of fossil fuel, domestic and industrial waste, manufacturing, and agricultural systems.

It is difficult to define the levels which constitute pollution. Much depends on the nature of the environment. For example, decomposition is much slower in cold environments – so oil slicks pose a greater threat in Arctic areas than in tropical ones. Similarly, levels of air quality which do not threaten healthy adults may affect young children, the elderly, or asthmatics.

Pollution costs

The costs of pollution are widespread and difficult to quantify. They include death, decreased levels of health, declining water resources, reduced soil quality, and poor air quality. It is vital to control and manage pollution. To be effective, pollution treatment must be applied at source. However, unless point sources can be targeted, this may be impossible. There is no point treating symptoms (e.g. treating acidified lakes with lime) if the cause is not tackled (e.g. emission of acid materials).

To learn more about pollution, go to www.pearsonhotlinks.co.uk, enter the book title or ISBN, and click on weblinks 1.8 and 1.9.

Many rich countries have knowingly polluted the environment, in return for the economic benefits they gain (e.g. energy production). Much of the cost of this pollution is borne by other countries – is this moral?

CONCEPTS: Environmental value systems

Most industrial nations adopt a cornucopian approach to the environment, believing that people can find a solution to the problems created by human (mis-)use of the environment, such as those caused by pollution.

Some forms of pollution cannot be contained by national boundaries and therefore can act either locally, regionally, or globally (e.g. acid deposition, page 340).

You are expected to be able to construct systems diagrams to show the impact of pollutants. Examples of pollution are explored in subsequent chapters and systems diagrams can be applied to each of these cases. Figure 1.32 is a general systems diagram showing the effects of pollution.

Systems diagram showing the effect of pollution

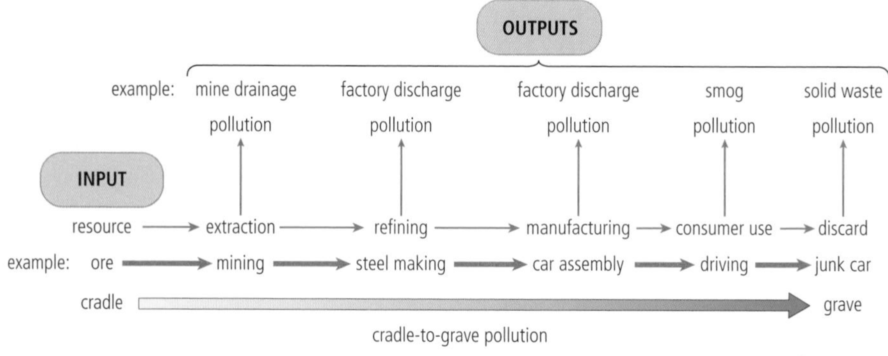

Figure 1.32 A systems diagram showing the pollution produced over a lifetime of a single car (cradle-to-grave pollution). Nearly all products, including food and other agricultural products, create such stepwise pollution.

Point-source pollution and non-point source pollution

Point-source pollution refers to discrete sources of contaminants that can be represented by single points on a map and the source of the pollution can be tracked. The nuclear explosions at Chernobyl, Ukraine, and Fukushima, Japan (pages 5–6), and the industrial pollution at Bhopal, India (page 5) are good examples of point-source pollution.

Point-source pollution – sediment is coming from a single identifiable source, a drain mid-way along a stream.

Non-point source pollution refers to more dispersed sources from which pollutants originate and enter the natural environment. A good example is the release of air pollutants from numerous, widely dispersed origins (e.g. vehicles and industries).

Point-source pollution is generally the more easily managed. Its point source is localized, making it easier to control emissions, apportion responsibility, and take legal action, if necessary. The localized impact is also easier to manage.

Pollution which arises from numerous widely dispersed origins is described as non-point source. Point-source pollution arises from a single clearly identifiable site.

Power stations may emit sulfur dioxide and nitrous oxide into the air: the deposition affects a wide area, and combines with the pollution from a variety of sources, and so is an example of non-point source pollution.

You should be aware that for some pollutants, there may be a time lag before an appreciable adverse effect on organisms is evident. An example here is DDT (see below).

Primary vs. secondary pollution

Pollutants may be primary (active on emission) or secondary (arising from primary pollutants undergoing physical or chemical change).

An example of primary pollution is the gas released from burning coal and other fossil fuels (e.g. SO_2, CO_2). An example of a secondary pollutant is ground level ozone, where the NO_x from car exhausts reacts with sunlight to form tropospheric (ground level) ozone (Chapter 6, page 333).

Acute vs. chronic effects of pollution

The effects of pollution can be acute or chronic.

- Acute – occurring after a short, intense exposure. Symptoms are usually experienced within hours.
- Chronic – occurring after low-level, long-term exposure. Disease symptoms develop up to several decades later.

Air pollution can have acute and chronic effects: acute effects include asthma attacks; chronic effects include lung cancer, chronic obstructive pulmonary disease (COPD), and heart disease. The acute and chronic effects of exposure to UV light are examined in Chapter 6 (page 325).

Persistent vs. biodegradable pollutants

Biodegradable means capable of being broken down by natural biological processes.

Persistent pollutants are ones that cannot be broken down by living organisms and so are passed along food chains (Chapter 2, page 86). Persistent organic pollutants (POPs) are organic compounds that are resistant to environmental breakdown through biological, chemical, or photolytic (i.e. broken down by light) processes. **Biodegradable pollutants** are ones that are not stored in biological matter or passed along food chains. Most modern pesticides, used to treat crops to as to ensure maximum yield, are biodegradable (e.g. Bt proteins that are rapidly decomposed by sunlight), although earlier chemicals were persistent (e.g. DDT).

DDT (dichlorodiphenyl-trichloroethane) is a synthetic pesticide with a controversial history. DDT exemplifies a conflict between the utility of a pollutant and its effect on the environment.

Costs and benefits of the ban on DDT

DDT was used extensively during the Second World War to control the lice that spread typhus and the mosquitoes that spread malaria. Its use led to a huge decrease in both diseases. After the war, DDT was used as an insecticide in farming, and its production soared.

In 1955, the World Health Organization (WHO) began a programme to eradicate malaria worldwide. This relied heavily on DDT. The programme was initially successful, but resistance evolved in many insect populations after only 6 years, largely because of the widespread agricultural use of DDT. In many parts of the world including Sri Lanka, Pakistan, Turkey, and central America, DDT has lost much of its effectiveness.

Between 1950 and 1980, DDT was used extensively in farming, and over 40 000 tonnes were used each year worldwide. Up to 1.8 million tonnes of DDT have been produced globally since the 1940s. About 4000–5000 tonnes of DDT are still produced and used each year for the control of malaria and other diseases. DDT is applied to the inside walls of homes (a process known as indoor residual spraying, IRS) to kill or repel mosquitoes entering the home. India is the largest consumer. The main producers are India, China, and North Korea.

In 1962, American biologist Rachel Carson published her hugely influential book *Silent Spring* (page 6) in which she claimed that the large-scale spraying of pesticides, including DDT, was killing wildlife. Top carnivores such as birds of prey were declining in numbers. Moreover, DDT could cause cancer in humans. Public opinion turned against DDT (see below).

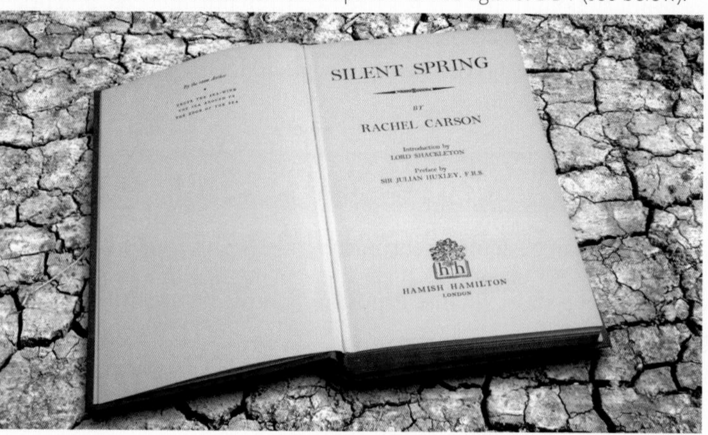

Rachel Carson's famous book became a *cause célèbre* and marked a turning point in attitudes to DDT.

Restrictions on the use of DDT

In the 1970s and 1980s, agricultural use of DDT was banned in most developed countries. DDT was first banned in Hungary (1968) followed by Norway and Sweden (1970), USA in 1972, and the UK in 1984. The use of DDT in vector control has not been banned, but it has been largely replaced by less persistent alternative insecticides.

The Stockholm Convention banned several POPs, and restricted the use of DDT to disease control. The Convention was signed by 98 countries and is endorsed by most environmental groups (pages 7–8). Despite the worldwide ban on agricultural use of DDT, its use in this context continues in India and North Korea.

Environmental impacts of DDT

DDT is a POP that is extremely hydrophobic and strongly absorbed by soils. DDT is not very soluble in water but is very soluble in lipids (fats). This means it can build up in fatty tissue. Its soil half-life can range from 22 days to 30 years.

Bioaccumulation is the retention or build-up of non-biodegradable or slowly biodegradable chemicals in the body. **Biomagnification** or biological amplification is the process whereby the concentration of a chemical increases at each trophic level. The end result is that top predators may have in their bodies concentrations of a chemical several million times higher than the same chemical's concentration in water and primary producers (Chapter 2, page 86).

DDT, and its breakdown products DDE and DDD, all biomagnify through the food chain (Figure 1.33). DDT is believed to be a major reason for the decline of the bald eagle in North America in the 1950s and 1960s. Other species affected included the brown pelican and peregrine falcon. Recent studies have linked the thinning of the birds' egg shells with high levels of DDE in particular, resulting in eggs being crushed by parents when incubating.

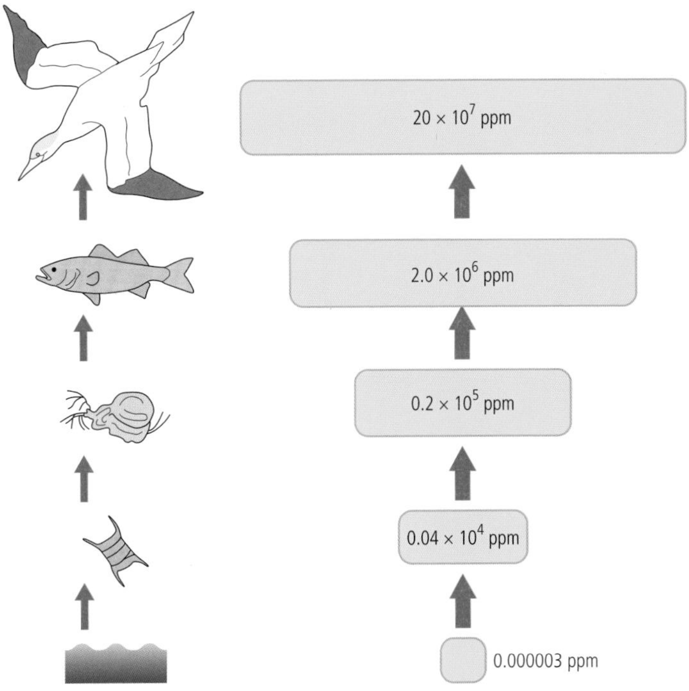

20×10^7 ppm

2.0×10^6 ppm

0.2×10^5 ppm

0.04×10^4 ppm

0.000003 ppm

Figure 1.33 Biomagnification of DDT along a food chain

Effects on human health

The effects of DDT on human health are disputed and conflicting. For example, some studies have shown that:

- farmers occupationally exposed to DDT had an increased incidence of asthma and/or diabetes
- some people exposed to DDT had a higher risk of liver, breast, and/or pancreatic cancer
- DDT exposure is a risk factor for early pregnancy loss, premature birth, and/or low birth weight
- a 2007 study found increased infertility among South African men from communities where DDT is used to combat malaria.

To learn more about the use and impact of DDT, go to www.pearsonhotlinks.co.uk, enter the book title or ISBN, and click on weblink 1.10.

Use of DDT against malaria

Malaria remains a major public health challenge in many parts of the world. WHO estimates that there are 250 million cases every year, resulting in almost 1 million deaths. About 90 per cent of these deaths occur in Africa. In 2006, only 13 countries were still using DDT.

Nevertheless, WHO is 'very much concerned with health consequences from use of DDT' and it has reaffirmed its commitment to eventually phase it out. In South America, malaria cases increased after countries stopped using DDT. In Ecuador, between 1993 and 1995, the use of DDT increased and there was a 61 per cent reduction in malaria rates.

Some donor governments and agencies have refused to fund DDT spraying, or made aid contingent on not using DDT. Use of DDT in Mozambique was stopped because 80 per cent of the country's health budget came from donor funds, and donors refused to allow the use of DDT.

You need to be able to demonstrate knowledge of both the anti-malarial and agricultural use of DDT, and evaluate the use of DDT.

Pollution management

Human causes of pollution are widespread and include farming and industrial practices, urbanization, development of transport, and the transport and burning of energy sources. The result depends on the amount of material released into the environment.

Figure 1.34 shows the stages leading to the impact of pollutants on the environment.

Figure 1.34 Model demonstrating the stages leading to pollutants having an impact on the environment.

Modern technology can reduce the impact of pollution. For example, applying the model in Figure 1.35 to cars and chemical factories, the impact of stage A could be managed by introducing electric and hybrid cars which use less fossil fuel; the impact of stage B could be minimized by fitting catalytic converters (which reduce atmospheric pollutants) to car exhaust systems, or adding scrubbers to industrial chimneys to remove toxic chemicals and allow for their reuse; stage C could be managed by using synthetic membranes to capture chemical spills (e.g. mats designed to capture and hold hydrocarbons).

Ideally, human behaviour should be changed to ensure that the pollution does not occur in the first place (this is called preventive action). If pollutants are released, the pollution should be regulated to ensure minimum exposure; if the release is such that it has an impact on the environment, clean-up and restoration must occur (reactive actions). There are, therefore, a number of ways in which the impacts of pollution can be managed (Figure 1.35):

- changing human activity
- regulating and reducing quantities of pollutants released at the point of emission
- cleaning up the pollutant and restoring the ecosystem after pollution has occurred.

Each of these strategies has advantages and limitations.

Figure 1.35 Pollution management targeted at three different levels

Process of pollution

A: human activity producing pollutant

B: release of pollutant into environment

C: impact of pollutant on ecosystems

Level of pollution management

Altering Human Activity

Most fundamental level of pollution management is to change the human activity that leads to production of pollutant in the first place, by promoting alternative technologies, lifestyles, and values through:

- campaigns, education, community groups
- governmental legislation, economic incentives/disincentives

Controlling Release of Pollutant

Where the activity/production is not completely stopped, strategies can be applied at the level of regulating or preventing the release of pollutants by:

- legislating and regulating standards of emission
- developing/applying technologies for extracting pollutant from emissions

Clean-up and Restoration of Damaged Systems

Where both the above levels of management have failed, strategies may be introduced to recover damaged ecosystems by:

- extracting and removing pollutant from ecosystem
- replanting/restocking lost or depleted populations and communities

You need to be able to evaluate the effectiveness of each of the three different levels of intervention in Figure 1.35. The principles of this figure should be used throughout the course when addressing issues of pollution. You should appreciate the advantages of employing the earlier strategies of pollution management over the later ones, and the importance of collaboration.

Changing human activities

The main advantage of changing human activities is that it may prevent pollution from happening. For example, if more societies were to use solar, hydro- or wind power there would be reduced emissions of greenhouse gases, and less risk of global warming. However, there are major limitations. Alternative technologies are expensive to develop and may only work in certain environments (Chapter 7, pages 356–360). Solar power is most effective in areas which have reliable hours of sunshine. Wind energy requires relatively high wind speeds and is best suited to coastal areas and high ground.

Reusing and recycling materials has reduced consumption of resources. Many items can be recycled such as newspapers, cans, glass, aluminium and plastics. However, there are certain goods which can only be recycled under special conditions. The increasing volume of electronic waste (e-waste) is creating major problems for its disposal and recycling.

Computer equipment contains toxic substances and is effectively hazardous waste. Much e-waste ends up in the developing world, and there is increasing concern about the pollution caused by hazardous chemicals and heavy metals there. A single computer can contain up to 2 kg of lead, and the complex mixture of materials make computers very difficult to recycle. New legislation in the European Union came into force in 2007 to cover waste electrical and electronic equipment (WEEE).

In the USA, up to 20 million 'obsolete' computers are discarded annually.

Increasingly, manufacturers of electronic goods incorporate e-waste management into their environmental policies and operate consumer recycling schemes. Dell, for example, cover the cost of home pick-up, shipping to the recycling centre, and recycling of any obsolete equipment. Hewlett–Packard (HP) recycled over 74 million kg of electronics in 2005. HP has recycling operations in 40 world regions. These schemes help to:

- reduce the volume of waste that ends up in landfill sites
- cut down on the amount of raw materials needed for the manufacture of new products
- make recycling convenient for the consumer.

Regulating activities

The next easiest way of reducing pollution is to reduce the amount of pollution at the point of emission. This may be done by having measures for extracting the pollutant from the waste emissions. A good example is the use of flue gas desulfurization (FGD). FGD is widely used to control the emissions of sulfur dioxide (SO_2) from coal- and oil-fired power stations and refineries. There are a variety of FGD processes available; most use an alkali to extract the acidic sulfur compounds from the flue gas. Flue gas treatment (FGT) is the process used for removing pollutants from waste incinerators. Such treatments are expensive and it is difficult to enforce such measures in the unregulated part of the economy (the informal sector).

It may be possible for people to adopt alternative, less-polluting lifestyles. During the period of high oil prices in 2008, more people than usual travelled by public transport, cycled, or walked to work or school. Since the 1990s, the Living Streets Walk to School Campaign has encouraged over 1 million primary school children to walk to school in the UK.

SYSTEMS APPROACH

There are a number of human factors that influence the choice and implementation of pollution management strategy. These include economic systems, EVSs, and political systems.

Countries at different stages of development place different sets of values on the natural environment. Many developing countries wish to use their resources for economic development. They argue that they are only doing the same as the rich countries, albeit many decades later.
Are they justified in this argument?

Levels of pollution can also be controlled by setting standards for air or water quality. For example, in 2008 the Environmental Protection Agency (EPA) in the USA improved air quality standards in an effort to help improve public health. It lowered the amount of ground-level ozone permitted in the atmosphere from 80 parts per billion (ppb) to 75 ppb. The EPA claimed the change could save 4000 lives each year. However, standards are not imposed to the same levels in all countries. Many developing countries need to develop their industries in order to improve their wealth. LEDCs are anxious not to be regulated by strict controls that would slow down their development. Indeed, some companies from rich countries locate in poor countries as the environmental legislation there is weaker or not enforced. US companies locating across the border in Mexico, the maquiladora industries, are a good example of this practice.

Cleaning up afterwards

The most expensive option (in terms of both time and money) is to clean up the environment after it has been polluted. Under natural conditions, bacteria take time to break down pollutants before the ecosystem recovers through secondary succession. In cold conditions, bacterial activity is reduced so pollutants in colder environments persist for longer than in warm environments. When people are employed in the clean-up process, it is often labour-intensive and, therefore, expensive.

Integration of policies

It is increasingly likely that integrated pollution-management schemes will employ aspects of each of the three approaches. It is unrealistic to expect human activities to cease to pollute the environment. However, any reduction will be beneficial. If the pollutants can be captured at the source of pollution, it will be cheaper in the long term because they will not have polluted the environment at that stage, so no clean-up will be required. Cleaning up widespread pollution is necessary, but it is the least effective option.

Exercises

1. Outline the processes of pollution.
2. Outline strategies for reducing the impacts of pollution.
3. Discuss the human factors that affect the approaches to pollution management.
4. Describe the variations in the level of DDT along a food chain.
5. Outline the main uses (past and present) of DDT.
6. Comment on the risks of using DDT.

Big questions

Having read this section, you can now discuss the following big questions:

- What strengths and weaknesses of the systems approach and the use of models have been revealed through this topic?
- What value systems can you identify at play in the causes and approaches to resolving the issues addressed in this topic?

Points you may want to consider in your discussions:

- How can systems diagrams be used to show the impact of pollution on environmental and social systems?
- How do EVSs influence the choice and implementation of pollution management strategy?

Practice questions

1 a Explain whether a small lake should be considered an open, closed, or isolated system. [2]

b Distinguish between *transfer* processes and *transformation* processes. [2]

c Annotate the diagram below to show the natural transfer and transformation processes which move water from the ocean to the lake. [2]

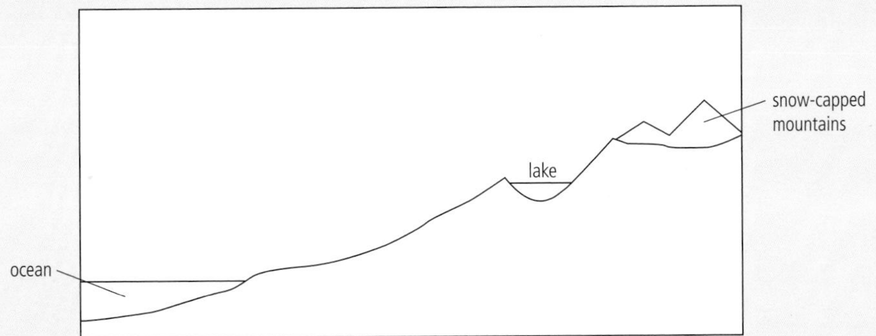

2 a Distinguish between *negative feedback* and *positive feedback*. [2]

b Construct a diagram to show how a positive feedback process involving methane may affect the rate of global warming. [2]

c Outline the basic components of an ecosystem using the systems approach. [4]

3 Using the figure below, construct a quantitative model to show the storages and flows in forest carbon cycling. [3]

- global forest biomass contains 283 gigatonnes* of carbon (GtC)
- dead wood, litter and soil contain 520 GtC
- in the atmosphere there are approximately 750 GtC
- it is estimated that forests release 60 GtC per year into the atmosphere
- worldwide deforestation releases approvimately 1.6 GtC per year (most in the tropics)
- some carbon is captured from the atmosphere when other crops are planted in the place of forests

* 1 gigatonne = 1 billion tonnes

4 Define the term *pollution*, and distinguish between *point-source* and *non-point source pollution*. [4]

5 Some mosquitoes may carry *Plasmodium*, so they are considered to be a disease vector. One controversial strategy for the control of malaria is to use the pesticide DDT (dichlorodiphenyltrichloroethane) to kill the mosquito.

The figure below shows the relationship between DDT use in Latin American countries and percentage change in the number of cases of malaria.

a Identify four countries on the graph where DDT is still in use. [1]

b The World Health Organization (WHO) has categorized DDT as a persistent organic pollutant (POP). The Stockholm Convention on Persistent Organic Pollutants is an international treaty that aims to eliminate or restrict the production and use of POPs. Within the Convention is the following provision:

> WHO recommends only indoor residual spraying (spraying only on the inside walls of buildings) of DDT for disease vector control.

With reference to Figure 3, evaluate this provision. [3]

c Suggest why an ecocentrist position might be opposed to indoor residual spraying. [2]

d Rachel Carson's book *Silent Spring* drew attention to the environmental impact of DDT on top carnivores. Explain the vulnerability of top carnivores to non-biodegradable toxins, such as DDT, entering food chains. [2]

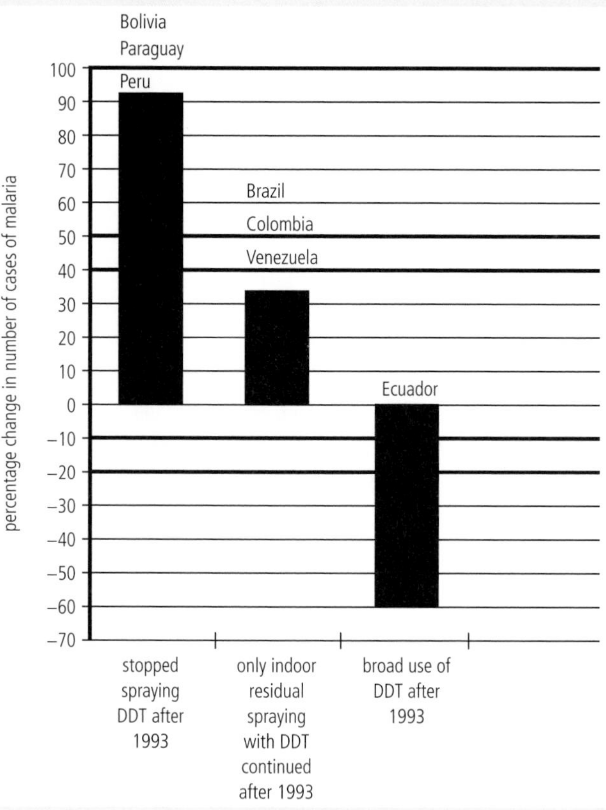

6 The figure below shows that sustainable development may depend on the interaction between three different priorities.

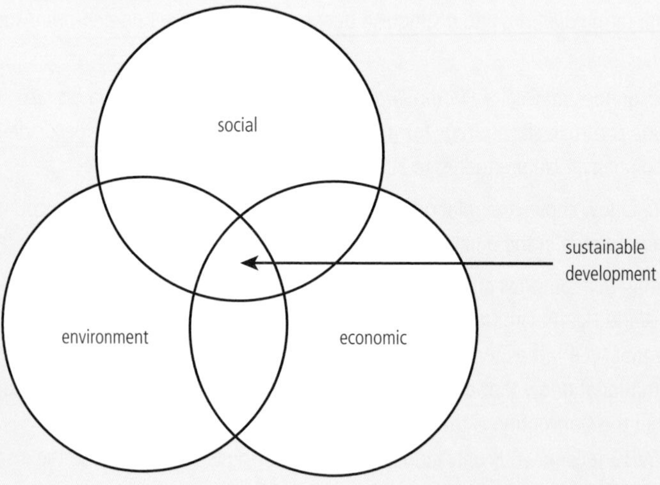

a State what is meant by the term *environmental value system*. [1]

b With reference to this figure, complete the table below to:

i identify the priority for each sector of society [1]

ii describe an example of how a conservation biologist and banker may support sustainable development [2]

	Priority	Example
Self-reliance soft ecologist	....	Community cooperative set up to sell local produce and share production costs to increase profits
Conservation biologist	....	....
Banker	....	....

c Explain why sustainable energy sources are not always adopted by societies. [2]

d Discuss the potential ecological services and goods provided by a named ecosystem. [6]

7 **a** Distinguish between the terms *sustainability* and *sustainable development*. [4]

b Explain why attitudes towards the environment change over time. Refer to named historical influences in your answer. [6]

02 Ecosystems and ecology

Opposite: Studying animals, such as the red-eyed tree frog in the Costa Rican rainforest, gives us a better understanding of their ecology and the ecosystems they live in.

2.1 Species and populations

Significant ideas

A species interacts with its abiotic and biotic environment, and its niche is described by these interactions.

Populations change and respond to interactions with the environment.

All systems have a carrying capacity for a given species.

Big questions

As you read this section, consider the following big question:

- What strengths and weaknesses of the systems approach and the use of models have been revealed through this topic?

Knowledge and understanding

- A species is a group of organisms sharing common characteristics that interbreed and produce fertile offspring.
- A habitat is the environment in which a species normally lives.
- A niche describes the particular set of abiotic and biotic conditions and resources to which an organism or population responds.
- The fundamental niche describes the full range of conditions and resources in which a species could survive and reproduce. The realized niche describes the actual conditions and resources in which a species exists due to biotic interactions.
- The non-living, physical factors that influence the organisms and ecosystem (e.g. temperature, sunlight, pH, salinity, precipitation) are termed abiotic factors.
- The interactions between the organisms (e.g. predation, herbivory, parasitism, mutualism, disease, competition) are termed biotic factors.
- Interactions should be understood in terms of the influences each species has on the population dynamics of others, and on the carrying capacity of the others' environment.
- A population is a group of organisms of the same species living in the same area at the same time, and which are capable of interbreeding.
- S and J population curves describe a generalized response of populations to a particular set of conditions (abiotic and biotic factors).
- Limiting factors will slow population growth as it approaches the carrying capacity of the system.

Species, habitat, and niche

Ecological terms are precisely defined and may vary from the everyday use of the same words.

Species

A **species** is a group of organisms sharing common characteristics that can interbreed and produce offspring that can also interbreed and produce young. Sometimes, two species breed together to produce a hybrid offspring, which may survive to adulthood

A snow leopard (*Panthera uncia*) – an example of a species. Species names have two parts – the genus name (in this case, *Panthera*) and a specific name (*uncia*). Species names are always written in italics, or underlined.

but cannot produce viable gametes and so is sterile. For example, a horse (*Equus caballus*) can breed with a donkey (*Equus asinus*) to produce a mule.

The species concept cannot:

- identify whether geographically isolated **populations** belong to the same species
- classify species in extinct populations
- account for asexually reproducing organisms
- clearly define species when barriers to reproduction are incomplete (Figure 2.1).

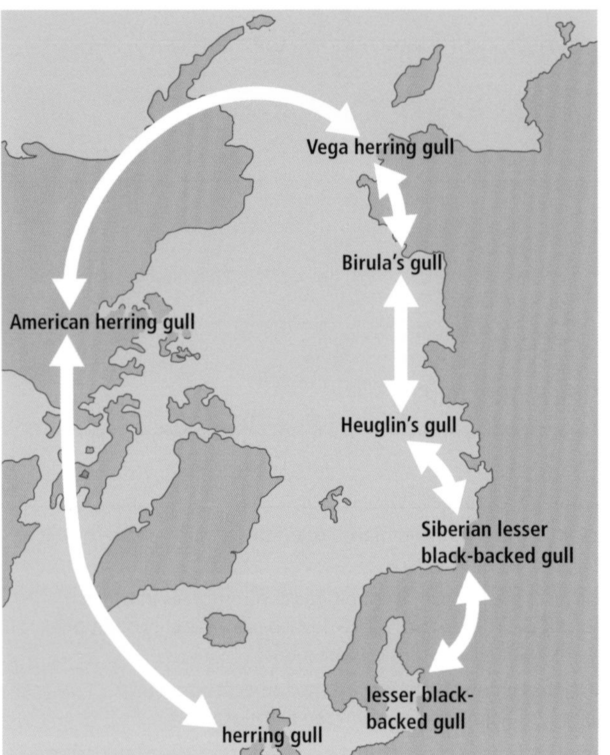

Figure 2.1 Gulls interbreeding in a ring around the Arctic are an example of ring species. Neighbouring species can interbreed to produce viable hybrids but herring gulls and lesser black-backed gulls, at the ends of the ring, cannot interbreed.

TOK

The species concept is sometimes difficult to apply: for example, can it be used to accurately describe extinct animals and fossils? The term is also sometimes loosely applied to what are, in reality, sub-species that can interbreed. This is an example of an apparently simple term that is difficult to apply in practical situations.

Species interact with their environment. The environment is the external surroundings that act on a species, influencing its survival and development. The environment contains living and non-living components, and an ecosystem includes both types of component. The non-living, physical factors that influence the organisms and ecosystem are termed abiotic factors (page 65). The living parts of the ecosystem, the organisms (animals, plants, algae, fungi, and bacteria) that live within it, are termed biotic factors.

When discussing examples of species, habitat, niche, and so on, you should use specific examples. For example, when referring to species, use Atlantic salmon rather than fish, Kentucky Bluegrass rather than grass, and silver birch rather than tree.

Habitat

A **habitat** is the environment in which a species normally lives. The habitat of the African elephant, for example, includes savannahs, forests, deserts, and marshes.

Elephant family in front of Mt Kilimanjaro, in the Amboseli National Park. These elephants live in an environment with open savannah grassland, acacia woodland, swamps, and marshlands.

You should be aware that for some organisms, habitats can change over time as a result of migration.

Niche

An ecological **niche** is best described as where, when, and how an organism lives. An organism's niche depends not only on where it lives (its habitat) but also on what it does. For example, the niche of an elephant includes everything that defines this species: its habitat, interactions between members of the herd, what it feeds on and when it feeds, and so on. No two species can have the same niche because the niche completely defines a species.

There are usually differences between the niche that a species can theoretically occupy and one that it actually occupies. Factors affecting how a species disperses itself and interacts with other species restrict the actual niche. The theoretical niche, which describes the full potential of where, when, and how a species can exist, is known as its **fundamental niche**. Where the species actually exists is known as its **realized niche**. The fundamental niche can, therefore, be simply defined as where and how an organism could live, and the realized niche as where and how an organism does live (e.g. Figure 2.2).

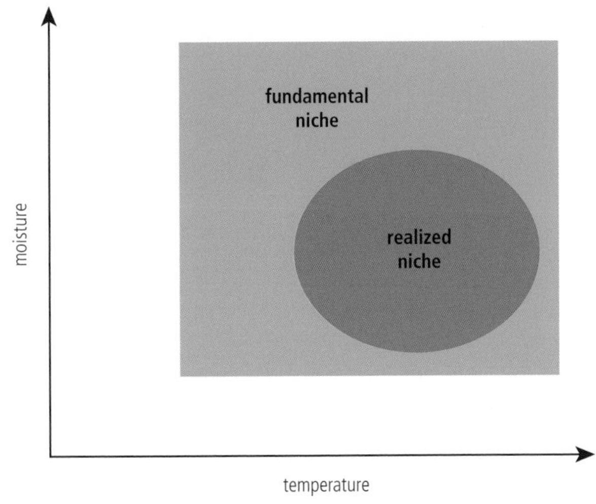

Figure 2.2 The distribution of a plant species is primarily determined by two factors – temperature and moisture. The fundamental niche includes all the areas where the species could live. The realized niche includes all the areas where the species does live – interaction with other species limits the niche in this way.

A niche describes the particular set of abiotic and biotic conditions and resources to which an organism or population responds.

The fundamental niche describes the full range of conditions and resources in which a species could survive and reproduce.

The realized niche describes the actual conditions and resources in which a species exists due to biotic interactions.

Case study

Fundamental and realized niches of barnacles

A colony of barnacles on the rocky shore.

American ecologist Joseph Connell investigated the realized and fundamental niches of two species of barnacle – a common animal on rocky shores in the UK. Connell had observed that one of the species, *Semibalanus (Balanus) balanoides*, was most abundant on the middle and lower intertidal area and that the other species, *Chthamalus stellatus*, was most common on the upper intertidal area of the shore. When he removed *Chthamalus* from the upper area of shore, he found that no *Semibalanus* replaced it: his explanation was that *Semibalanus* could not survive in an area that regularly dried out due to low tides. He concluded that *Semibalanus'* realized niche was the same as its fundamental niche.

In another experiment he removed *Semibalanus* from the upper and middle areas. He found that over time *Chthamalus* replaced it in the middle intertidal zone: his explanation was that *Semibalanus* was a more successful competitor in the middle intertidal zone and usually excluded *Chthamalus*. He concluded that the fundamental niche and realized niche for *Chthamalus* were not the same (Figure 2.3), and that its realized niche was smaller due to interspecific competition (i.e. competition between species) leading to competitive exclusion (when one species outcompetes and excludes another when their niches overlap, page 69).

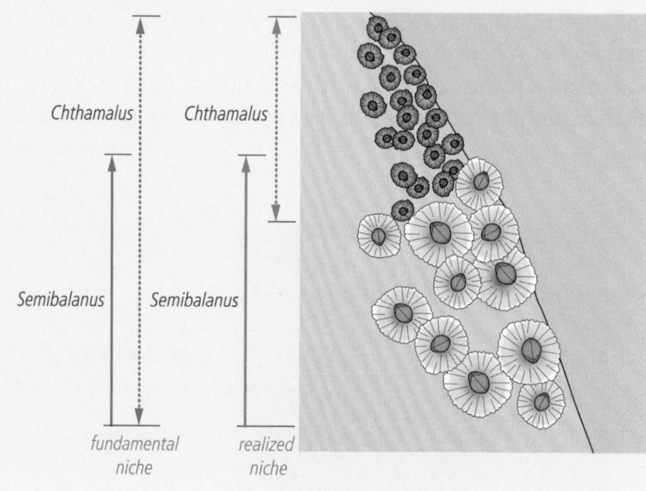

Figure 2.3 The fundamental and realized niches of two species of barnacle, *Chthamalus stellatus* (dotted arrows) and *Semibalanus balanoides*.

Abiotic factors

The non-living, physical factors that influence the organisms and ecosystem (e.g. temperature, sunlight, acidity/alkalinity (pH), rainfall (precipitation), and salinity) are termed abiotic factors. Abiotic factors also include the soil (edaphic factors) and topography (the landscape).

Abiotic factors are non-living parts of the environment. Such factors determine the fundamental and realized niche of species.

There are upper and lower levels of environmental factors beyond which a population cannot survive, and there is an optimum range within which species can thrive (Figure 2.4). These 'tolerance limits' exist for all important environmental factors. For some species, one factor may be most important in regulating distribution and abundance but, in general, many factors interact to affect species distribution.

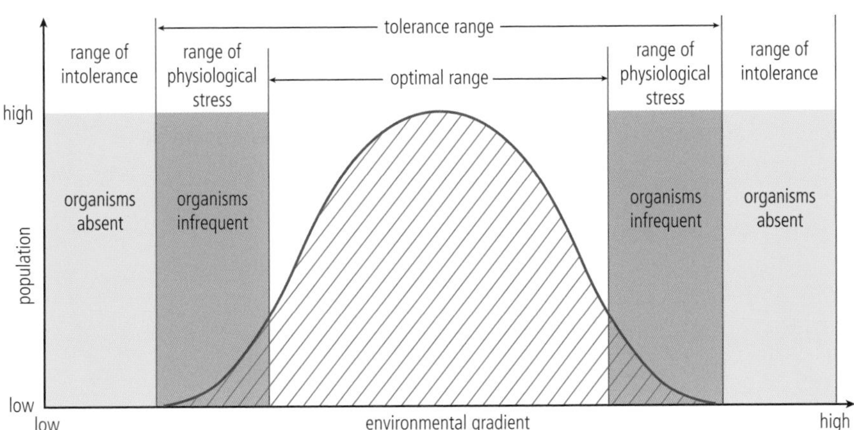

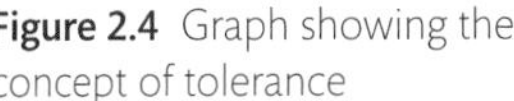

Figure 2.4 Graph showing the concept of tolerance

All plants and animals need water to survive. For plants, water stress (too little water) may cause germination to fail, seedlings to die, and seed yield to be reduced. Plants are extremely sensitive to water level. Categories of water-tolerant plants include:

- hydrophytes – water-tolerant plants which can root in standing water
- mesophytes – plants that inhabit moist but not wet environments
- xerophytes – plants that live in dry environments.

Xerophyte adaptations to avoid water shortages include remaining as seeds until rain stimulates germination, and storing water in stems, leaves, or roots. Plants that store water are called succulents. Many succulents have a crassulacean acid metabolism (CAM) which allows them to take in carbon dioxide at night when their stomata are open, and use it during the day when the stomata are closed. Other xerophytes have thick, waxy cuticles; small, sunken stomata; and drop their leaves in dry periods.

Population interactions

Carrying capacity refers to the number of organisms – or size of population – that an area or ecosystem can support sustainably over a long period of time.

Biotic factors are the living part of the environment. Interactions between organisms are also biotic factors. Ecosystems contain numerous populations with complex interactions between them. The nature of the interactions varies and can be broadly divided into specific types (predation, herbivory, parasitism, mutualism, disease, and competition). These are discussed below.

Predation

Predation occurs when one animal (or, occasionally, a plant) hunts and eats another organism.

Interactions should be understood in terms of the influences each species has on the population dynamics of others, and on the carrying capacity (page 72) of the others' environment.

Female snowy owl swoops down to catch a lemming on top of the snow

Predator–prey relationships are seen, for example, in lemming and snowy owl populations in the northern polar regions. The graph in Figure 2.5 shows fluctuations in the population sizes of lemming (the prey) and snowy owl (the predator) over several years.

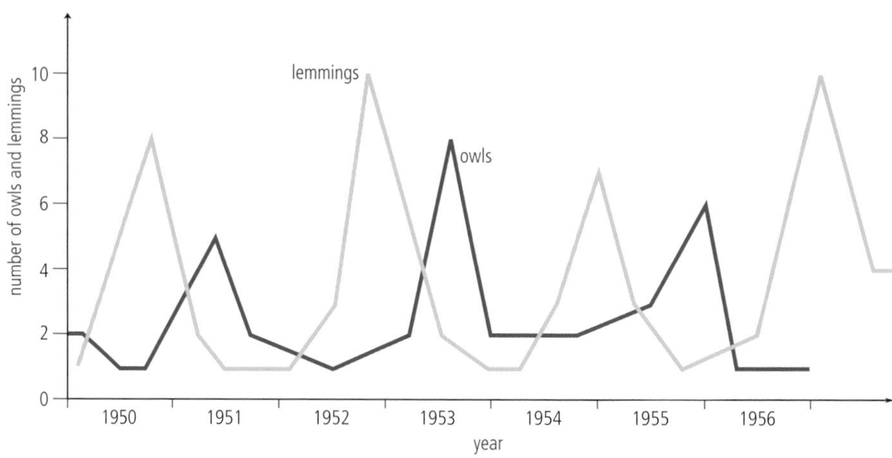

Figure 2.5 Variations in the populations of lemmings and snowy owls

These predator–prey interactions are often controlled by negative feedback mechanisms that control population densities (Figure 2.6). In the relative absence of the predatory snowy owl (due to a limited prey population), the population of lemmings begins to increase in size. As the availability of prey increases, there is an increase in predator numbers, after a time-lag. As the number of predators increases, the population size of the prey begins to decrease, again after a time-lag. With fewer prey, the number of predators decreases again. With fewer predators, the number of prey may begin to increase again and the cycle continues. Nevertheless, predation may be good for the prey: it removes old and sick individuals first as these are easier to catch. Those remaining are healthier and form a superior breeding pool.

Figure 2.6 Predator–prey relationships show negative feedback.

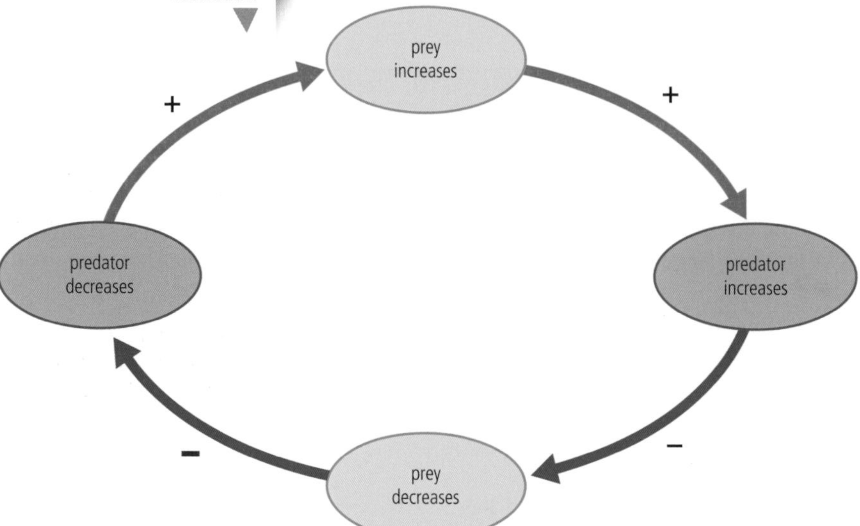

Herbivory

Herbivory is an interaction where an animal feeds on a plant. The animal that eats the plant is called a **herbivore**. An example of herbivory is provided by the hippo, which eats vegetation on the land during the coolness of the night. Hippos spend the day in rivers so they do not overheat.

The carrying capacity of a herbivore's environment is affected by the quantity of the plant it feeds on. An area with more abundant plant resources has a higher carrying capacity than an area that has less plant material available as food for the herbivore.

Not all predators are animals. Insectivorous plants, such as the Venus fly traps and pitcher plants trap insects and feed on them. Such plants often live in areas with nitrate-poor soils and obtain much of their nitrogen from animal protein. These plants still obtain energy from photosynthesis.

Nepenthes rajah, the largest pitcher plant, can hold up to 3.5 litres of water in the pitcher and has been known to trap and digest small mammals such as rats. *Nepenthes rajah* is endemic to Mount Kinabalu in Sabah, Malaysia, where it lives between 1500 and 2650 m above sea level.

A hippo has a specialized stomach to enable it to eat vegetation – its four chambers are the same as those found in other herbivores such as cows and deer.

Parasitism

A parasite is an organism that benefits at the expense of another (the host) from which it derives food. Ectoparasites live on the surface of their host (e.g. ticks and mites); endoparasites live inside their host (e.g. tapeworms).

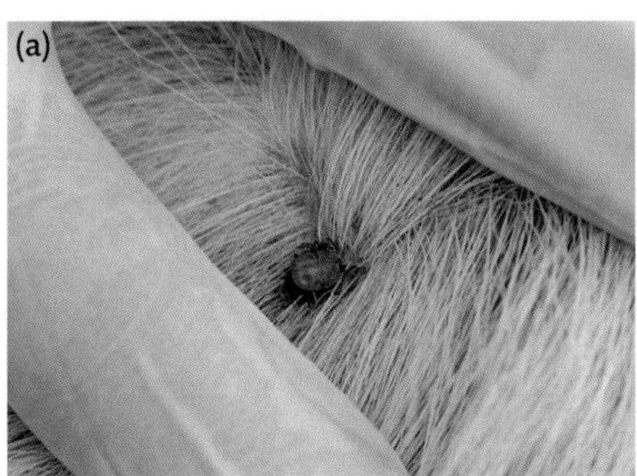

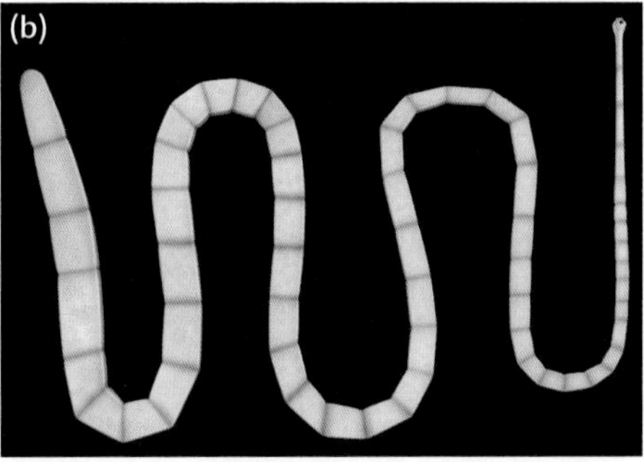

(a) A tick feeding on a dog
(b) A tapeworm – a parasite that lives in the intestines of mammals

The carrying capacity of the host may be reduced because of the harm caused by the parasite. Some plant parasites draw food from the host via their roots.

Parasitism is a symbiotic relationship in which one species benefits at the expense of the other. Mutualism is a symbiotic relationship in which both species benefit.

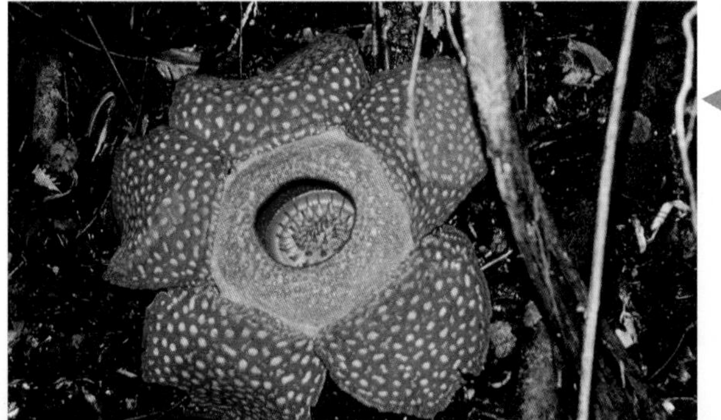

Rafflesia have the largest flowers in the world but no leaves. Without leaves, they cannot photosynthesize, so they grow close by South East Asian vines (*Tetrastigma* spp.) from which they draw the sugars they need for growth.

Mutualism

Symbiosis is a relationship in which two organisms live together. Parasitism is a form of symbiosis where one of the organisms is harmed. When both species benefit, the relationship is called mutualism. Examples include coral reefs and lichens. Coral reefs show a symbiotic relationship between the coral animal (polyp) and zooxanthellae (unicellular brown algae or dinoflagellates) that live within the coral polyp (Figure 2.7).

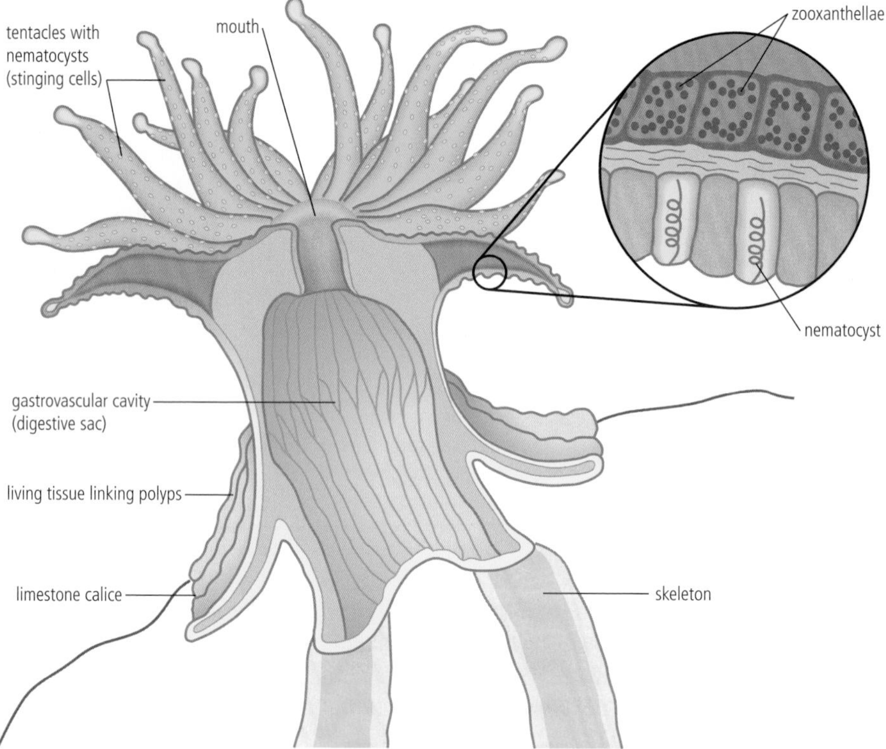

Figure 2.7 The zooxanthellae living within the polyp animal photosynthesize to produce food for themselves and the coral polyp, and in return are protected.

Disease

An organism that causes disease is known as a **pathogen**. Pathogens include bacteria, viruses, fungi, and single-celled animals called protozoa. The disease-causing species may reduce the carrying capacity of the organism it is infecting. Changes in disease can also cause populations to increase and decrease around the carrying capacity (Figure 2.11, page 71).

Dutch elm disease is caused by fungus (*Ascomycota*) that affects elm trees (see below). The fungus is spread by the elm bark beetle.

In Dutch elm disease, infection by a fungus results in clogging of vascular tissues. This prevents movement of water around the tree from the roots to the leaves and results in wilting and death.

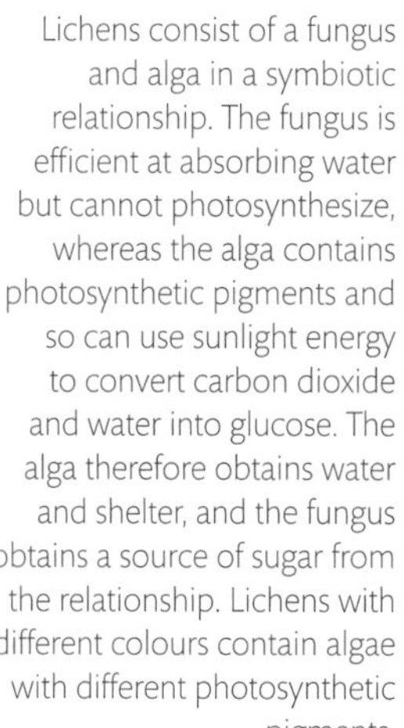

Lichens consist of a fungus and alga in a symbiotic relationship. The fungus is efficient at absorbing water but cannot photosynthesize, whereas the alga contains photosynthetic pigments and so can use sunlight energy to convert carbon dioxide and water into glucose. The alga therefore obtains water and shelter, and the fungus obtains a source of sugar from the relationship. Lichens with different colours contain algae with different photosynthetic pigments.

Competition

When resources are limiting, populations compete in order to survive. **Competition** is the demand by individuals for limited environmental resources. It may be either within a species (**intraspecific competition**) or between different species (**interspecific competition**). Interspecific competition exists when the niches of different species overlap (Figure 2.8).

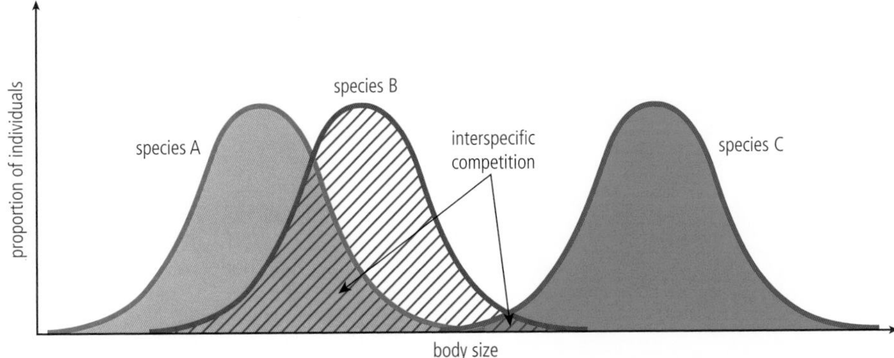

Figure 2.8 The niches of species A and species B, based on body size, overlap with each other to a greater extent than with species C. Strong interspecific competition will exist between species A and B but not with species C.

No two species can occupy the same niche, so the degree to which niches overlap determines the degree of interspecific competition. In this relationship, neither species benefits, although better competitors suffer less.

Experiments with single-celled animals have demonstrated the principle of competitive exclusion: if two species occupying similar niches are grown together, the poorer competitor will be eliminated (Figure 2.9).

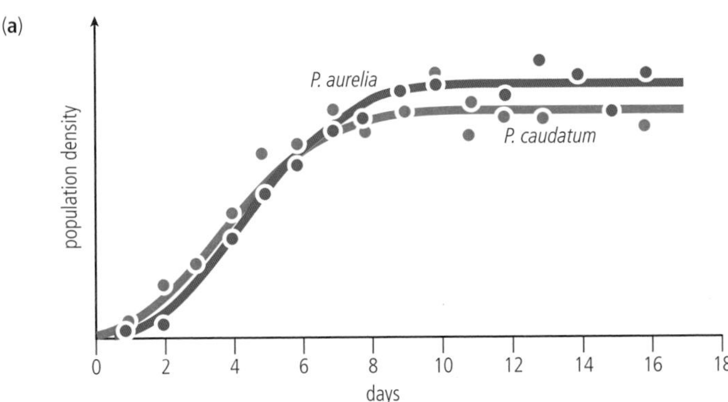

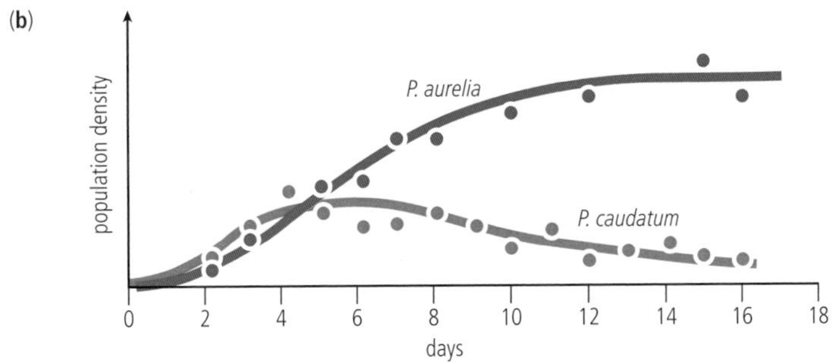

Figure 2.9 Species of *Paramecium* (a single-celled organism) can easily be grown in the laboratory. (a) If two species with very similar resource needs (i.e. similar niches) are grown separately, both can survive and flourish. (b) If the two species are grown in a mixed culture, the superior competitor – in this case *P. aurelia* – eliminates the other (this is known as competitive exclusion).

You need to be able to interpret graphical representations or models of factors that affect an organism's niche. Examples include predator–prey relationships, competition, and organism abundance over time.

Individuals within the same species occupy the same niche. Thus, if resources become limiting for a population, intraspecific competition becomes stronger. In this interaction, the stronger competitor (i.e. the one better able to survive) will reduce the carrying capacity of the other's environment.

CHALLENGE YOURSELF

ATL Thinking skills

This topic provides lots of opportunities for use of simulations and data-based analysis. Use the hotlink below to carry out a Virtual Lab experiment on population biology. To access worksheets and questions that accompany the experiment, use the second hotlink.

Population growth

A population of meerkats

Population growth curves

A population is a group of organisms of the same species living in the same area at the same time, and which are capable of interbreeding.

If a population is introduced into a new environment, such as that seen in the re-establishment of vegetation after the eruption of Krakatau in 1883 or the Mt St Helens eruption in 1980, specific population growth curves occur. Imagine rabbits are introduced into a new meadow. After an initial rapid (exponential) growth, the rabbit population will eat the vegetation faster than it can grow, because of the large numbers of rabbits. Further increases in population will stop. In this situation, the food supply has become a limiting factor in the growth of the rabbit population. Eventually, the rabbit population will reach the carrying capacity of the meadow (i.e. the size of rabbit population that the meadow can support).

To carry out an experiment in population biology by simulating growth in *Paramecium*, go to www.pearsonhotlinks.co.uk, enter the book title or ISBN, and click on weblink 2.1.

To access questions that accompany the experiment, go to www.pearsonhotlinks.co.uk, enter the book title or ISBN, and click on weblink 2.2.

S population curve

When a graph of population growth for such species is plotted against time, an **S-curve** is produced. This is also known as a sigmoid growth curve. An S-shaped population curve shows an initial rapid growth (exponential growth) and then slows down as the carrying capacity is reached (Figure 2.10).

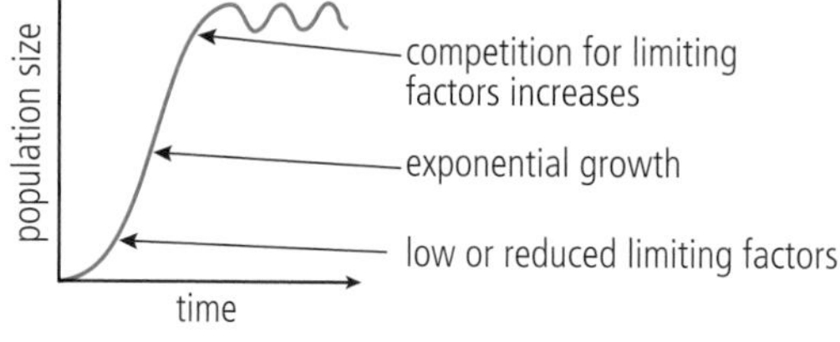

Figure 2.10 An S-shaped population growth curve

The graph shows slow growth at first when the population is small and there is a lack of mature adults. Early in the population growth curve there are few limiting factors and the population can expand exponentially. Competition between the individuals of the same species increases as a population grows. Competition increases because individuals are competing for the same limiting factors, such as resources (e.g. space on a rock for barnacles to attach). Competition for limiting factors, known as environmental resistance, results in a lower rate of population increase later on in the curve. The population eventually reaches its carrying capacity. Changes in the limiting factors cause the population size to increase and decrease (i.e. fluctuate) around the carrying capacity. Increases and decreases around the carrying capacity are controlled by negative feedback mechanisms.

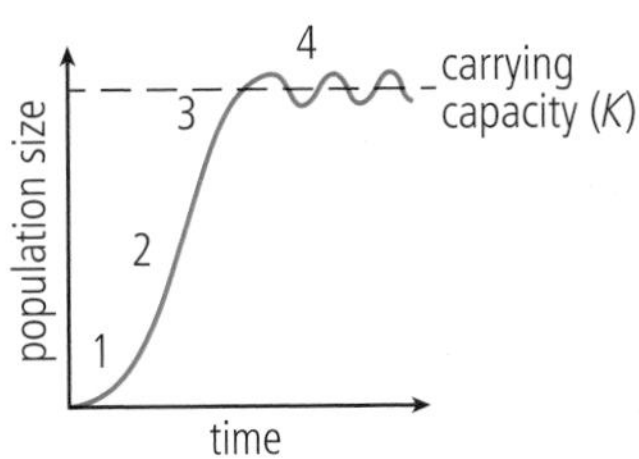

Figure 2.11 The four stages of an S-curve

The S-shaped population curve can be divided into four stages (Figure 2.11 and Table 2.1).

Table 2.1 The stages of an S-curve

Number of stage	Name of stage	Description	Explanation
1	lag phase	population numbers are low leading to low birth rates	• few individuals colonize a new area • because numbers of individuals are low, birth rates are also low
2	exponential growth phase	population grows at an increasingly rapid rate	• limiting factors are not restricting the growth of the population • there are favourable abiotic components, such as temperature and rainfall, and a lack of predators or disease • the numbers of individuals rapidly increases as does the rate of growth
3	transitional phase	population growth slows down considerably although continuing to grow	• limiting factors begin to affect the population and restrict its growth • there is increased competition for resources • an increase in predators and an increase in disease and mortality due to increased numbers of individuals living in a small area also cause a slowdown in growth rate
4	stationary phase	population growth stabilizes (the graph 'flattens') and then population fluctuates around a level that represents the carrying capacity	• limiting factors restrict the population to its carrying capacity (*K*) • changes in limiting factors, predation, disease, and abiotic factors cause populations to increase and decrease around the carrying capacity

Some factors that limit the size of populations depend on the density of the population, whereas others do not. **Density-dependent factors** are those that lower the birth rate or raise the death rate as a population grows. In contrast, **density-independent factors** are those which affect a population irrespective of population density. Factors affected by population density include supply of food and water, predation, parasites, and communicable disease (e.g. influenza). Factors not related to population density include climate (e.g. precipitation and humidity) and natural disasters (e.g. fire and flood).

You need to be able to explain population growth curves in terms of numbers and rates.

Sigmoidal curves (S-curves) are population growth curves which show an initial rapid growth (exponential growth) and then slow down as the carrying capacity is reached. Population size fluctuates around a set point (carrying capacity) (Figure 2.11). In contrast, a J-curve is a population growth curve which shows only exponential growth. Growth is initially slow and becomes increasingly rapid; it does not slow down (Figure 2.12).

J population curve

Exponential growth is an increasing or accelerating rate of growth, sometimes referred to as a J-shaped population curve or a **J-curve**. Growth is initially slow but becomes increasingly rapid, and does not slow down as population increases. Many populations show J-shaped rather than S-shaped population growth curves. Organisms showing J-shaped curves tend to produce many offspring rapidly and have little parental care (e.g. insects such as locusts).

Exponential growth occurs when:

- limiting factors are not restricting the growth of the population
- there are plentiful resources such as light, space, and food
- there are favourable abiotic components, such as temperature and rainfall.

Abiotic components can affect population growth (e.g. the carrying capacity of an environment for locusts can be raised due to rain). The sudden decrease in the population is called a population crash. A sudden decrease is shown in the Figure 2.12:

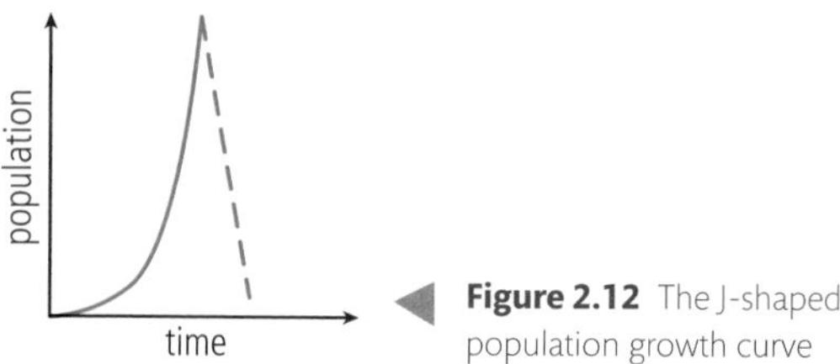

Figure 2.12 The J-shaped population growth curve

Populations showing J-shaped curves are generally controlled by abiotic but not biotic components, although lack of food can also cause populations to crash.

S and J population curves describe a generalized response of populations to a particular set of conditions (abiotic and biotic factors).

Limiting factors slow population growth as it approaches the carrying capacity of the system.

Limiting factors

Limiting factors include:

- for plants: light, nutrients, water, carbon dioxide, and temperature
- for animals: space, food, mates, nesting sites, and water.

Populations have an upper level or extent to the numbers that can be sustained in a given environment – carrying capacity is the term used to describe the maximum number of individuals of a species that can be sustained by an environment. The carrying capacity represents the population size at which environmental limiting factors limit further population growth.

The carrying capacity of a population is affected by various limiting factors, such as:

- the availability of food and water
- territorial space
- predation
- disease
- availability of mates.

The term *limiting factor* was first used by the German agricultural chemist Justus von Liebig (1803–73) who noted that the growth of crops was limited by the shortage of certain minerals. Liebig established the 'law of minimum' – the idea that the productivity, reproduction, and growth of organisms will be limited if one or more environmental factors is below its limiting level. Equally, there can be too much of a factor (i.e. there is an upper limit to how much of a particular nutrient plants can tolerate).

Exercises

1. Define the terms *species*, *population*, *habitat*, and *niche*. What is the difference between a habitat and a niche? Can different species occupy the same niche?
2. Explain the difference between fundamental and realized niche, using an example to illustrate your answer.
3. What is the difference between mutualism and parasitism? Give examples of each.
4. Describe and explain an S-shaped population curve.
5. The abundance of one species can affect the abundance of another. Give an ecological example of this, and explain how the predator affects the abundance of the prey, and vice versa. Are population numbers generally constant in nature? If not, what implications does this have for the measurement of wild population numbers?
6. Explain the concepts of limiting factors and carrying capacity in the context of population growth.
7. The data below show rates of growth in ticklegrass (as above-ground biomass in g m^{-2}) in soils with low or high nitrogen content and using high or low seed density.

Year	Low nitrogen, high seed density: above-ground biomass / $g m^{-2}$	Low nitrogen, low seed density: above-ground biomass / $g m^{-2}$	High nitrogen, high seed density: above-ground biomass / $g m^{-2}$	Low nitrogen, low seed density: above-ground biomass / $g m^{-2}$
1	0	0	0	0
2	420	60	500	30
3	780	80	1050	100
4	0	70	0	90
5	50	100	160	80
6	180	110	600	70

   a. Plot the data showing the growth rates among ticklegrass depending on nitrogen availability and density of seeds.
   b. Describe the results you have produced.
   c. Suggest reasons for these results.

8. The table below shows population growth in a population with discrete generations, starting with a population of 1000 and increasing at a constant reproductive rate of 1.2 per cent per generation.

Generation number	Population, *N*	Increase in population
*N*0	1000	-
*N*1	1200	200
*N*2	1440	240
*N*3	1728	288
*N*4	2074	346
*N*5	2488	414
*N*6	2986	498
*N*7	3583	597
*N*8		
*N*9		
*N*10		
*N*11		
*N*12		
*N*13		
*N*14		
*N*15		

   a. Complete the table by working out total population (column 2) and working out the increase in population size from generation to generation (column 3).
   b. Plot the graph of total population size.
   c. Describe the graph and identify the type of population growth that it shows.

- **Limiting factors are the factors that limit the distribution or numbers of a particular population.**
- **Limiting factors are environmental factors which slow down population growth.**

The figures in column 2 have been rounded to a whole number, but the real number for each generation has been multiplied by 1.2 to get the answer for the next generation (e.g. *N*4 has a population of 2073.6. This has been rounded up to 2074. However, to find *N*5, 2037.6 has been multiplied by 1.2 to make 2488.32 which is rounded down to 2488).

Big questions

Having read this section, you can now discuss the following big questions:

- What strengths and weaknesses of the systems approach and the use of models have been revealed through this topic?

Points you may want to consider in your discussions:

- Models to discuss include:
  - niche theory (fundamental vs. realized niche)
  - limits of tolerance
  - population growth curves (S-curve and J-curve).

2.2 Communities and ecosystems

Significant ideas

The interactions of species with their environment result in energy and nutrient flow.

Photosynthesis and respiration play a significant role in the flow of energy in communities.

The feeding relationships in a system can be modelled using food chains, food webs, and ecological pyramids.

Big questions

As you read this section, consider the following big question:

- What strengths and weaknesses of the systems approach and the use of models have been revealed through this topic?

Knowledge and understanding

- A community is a group of populations living and interacting with each other in a common habitat.
- An ecosystem is a community and the physical environment it interacts with.
- Respiration and photosynthesis can be described as processes with inputs, outputs, and transformations of energy and matter.
- Respiration is the conversion of organic matter into carbon dioxide and water in all living organisms, releasing energy. Aerobic respiration can simply be described as:

  glucose + oxygen → carbon dioxide + water
- During respiration large amounts of energy are dissipated as heat, increasing the entropy in the ecosystem while enabling the organisms to maintain relatively low entropy/high organization.
- Primary producers in the majority of ecosystems convert light energy into chemical energy in the process of photosynthesis.

- The photosynthesis reaction is:

  carbon dioxide + water → glucose + oxygen
- Photosynthesis produces the raw material for producing biomass.
- The trophic level is the position that an organism occupies in a food chain, or a group of organisms in a community that occupy the same position in food chains.
- Producers (autotrophs) are typically plants or algae that produce their own food using photosynthesis and form the first trophic level in a food chain. Exceptions include chemosynthetic organisms which produce food without sunlight.
- Feeding relationships involve producers, consumers and decomposers. These can be modelled using food chains, food webs, and using ecological pyramids.
- Ecological pyramids include pyramids of numbers, biomass, and productivity and are quantitative models and are usually measured for a given area and time.
- In accordance with the second law of thermodynamics, there is a tendency for numbers and quantities of biomass and energy to decrease along food chains; therefore the pyramids become narrower towards the apex.
- Bioaccumulation is the build-up of persistent/non-biodegradable pollutants within an organism or trophic level because they cannot be broken down.
- Biomagnification is the increase in concentration of persistent/non-biodegradable pollutants along a food chain.
- Toxins such as DDT and mercury accumulate along food chains due to the decrease of biomass and energy.
- Pyramids of numbers can sometimes display different patterns, for example, when individuals at lower trophic levels are relatively large (inverted pyramids).
- A pyramid of biomass represents the standing stock/storage of each trophic level measured in units such as grams of biomass per square metre (g m^{-2}) or joules per square metre (J m^{-2}) (units of biomass or energy).
- Pyramids of biomass can show greater quantities at higher trophic levels because they represent the biomass present at a given time, but there may be marked seasonal variations.
- Pyramids of productivity refer to the flow of energy through a trophic level, indicating the rate at which that stock/storage is being generated.
- Pyramids of productivity for entire ecosystems over a year always show a decrease along the food chain.

Communities and ecosystems

Community

A **community** is many species living together, whereas the term *population* refers to just one species. The savannah grasslands and lakeland ecosystems of Africa contain wildebeest, lions, hyenas, giraffes, and elephants as well as zebras. Communities include all biotic parts of the ecosystem, both plants and animals.

An animal community in the Ngorongoro Conservation Area, Tanzania

- A community is a group of populations living and interacting with each other in a common habitat.
- An ecosystem is a community and the physical environment it interacts with.

Ecosystem

An ecosystem is a community of interdependent organisms (the biotic component) and the physical environment (the abiotic component) they inhabit.

Ecosystems can be divided into three types: terrestrial, marine, and freshwater. Marine ecosystems include the sea, estuaries, salt marshes, and mangroves. Marine ecosystems all have a high concentration of salt in the water. Estuaries are included in the same group as marine ecosystems because they have high salt content compared to freshwater ecosystems. Freshwater ecosystems include rivers, lakes, and wetlands. Terrestrial ecosystems include all land-based ecosystems.

Taiga forest (Chapter 1, page 35) is an example of a terrestrial ecosystem.

Coral reef (page 68) is an example of a marine ecosystem.

The Orinoco River, Venezuela, is an example of a freshwater ecosystem.

Ecosystems such as the northern coniferous forest (Taiga, page 76) cross several countries and so their conservation and ecology has an international dimension.

Each type of ecosystem has specific abiotic factors which characterize and define the ecosystem. Each ecosystem also has abiotic factors shared with other types of ecosystem. Measuring different abiotic factors is discussed later in this chapter (pages 128–132). The biotic component of an ecosystem (i.e. the species found there) depends on the abiotic factors that define the ecosystem.

Photosynthesis and respiration

Continual inputs of energy and matter are essential in the support of ecosystems. Two processes control the flow of energy through ecosystems: photosynthesis and respiration. **Photosynthesis** converts light energy to chemical energy, which is stored in biomass. **Respiration** releases this energy so that it can be used to support the life processes (e.g. movement) of organisms.

Respiration and photosynthesis can be described as processes with inputs, outputs, and transformations of energy and matter.

SYSTEMS APPROACH

Respiration and photosynthesis can be represented as systems diagrams, with inputs, outputs, storages, and processes.

Photosynthesis

Photosynthesis requires carbon dioxide, water, chlorophyll, and light, and is controlled by enzymes. Oxygen is produced as a waste product in the reaction.

The photosynthesis reaction is:

$$\text{carbon dioxide} + \text{water} \xrightarrow[\text{chlorophyll}]{\text{light}} \text{glucose} + \text{oxygen}$$

Photosynthesis is the process by which green plants convert light energy from the Sun into useable chemical energy stored in organic matter.

Photosynthesis produces the raw material for producing biomass.

In terms of inputs, outputs, and energy transformations, photosynthesis can be summarized as follows.

- Inputs – sunlight as energy source, carbon dioxide, and water.
- Outputs – glucose, used as an energy source for the plant and as the basic starting material for other organic molecules (e.g. cellulose, starch);
  – oxygen, released to the atmosphere through stomata.
- Transformations – the energy change is from light energy into stored chemical energy, and thus the chemical energy is stored in organic matter (i.e. carbohydrates, fats, and proteins). Chlorophyll is needed to capture certain visible wavelengths of sunlight energy and allow this energy to be transformed into chemical energy.

Respiration

Respiration is the conversion of organic matter into carbon dioxide and water in all living organisms, releasing energy.

Respiration releases energy from glucose and other organic molecules inside all living cells. It begins as an anaerobic process in the cytoplasm of cells, and is completed inside mitochondria with aerobic chemical reactions occurring. The process is controlled by enzymes. The energy released is in a form available for use by living organisms, but is ultimately lost as heat (Chapter 1).

Aerobic respiration can simply be described as:

glucose + oxygen → carbon dioxide + water

All organisms respire: bacteria, algae, plants, fungi, and animals. Only plants, algae and cyanobacteria photosynthesize.

Respiration can be summarized as follows (Figure 2.13).

- Inputs – organic matter (glucose) and oxygen.
- Processes – oxidation processes inside cells.
- Outputs – release of energy for work and heat.
- Transformations – the energy transformation is from stored chemical energy into kinetic energy and heat. Energy is released in a form available for use by living organisms, but much is also eventually lost as heat (the second law of thermodynamics, page 27).

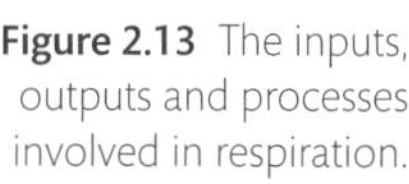

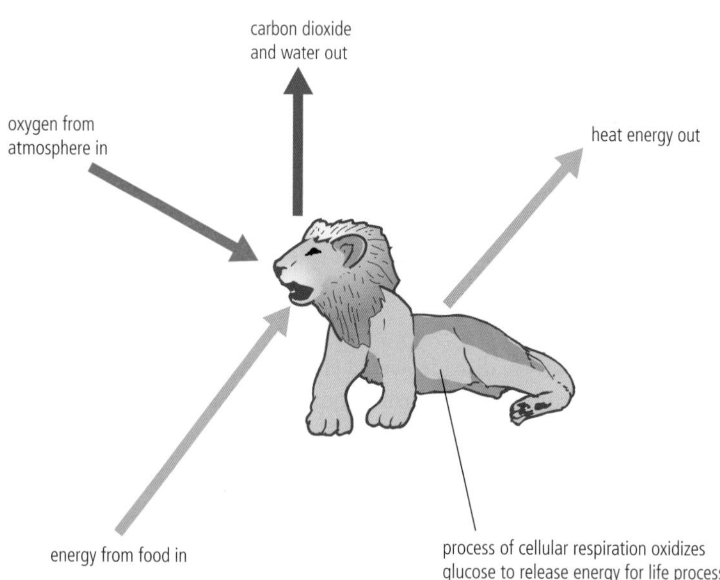

Figure 2.13 The inputs, outputs and processes involved in respiration.

You need to be able to construct system diagrams representing photosynthesis and respiration.

During respiration large amounts of energy are dissipated as heat, increasing the entropy in the ecosystem (Chapter 1, page 28) while enabling the organisms to maintain relatively low entropy (i.e. high organization).

Photosynthesis involves the transformation of light energy into the chemical energy of organic matter. Respiration is the transformation of chemical energy into kinetic energy with, ultimately, heat lost from the system.

Feeding relationships

Producers

Certain organisms in an ecosystem convert abiotic components into living matter. These are the **producers**; they support the ecosystem by constant input of energy and new biological matter (biomass) (Figure 2.14). Producers are also known as **autotrophs**.

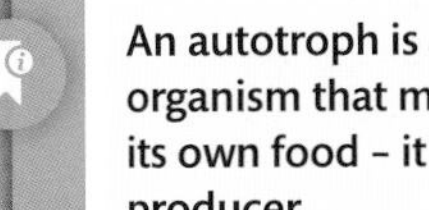

An autotroph is an organism that makes its own food - it is a producer.

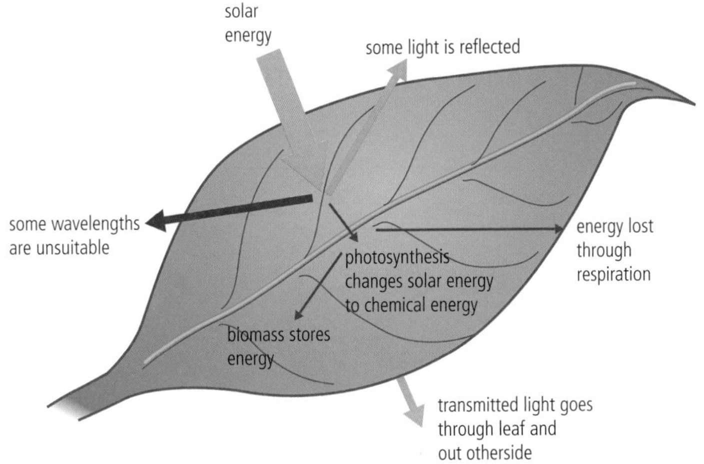
solar energy
some light is reflected
some wavelengths are unsuitable
energy lost through respiration
photosynthesis changes solar energy to chemical energy
biomass stores energy
transmitted light goes through leaf and out otherside

Figure 2.14 Producers convert sunlight energy into chemical energy using photosynthetic pigments (e.g. chlorophyll). The food produced supports the rest of the food chain.

Plants, algae, and some bacteria are producers. Organisms that use sunlight energy to create their own food are called photoautotrophs; all green plants are photoautotrophs. Not all producers use sunlight to make food. For example, some bacteria use chemical energy rather than sunlight to make sugars; chemosynthetic bacteria are part of the nitrogen cycle (page 98). Giant tube worms (*Riftia pachyptila*) live on or near deep-sea hydrothermal vents (page 111); they have a symbiotic relationship with chemosynthetic bacteria using hydrogen sulfide and carbon dioxide to produce sugars.

Giant tube worms at a hydrothermal vent

Consumers

Organisms that cannot make their own food eat other organisms to obtain energy and matter: they are **consumers**. Consumers do not contain photosynthetic pigments such as chlorophyll so they cannot make their own food. They must obtain their energy, minerals and nutrients by eating other organisms – they are **heterotrophs**. Herbivores feed on autotrophs, **carnivores** feed on other heterotrophs, and omnivores feed on both.

Consumers pass energy and biomass from producers through to the top carnivores.

Decomposers

Decomposers obtain their food and nutrients from the breakdown of dead organic matter. When they break down tissue, they release nutrients ready for reabsorption by producers. They form the basis of a decomposer food chain. Decomposers also contribute to the build-up of humus in soil. Humus is organic material in soil made by the decomposition of plant or animal matter. It improves the ability of soil to retain nutrients. Decomposers are essential for cycling matter in ecosystems. Matter that is cycled by decomposers in ecosystems includes elements such as carbon and nitrogen.

Trophic levels, food chains and food webs

The flow of energy and matter from organism to organism can be shown in a food chain. The position that an organism occupies in a food chain is called the **trophic level** (Figure 2.15). Trophic level can also mean the position in the food chain occupied by a group of organisms in a community. Producers form the first trophic level in a food chain.

western wheat grass →	club-horned grasshopper →	Great Plains toad →	garter snake →	Swainson's hawk
producer	primary consumer	secondary consumer	tertiary consumer	quaternary consumer
autotroph	herbivore	omnivore/ carnivore	carnivore	carnivore

Figure 2.15 A food chain. Ecosystems contain many food chains.

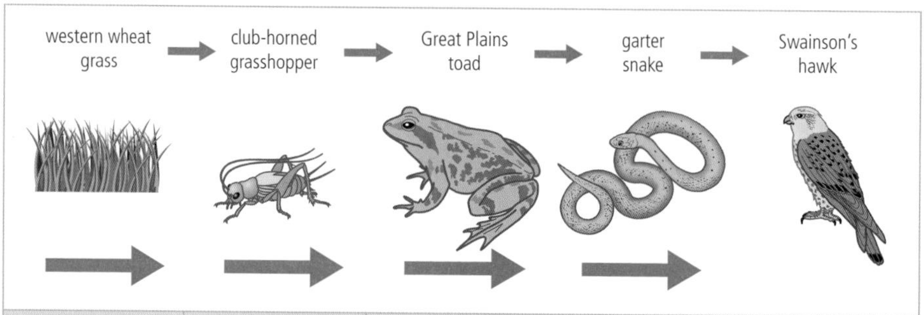

If you are asked to draw a food chain, you do not need to draw the animals and plants involved. You do need to give specific names for the different organisms (e.g. salmon rather than fish). Arrows show the flow of energy from one organism to the next and should be in the direction of energy flow.

Ecosystems contain many interconnected food chains that form **food webs** (Figure 2.16).

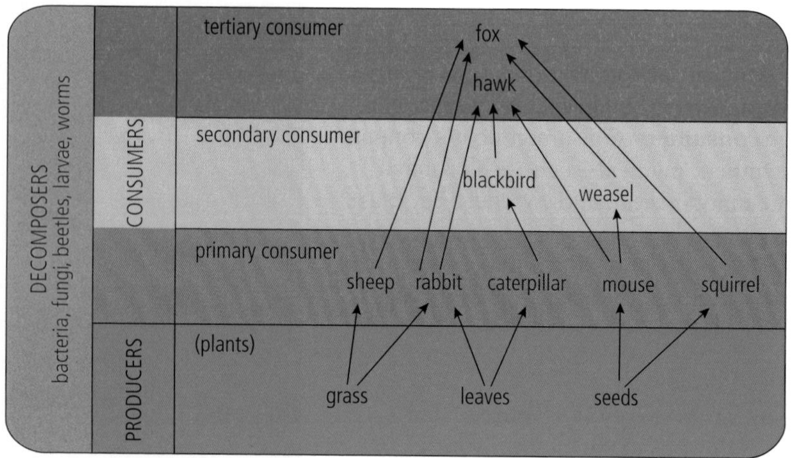

Figure 2.16 A food web showing its trophic levels

One species may occupy several different trophic levels depending on which food chain it is present in. In Figure 2.16, foxes and hawks are both secondary and tertiary consumers depending on which food chain they are in. Decomposers feed on dead organisms at each trophic level.

Case study

The effect of harvesting on food webs in the North Sea

Diagrams of food webs can be used to estimate the knock-on effects of changes to an ecosystem. Figure 2.17 shows a food web for the North Sea. In the figure, the producer is phytoplankton (microscopic algae), the primary consumers (herbivores) are zooplankton (microscopic animal life), the secondary consumers (carnivores) include jellyfish, sand eels, and herring (each on different food chains), and the tertiary consumers (top carnivores) are mackerel, seals, seabirds, and dolphins (again, on different food chains).

During the 1970s, sand eels were harvested and used as animal feed, for fishmeal and for oil and food on salmon farms: Figure 2.17 can be used to explain what impacts a dramatic reduction in the number of sand eels might have on the rest of the ecosystem. Sand eels are the only source of food for mackerel, puffin, and gannet, so populations of these species may decline or they may have to switch food source. Similarly, seals will have to rely more on herring, possibly reducing their numbers or they may also have to switch food source. The amount of zooplankton may increase, improving food supply for jellyfish and herring.

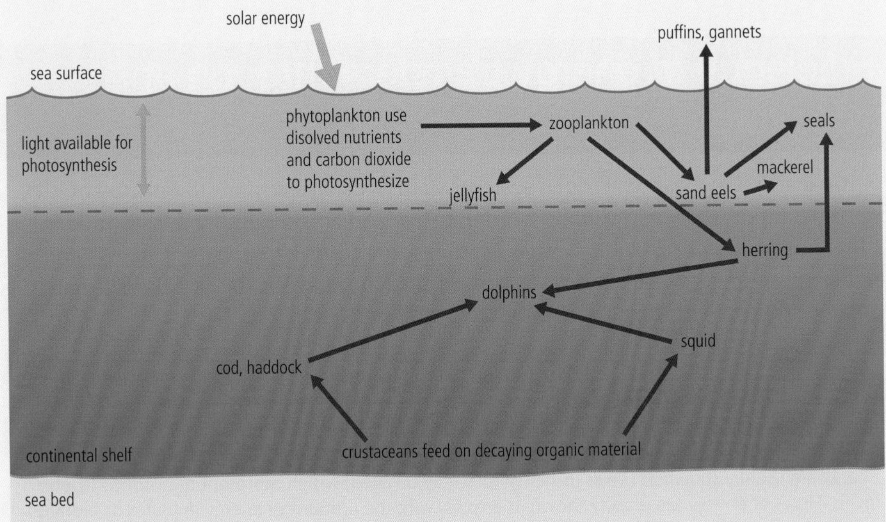

Figure 2.17 A simplified food web for the North Sea in Europe

An estimated 1000 kg of phytoplankton (plant plankton) are needed to produce 100 kg of zooplankton (animal plankton). The zooplankton is in turn consumed by 10 kg of fish, which is the amount needed by a person to gain 1 kg of body mass. Biomass and energy decline at each successive trophic level so there is a limit to the number of trophic levels which can be supported in an ecosystem. Energy is lost as heat (produced as a waste product of respiration) at each stage in the food chain, so only energy stored in biomass is passed on to the next trophic level. Thus, after 4 or 5 trophic stages, there is not enough energy to support another stage.

Food chains always begin with the producers (usually photosynthetic organisms), followed by primary consumers (herbivores), secondary consumers (omnivores or carnivores), and then higher consumers (tertiary, quaternary, etc.). Decomposers feed at every level of the food chain.

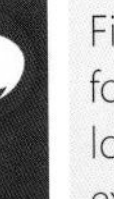

Find an example of a food chain from your local area, with named examples of producers, consumers, decomposers, herbivores, carnivores, and top carnivores.

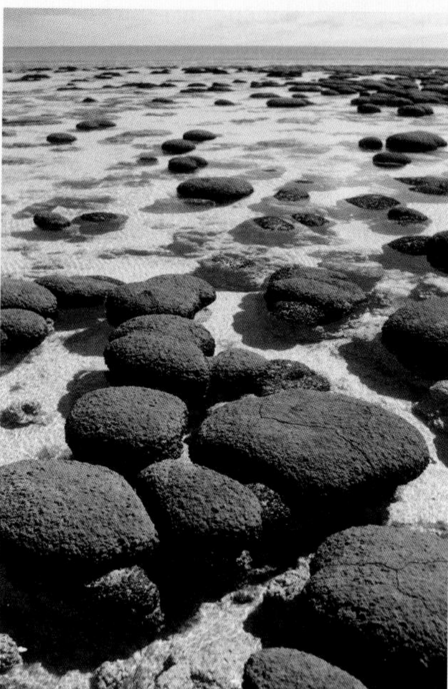

The earliest forms of life on Earth, 3.8 billion years ago, were consumers feeding on organic material formed by interactions between the atmosphere and the land surface. Producers appeared around 3 billion years ago – these were photosynthetic bacteria. Because oxygen is a waste product of photosynthesis, these bacteria eventually brought about a dramatic increase in the amount of oxygen in the atmosphere. The oxygen enabled organisms using aerobic respiration to evolve and generate the large amounts of energy they needed. And eventually, complex ecosystems followed.

Stromatolites were the earliest producers on the planet and are still here. These large aggregations of cyanobacteria can be found in the fossil record and alive in locations such as Western Australia and Brazil.

You need to actively link this topic with what you have learned in Chapter 1 – questions will arise requiring you to use your knowledge of thermodynamics to explain energy flow in ecosystems.

Efficiency of energy transfers through an ecosystem

As you learned in Chapter 1, open systems such as ecosystems are supported by continual input of energy, usually from the Sun. You will recall that the second law of thermodynamics (Chapter 1, page 27) states that transformations of energy are inefficient, so energy is lost from the system at each stage of a food chain, ultimately as heat energy. You will learn more about this later (page 89).

SYSTEMS APPROACH

Systems diagrams can be used to show the flow of energy through ecosystems (Figure 2.18). Stores of energy are usually shown as boxes (other shapes may be used) which represent the various trophic levels. Flows of energy are usually shown as arrows (with the amount of energy in joules or biomass per unit area represented by the thickness of the arrow).

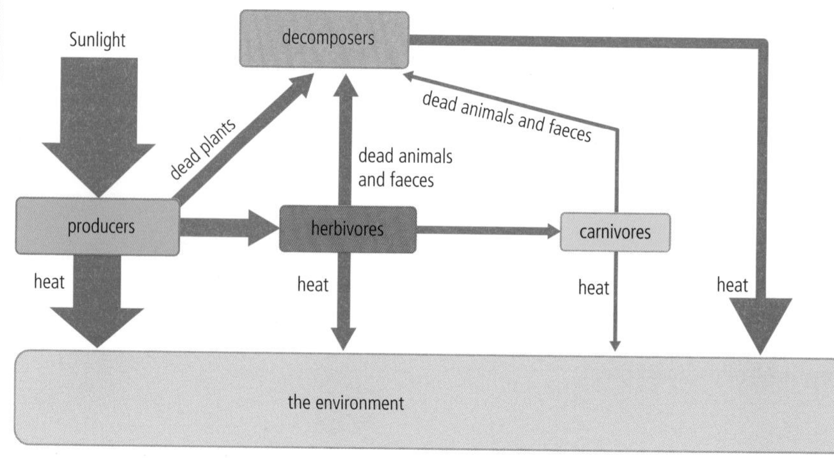

Figure 2.18 Energy flow in an ecosystem. The width of the arrows is proportional to the quantity of energy transferred. Producers convert energy from sunlight into new biomass through photosynthesis. Heat is released to the environment through respiration.

You need to be able to analyse the efficiency of energy transfers through a system.

Pyramids of numbers, biomass, and productivity

Pyramids are graphical models of the quantitative differences (e.g. differences in numbers) that exist between the trophic levels of a single ecosystem, and are usually measured for a given area and time. These models provide a better understanding of the workings of an ecosystem by showing the feeding relationships in a community. There are three types of pyramid: pyramid of numbers, pyramid of biomass, and pyramid of productivity.

Pyramids are graphical models showing the quantitative differences between the trophic levels of an ecosystem and are usually measured for a given area and time. There are three types.

- **Pyramid of numbers records the number of individuals at each trophic level coexisting in an ecosystem. Quantitative data for each trophic level are drawn to scale as horizontal bars arranged symmetrically around a central axis.**
- **Pyramid of biomass represents the biological mass of the standing stock at each trophic level at a particular point in time measured in units such as grams of biomass per square metre (g m^{-2}). Biomass may also be measured in units of energy, such as J m^{-2}.**
- **Pyramid of productivity shows the flow of energy (i.e. the rate at which the stock is being generated) through each trophic level of a food chain over a period of time. Productivity is measured in units of flow (g m^{-2} yr^{-1} or J m^{-2} yr^{-1}).**

How are pyramids constructed?

Quantitative data for a food chain are shown in Table 2.2.

Species	Number of individuals
leaves	40
caterpillar	20
blackbird	14
hawk	16

Table 2.2 Data for a terrestrial food chain

To construct a pyramid of numbers for these data, first draw two axes on graph paper. Draw the horizontal axis along the bottom of the graph paper and the vertical axis in the centre of the graph paper. Plot data from the table symmetrically around the vertical axis. As there are 40 leaves, the producer trophic level is drawn with 20 units to the left and 20 units to the right of the vertical axis. The height of the bars is kept the same for each trophic level. Each trophic level is labelled with the appropriate organism. Figure 2.19 shows the result.

Pyramids of biomass and pyramids of productivity are constructed in the same way.

Figure 2.19 Pyramid of numbers for given data

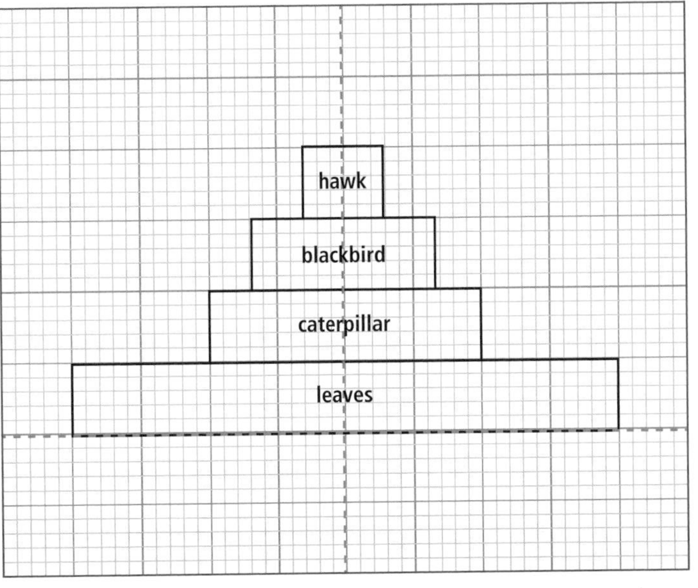

Pyramids of numbers

The numbers of producers and consumers coexisting in an ecosystem can be shown by counting the numbers of organisms in an ecosystem and constructing a pyramid. Sometimes, rather than counting every individual in a trophic level, limited collections may be done in a specific area and this multiplied up to the total area of the ecosystem. In accordance with the second law of thermodynamics, there is a tendency for numbers to decrease along food chains, and so graphical models tend to be pyramids – they are narrower towards the apex (Figure 2.20a). However, pyramids of numbers are not always pyramid shaped. For example, in a woodland ecosystem with many insect herbivores feeding on trees, there are fewer trees than insects. This means the pyramid is inverted (upside-down) as in Figure 2.20b. This situation arises when the size of individuals at lower trophic levels are relatively large. Pyramids of numbers, therefore, have limited use in representing the flow of energy through food chains.

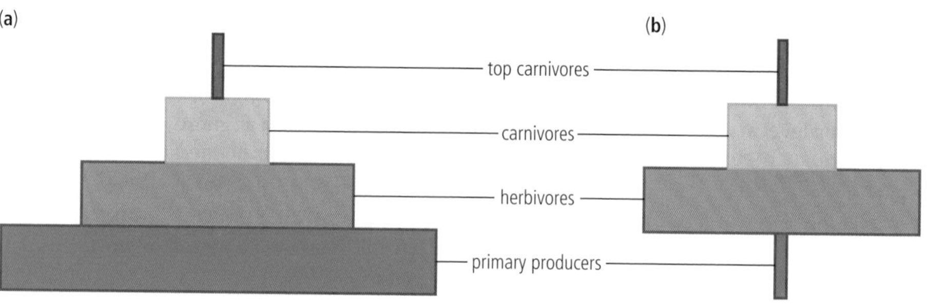

Figure 2.20 Pyramids of numbers. (a) A typical pyramid where the number of producers is high. (b) A limitation of number pyramids is that they are inverted when the producers are outnumbered by the herbivores.

Pyramids of biomass can show greater quantities at higher trophic levels because they represent the biomass present at a given time, but there may be marked seasonal variations. For example, phytoplankton vary in productivity (and therefore biomass) depending on sunlight intensity. The biomass present in an area also depends on the quantity of zooplankton consuming the phytoplankton.

Pyramids of biomass

A pyramid of biomass quantifies the amount of biomass present at each trophic level at a certain point in time, and represents the standing stock of each trophic level. Biomass may be measured in grams of biomass per metre squared (g m^{-2}) or units of energy, such as joules per metre squared (J m^{-2}). Following the second law of thermodynamics, there is a tendency for quantities of biomass (like numbers) to decrease along food chains, so the pyramids become narrower towards the top.

Although pyramids of biomass are usually pyramid shaped, they can sometimes be inverted and show greater quantities at higher trophic levels. This is because, as with pyramids of numbers, they represent the biomass present at a given time (i.e. they are a snap-shot of the ecosystem). The standing crop biomass (the biomass taken at a certain point in time) gives no indication of productivity over time. For example, a fertile intensively grazed pasture may have a lower standing crop biomass of grass but a higher productivity than a less fertile ungrazed pasture (because the fertile pasture has biomass constantly removed by herbivores). This results in an inverted pyramid of biomass. In a pond ecosystem, the standing crop of phytoplankton (the major producers) at any given point will be lower than the mass of the consumers, such as fish and insects. This is because phytoplankton reproduce very quickly. Inverted pyramids sometimes result from marked seasonal variations.

Pyramids of productivity refer to the flow of energy through a trophic level, indicating the rate at which that stock/storage is being generated.

Pyramids of productivity

Pyramids of biomass represent the momentary stock, whereas pyramids of productivity show the rate at which that stock is being generated. You cannot compare the turnover of two shops by comparing the goods displayed on the shelves, because you also need to know the rates at which the goods are sold and the shelves are restocked. The same is true of ecosystems.

Pyramids of productivity take into account the rate of production over a period of time because each level represents energy per unit area per unit time. Productivity is measured in units of flow – mass or energy per metre squared per year (g m^{-2} yr^{-1} or J m^{-2} yr^{-1}). This is a more useful way of measuring changes along a food chain than looking at either biomass (measured in g m^{-2}) or energy (measured in J m^{-2}) at one moment in time. Pyramids of productivity show the flow of energy through an entire ecosystem over a year. This means they invariably show a decrease along the food chain. There are no inverted pyramids of productivity. The relative energy flow within an ecosystem can be studied, and different ecosystems can be compared. Pyramids of productivity also overcome the problem that two species may not have the same energy content per unit weight: in these cases, biomass is misleading but energy flow is directly comparable.

You need to be able to explain the relevance of the laws of thermodynamics to the flow of energy through ecosystems.

CHALLENGE YOURSELF

Thinking skills 

Use the hotlink below to choose organisms and construct pyramids of numbers, biomass and productivity for five different ecosystems.

Pyramids of biomass refers to a standing crop (a fixed point in time) and pyramids of productivity refer to the rate of flow of biomass or energy, as shown in Table 2.3.

Table 2.3 Units for pyramids of biomass and productivity

Pyramid	Units
biomass (standing crop)	g m^{-2}
productivity (flow of biomass/energy)	g m^{-2} yr^{-1} J m^{-2} yr^{-1}

To learn more about constructing pyramids of numbers, biomass and productivity, go to www.pearsonhotlinks.co.uk, enter the book title or ISBN, and click on weblink 2.3.

Pyramid structure and ecosystem functioning

Because energy is lost through food chains, top carnivores are at risk from disturbance further down the food chain. A reduction in the numbers of producers or primary consumers can threaten the existence of the top carnivores when there are not enough of the producers or primary consumers (and therefore energy and biomass) to support the top carnivores. Because of their relatively small populations, top carnivores may be the first population we notice to suffer through ecosystem disruption.

Feeding relationships can be represented by different models (food chain, food webs, and ecological pyramids). How can we decide when one model is better than another?

Case study

A snow leopard hunting

Snow leopards are found in the mountain ranges of Central Asia. They feed on wild sheep and goats. Effects lower down the food chain threaten this top carnivore. Overgrazing of the mountain grasslands by farmed animals leaves less food for the snow leopard's main prey. Less food for the wild sheep and goats means fewer of these animals are available for the snow leopard, so its existence is at risk. The snow leopard has little choice but to prey on the domestic livestock in order to survive. But this leads the herdsmen to attack and kill the snow leopards.

The total wild population of the snow leopard is estimated at between 4100 and 6600 individuals. They have now been designated as endangered by the International Union for Conservation of Nature (IUCN).

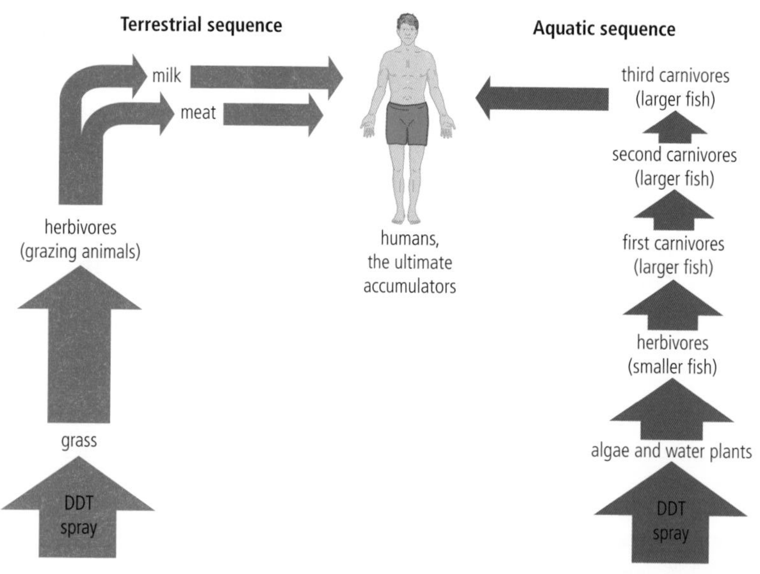

Figure 2.21 Simple food chains showing the accumulation of the non-biodegradable pesticide, DDT

Top carnivores can also be put at risk through other interferences in the food chain. Farmers often use pesticides to improve crop yield and to maximize profits. Today's pesticides break down naturally and lose their toxic properties (i.e. they are biodegradable), but this was not always the case (Chapter 1, page 52). In the past, pesticides weren't biodegradable, and their use had serious knock-on effects for ecosystems. Figure 2.21 shows the effect of the very effective non-biodegradable pesticide DDT on food chains. The producers, algae and plants or grass (first accumulators) take in the DDT. Organisms in the second trophic level (the primary consumers) eat the DDT-containing producers and retain the pesticide in their body tissue (mainly in fat) – this is **bioaccumulation**. The process continues up the food chain with more and more DDT being accumulated at each level. The top carnivores (humans, at level 6 in the aquatic food chain or level 4 in the terrestrial chain) are the final destination of the pesticide (ultimate accumulators).

You need to be able to explain the impact of a persistent or non-biodegradable pollutant in an ecosystem.

The pesticide accumulates in body fat and is not broken down. Each successive trophic level supports fewer organisms, so the pesticide becomes increasingly concentrated in the tissues – this is **biomagnification**. Organisms higher in the food chain have progressively longer life spans, so they have more time to accumulate more of the toxin by eating many DDT-containing individuals from lower levels. Top carnivores such as the bald eagle are, therefore, at risk from DDT poisoning (pages 53, 183).

- **Bioaccumulation is the build-up of persistent/non-biodegradable pollutants within an organism or trophic level because they cannot be broken down.**
- **Biomagnification is the increase in concentration of persistent/non-biodegradable pollutants along a food chain.**
- **Toxins such as DDT and mercury accumulate along food chains due to the decrease of biomass and energy.**

Exercises

1. Define the terms *community* and *ecosystem*.
2. Explain the role of producers, consumers, and decomposers in the ecosystem.
3. Summarize photosynthesis in terms of inputs, outputs, and energy transformations. Now do the same for respiration.
4. **a.** Why is not all available light energy transformed into chemical energy in biomass?
   b. Why is not all of the energy in biomass made available to the next tropic level?
5. Construct an energy-flow diagram illustrating the movement of energy through ecosystems, including the productivity of the various trophic levels.
6. What are the differences between a pyramid of biomass and a pyramid of productivity? Which is always pyramid shaped, and why? Give the units for each type of pyramid.
7. Explain the impact of a persistent/non-biodegradable pollutant on a named ecosystem.

Big questions

Having read this section, you can now discuss the following big questions:

- What strengths and weaknesses of the systems approach and the use of models have been revealed through this topic?

Points you may want to consider in your discussions:

- What are the strengths and weaknesses of models of food chains, food webs, and ecological pyramids?
- How can pyramids of productivity be used to predict the effect of human activities on ecosystems?
- How can systems diagrams be used to show energy flow through ecosystems? What are the strengths and weaknesses of such diagrams?

2.3 Flows of energy and matter

Significant ideas

Ecosystems are linked together by energy and matter flows.

The Sun's energy drives these flows. Humans are impacting the flows of energy and matter both locally and globally.

Big questions

As you read this section, consider the following big questions:

- What strengths and weaknesses of the systems approach and the use of models have been revealed through this topic?
- How are the issues addressed in this topic of relevance to sustainability or sustainable development?

Knowledge and understanding

- As solar radiation (insolation) enters Earth's atmosphere some energy becomes unavailable for ecosystems as the energy is absorbed by inorganic matter or reflected back into the atmosphere.
- Pathways of radiation through the atmosphere involve a loss of radiation through reflection and absorption.
- Pathways of energy through an ecosystem include:
  - conversion of light energy to chemical energy
  - transfer of chemical energy from one trophic level to another with varying efficiency
  - overall conversion of ultraviolet and visible light to heat energy by an ecosystem
  - re-radiation of heat energy to the atmosphere.
- The conversion of energy into biomass for a given period of time is measured as productivity.
- Net primary productivity (NPP) is calculated by subtracting respiratory losses (R) from gross primary productivity (GPP).

$$NPP = GPP - R$$

- Gross secondary productivity (GSP) is the total energy/biomass assimilated by consumers and is calculated by subtracting the mass of faecal loss from the mass of food eaten.

  GSP = food eaten – faecal loss
- Net secondary productivity (NSP) is calculated by subtracting respiratory losses (R) from GSP.

  NSP = GSP – R
- Maximum sustainable yields are equivalent to the net primary or net secondary productivity of a system.
- Matter also flows through ecosystems linking them together. This flow of matter involves transfers and transformations.
- The carbon and nitrogen cycles are used to illustrate this flow of matter using flow diagrams. These cycles contain storages (sometimes referred to as sinks) and flows that move matter between storages.
- Storages in the carbon cycle include organisms, including forests (organic), atmosphere, soil, fossil fuels, and oceans (all inorganic).
- Flows in the carbon cycle include consumption (feeding), death, and decomposition, photosynthesis, respiration, dissolving, and fossilization.
- Storages in the nitrogen cycle include organisms (organic), soil, fossil fuels, atmosphere, and water bodies (all inorganic).
- Flows in the nitrogen cycle include nitrogen fixation by bacteria and lightning, absorption, assimilation, consumption (feeding), excretion, death, and decomposition, and denitrification by bacteria in water-logged soils.
- Human activities such as burning fossil fuels, deforestation, urbanization, and agriculture impact energy flows as well as the carbon and nitrogen cycles.

Transfer and transformation of energy

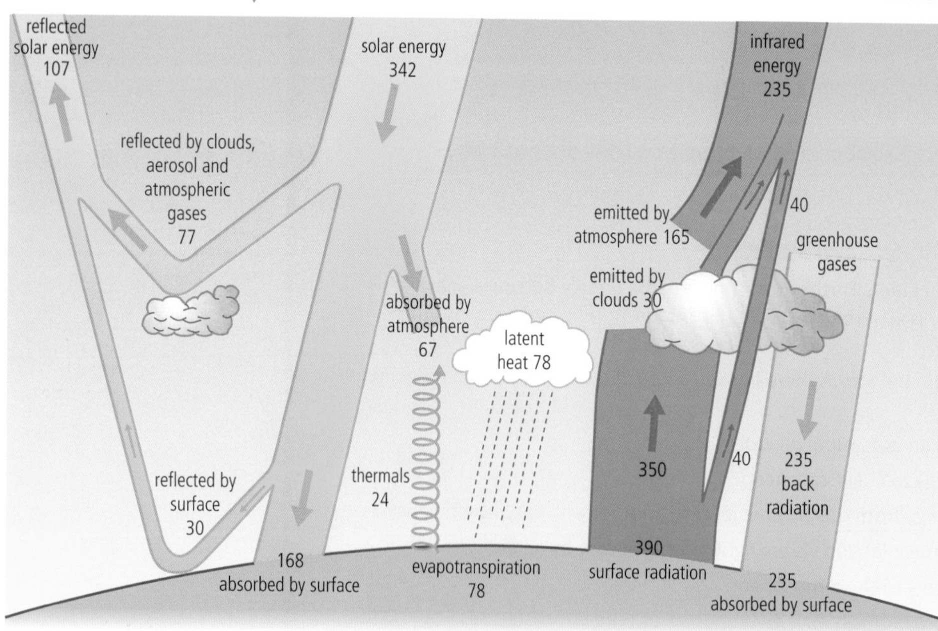

Figure 2.22 The Earth's energy budget. Mean vertical energy flows in the terrestrial system (atmosphere and surface), in watts per square metre. Most important are the 342 W m^{-2} of solar energy which enter the outer atmospheric layer and the approximately 390 W m^{-2} which are reradiated from the soil surface in the form of infrared waves.

The pathway of sunlight entering the Earth's atmosphere is complex (Figure 2.22). Sunlight contains a broad spectrum of wavelengths from X-rays to radio waves, though most exists as ultraviolet, visible light and infrared radiation. Almost half of the Sun's total radiation is visible light.

As solar radiation (insolation) enters the Earth's atmosphere some energy becomes unavailable for ecosystems as the energy is absorbed by inorganic matter or reflected back into the atmosphere. Very little of the sunlight available from the Sun ends up as biomass in ecosystems. Around 51 per cent of the available energy from the Sun does not reach producers. Figure 2.22 shows that the pathways of radiation through the atmosphere involve a loss of radiation through reflection and absorption.

Percentage losses include:

- reflection from clouds – 19 per cent
- absorption of energy by clouds – 3 per cent
- reflection by scatter from aerosols and atmospheric particles – 3 per cent
- absorption by molecules and dust in the atmosphere – 17 per cent
- reflection from the surface of the Earth – 9 per cent.

Of the 49 per cent that is absorbed by the ground, only a small proportion ends up in producers. First, much of the incoming solar radiation fails to enter the chloroplasts of leaves because it is reflected, transmitted or is the wrong wavelength to be absorbed (Figure 2.14, page 79). Of the radiation captured by leaves, only a small percentage ends up as biomass in growth compounds as the conversion of light to chemical energy is inefficient. Overall, only around 0.06 per cent of all the solar radiation falling on the Earth is captured by plants.

Once producers have converted energy into a chemical store, energy is available in useable form both to the producers and to organisms higher up the food chain. As you have learned (Chapter 1), there is loss of chemical energy from one trophic level to another (Figure 2.22). The percentage of energy transferred from one trophic level to the next is called the ecological efficiency.

$$\text{ecological efficiency} = \left(\frac{\text{energy used for growth (new biomass)}}{\text{energy supplied}}\right) \times 100$$

Consider Figure 2.23: if energy used for new growth (0.1 J converted to new biomass in the blackbird) and 1 J of energy is available (the amount of energy consumed by the blackbird) then the ecological efficiency = (0.1/1) × 100 = 10%.

Efficiencies of transfer are low and they account for the energy loss (Figure 2.23). Ecological efficiency varies between 5 and 20 per cent with an average of 10 per cent: on average, one tenth of the energy available to one trophic level becomes available to the next.

Ultimately all energy lost from an ecosystem is in the form of heat, through the inefficient energy conversions of respiration so, overall, there is a conversion of light energy to heat energy by an ecosystem. Heat energy is re-radiated into the atmosphere.

You need to understand the difference between storages and flows of energy. Storages of energy are shown as boxes that represent the trophic level. Storages are measured as the amount of energy or biomass per unit area. Flows of energy or productivity are given as rates, for example $J\ m^{-2}\ day^{-1}$.

Systems diagrams showing energy flow through ecosystems need to show the progressive loss of energy in both storages and flows. Boxes and storages are drawn in proportion to the amount of energy they represent (Figure 2.24). Boxes show storages of energy. Storages of energy are measured as the amount of energy or biomass in a specific area. The flows of energy are shown as arrows. Arrows also represent flows of productivity. Flows are measured as rates; for example, $J\ m^{-2}\ day^{-1}$.

Figure 2.23 Loss of energy in food chains

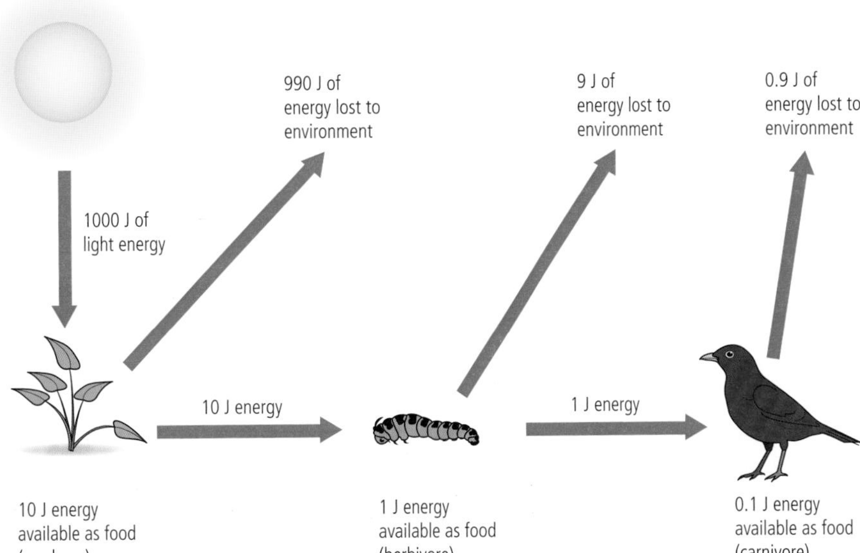

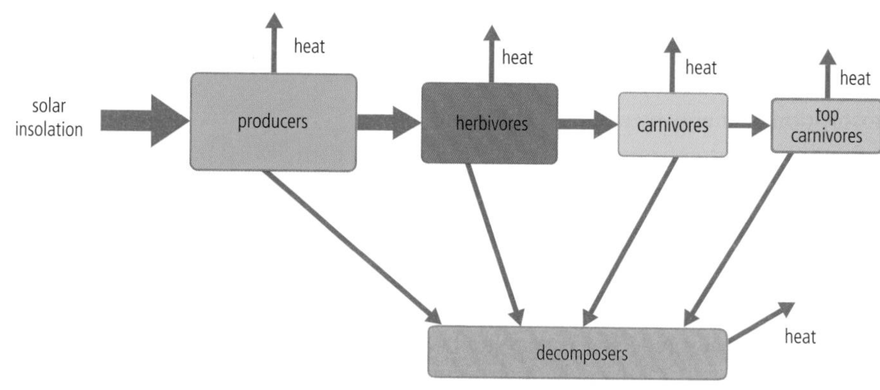

Figure 2.24 An energy-flow diagram showing the flow of energy through an ecosystem. Storages (boxes) and flows (arrows) vary in width and are proportional to the amount of energy being transferred.

Pathways of energy through an ecosystem include:

- **conversion of light energy to chemical energy**
- **transfer of chemical energy from one trophic level to another with varying efficiency**
- **overall conversion of ultraviolet and visible light to heat energy by an ecosystem**
- **re-radiation of heat energy to the atmosphere.**

Productivity diagrams were pioneered by American scientist Howard Odum in the 1950s. He carried out the first complete analysis of a natural ecosystem (a spring-fed stream) at Silver Spring in Florida. He mapped in detail all the flow routes to and from the stream, and measured the energy and organic matter inputs and outputs, and from these calculated productivity for each trophic level and the flows between them. Productivity was calculated in kcal m^{-2} yr^{-1}. The information from the Silver Spring study was simplified as a productivity diagram (Figure 2.25): such diagrams are useful as they give an indication of turnover in ecosystems by measuring energy flows per unit time as well as area.

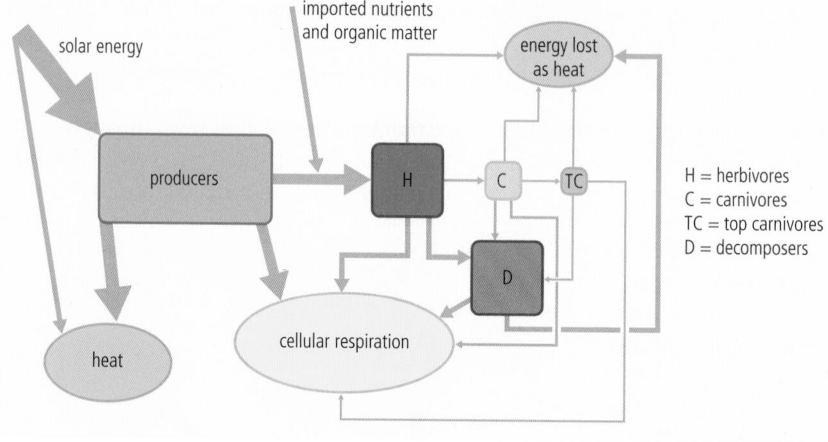

Figure 2.25 Energy flows through an ecosystem drawn from Odum's Silver Spring data. Rectangles represent stores of biomass and ovals are movements of energy from the system. Arrows are approximately proportional to each other and indicate differences in energy flow between different parts of the system.

Primary and secondary productivity

The conversion of energy into biomass for a given period of time is measured as productivity.

You have just learned that **productivity** in ecosystems can be described as production of biomass per unit area per unit time. Productivity occurs at each level of a food chain, and depending on where productivity occurs, it is referred to *primary* or *secondary* productivity.

- **Primary productivity** – the gain by producers (autotrophs) in energy or biomass per unit area per unit time.

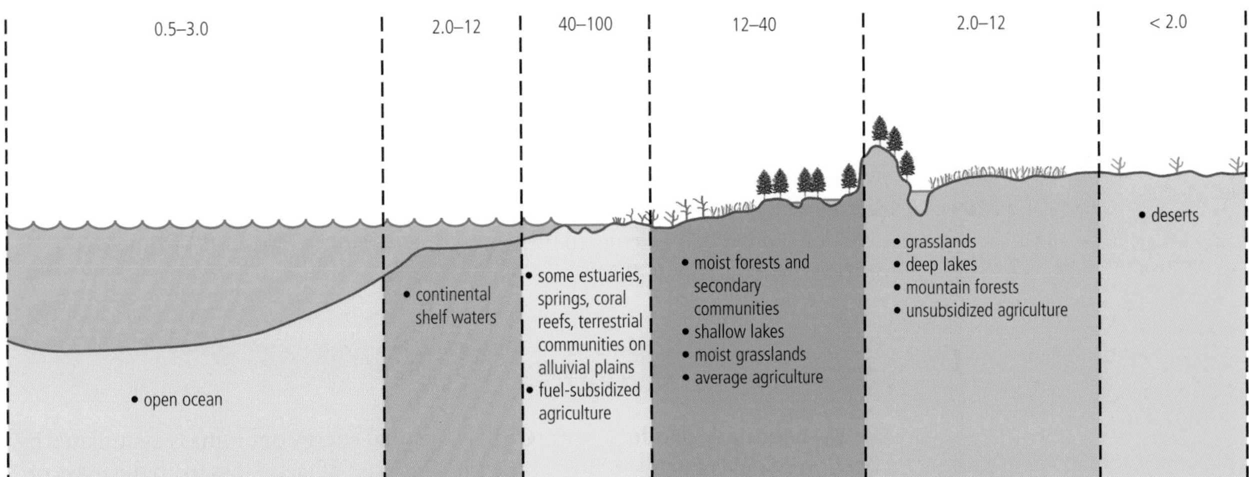

Figure 2.26 Comparison of biomes in terms of primary production / 103 kJ m^{-2} yr^{-1}

- **Secondary productivity** – the biomass gained by heterotrophic organisms, through feeding and absorption, measured in units of mass or energy per unit area per unit time.

Primary productivity is the conversion of solar energy into chemical energy whereas secondary productivity involves feeding and absorption. Primary productivity depends on the amount of sunlight, the ability of producers to use energy to synthesize organic compounds, and the availability of other factors needed for growth (e.g. minerals and nutrients) (Figure 2.26).

Secondary productivity depends on the amount of food present and the efficiency of consumers turning this into new biomass.

Primary production is highest where conditions for growth are optimal – where there are high levels of insolation, a good supply of water, warm temperatures, and high nutrient levels. For example, tropical rainforests have high rainfall and are warm throughout the year so they have a constant growing season and high productivity. Deserts have little rain which is limiting to plant growth. Estuaries receive sediment containing nutrients from rivers, they are shallow and therefore light and warm and so have high productivity. Deep oceans are dark below the surface and this limits productivity of plants (nutrients are the limiting factors at the surface). The productivity in different biomes is examined in detail later in this chapter (pages 104–111).

Productivity can further be divided into gross and net productivity, in the same way that monetary income can be divided into gross and net profits. Gross income is the total monetary income, and net income is gross income minus costs. Similarly, **gross productivity (GP)** is the total gain in energy or biomass per unit area per unit time. **Net productivity (NP)** is the gain in energy or biomass per unit area per unit time remaining after allowing for respiratory losses (R). NP represents the energy that is incorporated into new biomass and is therefore available for the next trophic level. It is calculated by taking away from gross productivity the energy lost through respiration (other metabolic process may also lead to the loss of energy but these are minor and are discarded).

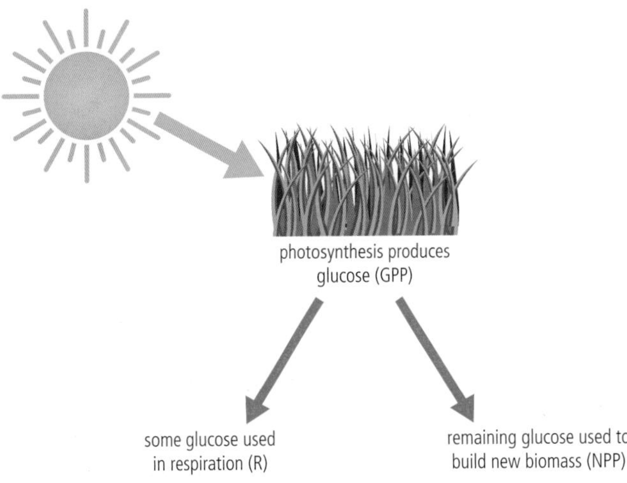

Figure 2.27 NPP is the rate at which plants accumulate new dry mass in an ecosystem. It is a more useful value than GPP as it represents the actual store of energy contained in potential food for consumers rather than just the amount of energy fixed into sugar initially by the plant through photosynthesis. The accumulation of dry mass is more usually termed biomass, and has a key part in determining the structure of an ecosystem.

To learn more about primary and secondary productivity online, go to www.pearsonhotlinks.co.uk, enter the book title or ISBN, and click on weblink 2.4.

Primary productivity

- **Gross primary productivity (GPP)** is equivalent to the mass of glucose created by photosynthesis per unit area per unit time in primary producers.
- **Net primary productivity (NPP)** is the gain by producers in energy or biomass per unit area per unit time remaining after allowing for respiratory losses (R). This is potentially available to consumers in an ecosystem (Figure 2.27). Net primary productivity (NPP) is calculated by subtracting respiratory losses (R) from gross primary productivity (GPP):

$$NPP = GPP - R$$

Secondary productivity

- **Gross secondary productivity (GSP)** is the total energy or biomass assimilated by consumers and is calculated by subtracting the mass of faecal loss from the mass of food consumed:

$$GSP = \text{food eaten} - \text{faecal loss}$$

GSP is the total energy gained through absorption in consumers (Figure 2.28).

- **Net secondary productivity (NSP)** is calculated by subtracting respiratory losses (R) from GSP:

$$NSP = GSP - R$$

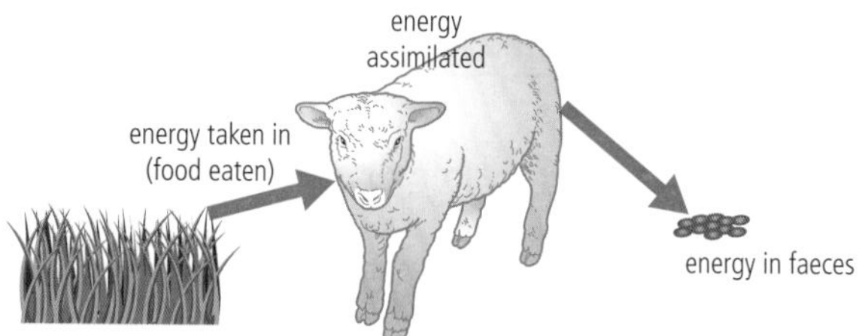

Figure 2.28 Animals do not use all the biomass they consume. Some of it passes out in faeces and excretion. Gross production in animals (GSP) is the amount of energy or biomass assimilated minus the energy or biomass of the faeces (i.e. the amount of energy absorbed by the body).

- **Net secondary production (NSP)** is the gain by consumers in energy or biomass per unit area per unit time remaining after allowing for respiratory losses (R) (Figure 2.29).

$$NSP = GSP - R$$

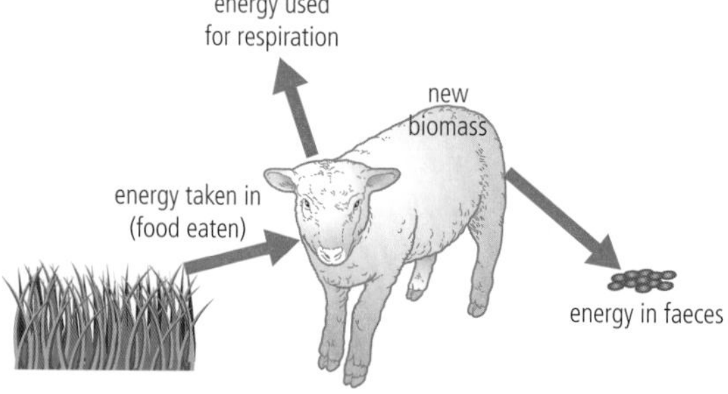

Figure 2.29 Some of the energy assimilated by animals is used in respiration, to support life processes, and the remainder is available to form new biomass (NSP). It is this new biomass that is then available to the next trophic level.

The term *assimilation* is sometimes used instead of *secondary productivity*.

Experiment to calculate gross primary productivity (GPP) and net primary productivity (NPP)

The easiest way to measure gross primary productivity (GPP) and net primary productivity (NPP) is by using aquatic plants. To calculate GPP and NPP, measurements of photosynthesis and respiration need to be taken. Photosynthesis and respiration either produce or use oxygen. Measuring dissolved oxygen will, therefore, give an indirect measurement of the amounts of photosynthesis and respiration in aquatic plants (but not a direct measure of the amount of energy fixed).

Net primary productivity can be estimated by measuring the increase in dissolved oxygen when aquatic plants are put in the light. In the light, both photosynthesis and respiration will be occurring but photosynthesis is the bigger process, and it produces more oxygen than the plant uses in respiration.

Gross primary productivity can be calculated using the equation:

$$\text{NPP} = \text{GPP} - \text{R, where R} = \text{respiratory loss}$$

Respiration can be calculated by measuring the decrease in dissolved oxygen when aquatic plants are put in the dark. In the dark, only respiration will occur and not photosynthesis. The equation can be rearranged to calculate GPP:

$$\text{GPP} = \text{NPP} + \text{R}$$

An aquatic plant was put in light and dark conditions. Dissolved oxygen was measured before and after the plant was put in light and in the dark. In this experiment, gross primary productivity (GPP) and net primary productivity (NPP) were measured by using changes in dissolved oxygen in milligrams of oxygen per litre per hour.

Plant in the light

Amount of dissolved oxygen at the start of the experiment = 10 mg of oxygen per litre

Amount of dissolved oxygen at the end of the experiment = 12 mg of oxygen per litre

Increase in dissolved oxygen = 2 mg of oxygen per litre

The increase in dissolved oxygen is a measure of NPP. The experiment lasted 1 hour and so the indirect measurement of NPP = 2 mg of oxygen per litre per hour (this could be used to estimate the amount of new biomass produced).

Plant in the dark

Amount of dissolved oxygen at the start of the experiment = 10 mg of oxygen per litre

Amount of dissolved oxygen at the end of the experiment = 7 mg of oxygen per litre

Loss of dissolved oxygen = 3 mg of oxygen per litre per hour.

The loss of dissolved oxygen is a measure of respiration (R).

$$\text{NPP} = \text{GPP} - \text{R}, \quad \text{so} \quad \text{GPP} = \text{NPP} + \text{R}$$

Therefore indirect estimation of GPP = 2 + 3 = 5 mg of oxygen per litre per hour (this could be used to estimate the amount of glucose produced).

You need to be able to calculate the values of both gross primary productivity (GPP) and net primary productivity (NPP) from given data.

The definitions of productivity must include units (i.e. the gain in biomass per unit area per unit time).

Experiment to calculate gross secondary productivity (GSP) and net secondary productivity (NSP)

You need to be able to calculate the values of both gross secondary productivity (GSP) and net secondary productivity (NSP) from given data, as in Table 2.4. A total of 10 stick insects were fed privet leaves for 5 days.

Table 2.4 Data collected from an experiment using stick insects

	Start of experiment	End of experiment
mass of leaves / g	29.2	26.3
mass of stick insect / g	8.9	9.2
mass of faeces / g	0.0	0.5

Calculating NSP

NSP can be calculated by measuring the increase in biomass in stick insects over a specific amount of time. The increase in biomass in stick insects (NSP) is equal to the mass of food eaten minus biomass lost through respiration and faeces.

In this experiment NSP = mass of stick insects at end of experiment – mass of stick insects at start of experiment.

Over a 5-day period: NSP = 9.2 – 8.9 = 0.3 g

Therefore, NSP = 0.3/5 = 0.06 g per day.

Calculating GSP

GSP can be calculated using the following equation:

GSP = food eaten – faecal loss

Food eaten = mass of leaves at start of the experiment – mass of leaves at end of the experiment.

Food eaten = 29.2 – 26.3 = 2.9 g

Also, faecal loss = mass of faeces at end of experiment = 0.5 g

Therefore, over a 5-day period: GSP = 2.9 – 0.5 = 2.4 g

Therefore, GSP = 2.4/5 = 0.48 g per day.

GSP represents the amount of food absorbed by the consumer.

Calculating respiration

Respiration (R, the loss of glucose as respiration breaks it down) can be calculated from the equation:

NSP = GSP – R

The equation can be rearranged:

R = GSP – NSP

Therefore, R = 0.48 – 0.06 = 0.42 g per day.

Maximum sustainable yields

Sustainable yield (SY) is the rate of increase in natural capital (i.e. natural income) that can be exploited without depleting the original stock or its potential for replenishment.

Sustainable yield means that a natural resource can be harvested at a rate equal to or less than their natural productivity so the natural capital is not diminished. The annual sustainable yield for a given natural resource such as a crop is the annual gain in biomass or energy through growth and recruitment. **Maximum sustainable yield** is the maximum flow of a given resource such that the stock does not decline over time (i.e. highest rate of harvesting that does not lead to a reduction in the original natural capital). In Chapter 4, you will explore maximum sustainable yield (MSY) as applied to fish stocks (page 240).

Maximum sustainable yields are equivalent to the net primary or net secondary productivity of a system. Net productivity is measured in the amount of energy stored as new biomass per year, and so any removal of biomass at a rate greater than this rate means that NPP or NSP would not be able to replace the biomass that had been extracted. Any harvesting that occurs above these levels is unsustainable and will lead to a reduction in the natural capital.

CONCEPTS: Sustainability

You need to understand the link between sustainable yields and productivity. Maximum sustainable yields are equivalent to the net primary or net secondary productivity of a system. Harvesting above maximum sustainable yields leads to a reduction in the natural capital and is unsustainable.

Nutrient cycles

Energy flows through ecosystems. For example, it may enter as sunlight energy and leave as heat energy. Matter cycles between the biotic and abiotic environment. Nutrient cycles can be shown in simple diagrams which show stores and transfers of nutrients (Figure 2.30).

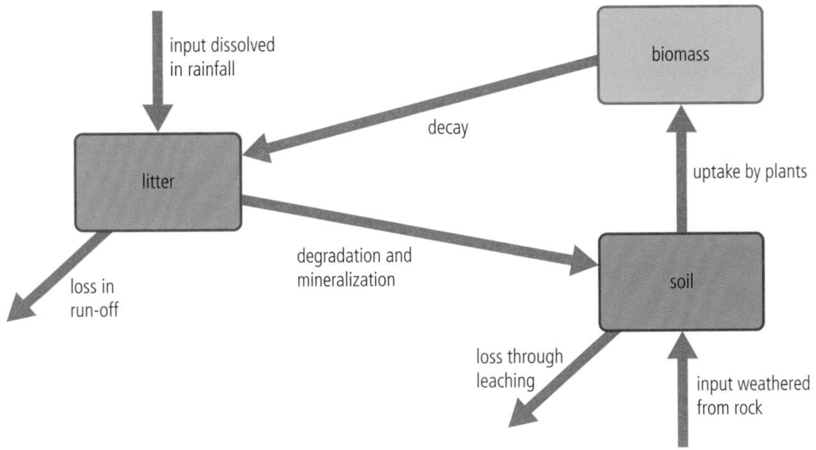

Figure 2.30 Systems diagram (developed by Gersmehl) showing nutrient cycles

The factors that affect the store of nutrients and their transfer are those that affect:

- the amount and type of weathering
- overland run-off and soil erosion
- the amount of rainfall

- rates of decomposition
- the type of vegetation (woody perennial species hold onto nutrients for much longer than annuals)
- the age and health of plants
- plant density
- fire.

Hence, explaining the differences between nutrient cycles in different ecosystems involves consideration of many processes.

Nutrients are circulated and reused frequently. Natural elements are capable of being absorbed by plants, either as gases or as soluble salts. Only oxygen, carbon, hydrogen, and nitrogen are needed in large quantities. These are known as macronutrients. The rest are trace elements or micronutrients and are needed only in small quantities (e.g. magnesium, sulfur, and phosphorus). Nutrients are taken in by plants and built into new organic matter. When animals eat the plants, they take up the nutrients. The nutrients eventually return to the soil when the plants and animals die and are broken down by the decomposers, and when animals defecate and excrete.

Organic – made from living matter (e.g. plants and animals); inorganic – made from non-living matter (e.g. rocks).

All nutrient cycles involve interaction between soil and the atmosphere, and many food chains. Nevertheless, there is great variety between the cycles. Nutrient cycles can be sedimentary based, in which the source of the nutrient is from rocks (e.g. the phosphorus cycle), or they can be atmospheric based, as in the case of the nitrogen cycle.

Matter flows through and between ecosystems, linking them together. This flow of matter involves transfers and transformations.

Figure 2.31 The carbon cycle

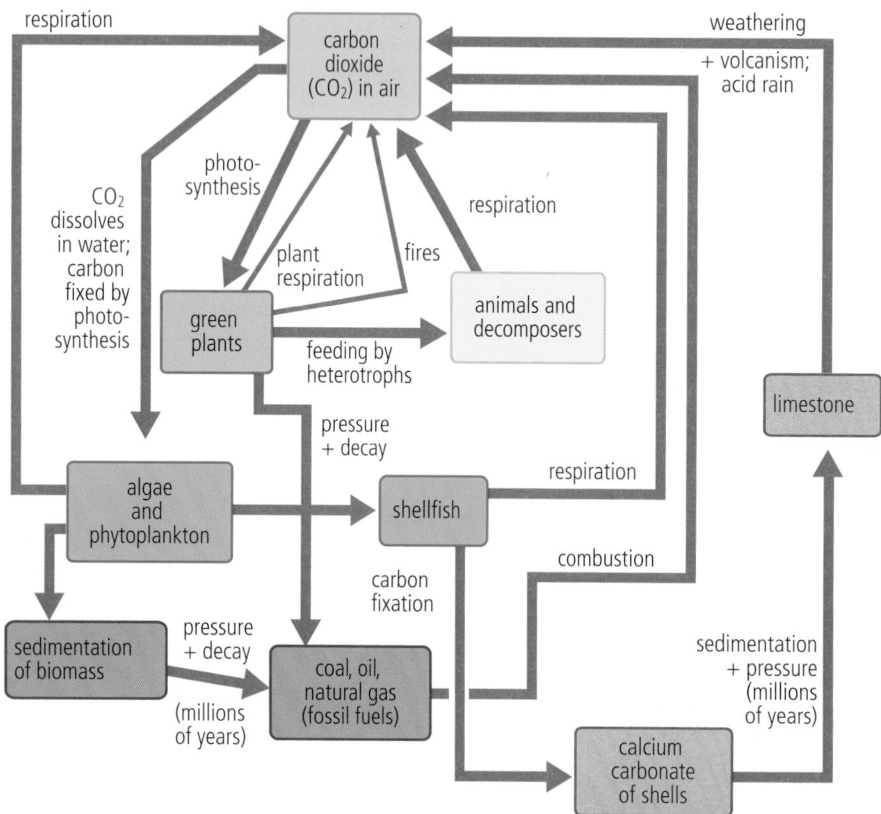

Carbon cycle

As you have seen above, unlike energy, nutrients are recycled and reused in ecosystems. Without this recycling, Earth would be covered with detritus and the availability of nutrients would decline. Decomposition is at the centre of these nutrient cycles, but other processes play their part as well.

Carbon is an essential element in ecosystems as it forms the key component of biological molecules (e.g. carbohydrates, fats, and protein). Although ecosystems form an important store of carbon (especially trees), it is also stored in fossil fuels (coal, gas, peat, and oil) and in limestone, and can remain in these forms for very long periods of time (Figure 2.31).

Storages in the carbon cycle include:

- organic storage:
  - organisms, including forests
- inorganic storages:
  - atmosphere
  - soil
  - fossil fuels
  - oceans.

Flows in the carbon cycle can be divided into transfers and transformations.

Transfers in the carbon cycle include:

- herbivores feeding on producers
- carnivores feeding on herbivores
- decomposers feeding on dead organic matter
- carbon dioxide from the atmosphere dissolving in rainwater oceans.

Transformations in the carbon cycle include the following.

- Photosynthesis, which converts inorganic materials into organic matter. Photosynthesis transforms carbon dioxide and water into glucose using sunlight energy trapped by chlorophyll.
- Respiration converts organic storage into inorganic matter. Respiration transforms organic matter such as glucose into carbon dioxide and water.
- Combustion transforms biomass into carbon dioxide and water.
- Fossilization transforms organic matter in dead organisms into fossil fuels through incomplete decay and pressure.

Carbon dioxide is fixed (i.e. converted from a simple inorganic molecule into a complex organic molecule – glucose) by autotrophs in either aquatic or terrestrial systems. These organisms respire and return some carbon to the atmosphere as carbon dioxide, or assimilate it into their bodies as biomass. When the organisms die, they are consumed by decomposers which use the dead tissue as a source of food, returning carbon to the atmosphere when they respire.

Oil and gas were formed millions of years ago when marine organisms died and fell to the bottom of the ocean, where anaerobic conditions slowed the decay process. Burial of the organisms followed by pressure and heat over long periods of time created these fuels. Coal was formed largely by similar processes acting on land vegetation. Limestone (calcium carbonate) was formed by the shells of ancient organisms and corals being crushed and compressed into sedimentary rock. Weathering of limestone, acid rain, and the burning of fossil fuels, returns carbon to the atmosphere.

Nitrogen cycle

Nitrogen is an essential building block of amino acids (which link together to make proteins) and DNA. It is a vital element for all organisms.

Nitrogen is the most abundant gas in the atmosphere (80 per cent) but because it is very stable it is not directly accessible by animals or plants. Only certain species of bacteria (nitrogen-fixing bacteria) can generate the energy needed to convert nitrogen gas into ammonia (Figure 2.32).

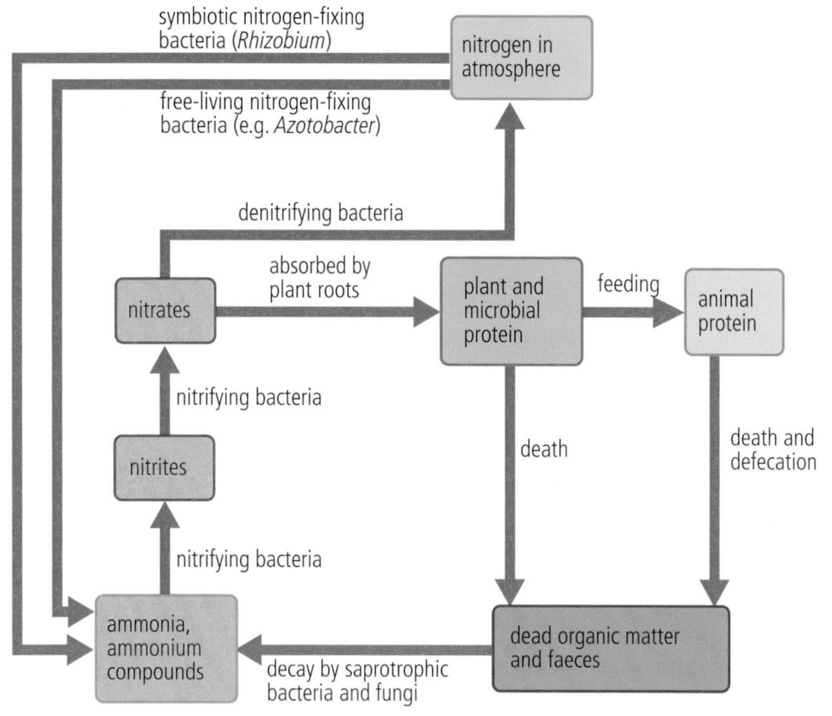

Figure 2.32 The nitrogen cycle

Storages in the nitrogen cycle include:

- organic storage:
  - organisms
- inorganic storages:
  - soil
  - fossil fuels
  - atmosphere
  - water bodies.

Flows in the nitrogen cycle can be divided into transfers and transformations.

Transfers in the nitrogen cycle include:

- herbivores feeding on producers
- carnivores feeding on herbivores
- decomposers feeding on dead organic matter
- plants absorbing nitrates through their roots
- removal of metabolic waste products from an organism (excretion).

Transformations in the nitrogen cycle include the following.

- Lightening transforms nitrogen in the atmosphere into NO_3. This is called nitrogen fixation.
- Nitrogen-fixing bacteria transform nitrogen gas in the atmosphere into ammonium ions.
- Nitrifying bacteria transform ammonium ions into nitrite and then nitrate.
- Denitrifying bacteria transform nitrates into nitrogen.
- Decomposers break down organic nitrogen (protein) into ammonia. The breakdown of organic nitrogen into ammonia is called deamination.
- Nitrogen from nitrates is used by plants to make amino acids and protein (assimilation).

Mycorrhizae attached to plant roots, form a thread-like network, extending beyond the roots. This extra network takes up additional water and nutrients and supplies them to the plant (see International mindedness, top of page 99).

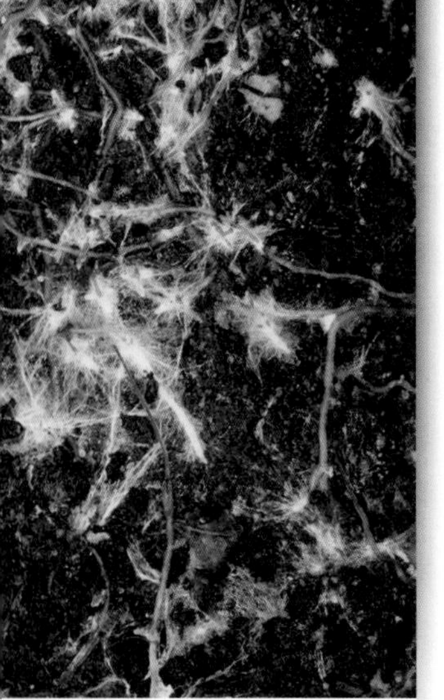

Lightning can fix atmospheric nitrogen into ammonia. Decomposers produce ammonia and ammonium compounds. Ammonia is also present in excretory products.

Nitrogen-fixing bacteria are found either free-living in the soil (e.g. *Azotobacter*) or living within root nodules of leguminous plants (*Rhizobium*). Species in root nodules are symbiotic with the plant – they derive the sugars they need for respiration from the plant (a lot of energy is needed to split the nitrogen molecule) and the plants gain a useable form of nitrogen. These bacteria fix atmospheric nitrogen into ammonium ions.

Nitrifying bacteria found in the soil oxidize ammonium ions first into nitrites and then into nitrates. These chemosynthetic organisms (page 79) convert inorganic materials into organic matter. The bacteria gain energy from this reaction to form food (glucose). Ammonia and nitrites are toxic to plants but the nitrates are taken up with water into plant roots and used to create amino acids and other organic chemicals.

Nitrogen is returned to the atmosphere by denitrifying bacteria, which remove oxygen from nitrates for use in respiration (they live in oxygen-poor soils where free oxygen is not readily available). Nitrogen gas is released as a by-product. The reason that waterlogged soils are not good for farmers is that denitrifying bacteria enjoy these conditions and dramatically reduce the quantity of nitrates available for crop growth.

The breakdown of organic matter is higher in tropical forest than in temperate woodland because high temperatures and year-round availability of water in tropical forests allow for continuous breakdown of nitrogen-containing compounds. This results in very rapid cycling and reabsorption. In temperate woodland, the breakdown of organic matter slows down significantly during winter months, causing nitrogen build-up in soil.

Some tropical forest trees have specific species of mycorrhizal fungi associated with their roots that increase rate of organic matter breakdown leading to rapid reabsorption of nitrogen. Although soils are nutrient-poor in tropical forests, the rapid recycling of nitrogen compared to temperate forests allows for more rapid growth to occur.

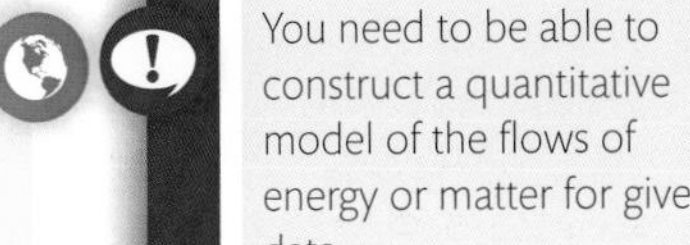

You need to be able to construct a quantitative model of the flows of energy or matter for given data.

The impact of human activities on energy flows and matter cycles

You need to be able to discuss human impacts on energy flows, and on the carbon and nitrogen cycles.

Human activities such as burning fossil fuels, deforestation, urbanization, and agriculture impact energy flows as well as the carbon and nitrogen cycles.

Energy flows and nutrient cycles occur at a global level – so human impacts have worldwide implications.

Energy flows

For thousands of years, humankind's only source of energy was radiation from the Sun. Sunlight energy, trapped by producers through photosynthesis, provided energy for food. This limited population growth as only limited amounts of food were available (i.e. that which occurred naturally). With the advent of the industrial revolution and the increased use of fossil fuels, industry could harness the sunlight energy trapped in coal and oil. Energy trapped by plants millions of years ago could be released: the amount of energy available to humans increased hugely, enabling industry to be powered and agricultural output, through the use of machinery, to increase. Population growth, through increased food output, increased rapidly. This change in the Earth's energy budget has ultimately led to many of the environmental issues covered in this course – habitat destruction, climate change, the reduction of non-renewable resources, acid deposition, and so on.

The combustion of fossil fuels has altered the way in which energy from the Sun interacts with the atmosphere and the surface of our planet. Increased carbon dioxide levels, and the corresponding increase in temperatures (Chapter 7) have led to the reduction in Arctic land and sea ice, reducing the amount of reflected sunlight energy (Chapter 7, page 377). Changes in the atmosphere through pollution (Chapters 6 and 7) have led to increased interception of radiation from the Sun, through changes in reflection by scatter from tiny atmospheric particles, and absorption by molecules and dust in the atmosphere (Figure 7.26, page 381).

Matter cycles

Timber harvesting (i.e. logging) interferes with nutrient cycling. This is especially true in tropical rainforests, where soils have low fertility and nutrients cycle between the leaf litter and tree biomass. Rapid decomposition, due to warm conditions and high rainfall, leads to the breakdown of the rich leaf litter throughout the year: once the trees have been removed, the canopy no longer intercepts rainfall, and the soil and leaf litter is washed away, and with it much of the available nutrients. In South East Asia, large areas have been cleared to grow oil palm. Oil palm is used in food production, domestic products, and to provide biofuel. Once the original forest has been removed, natural nutrient recycling is also lost. The soils, as you have seen, are generally nutrient poor, so oil-palm trees require fertilizer to produce yields that return a reasonable profit. Fertilizers can have various negative environmental impacts. Adding fertilizers, such as those containing nitrates, can cause **eutrophication** in nearby bodies of water (Chapter 4, page 255) when nitrates run-off from soils, causing disruption to ecosystems.

When crops are harvested and transported to be sold at a market usually some distance away, the nitrogen they contain is also transported. These changes to the location of the nitrogen storages alter the nitrogen cycle and can cause disruption to ecosystems.

Burning fossil fuels increases the amount of carbon dioxide in the atmosphere, leading to global warming and climate change (Chapter 7). Mining and burning of fossil fuels reduces the storages of these non-renewable sources of energy and increases the storage of carbon in the atmosphere. Increased carbon dioxide levels in the atmosphere can lead to increased vegetation growth, because there is more carbon dioxide available for photosynthesis, again altering the carbon cycle.

Exercises

1. Define the terms *gross productivity*, *net productivity*, *primary productivity*, and *secondary productivity*.
2. How is NPP calculated from GPP? Which figure represents the biomass available to the next trophic level?
3. Define the terms *gross secondary productivity (GSP)* and *net secondary productivity (NSP)*. Write the formula for each.
4. NPP, mean biomass, and NPP per kg biomass vary in different biomes, depending on levels of insolation, rainfall, and temperature. Mean NPP for tropical rainforest is greater than tundra because rainforest is hot and wet, so there is more opportunity to develop large biomass than in tundra. However, NPP per kg biomass is far lower in rainforest than tundra because rainforest has a high rate of both photosynthesis and respiration, so NPP compared to total biomass is low. Tundra are cold and dry and have low rates of photosynthesis and respiration; plants are slow growing with a gradual accumulation of biomass but relatively large growth in biomass per year.

   The table below shows values for these parameters for different biomes.

Biome	Mean net primary productivity (NPP) / kg m^{-2} yr^{-1}	Mean biomass / kg m^{-2}	NPP per kg biomass per year
desert	0.003	0.002	
tundra	0.14	0.60	0.233
temperate grassland	0.60	1.60	0.375
savannah (tropical) grassland	0.90	4.00	0.225
temperate forest	1.20	32.50	0.037
tropical rainforest	2.20	45.00	0.049

   a. Calculate the NPP per kg of biomass per year for the desert biome.
   b. How does this figure compare those for other biomes? Explain the figure you have calculated in terms of NPP, and NPP per kg biomass.
   c. Compare the figures for NPP in temperate and tropical grassland. Explain the difference.

You need to be able to analyse data for a range of biomes.

5. Draw systems diagrams for each of the following cycles:
   - the carbon cycle
   - the nitrogen cycle.

   Each should contain storages, flows, transfers, and transformations.
6. Outline the effect that human activities have had on
   - energy flows
   - matter cycles.

Big questions

Having read this section, you can now discuss the following big questions:

- What strengths and weaknesses of the systems approach and the use of models have been revealed through this topic?
- How are the issues addressed in this topic of relevance to sustainability or sustainable development?

Points you may want to consider in your discussions:

- Why are maximum sustainable yields equivalent to the net primary or net secondary productivity of a system? Why would harvesting biomass at a rate greater than NPP or GPP be unsustainable?
- How could systems diagrams of carbon and nitrogen cycles be used to show the effect of human activities on ecosystems? What are the strengths and weaknesses of such diagrams?

2.4 Biomes, zonation, and succession

Significant ideas

Climate determines the type of biome in a given area although individual ecosystems may vary due to many local abiotic and biotic factors.

Succession leads to climax communities that may vary due to random events and interactions over time. This leads to a pattern of alternative stable states for a given ecosystem.

Ecosystem stability, succession, and biodiversity are intrinsically linked.

Big questions

As you read this section, consider the following big questions:

- What strengths and weaknesses of the systems approach and the use of models have been revealed through this topic?
- How are the issues addressed in this topic of relevance to sustainability or sustainable development?

Knowledge and understanding

- Biomes are collections of ecosystems sharing similar climatic conditions which can be grouped into five major classes - aquatic, forest, grassland, desert, and tundra. Each of these classes will have characteristic limiting factors, productivity, and biodiversity.
- Insolation, precipitation, and temperature are the main factors governing the distribution of biomes.
- The tricellular model of atmospheric circulation explains the distribution of precipitation and temperature and how they influence structure and relative productivity of different terrestrial biomes.
- Climate change is altering the distribution of biomes and causing biome shifts.
- Zonation refers to changes in community along an environmental gradient due to factors such as changes in altitude, latitude, tidal level, or distance from shore or coverage by water.

- Succession is the process of change over time in an ecosystem involving pioneer, intermediate, and climax communities.
- During succession, the patterns of energy flow, gross and net productivity, diversity, and mineral cycling change over time.
- Greater habitat diversity leads to greater species and genetic diversity.
- *r*- and *K*-strategist species have reproductive strategies that are better adapted to pioneer and climax communities, respectively.
- In early stages of succession, gross productivity is low due to the unfavourable initial conditions and low density of producers. The proportion of energy lost through community respiration is relatively low too, so net productivity is high; that is, the system is growing and biomass is accumulating.
- In later stages of succession, with an increased consumer community, gross productivity may be high in a climax community. However, this is balanced by respiration, so net productivity approaches zero and the productivity : respiration (P : R) ratio approaches 1.
- In a complex ecosystem, the variety of nutrient and energy pathways contributes to its stability.
- There is no one climax community but rather a set of alternative stable states for a given ecosystem. These depend on the climatic factors, the properties of the local soil and a range of random events which can occur over time.
- Human activity is one factor which can divert the progression of succession to an alternative stable state, by modifying the ecosystem, for example the use of fire in an ecosystem, use of agriculture, grazing pressure, or resource use such as deforestation. This diversion may be more or less permanent depending on the resilience of the ecosystem.
- An ecosystem's capacity to survive change may depend on its diversity and resilience.

Biomes are collections of ecosystems sharing similar climatic conditions. They can be grouped into five major classes - aquatic, forest, grassland, desert, and tundra.

Biomes

Biomes have distinctive abiotic factors and species which distinguish them from other **biomes** (Figure 2.33). Water (rainfall), insolation (sunlight), and temperature are the climate controls that determine how biomes are structured, how they function, and where they are found round the world.

Water is needed for photosynthesis, transpiration, and support (cell turgidity). Sunlight is needed for photosynthesis. Photosynthesis is a chemical reaction, so temperature affects the rate at which it progresses. Rates of photosynthesis determine the productivity of an ecosystem (net primary productivity, NPP, pages 91–92) – the

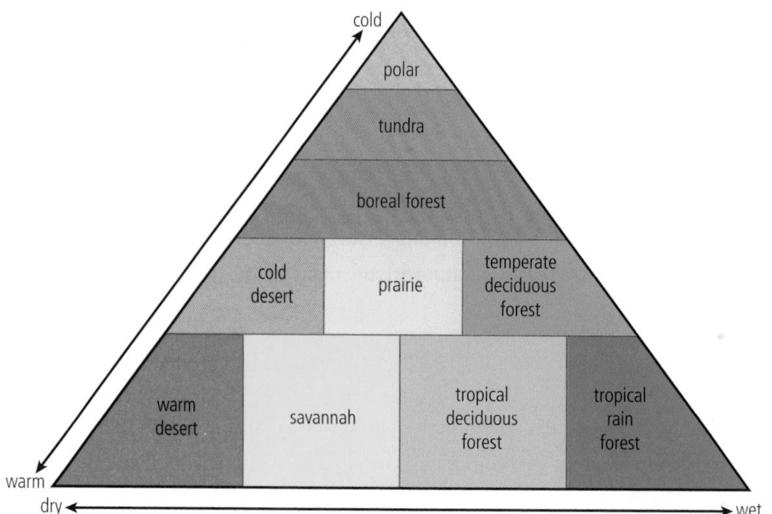

Figure 2.33 Temperature and precipitation determine biome distribution around the globe. Levels of insolation also play an important role, which correlates broadly with temperature (areas with higher levels of light tend to have higher temperatures).

more productive a biome, the higher its NPP. So, rainfall, temperature, and insolation determine rates of photosynthesis – and this is what determines the structure, function, and distribution of biomes.

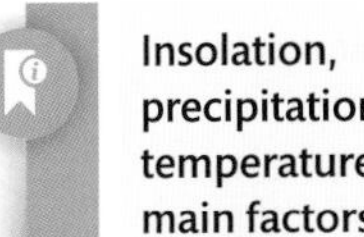

Insolation, precipitation, and temperature are the main factors governing the distribution of biomes.

You have already learned that ecosystems can be divided into terrestrial, freshwater, and marine (page 76). Similarly, biomes can be grouped into five major classes:

- forest, desert, tundra, and grassland (terrestrial ecosystems)
- aquatic (marine and freshwater ecosystems).

Each of these classes has characteristic limiting factors, productivity, and biodiversity. As you have just seen, insolation, precipitation, and temperature are the main factors governing the distribution of biomes.

Tricellular model of atmospheric circulation

As well as the differences in insolation and temperature found from the equator to higher (more northern) **latitudes**, the distribution of biomes can be understood by looking at patterns of atmospheric circulation. The **tricellular model** of atmospheric circulation is a way of explaining differences in atmospheric pressure belts, temperature, and precipitation that exist across the globe (Figure 2.34).

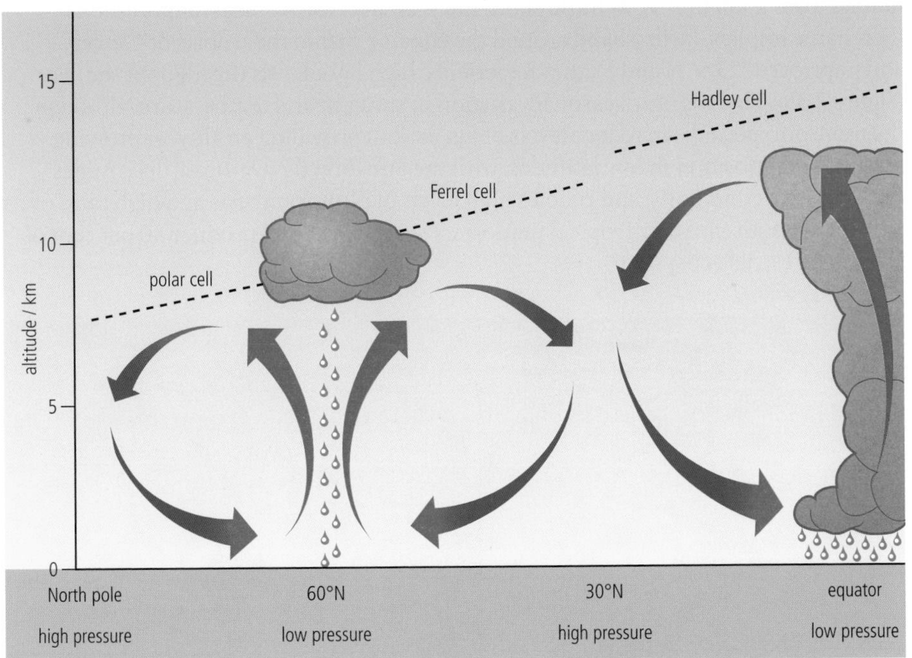

Figure 2.34 The tricellular model is made up of the polar cell, the Ferrel cell in mid-latitudes and the Hadley cell in the tropics. Downward air movement creates high pressure. Upward air movement creates low pressure and cooling air that leads to increased cloud formation and precipitation.

Atmospheric movement can be divided into three major cells: Hadley, Ferrel, and polar, with boundaries coinciding with particular latitudes (although they move on a seasonal basis). The Hadley cell controls weather over the tropics, where the air is warm and unstable. The equator receives most insolation per unit area of Earth: this heats up the air which rises, creating the Hadley cell. As the air rises, it cools and condenses, forming large cumulonimbus clouds that create the thunderstorms characteristic of tropical rainforest. These conditions provide the highest rainfall on the planet. The pressure at the equator is low as air is rising. Eventually, the cooled air begins to spread out, and descends at approximately 30° north and south of the equator. Pressure here is therefore high (because air is descending). This air is dry, so it is in these locations that the desert biome is found. Air then either returns to the

The tricellular model of atmospheric circulation explains the distribution of precipitation and temperature, and how these influence structure and relative productivity of different terrestrial biomes.

equator at ground level or travels towards the poles as warm winds. Where the warm air travelling north and south hits the colder polar winds, at approximately 60° north and south of the equator, it rises because it is less dense. This creates an area of low pressure. As the air rises, it cools and condenses, forming clouds. Precipitation results, so this is where temperate forest biomes are found. The model explains why rainfall is high at the equator and at 60° north and south.

Biomes cross national boundaries. In Borneo, for example, the rainforest crosses three countries: Indonesia, Malaysia, and Brunei. Studying biomes may therefore require investigations to be carried out across national frontiers – this can sometimes be politically as well as logistically difficult.

You need to be able to explain the distribution, structure, biodiversity and relative productivity of contrasting biomes. Climate should be explained in terms of temperature, precipitation, and insolation only.

Investigating different biomes

Different biomes have characteristic limiting factors, productivity, and biodiversity.

Tropical rainforest

Tropical rainforests have constant high temperatures (typically 26 °C) and high rainfall (over 2500 mm yr^{-1}) throughout the year. Because tropical rainforests, as their name implies, lie in a band around the equator within the tropics of Cancer and Capricorn (23.5° N and S), they experience high light levels throughout the year (Figure 2.35). There is little seasonal variation in sunlight and temperature (although the monsoon period can reduce levels of insolation) providing an all-year growing season. Their position in low latitudes, with the Sun directly overhead, determines their climatic conditions, and enables high levels of photosynthesis and high rates of NPP throughout the year. Tropical rainforests are estimated to produce 40 per cent of NPP of terrestrial ecosystems.

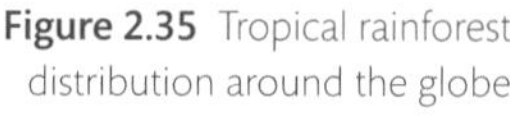

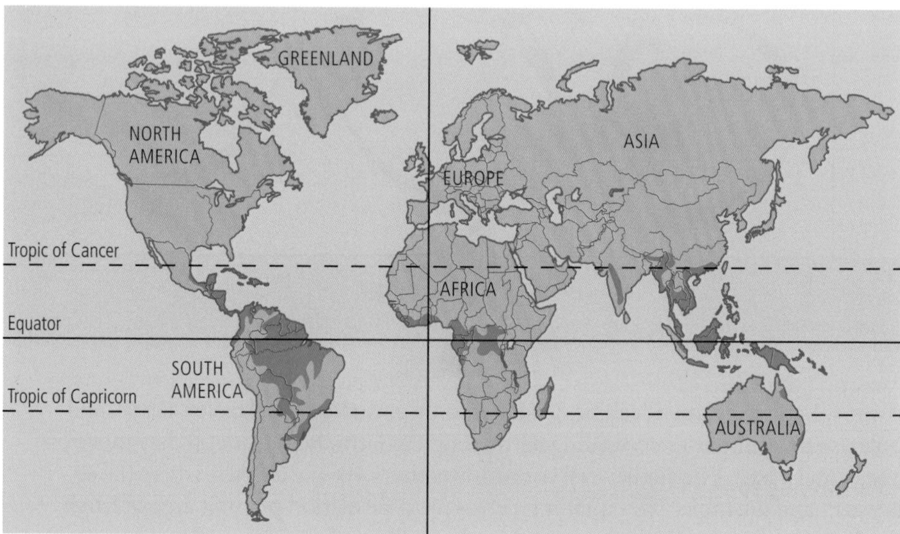

Figure 2.35 Tropical rainforest distribution around the globe

Rainforests are broad-leaved evergreen forests with a very high diversity of animals and plants. A rainforest may have up to 480 species of tree in a single hectare (2.5 acres), whereas temperate forest may only have six tree species making up the majority of the forest. The high diversity of plants is because of the high levels of productivity resulting from year-round high rainfall and insolation. The high diversity of animals follows from the complexity of the forests: they are multilayered and provide many different niches allowing for an enormous variety of different organisms (Figure 2.36).

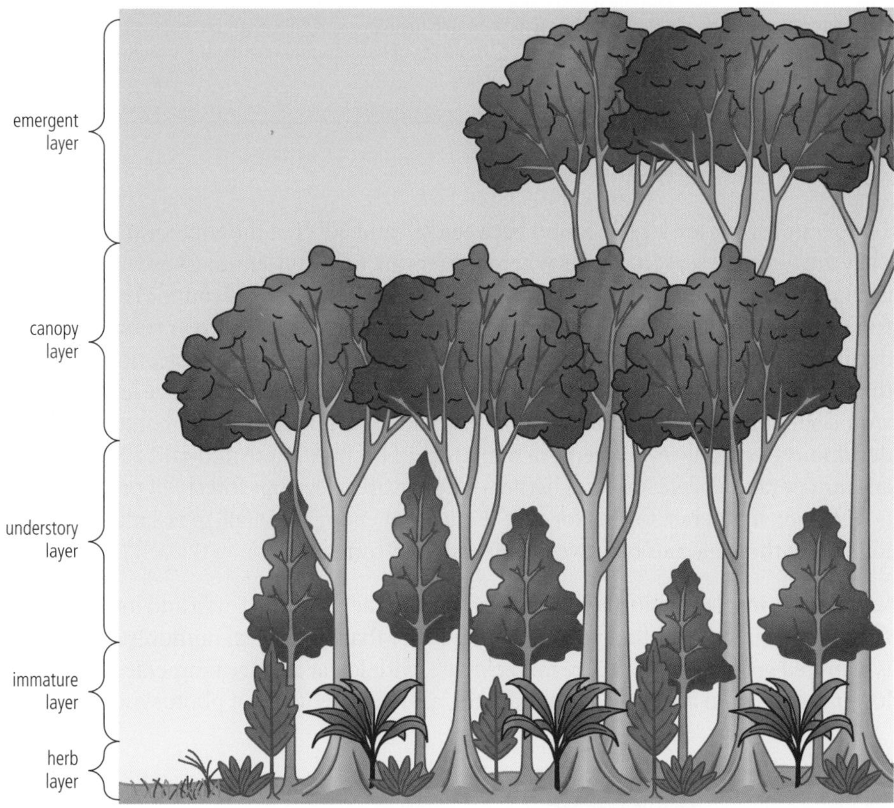

Figure 2.36 Rainforests show a highly layered, or stratified, structure. Emergent trees can be up to 80 m high, although overall structure depends on local conditions and varies from forest to forest. Only about 1 per cent of light hitting the canopy layer reaches the floor, so the highest levels of NPP are found in the canopy – one of the most productive areas of vegetation in the world. High productivity in the canopy results in high biodiversity, and it is believed that half of the world's species could be found in rainforest canopies.

Although rainforests are highly productive, much of the inorganic nutrients needed for growth are locked up in the trees. The soil is low in nutrients. Trees obtain their nutrients from rapid recycling of detritus that occurs on the forest floor. If rates of decay are high enough, the forest can maintain levels of growth. However, heavy rainfall can cause nutrients to be washed from soils (leaching) resulting in an increased lack of inorganic nutrients that could limit primary production. Because soils in rainforest are thin, trees have shallow root systems with one long tap root running from the centre of the trunk into the ground plus wide buttresses to help support

Buttress roots grow out from the base of the trunk, sometimes as high as 5 m above the ground, and provide support for the tree on thin soils. These also allow roots to extend out from the tree increasing the area over which nutrients can be absorbed from the soil.

the tree. The forest canopy provided by the trees protects the soils from heavy rainfall – but once areas have been cleared through logging, the soils are quickly washed away (eroded) making it difficult for forests to re-establish (it may take about 4000 years for a logged area to recover its original biodiversity.

CONCEPTS: Biodiversity

Biomes such as tropical rainforest and coral reef are found in equatorial areas with high light intensity all year round and warm temperatures which enable high levels of NPP. High productivity leads to high levels of resources (food, etc.), high complexity of habitats and niches, and therefore high biodiversity.

Temperate forest

Temperate forests are largely found between 40° and 60° N of the equator (Figure 2.37). They are found in seasonal areas where winters are cold and summers are warm, unlike tropical rainforests which enjoy similar conditions all year round. Two different tree types are found in temperate forest – evergreen (which leaf all year round) and deciduous (which lose their leaves in winter). Evergreen trees have protection against the cold winters (thicker leaves or needles), whereas deciduous trees have leaves that would suffer frost damage, so they shut down in winter. Forests might contain only deciduous trees, only evergreens, or a mixture of both. At these mid-latitudes the amount of rainfall determines whether or not an area develops forest – if precipitation is sufficient, temperate forests form; if there is not enough rainfall, grasslands develop. Rainfall in these biomes is between 500 and 1500 mm yr^{-1}.

Variation in insolation during the year, caused by the tilt of the Earth and its rotation around the Sun, means that productivity is lower than in tropical rainforests as there is a limited growing season. The mild climate, with lower average temperatures and lower rainfall than are found at the equator, also reduces levels of photosynthesis and

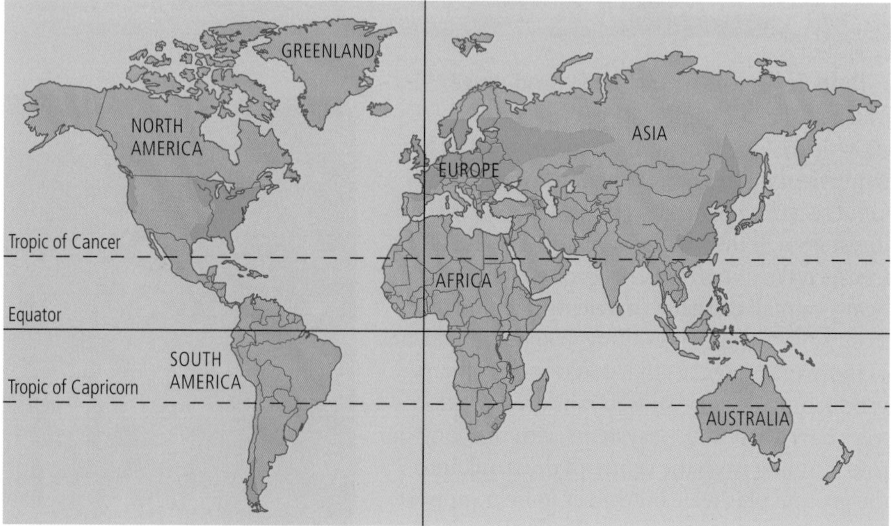

Figure 2.37 Distribution of temperate forest around the globe

productivity compared to tropical rainforest. But temperate forests have the second highest NPP (after rainforests) of all biomes. Diversity is lower than in rainforest and the structure of temperate forest is simpler. These forests are generally dominated by one species and 90 per cent of the forest may consist of only six tree species. There is some layering of the forest, although the tallest trees generally do not grow more than 30 m, so vertical stratification is limited. The less complex structure of temperate forests compared to rainforest reduces the number of available niches and therefore species diversity is much less. The forest floor has a reasonably thick leaf layer that is rapidly broken down when temperatures are higher, and nutrient availability is in general not limiting. The lower and less dense canopy means that light levels on the forest floor are higher than in rainforest, so the shrub layer can contain many plants such as brambles, grasses, bracken, and ferns.

The loss of leaves from deciduous trees in temperate forests over winter allows increased insulation of the forest floor, enabling the seasonal appearance of species such as bluebells.

Deserts

The Sahara Desert in northern Africa is the world's largest desert. Covering more than 9 million square kilometres (3.5 million square miles), it is slightly smaller than the USA. However, it is not the site of the world's lowest rainfall – that occurs in Antarctica, which receives less than 50 mm of precipitation annually.

Deserts are found in bands at latitudes of approximately 30° N and S (Figure 2.38). They cover 20–30 per cent of the land surface. It is at these latitudes that dry air descends having lost water vapour over the tropics. Hot deserts are characterized by high temperatures at the warmest time of day (typically 45–49 °C) in early afternoon and low precipitation (typically under 250 mm yr^{-1}). Rainfall may be unevenly distributed. The lack of water limits rates of photosynthesis and so rates of NPP are very low. Organisms also have to overcome fluctuations in temperature (night temperatures, when skies are clear, can be as low as 10 °C, sometimes as low as 0 °C), which make survival difficult.

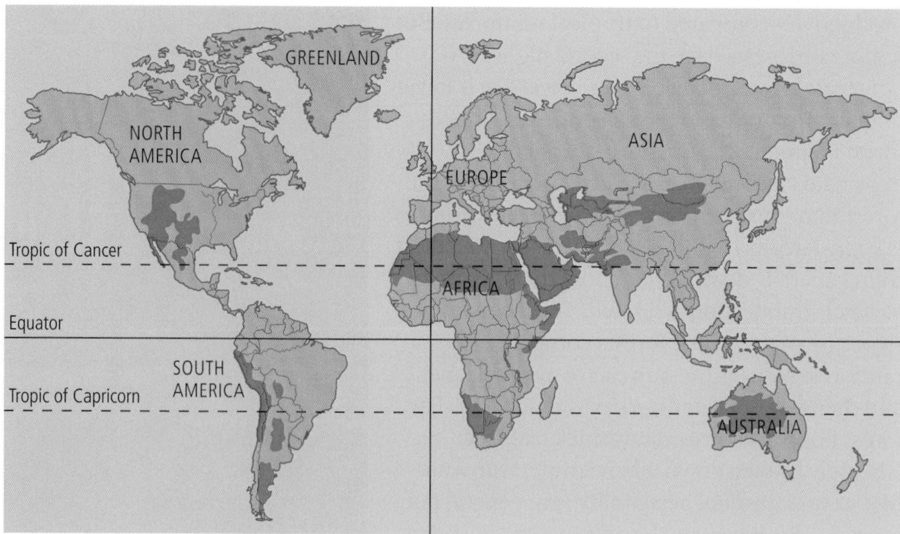

Figure 2.38 The distribution of deserts around the globe

Low productivity means that vegetation is sparse. Soils can be rich in nutrients as they are not leached away; this helps to support the plant species that can survive there. Decomposition levels are low because of the dryness of the air and lack of water. The species that can exist in deserts are highly adapted, showing many xerophytic adaptations (i.e. adaptations to reduce water loss in dry conditions). Cacti (a group restricted to the Americas) have reduced their surface area for transpiration by converting leaves into spines. They store water in their stems, which have the ability to expand, enabling more water to be stored and decreasing the surface area : volume ratio thus further reducing water loss from the surface. The spiny leaves deter animals from eating the plants and accessing the water. Xerophytes have a thick cuticle that also reduces water loss. Roots can be both deep (to access underground sources of water) and extensive near the surface (to quickly absorb precipitation before it evaporates).

Animals have also adapted to desert conditions. Snakes and reptiles are the commonest vertebrates – they are highly adapted to conserve water and their cold-blooded metabolism is ideally suited to desert conditions. Mammals have adapted to live underground and emerge at the coolest parts of day.

Elk crossing frozen tundra

Tundra

Tundra is found at high latitudes where insolation is low (Figure 2.39). Short day length also limits levels of sunlight. Water may be locked up in ice for months at a time and this combined with little rainfall means that water is also a limiting factor. Low light intensity and rainfall mean that rates of photosynthesis and productivity are low. Temperatures are very low for most of the year; temperature is also a limiting factor because it affects the rate of photosynthesis, respiration, and decomposition (these enzyme-driven chemical reactions are slower in colder conditions). Soil may be permanently frozen (permafrost) and nutrients are limiting. Low temperature means that the recycling of nutrients is low, leading to the

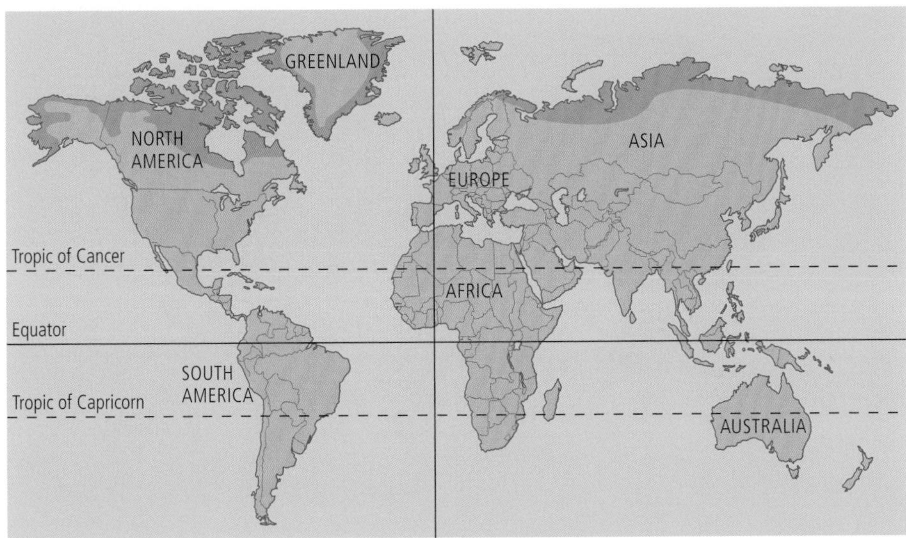

Figure 2.39 The distribution of tundra around the globe

formation of peat bogs where much carbon is stored. The vegetation consists of low scrubs and grasses.

Most of the world's tundra is found in the north polar region (Figure 2.39), and so is known as Arctic tundra. There is a small amount of tundra in parts of Antarctica that are not covered with ice, and in lower latitudes on high altitude mountains (alpine tundra).

During winter months, temperatures can reach −50 °C: all life activity is low in these harsh conditions. In summer, the tundra changes: the Sun is out almost 24 hours a day, so levels of insolation and temperature both increase leading to plant growth. Only small plants are found in this biome because there is not enough soil for trees to grow and, even in the summer, the permafrost drops to only a few centimetres below the surface.

In the summer, animal activity increases, due to increased temperatures and primary productivity. The growing period is limited to 6 weeks of the year, after which temperatures drop again and hours of sunlight decline. Plants are adapted with leathery leaves or underground storage organs, and animals with thick fur. Arctic animals are, on average, larger than their more southerly relations, which decreases their surface area relative to their size enabling them to reduce heat loss (e.g. the Arctic fox is larger than the European fox).

Tundra is the youngest of all biomes as it was formed after the retreat of glaciers from 15 000 to 10 000 years ago.

▲ Bison roam on mixed grass prairie

Grasslands

Grasslands are found on every continent except Antarctica, and cover about 16 per cent of the Earth's surface (Figure 2.40). They develop where there is not enough precipitation to support forests, but enough to prevent deserts forming. There are several types of grassland: the Great Plains and the Russian Steppes are temperate grasslands; the savannahs of east Africa are tropical grassland.

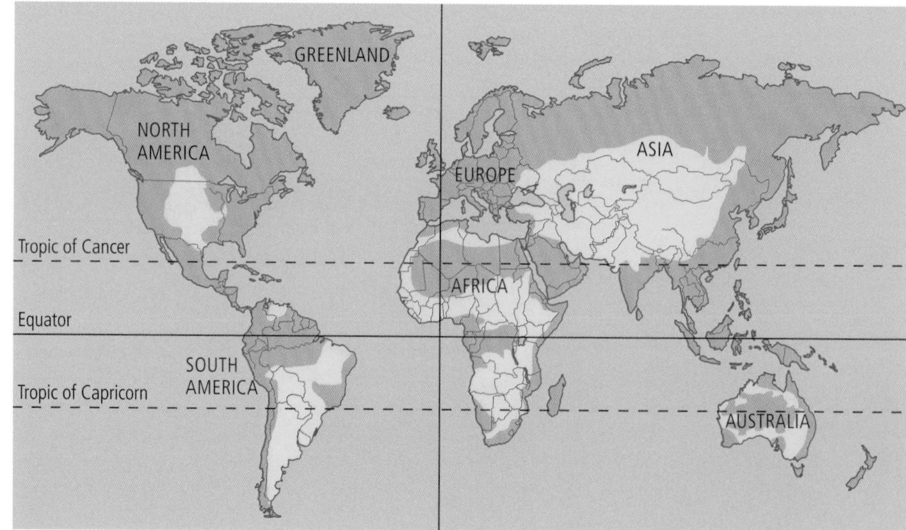

Figure 2.40 The distribution of grasslands around the globe

Grasses have a wide diversity but low levels of productivity. Grasslands away from the sea have wildly fluctuating temperatures which can limit the survival of animals and plants. They are found in the area where the polar and Ferrel cells meet (Figure 2.34, page 103), and the mixing of cold polar air with warmer southerly winds (in the northern hemisphere) causes increased precipitation compared to polar and desert regions. Rainfall is approximately in balance with levels of evaporation. Decomposing vegetation forms a mat containing high levels of nutrients, but the rate of decomposition is not high because of the cool climate. Grasses grow beneath the surface and during cold periods (more northern grasslands suffer a harsh winter) can remain dormant until the ground warms.

Tropical coral reefs

Tropical coral reefs (photo, page 76) are known as the rainforests of the ocean. This is because, like rainforests, they have high biodiversity and complex three-dimensional structure. Coral reefs are located near the equator (Figure 2.41) where seas are warm and there is strong sunlight throughout the year. Small animals called polyps (page 68) take carbon dioxide and calcium from seawater and transform it into calcium carbonate skeletons (which form the reef). Because of the warm sea temperatures,

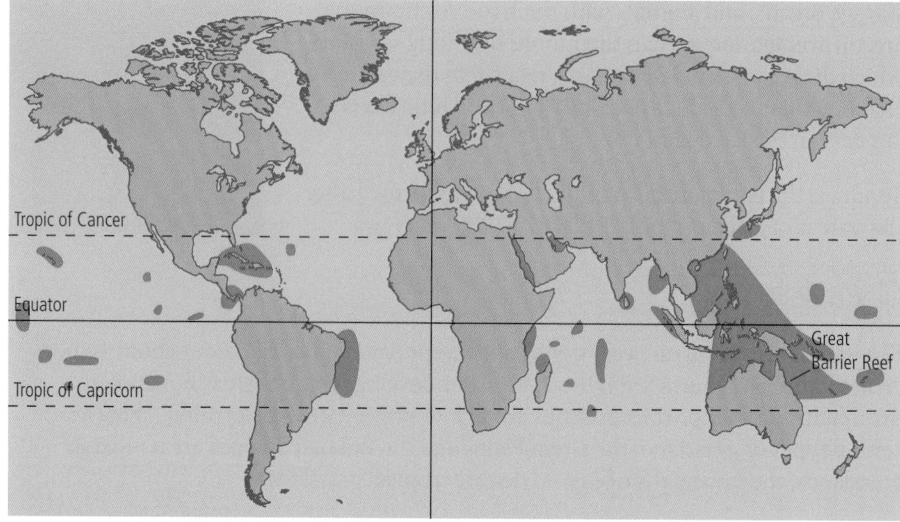

Figure 2.41 The distribution of coral reefs around the globe

productivity of the symbiotic algae that live within the polyps (page 68) is high, meaning that the production of new coral skeletons can also be high. The input of energy from the Sun via the symbiotic algae, and other producers such as seaweeds, provides a constant input of energy into the ecosystem, maintaining complex food webs. The complex structure of the reefs means that there are many niches and a corresponding high biodiversity. The Great Barrier Reef off the coast of Queensland, Australia, for example, has 1500 species of fish, 359 types of hard coral, a third of the world's soft corals, 6 of the world's 7 species of threatened marine turtle and more than 30 species of marine mammals including vulnerable dugongs (sea cows).

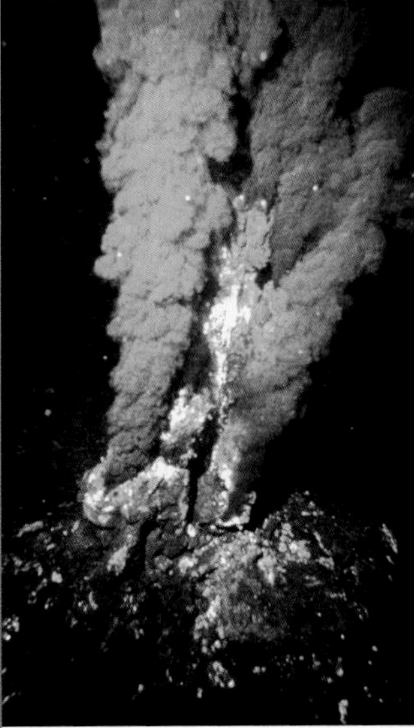

A deep-sea hydrothermal vent (also known as a black smoker)

Hydrothermal vents

Deep sea hydrothermal vents are one of the hottest environments on Earth. They are found in volcanically active areas along tectonic plate margins (Figure 2.42). They occur when cold-seawater penetrates the ocean crust and comes into contact with the hot rock below the surface.

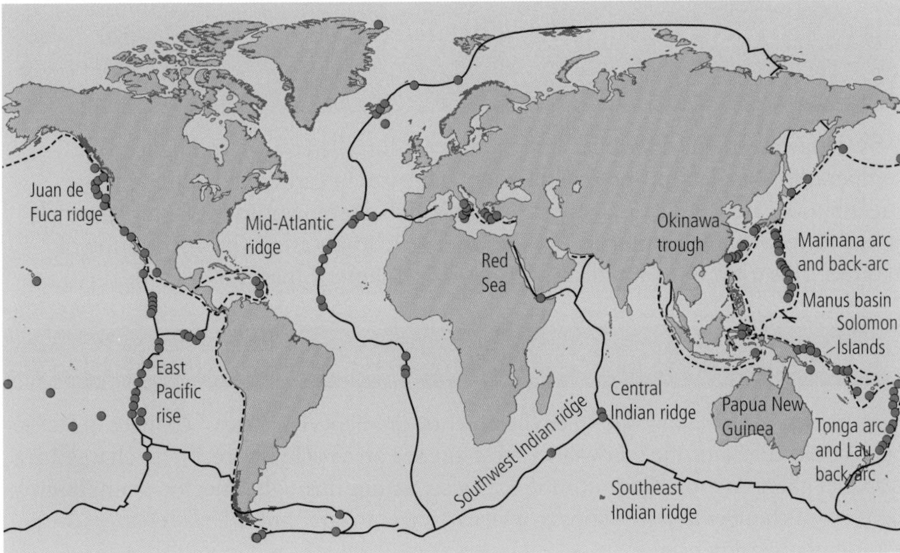

Figure 2.42 The distribution of active hydrothermal vents around the globe

The vent chimneys expel super-heated, metal-laden fluids, and so only extremely heat tolerant organisms (e.g. some chemosynthetic bacteria and some polychaete worms) can live there. Such organisms are called thermophilic species (extremophile organisms that thrive at the relatively high temperatures of 45–122 °C). The hot fluid released by the vents is enriched with chemicals, especially hydrogen sulfide, through which the chemosynthetic bacteria can obtain energy.

Sunlight does not penetrate to the depths of the sea where the hydrothermal vents are found, and so photosynthetic producers cannot exist there. Food chains in hydrothermal vents are based on chemosynthetic bacteria, in contrast to the food chains of coral reefs which are supported by photosynthetic algae within coral polyps (page 68) and other photosynthetic organisms.

The structure of hydrothermal vents is much simpler than that of coral reefs, with fewer niches available. Diversity at the hydrothermal vents is much lower than that found in coral reefs, although the productivity of the bacteria supports diverse communities of very specialized organisms not found elsewhere on the planet (Figure 2.43).

CHALLENGE YOURSELF

Research skills 

Research another pair contrasting biomes – temperate bog and tropical mangrove forest – and produce fact sheets on both. Draw up a table comparing four pairs of contrasting biomes (the ones you have researched and the examples in this book). How do their structure, biodiversity and relative productivity compare?

Now research the values of GPP for all the biomes you have studied. How do values for GPP and NPP compare? Why do these differences exist?

You should study at least four contrasting pairs of biomes. Examples of contrasting biomes include: temperate forests and tropical seasonal forests; tundra and deserts; tropical coral reefs and hydrothermal vents; temperate bogs and tropical mangrove forests.

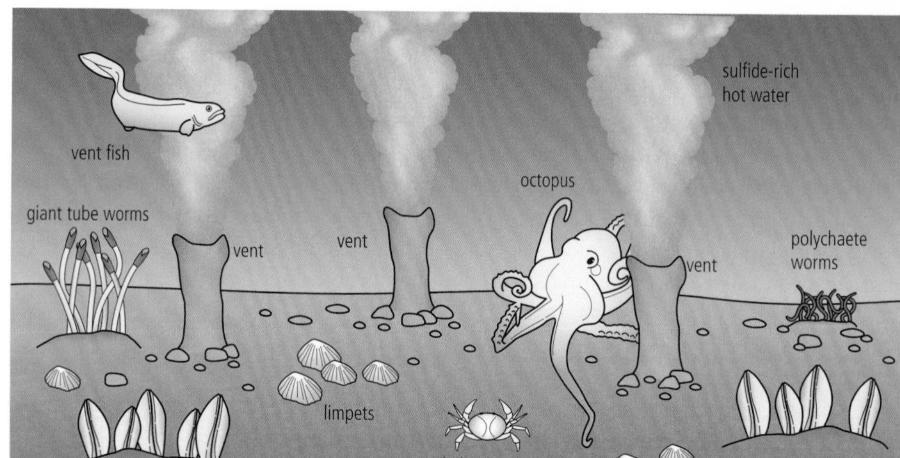

Figure 2.43 A hydrothermal vent community

Climate change is altering the distribution of biomes and causing biome shifts.

The effect of climate change on biome distribution

As you have seen, the distribution of biomes is controlled by a combination of temperature, insolation, and precipitation. Increases in carbon dioxide and other greenhouse gases lead to an increase in mean global temperature (Chapter 7), which in turn affects rainfall patterns. These changes in climate affect the distribution of biomes. This topic is explored in more detail in Chapter 7 (pages 373–375).

Spatial and temporal changes in communities

Communities can change along environmental gradients because of changes in factors such as altitude, latitude, or distance from the sea on a rocky shore. These changes are a spatial phenomenon. Communities can also change through time; for example, an ecosystem changes as it develops from early stages to later ones. These changes are a temporal phenomenon.

Zonation

Rocky shores can be divided into zones from the lower to upper shore. Each zone is defined by the spatial patterns of animals and plants. Seaweeds in particular show distinct **zonation** patterns, with species more resilient to water loss found on the upper shore (e.g. channel wrack) and those less resilient to water loss on the lower shore where they are not out of water for long (e.g. kelp).

Zonation is the arrangement or pattern of communities in bands in response to change in some environmental factor over a distance. For example, changes in ecosystems up a mountain as altitude increases.

Case study

Zonation on rocky shores

Figure 2.44 shows how abiotic factors vary along an environmental gradient on rocky shore. Organisms high on the rocky shore, and exposed to the air for long periods of time, have to withstand desiccation (drying out) and variations in temperature and salt concentration. Organisms that are lower on the shore are covered by seawater for much of the time and so are unlikely to dry out. They experience less variation in temperature and salt concentration in their environment, although wave action is greater.

Rocky shores provide an ideal location for studying zonation.

Because of the varying conditions, organisms can be expected to show zonation depending on their adaptations to abiotic factors. Seaweeds show marked zonation (Figure 2.45).

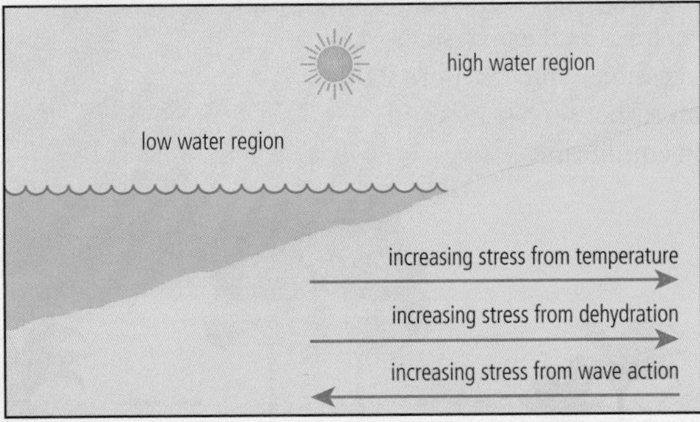

Figure 2.44 Variation in abiotic factors along a rocky shore

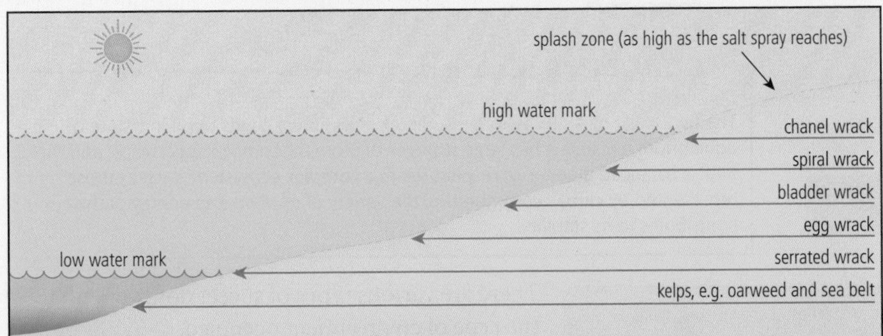

Figure 2.45 Zonation of seaweeds on a rocky shore

Variety of algae (seaweed) on a rocky shore. The area shown here is dominated by egg wrack, which has air bladders along its fronds that keep the seaweed afloat when the tide comes in, enabling them to obtain the maximum amount of sunlight for photosynthesis.

To learn more about rocky shores, go to www.pearsonhotlinks.co.uk, enter the book title or ISBN, and click on weblink 2.5.

Zonation occurs on different scales that can be both local and global. Rocky shores show local zonation whereas biome distribution is an example of zonation on a global scale.

Succession

Succession is the process of change over time in an ecosystem involving pioneer, intermediate, and climax communities.

The long-term change in the composition of a community is called **succession** (Figure 2.46). It explains how ecosystems develop from bare substrate over a period of time. The change in communities from the earliest community to the final community is called a **sere**. Successions can be divided into a series of stages, with each distinct community in the succession called a seral stage. The first seral stage of a succession is called the **pioneer community**. A pioneer community can be defined as the first stage of an ecological succession that contains hardy species able to withstand difficult conditions. The later communities in a sere are more complex than those that appear earlier. The final seral stage of a succession is called the **climax community**. A climax community can be defined as the final stage of a succession, which is more stable than earlier seral stages and is in equilibrium.

climax community
small trees
shrubs
pioneers (grasses)
bare rock
time

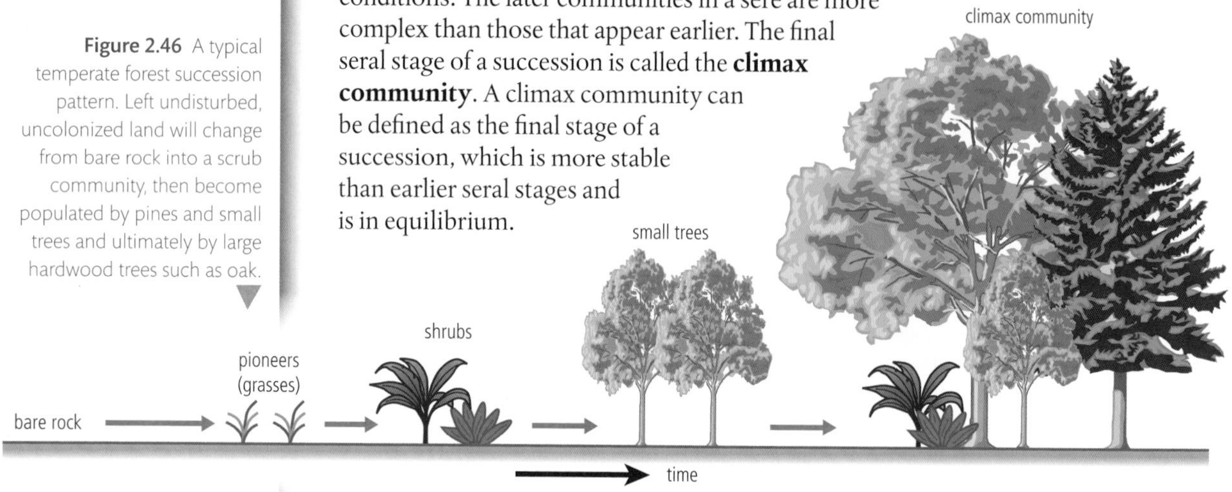

Figure 2.46 A typical temperate forest succession pattern. Left undisturbed, uncolonized land will change from bare rock into a scrub community, then become populated by pines and small trees and ultimately by large hardwood trees such as oak.

CONCEPTS: Equilibrium

The final stage of a succession, the climax community, tends to be in a state of equilibrium because it has large storages of biomass, complex food webs, and the NPP is balanced by rates of respiration. In a complex ecosystem, such as those represented by climax communities, the variety of nutrient and energy pathways contributes to its stability.

bare rock
colonization by lichens, weathering rock, and production of dead organic material
growth of moss, further weathering, and the beginnings of soil formation
growth of small plants such as grasses and ferns, further improvement in soil
larger herbaceous plants can grow in the deeper and more nutrient-rich soil
climax community dominated by shrubs and trees

Figure 2.47 A model of succession on bare rock

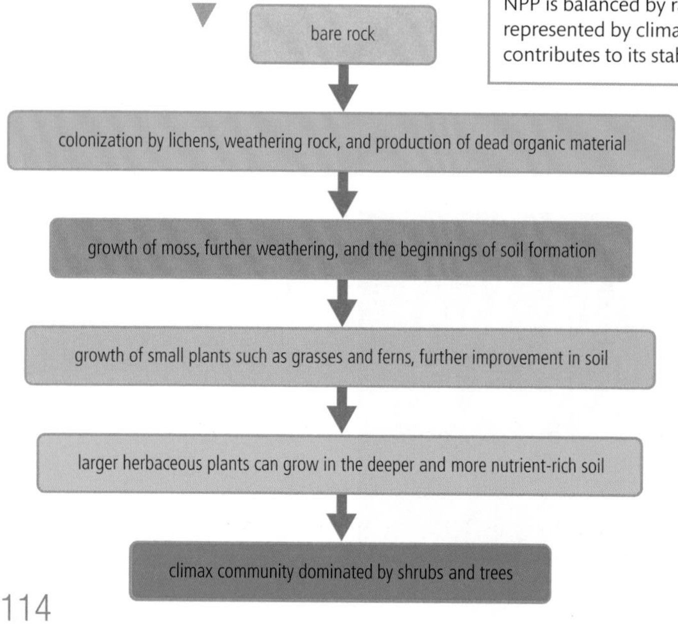

There are various types of succession, depending on the type of environment occupied:

- succession on bare rock is a lithosere (Figures 2.46 and 2.47)
- succession in a freshwater habitat is a hydrosere
- succession in a dry habitat (e.g. sand) is a xerosere.

Succession occurring on a previously uncolonized substrate (e.g. rock) is called a **primary succession**. **Secondary succession** occurs in places where a previous community has been destroyed (e.g. after forest fires). Secondary succession is faster than primary succession because of the presence of soil and a seed bank.

Succession happens when species change the habitat they have colonized and make it more suitable for new species. For example, lichens and mosses are typical pioneer species. Very few species can live on bare rock as it contains little water and has few available nutrients. Lichens and moss can photosynthesize and are effective at absorbing water, so they need no soil to survive and are excellent pioneers. Once established, they trap particles blown by the wind; their growth reduces wind speed and increases temperature close to the ground. When they die and decompose they form a simple soil in which grasses can germinate. The growth of pioneers also helps to weather parent rock adding still further to the soil. Other species, such as grasses and ferns that grow in thin soil, are now better able to colonize.

The new wave of species are better competitors than the earlier species; for example, grasses grow taller than mosses and lichens, so they get more light for photosynthesis. Their roots trap substrate (the thin soil) thereby reducing erosion, and they have a larger photosynthetic area, so they grow faster. The next stage involves the growth of herbaceous plants (e.g. dandelions and goosegrass), which require more soil to grow but which outcompete the grasses – they have wind-dispersed seeds and rapid growth, so they become established before larger plants. Shrubs then appear (e.g. bramble, gorse, and rhododendron); these larger plants grow in fertile soil, and are better competitors than the pioneers.

The final stage of a succession is the climax community. In this community, trees produce too much shade for the shrubs, which are replaced by shade-tolerant forest floor species (case study). The amount of organic matter present increases as succession progresses because as pioneer and subsequent species die out, their remains contribute to a build-up of litter from their biomass. Soil organisms such as earthworms move in and break down litter, leading to a build-up of organic matter in the soil making it easier for other species to colonize. Soil also traps water, and increasing amounts of moisture are, therefore, available to plants in the later stages of the succession.

You need to be able to describe the process of succession in a given example, and explain the general patterns of change in communities undergoing succession. Named examples of organisms from the pioneer, intermediate, and climax communities should be provided.

Case study

Succession on a shingle ridge

On a shingle ridge, lichens and mosses are pioneer species. Shingle has few available nutrients but lichens can photosynthesize and are effective at absorbing water. Once established, lichens and mosses trap particles blown by the wind, reduce wind speed, and increase temperature at the shingle surface. Their growth helps to weather the parent rock. When they die and decompose, a thin soil results. Grasses that grow in thin soil, such as red fescue, can now colonize the area. Grass roots trap soil and stop erosion, and have a larger photosynthetic area than pioneers and so can grow faster. Early in the succession xerophytic plants (page 65) are found, including the yellow-horned poppy and sea kale, which have thick, waxy leaves to prevent water loss and a bluish white colour that reflects sunlight and protects the plant. Plants with nitrogen-fixing bacteria in root nodules (page 98), such as rest harrow and bird's foot trefoil, enter the succession. These new species are better competitors than the pioneer species.

The next stage of the succession involves the growth of larger plants such as sea radish and then a shrub community dominated by bramble (*Rubus fruticosus*). The larger plants grow in deeper soil and are better competitors than the plants of the earlier seral stages.

Succession on a shingle ridge in Devon, UK. The community changes from a pioneer community of lichens and mosses through to a climax woodland community containing sycamore and oak trees.

The final stage of a succession is the climax community, a temperate forest ecosystem (page 106). Here, trees block sunlight to the shrub community and the shrubs are replaced by shade-tolerant forest floor species (species that can survive in shady conditions) such as ferns.

Succession occurs over time, whereas zonation refers to a spatial pattern.

You need to be able to interpret models or graphs related to succession and zonation.

The concept of succession must be carefully distinguished from the concept of zonation. Succession refers to changes over time, whereas zonation refers to spatial patterns. As you have seen, rocky shores can be divided into zones from lower to upper shore, with each zone defined by the spatial patterns of animals and plants. Succession, in contrast, is the orderly process of change over time in a community. Changes in the community of organisms frequently cause changes in the physical environment that allow another community to become established and replace the former through competition. Often, but not inevitably, the later communities in such a sequence or sere are more complex than those that appear earlier.

Changes through a succession

Gross productivity, net productivity, diversity, and mineral cycles change over time as an ecosystem goes through succession. During the course of a succession, greater habitat diversity leads to greater species and genetic diversity.

Changes in energy flow, gross productivity, and net productivity

In the early stages of a succession, the gross productivity is low because of the low density of producers. The density of producers in the early stages of succession is low because of the lack of soil, water, and nutrients. In the early stages of a succession, the proportion of energy lost through community respiration is relatively low and so net productivity is high. When net productivity is high, the ecosystem is growing and biomass is accumulating.

In later stages of a succession, the gross productivity is high in the climax community as there is an increased consumer community. The gross productivity is balanced by respiration in later stages of a succession (page 117), and so the net productivity approaches zero and the ratio of production to respiration approaches 1.

CONCEPTS: Equilibrium

In the later stages of succession, when the ratio of production to respiration approaches 1, the ecosystem is in steady-state equilibrium.

Changes in diversity

Early in the succession, there is low biomass and few niches. The plant community changes through each seral stage, leading to larger plants and greater complexity. As the plant community grows and complexity increases, the number of niches increases. As the number of niches increases, the food webs become more complex and both habitat diversity and species diversity increase.

Changes in mineral cycling

Mineral cycling forms an open system at early stages of succession. Elements such as carbon and nitrogen are introduced to the system from the surrounding area and can also leave the system. Later in the succession, mineral cycling forms a more closed system. Elements such as carbon and nitrogen can remain and cycle within the system. Minerals pass from the soil into living biomass. Minerals return to the soil when organisms die and decay.

Further changes in a succession are shown in Figure 2.48.

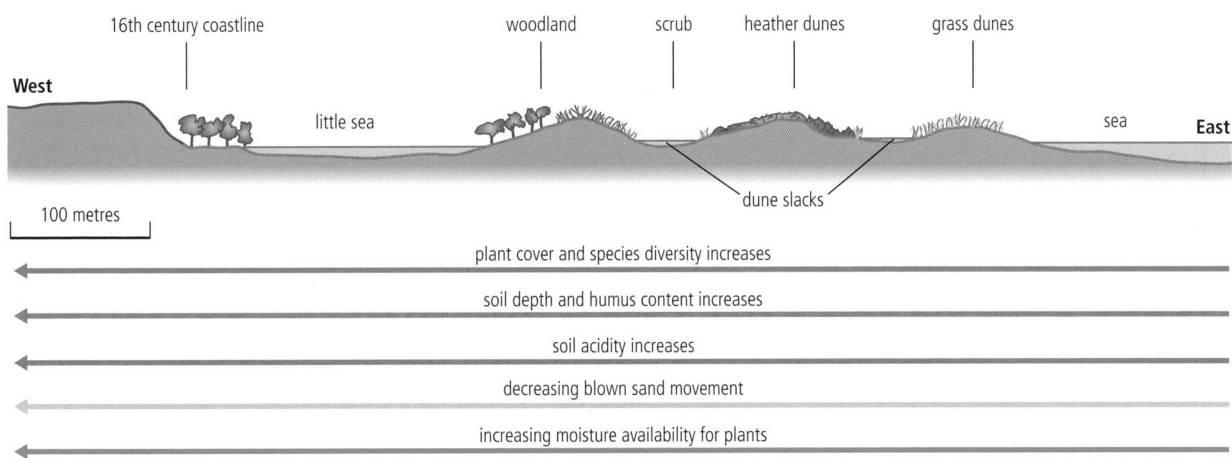

Figure 2.48 Changes in biotic and abiotic factors along a sand-dune succession

Table 2.5 summarizes differences in productivity, diversity, and mineral cycling between early and late stages of succession:

Table 2.5 Features of early and late succession

Feature	Pioneer community	Climax community
GPP	low	high
NPP	high	low
total biomass	low	high
niches	few	many
species richness	low	high
diversity	low	high
organic matter	small	large
soil depth	shallow	deep
minerals	external	internal
mineral cycles	open system	closed system
mineral conservation	poor	good
role of detritus	small	large

Production : respiration ratio

You have seen how the early stages of a succession have low GPP but high NPP because of the low overall rates of respiration. This relationship can be described as a ratio (production : respiration ratio or P/R ratio).

- If production is equal to rate of respiration, the value of P/R is 1.
- Where P/R is greater than 1, biomass accumulates.
- Where P/R is less than 1, biomass is depleted.
- Where P/R = 1 a steady-state community results.

In the later stages of a succession, with an increased consumer community, rates of community respiration are high. Gross productivity may be high in a climax community but, as this is balanced by respiration, the net productivity approaches zero (NP = GP – respiration), and the P/R ratio approaches 1.

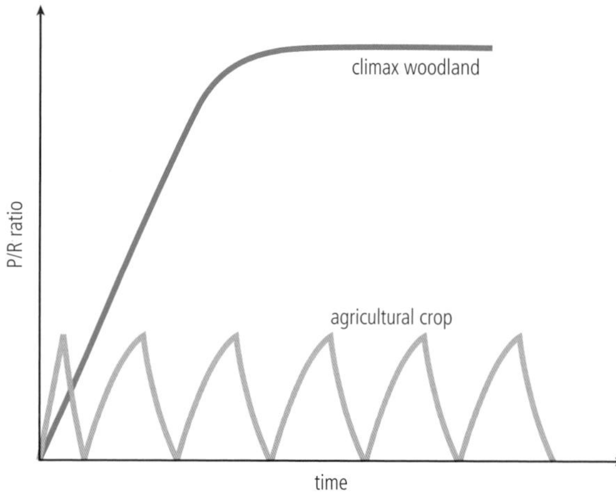

Figure 2.49 Difference in P/R ratios between natural and agricultural systems

If the P/R ratios of a food production system are compared to a natural ecosystem with a climax community, clear differences can be seen. Figure 2.49 compares intensive crop production with deciduous woodland. Fields and woodland both have low initial productivity, which increases rapidly as biomass accumulates. Farmers do not want the P/R ratio to reach 1 because, at that point, community respiration negates the high rates of gross productivity, which means that yields are not increased. The wheat is therefore harvested before P/R = 1. Community respiration is also controlled in the food production system by isolating herbivores and thereby increasing net productivity and growth. In natural woodland, the consumer community increases, so naturally high productivity is balanced by consumption and respiration. The woodland reaches its climax community when P/R = 1 (i.e. all woodland productivity is balanced by respiration).

- Early stages of succession
  - Gross productivity is low; the proportion of energy lost through community respiration is also low, so net productivity is high; the system is growing and biomass is accumulating.
- Later stages of succession
  - With an increased consumer community, gross productivity may be high; however, this is balanced by respiration, so net productivity approaches zero and the production : respiration (P/R) ratio approaches 1.

CONCEPTS: Equilibrium

Where P/R is greater than 1, biomass accumulates; where P/R is less than 1, biomass is reduced. Where P/R = 1, a community in steady-state equilibrium results.

- **In early stages of succession, gross productivity is low due to the unfavourable initial conditions and low density of producers. The proportion of energy lost through community respiration is relatively low too, so net productivity is high; that is, the system is growing and biomass is accumulating.**
- **In later stages of succession, with an increased consumer community, gross productivity may be high in a climax community. However, this is balanced by respiration, so net productivity approaches zero and the P/R ratio approaches 1.**

A climax community is a community of organisms that is more or less stable (i.e. in steady-state equilibrium), and is also in equilibrium with natural environmental conditions such as climate. It is the end-point of ecological succession.

Climax communities

Ecosystem stability refers to how well an ecosystem is able to cope with changes. As you saw in Chapter 1, most ecosystems are negative feedback systems – they contain inbuilt checks and balances without which they would spiral out of control, and no ecosystem would be self-sustaining. Ecosystems with more feedback mechanisms are more stable than simple ecosystems. Thus, ecosystems in the later stages of succession are likely to be more stable because food webs are more complex (because of high species diversity). This means that a species can turn to alternative food sources if its main food source is reduced. By late succession, large amounts of organic matter are available to provide a good source of nutrients. Nutrient cycles are more closed and self-sustaining; they are not dependent on external influences. This also contributes to stability.

In a climax community there are continuing inputs and outputs of matter and energy, but the system as a whole remains in a more-or-less constant state (steady-state equilibrium).

▲ Lowland tropical rainforest is a climax community in South East Asia. Hardwood trees of the family Dipterocarpaceae are dominant. They are often very tall and provide a rich three-dimensional structure to the forest.

The features of a climax community (compared to an early community) are:

- greater biomass
- higher levels of species diversity
- more favourable soil conditions (e.g. greater organic content and deeper soil)
- better soil structure (therefore greater water retention and aeration)
- taller and longer-living plant species
- greater community complexity and stability
- greater habitat diversity
- steady-state equilibrium.

There is no one climax community but rather a set of alternative stable states for a given ecosystem. These depend on the climatic factors, the properties of the local soil and a range of random events which can occur over time.

▲ Temperate forests are often dominated by a single tree species, such as the oak.

Redwood forests along the Pacific coast of the USA contain some of the tallest trees in the world. The dominant species in terms of biomass is *Sequoia sempervirens*. Trees can reach up to 115.5 m (379.1 feet) in height and 8 m (26 feet) in diameter. ▶

Species that are r-strategists grow and mature quickly and produce many, small offspring, whereas K-strategists are slow growing and produce few, large offspring that mature slowly. Pioneer communities are suited to r-strategists, whereas K-strategists are better adapted to climax communities.

r- and *K*-strategist species

Species can be classified according to how rapidly they reproduce, and the degree to which they give parental care. The type of species found along a succession, based on such criteria, varies. Species characterized by periods of rapid growth followed by decline, tend to inhabit unpredictable, rapidly changing environments (i.e. early seral stages) and are termed opportunistic species. They have a fast rate of increase (*r*) and are called **r-strategists** or *r*-species. Slow growing organisms tend to be limited by the carrying capacity of an environment (*K*), and so are known as **K-strategists** or *K*-species. They inhabit stable environments (i.e. later seral stages). *r*- and *K*-strategist species have reproductive strategies that are better adapted to pioneer and climax communities, respectively.

Species characterized as *r*-strategists produce many, small offspring that mature rapidly. They receive little or no parental care. Species producing egg-sacs are a good example. In contrast, species that are *K*-strategists produce very few, often very large offspring that mature slowly and receive much parental support. Elephants and whales are good examples. As a result of the low birth rate, *K*-strategists are vulnerable to high death rates and extinction.

Many species lie in between these two extremes and are known as *C*-strategists or *C*-species.

r- and *K*-selection theory

This theory states that natural selection (pages 156–157) may favour individuals with a high reproductive rate and rapid development (*r*-strategists) over those with lower reproductive rates but better competitive ability (*K*-strategists). Characteristics of the classes are shown in Table 2.6.

Table 2.6 A comparison of *r*- and *K*- species

r-strategists	*K*-strategists
initial colonizers	dominant species
large numbers of a few species	diverse range of species
highly adaptable	specialist
rapid growth and development	slow development
early reproduction	delayed reproduction
short life	long living
small size	large size
very productive	not very productive

Species can have traits of both *K*- and *r*-strategists. Studies showed dandelions in a disturbed lawn had high reproduction rates whereas those on an undisturbed lawn produced fewer seeds but were better competitors.

In predictable environments – those in which resources do not fluctuate – there is little advantage to rapid growth. Instead, natural selection favours species that can maximize use of natural resources and which produce only a few young with have a high probability of survival. These *K*-strategists have long life spans, large body size and develop slowly. In contrast, disturbed habitats with rapidly changing conditions favour *r*-strategists that can respond rapidly, develop quickly and have early reproduction. This leads to a high rate of productivity. Such colonizer species often have a high dispersal ability to reach areas of disturbance.

Survivorship curves

Rates of mortality vary with age, size and sometimes gender. Survivorship curves show changes in survivorship over the lifespan of a species (Figure 2.50).

Factors that influence survivorship rates include:

- competition for resources
- reproductive strategy
- adverse environmental conditions
- predator–prey relationships.

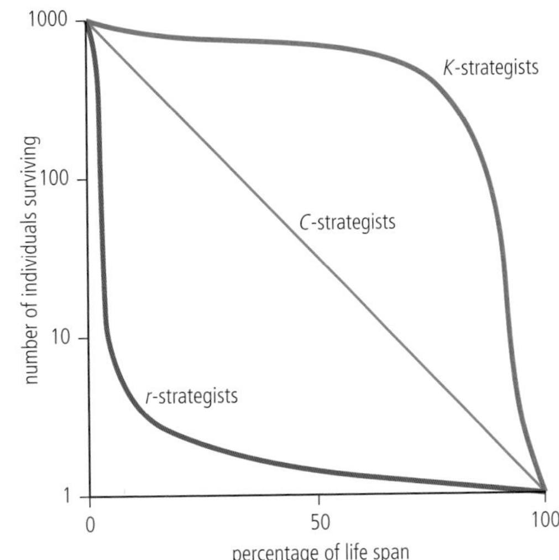

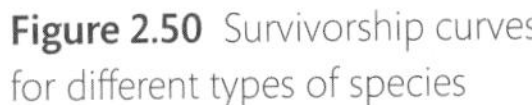

Figure 2.50 Survivorship curves for different types of species

The two extreme examples of a survivorship curve (as shown in Figure 2.50) are:

- the curve for a species where almost all individuals survive for their potential life span, and then die almost simultaneously (K-strategists) – salmon and humans are excellent examples
- the curve for a species where most individuals die at a very young age but those that survive are likely to live for a very long time (*r*-strategists) – turtles and oysters are very good examples.

Species that are *r*-strategists produce large numbers of offspring so they can colonize new habitats quickly and make use of short-lived resources (i.e. they make good pioneer species in a succession). Species that are *K*-strategists tend to produce a small number of offspring, so increasing their survival rate and enabling them to survive in long-term climax communities.

Note that the scale in Figure 2.50 is semi-logarithmic. This means that one axis (here, the horizontal or *x*-axis) is a normal scale and the gaps between each unit are regular: the distance between 0 and 50 is the same as between that between 50 and 100. In contrast, the vertical or *y*-axis is logarithmic.

Note that on the logarithmic scale:

- the scale does not start at 0 (here, it starts at 1)
- the scale goes up in logarithms (the first cycle goes up in ones, the second cycle in tens, the next in hundreds).

The reason for using a logarithmic scale is that it enables us to show very large values on the same graph as very small values.

You need to be able to distinguish the roles of *r*- and *K*-selected species in succession.

The impact of human activities on succession

Climatic and edaphic (i.e. relating to soil) factors determine the nature of a climax community. Human factors frequently affect this process through disturbance. The interference or disturbance halts the process of succession and diverts it so that a different stable state is reached rather than the climax community. This interrupted succession is known as **plagioclimax**. An example is the effect of footpath erosion caused by continued trampling by feet. Or consider a sand dune ecosystem, where walkers might trample plants to the extent that they are eventually destroyed. Human activity can affect the climax community through agriculture, hunting, forest clearance, burning, and grazing: all these activities divert the progression of succession to an alternative stable state so that the original climax community is not reached.

As you have learned in Chapter 1 (page 36), an ecosystem's capacity to survive change may depend on its diversity and resilience.

Human activity is one factor which can divert the progression of succession to an alternative stable state, by modifying the ecosystem. Examples include the use of fire, agriculture, grazing pressure, or resource use such as deforestation. This diversion may be more or less permanent depending on the resilience of the ecosystem.

You need to be able to discuss the factors which could lead to alternative stable states in an ecosystem, and discuss the link between ecosystem stability, succession, diversity, and human activity.

Controlled burning of heather also prevents the re-establishment of deciduous woodland. The heather is burned after 15 years, before it becomes mature. If the heather matured, it would allow colonization of the area by other plants. The ash adds to the soil fertility and the new heather growth that results increases the productivity of the ecosystem.

Burning and deforestation of the Amazon forest to make grazing land leads to loss of large areas of rainforest. Continued burning and clearance, and the establishment of grasslands, prevents succession occurring.

Large parts of the UK were once covered by deciduous woodland. Some heather would have been present in the north, but relatively little. From the Middle Ages onward, forests were cleared to supply timber for fuel, housing, construction of ships (especially oak), and to clear land for agriculture. As a result, soil deteriorated and heather came to dominate the plant community. Sheep grazing and associated burning has prevented the re-growth of woodland by destroying young saplings.

Deforestation is having a major impact on one of the most diverse biomes, tropical rainforest (Figures 2.51, 2.52). An area of rainforest the size of a football pitch is destroyed every 4 seconds. As well as loss of habitat and destruction of a complex climax community, the carbon dioxide released when the trees are burned returns to the atmosphere: this amount of carbon dioxide is more than that from the entire global transport sector.

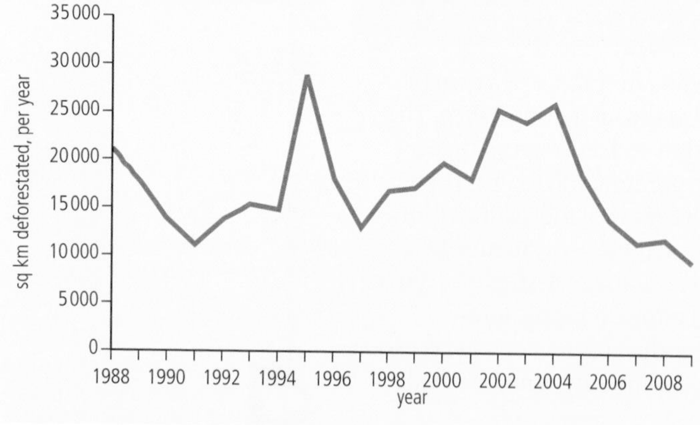

Figure 2.51 Deforestation in the Brazilian Amazon basin fluctuates but remains high despite warnings about the consequences for the planet. The loss of the highly diverse climax community and its replacement by agricultural or grazing ecosystems affects global biodiversity, regional weather, the water cycle and sedimentation patterns.

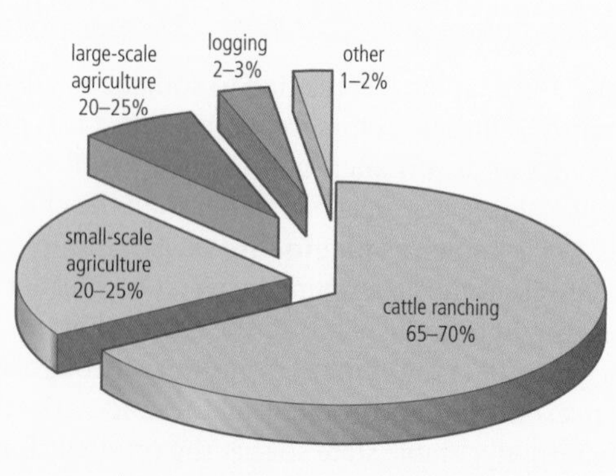

Figure 2.52 Deforestation in the Brazilian Amazon basin, 2000–05 – the reasons land is deforested. The high percentage of meat in western diet, and the increasing consumption of beef in the developing world, demand land for cattle ranching. This is the main driver of human impact on the ecology of the Amazon.

Case study

Deforestation in Borneo

Deforestation in Borneo has progressed rapidly in recent years (Figure 2.53). It affects people, animals, and the environment. A recent assessment by the United Nations Environment Program (UNEP) predicts that the Bornean orang-utan (endemic to the island) will be extinct in the wild by 2025 if current trends continue. Rapid forest loss and degradation threaten many other species, including the Sumatran rhinoceros and clouded leopard. The main cause of forest loss in Borneo is logging and the clearance of land for oil palm plantations.

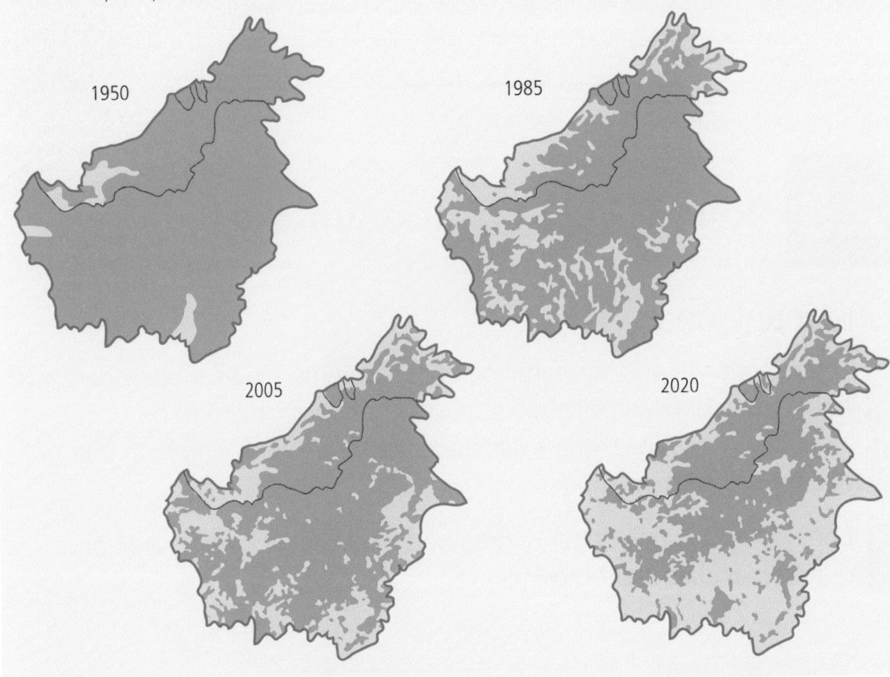

Figure 2.53 Loss of primary forest between 1950 and 2005, and a projection for forest cover in 2020 based on current trends

Exercises

1. Define the term *biome*. How does this differ from the term *ecosystem*?
2. Draw up a table listing the following biomes: tropical rainforest, temperate forest, hot desert, and tundra. The table should include information about the levels of insolation (sunlight), rainfall (precipitation), and productivity for each biome.
3. Which biome has the highest productivity? Why? Which has the lowest? Why?
4. What are the differences between succession and zonation? Give examples of each, using named examples.
5. What changes occur along a succession? What impact do these changes have on biodiversity?
6. What is the P/R ratio? What does it measure? How does the P/R ratio change from early to late succession?
7. What are the characteristics of climax communities? Your answer should include details of biomass levels, species diversity, soil conditions, soil structure, pH, community complexity, type of equilibrium, and habitat diversity.
8. What is a plagioclimax, and how is one formed? Give four examples of how human activities divert the progression of succession to an alternative stable state by modifying the ecosystem.
9. Draw a table with *r*-strategists in one column and *K*-strategists in the other. List the characteristics that apply to each.

Big questions

Having read this section, you can now discuss the following big questions:

- What strengths and weaknesses of the systems approach and the use of models have been revealed through this topic?
- How are the issues addressed in this topic of relevance to sustainability or sustainable development?

Points you may want to consider in your discussions:

- What are the strengths and weaknesses of models of succession and zonation?
- How could the P/R ratio be used to estimate whether the harvesting of a natural capital, such as trees, is sustainable or not?

2.5 Investigating ecosystems

Significant ideas

The description and investigation of ecosystems allows for comparisons to be made between different ecosystems and for them to be monitored, modelled, and evaluated over time, measuring both natural change and human impacts.

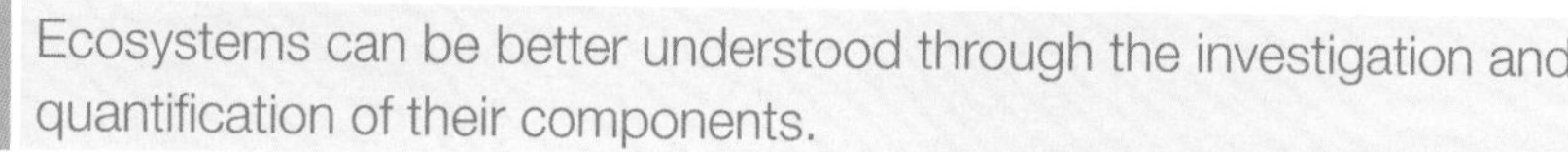

Ecosystems can be better understood through the investigation and quantification of their components.

Big questions

As you read this section, consider the following big questions:

- What strengths and weaknesses of the systems approach and the use of models have been revealed through this topic?
- How are the issues addressed in this topic of relevance to sustainability or sustainable development?

Knowledge and understanding

- The study of an ecosystem requires that it be named and located (e.g. Deinikerwald, Baar, Switzerland, a mixed deciduous-coniferous managed woodland).
- Organisms in an ecosystem can be identified using a variety of tools including keys, comparison to herbarium or specimen collections, technologies and scientific expertise.
- Sampling strategies may be used to measure biotic and abiotic factors and their change in space along an environmental gradient, or over time through succession or before and after a human impact (e.g. as part of an EIA).
- Measurements should be repeated to increase reliability of data. The number of repetitions required depends on the factor being measured.
- Methods for estimating the biomass and energy of trophic levels in a community include measurement of dry mass, controlled combustion, and extrapolation from samples. Data from these methods can be used to construct ecological pyramids.
- Methods for estimating the abundance of non-motile organisms include the use of quadrats for making actual counts, measuring population density, percentage cover, and percentage frequency.

- Direct and indirect methods for estimating the abundance of motile organisms can be described and evaluated. Direct methods include actual counts and sampling. Indirect methods include the use of capture–mark–recapture with the application of the Lincoln Index:

$$\text{Lincoln index} = \frac{n_1 \times n_2}{n_m}$$

where n_1 is the number caught in the first sample, n_2 is the number caught in the second sample and n_m is the number caught in the second sample that were previously marked.
- Species richness is the number of species in a community and is a useful comparative measure.
- Species diversity is a function of the number of species and their relative abundance and can be compared using an index. There are many versions of diversity indices but you are only expected to be able to apply and evaluate the result of the Simpson diversity index as shown below. Using this formula, the higher the result, the greater the species diversity. This indication of diversity is only useful when comparing two similar habitats or the same habitat over time.

$$D = \frac{N(N-1)}{\sum n(n-1)}$$

where D is the Simpson diversity index, N is the total number of organisms of all species found and n is the number of individuals of a particular species.

Studying ecosystems

As you have already seen, ecosystems are highly complex systems. You have looked at how abiotic factors such as temperature, insolation, and precipitation define where ecosystems are found, and how they influence the biotic components (i.e. the organisms found there). Flows of energy and cycles of matter support ecosystems (pages 87–100). You are now going to consider how ecosystems can be better understood through the investigation and quantification of their components. Given the complexity of ecosystems, standardized methods are needed to compare ecosystems with one another. Such studies also allow ecosystems to be monitored, modelled and evaluated over time, with both natural change and human impacts being measured. In order for human effects to be established, the undisturbed ecosystem must first be researched. Let's consider how such investigations can be done.

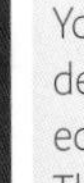

You need to be able to design and carry out ecological investigations. These make ideal studies for your Internal Assessment project. The study of an ecosystem requires that it be named and located (e.g. Deinikerwald, Baar, Switzerland, a mixed deciduous–coniferous managed woodland).

Identifying organisms in ecosystems

Ecology is the study of living organisms in relation to their environment. In any ecological study, it is important to correctly identify the organisms in question, otherwise results and conclusions will be invalid. A **dichotomous key** is a handy tool for identification of organisms that you are not familiar with.

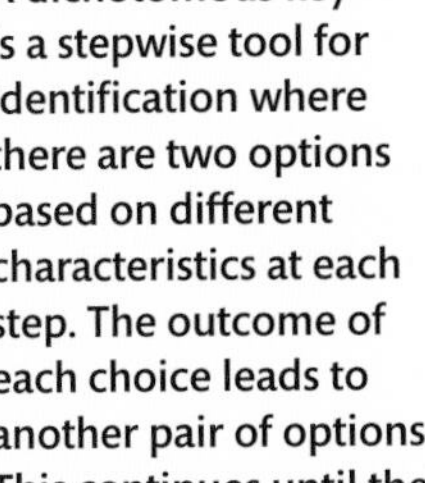

A dichotomous key is a stepwise tool for identification where there are two options based on different characteristics at each step. The outcome of each choice leads to another pair of options. This continues until the organism is identified.

Dichotomous means 'divided into two parts'. The key is written so that identification is done in steps. At each step, you have a choice of two options, based on different possible characteristics of the organism you are looking at. Sometimes, such keys are in written form, sometimes they are drawn as a tree diagram.

Suppose you want to identify one of the organisms or objects pictured below.

A random selection of animate and inanimate objects

You could use a written key such as the one below, or a tree diagram such as Figure 2.54.

1	**a**	Organism is living	go to 4
	b	Organism is non-living	go to 2
2	**a**	Object is metallic	go to 3
	b	Object is non-metallic	pebble

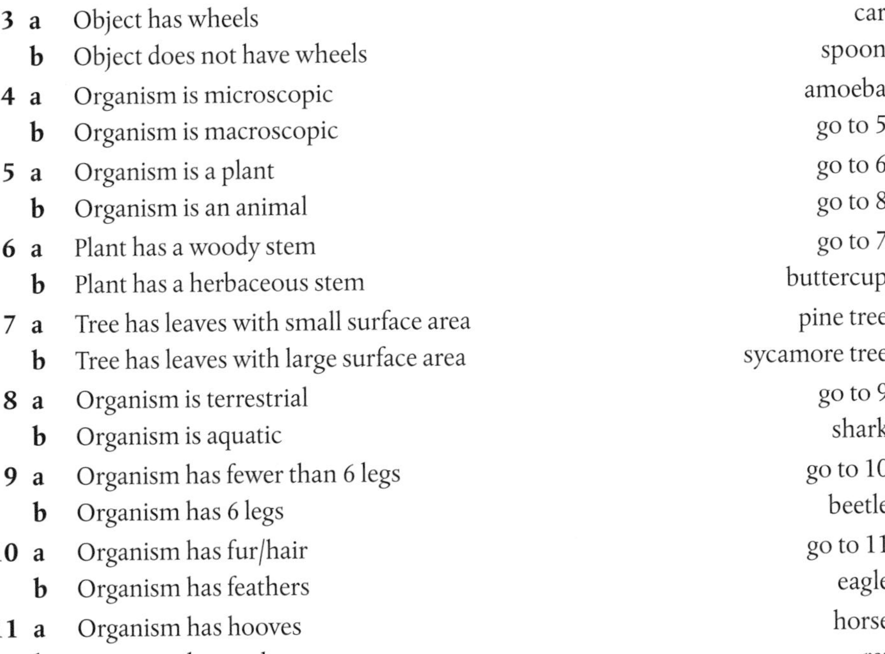

3	**a**	Object has wheels	car
	b	Object does not have wheels	spoon
4	**a**	Organism is microscopic	amoeba
	b	Organism is macroscopic	go to 5
5	**a**	Organism is a plant	go to 6
	b	Organism is an animal	go to 8
6	**a**	Plant has a woody stem	go to 7
	b	Plant has a herbaceous stem	buttercup
7	**a**	Tree has leaves with small surface area	pine tree
	b	Tree has leaves with large surface area	sycamore tree
8	**a**	Organism is terrestrial	go to 9
	b	Organism is aquatic	shark
9	**a**	Organism has fewer than 6 legs	go to 10
	b	Organism has 6 legs	beetle
10	**a**	Organism has fur/hair	go to 11
	b	Organism has feathers	eagle
11	**a**	Organism has hooves	horse
	b	Organism has no hooves	rat

There are limitations to the use of keys. For a start, keys tend to examine physical characteristics rather than behaviour – two species that appear very similar may have very different types of activity. Some keys use technical terms that only an expert would understand. It is also possible that there may not be a key available for the type of organisms you are trying to identify. In addition, some features of organisms cannot be easily established in the field. For example, whether or not an animal has a placenta; whether an animal is endothermic or ectothermic (warm- or cold-blooded). Some organisms significantly change their body shape during their lifetime (e.g. frogs have an aquatic tadpole juvenile form which is very different from the adult), which keys must take into account. Many insects, for example, show differences between male and females of the species which can cause difficulties when identifying species.

Figure 2.54 A dichotomous key for a random selection of animate and inanimate objects

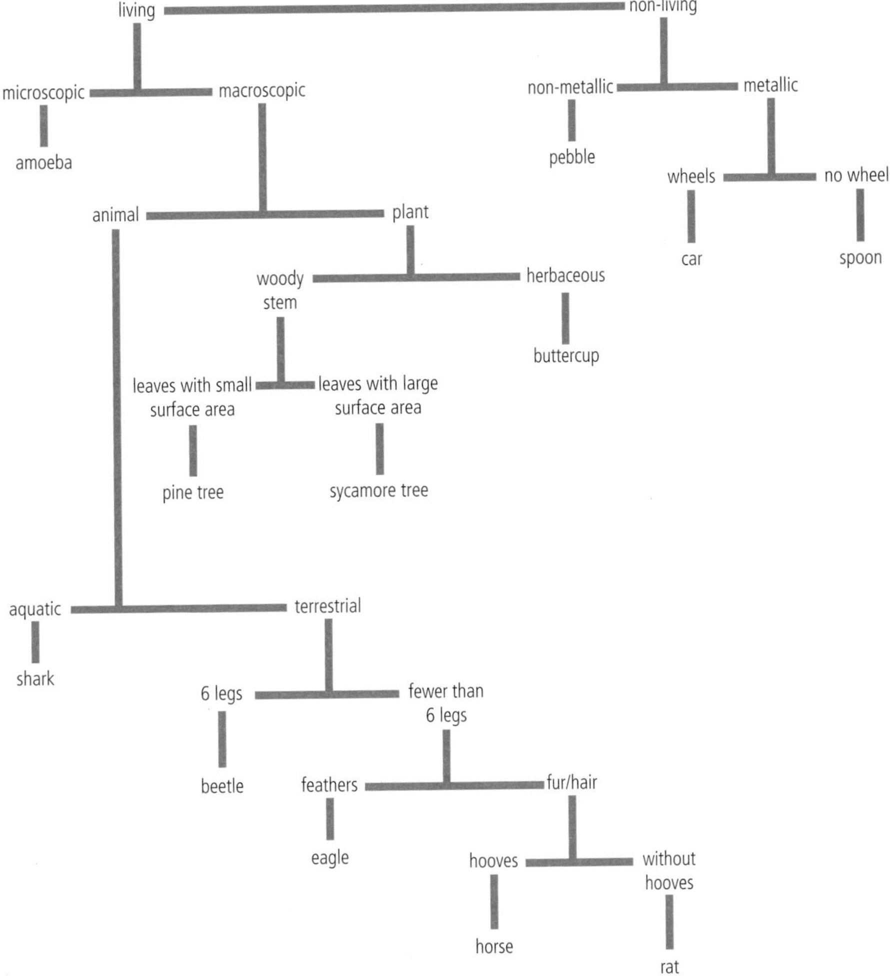

To learn more about using dichotomous keys, go to www.pearsonhotlinks.co.uk, enter the book title or ISBN, and click on weblink 2.6.

You need to be able to construct your own keys for up to eight species. When constructing an identification key, do not use terms such as big/small – these are not useful. You need to write quantitative descriptors (i.e. descriptions that allow numbers to be given to features, such as the number of legs) and simple presence/absence of external features.

Organisms in an ecosystem can also be identified by comparing specimens to those in a herbarium or to museum specimen collections or by using scientific expertise. Museums today use DNA profiling techniques to identify differences between specimens. This can be a more accurate way of determining the identity of an organism than its physical appearance.

Measuring abiotic components of the ecosystem

You have already seen how ecosystems can be broadly divided into three types (page 76). Here is a reminder:

- marine – the sea, estuaries, salt marshes, and mangroves, all characterized by high salt content of the water
- freshwater – rivers, lakes, and wetlands
- terrestrial – land based.

Each ecosystem has its own specific abiotic factors as well as the ones they share.

Abiotic factors of a marine ecosystem include:

- salinity
- pH
- temperature
- dissolved oxygen
- wave action.

Estuaries are classified as marine ecosystems because they have high salt content compared to freshwater. Mixing of fresh water and oceanic seawater leads to diluted salt content but it is still high enough to influence the distribution of organisms within it – salt-tolerant animals and plants have specific adaptations to help them cope with the osmotic demands of saltwater.

Only a small proportion of fresh water is found in ecosystems (Chapter 4, page 213). Abiotic factors of a freshwater ecosystem include:

- turbidity
- pH
- temperature
- dissolved oxygen
- flow velocity.

Abiotic factors of a terrestrial ecosystem include:

- temperature
- light intensity
- wind speed
- particle size
- slope/aspect
- soil moisture
- drainage
- mineral content.

You need to know methods for measuring each of the abiotic factors listed above and how they might vary in any given ecosystem with depth, time or distance. Abiotic factors are examined in conjunction with related biotic components (pages 132–140).

To learn more about sampling techniques, go to www.pearsonhotlinks.co.uk, enter the book title or ISBN, and click on weblink 2.7.

The Nevada desert, USA. Water supply in terrestrial ecosystems can be extremely limited, especially in desert areas, and is an important abiotic factor in controlling the distribution of organisms.

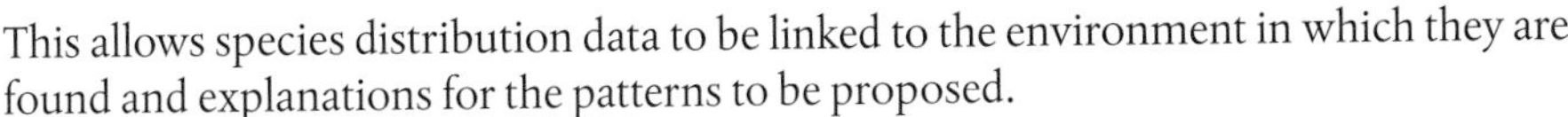

This allows species distribution data to be linked to the environment in which they are found and explanations for the patterns to be proposed.

Evaluating measures for describing abiotic factors

Abiotic factors that can be measured within an ecosystem include:

- **marine environment: salinity, pH, temperature, dissolved oxygen, wave action**
- **freshwater environment: turbidity, pH, temperature, dissolved oxygen, flow velocity**
- **terrestrial environment: temperature, light intensity, wind speed, particle size, slope, soil moisture, drainage, mineral content.**

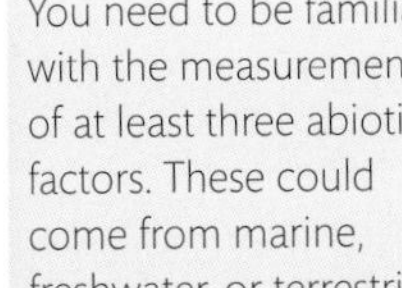

You need to be familiar with the measurement of at least three abiotic factors. These could come from marine, freshwater, or terrestrial ecosystems.

Let's consider the techniques used for measuring abiotic factors. An inaccurate picture of an environment may be obtained if errors are made in sampling, so possible sources of error are examined.

Measurements should be repeated to increase reliability of data. The number of repetitions required depends on the factor being measured.

Light

A light-meter can be used to measure the light intensity in an ecosystem. The meter should be held at a standard, fixed height above the ground and read when the value is steady and not fluctuating. Cloud cover and changes in light intensity during the day mean that values must be taken at the same time of day and same atmospheric conditions: this can be difficult if several repeats are taken. The direction of the light-meter also needs to be standardized so it points in the same direction at the same angle each time it is used. Care must be taken not to shade the light-meter during a reading.

Using a light meter to record light intensity falling on ivy.

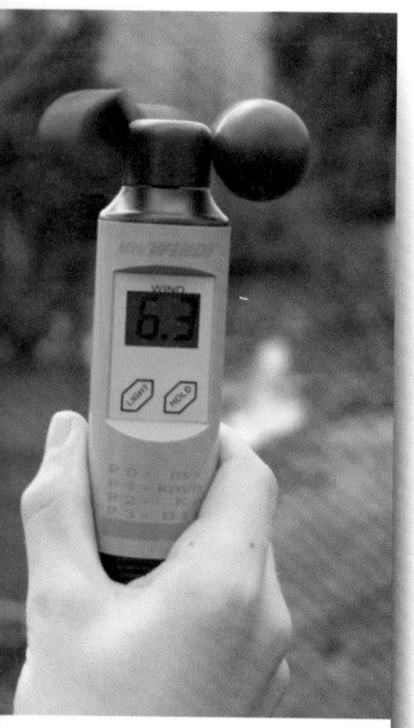

An anemometer measuring wind speed. It works by converting the number of rotations made by three cups at the top of the apparatus into wind speed.

Temperature

An electronic thermometer with probes (datalogger) allows temperature to be accurately measured in air, water, and at different depths in soil. The temperature needs to be taken at a standard depth. Problems arise if the thermometer is not buried deeply enough: the depth needs to be checked each time it is used. Temperature can only be measured for a short period of time using conventional digital thermometers: dataloggers can be used to measure temperature over long periods of time and take fluctuations in temperature into account.

pH

This can be measured using a pH meter or datalogging pH probe. Values in fresh water range from slightly basic to slightly acidic depending on surrounding soil, rock, and vegetation. Seawater usually has a pH above 7 (alkaline). The meter or probe must be cleaned between readings and each reading must be taken at the same depth. Soil pH can be measured using a soil test kit – indicator solution is added and the colour compared to a chart.

Wind

Measurements can be taken by observing the effects of wind on objects – these are then related to the Beaufort scale. Precise measurements of wind speed can be made with a digital anemometer. The device can be mounted or hand-held. Some use cups to capture the wind, whereas other smaller devices use a propeller. Care must be taken not to block the wind. Gusty conditions may lead to large variations in data.

Particle size

Soil can be made up of large, small, or intermediate particles. Particle size determines drainage and water-holding capacity (page 275). Large particles (pebbles) can be measured individually and the average particle size calculated. Smaller particles can be measured by using a series of sieves with increasingly fine mesh size. The smallest particles can be separated by sedimentation. Optical techniques (examining the properties of light scattered by a suspension of soil in water) can also be used to study the smallest particles.

Slope

Surface run-off is determined by slope, which can be calculated using a clinometer (Figure 2.55). Aspect can be determined using a compass. Care must be taken in interpreting results as the slope may vary in angle over its distance.

> If the slope is 10 degrees,
>
> percentage slope = tan(10) × 100 = 0.176 × 100 = 17.6%

Figure 2.55 The slope angle is taken by sighting along the protractor's flat edge and reading the degree aligned with the string. Percentage slope can be calculated by determining the tangent of the slope using a scientific calculator and multiplying by 100.

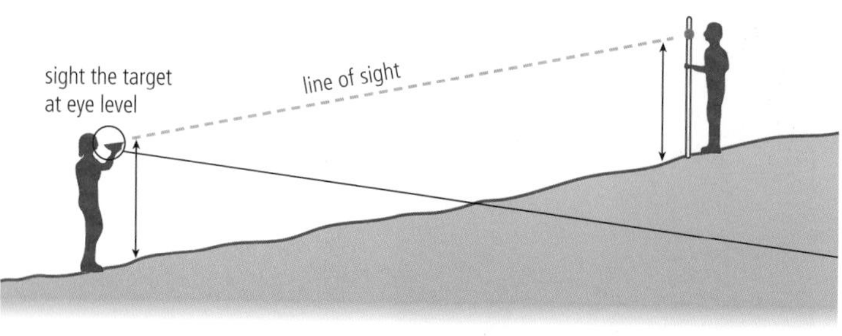

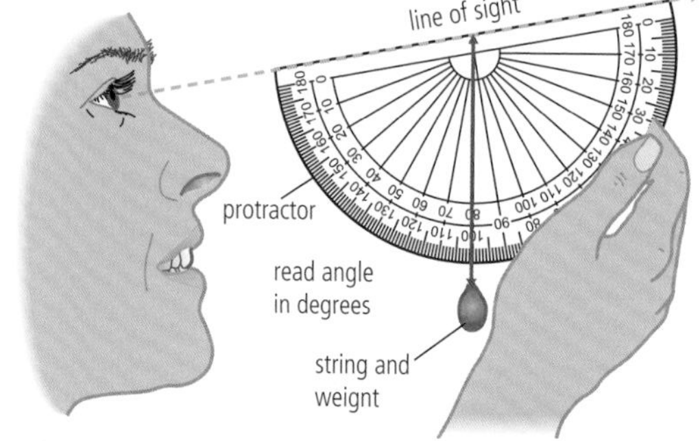

Soil moisture

Soils contain water and organic matter. Weighing samples before and after heating in an oven gives the weight of water evaporated and therefore moisture levels. If the oven is too hot when evaporating the water, organic content can also burn off further reducing soil weight and giving inaccurate readings. Repeated readings should be taken until no further weight loss is recorded – the final reading should be used. Loss of weight can be calculated as a percentage of the starting weight. Soil moisture probes are also available, which are simply pushed into the soil. These need to be cleaned between readings, and can be inaccurate.

Mineral content

The loss on ignition (LOI) test can determine mineral content. Soil samples are heated to high temperatures (500–1000 °C) for several hours to allow volatile substances (i.e. ones that can evaporate) to escape. The loss of mass is equivalent to the quantity of volatile substances present. The temperature and duration of heating depend on the mineral composition of the soil, but there are no standard methods. The same conditions should be used when comparing samples.

Flow velocity

Surface flow velocity can be measured by timing how long it takes a floating object to travel a certain distance. More accurate measurements can be taken using a flow-meter (a calibrated propeller attached to a pole). The impeller is inserted into water just below the surface and pointed into the direction of flow. A number of readings are taken to ensure accuracy. As velocity varies with distance from the surface, readings must be taken at the same depth. Results can be misleading if only one part of a stream is measured. Water flows can vary over time because of rainfall or glacial melting events.

▲ A flow-meter allows water velocity to be recorded at any depth.

Salinity

Salinity can be measured using electrical conductivity (with a datalogger) or by the density of the water (the higher the salt content, the higher the density). Salinity is most often expressed in parts per thousand (ppt); this means parts of salt per thousand parts of water. Seawater has an average salinity of 35 ppt, which is equivalent to 35 g dm^{-3} or 35‰.

Dissolved oxygen

Oxygen-sensitive electrodes connected to a meter can be used to measure dissolved oxygen. Readings may be affected by oxygen in the air, so care must be taken when using an oxygen meter to avoid contamination with oxygen in the air. A more labour-intensive method is Winkler titration. This is based on the principle that oxygen in the water reacts with iodide ions, and acid can be added to release iodine that can be quantitatively measured.

Wave action

Areas with high wave action have high levels of dissolved oxygen due to mixing of air and water in the turbulence. Wave action is measured using a dynamometer, which measures the force in the waves. Changes in tide and wave strength during the day and over monthly periods mean that average results must be used to take this variability into account.

Turbidity

Cloudy water is said to have high turbidity and clear water low turbidity. Turbidity affects the penetration of sunlight into water and therefore rates of photosynthesis.

Figure 2.56 A Secchi disc is mounted on a pole or line and is lowered into water until it is just out of sight. The depth is measured using the scale on the line or pole. The disc is raised until it is just visible again and a second reading is taken. The average depth calculated is known as the Secchi depth.

Turbidity can be measured using a Secchi disc (Figure 2.56). Problems may be caused by the Sun's glare on the water, or the subjective nature of the measure (one person may see the disc at one depth but someone with better eyesight may see it at a greater depth). Errors can be avoided by taking measures on the shady side of a boat. More sophisticated optical devices can also be used (e.g. a nephelometer or turbidimeter) to measure the intensity of light scattered at 90° as a beam of light passes through a water sample.

Short-term and limited field sampling reduces the effectiveness of the above techniques because abiotic factors may vary from day to day and season to season. The majority of these abiotic factors can be measured using datalogging devices. The advantage of dataloggers is that they can provide continuous data over a long period of time, making results more representative of the area. The results can also be made more reliable by taking many samples.

Abiotic data can be collected using instruments that avoid issues of objectivity as they directly record quantitative data. Instruments allow us to record data that would otherwise be beyond the limit of our perception.

TOK

The measurement of the biotic factors is often subjective, relying on your interpretation of different measuring techniques to provide data. It is rare in environmental investigations to be able to provide ways of measuring variables that are as precise and reliable as those in the physical sciences. Will this affect the value of the data collected and the validity of the knowledge?

Measuring biotic components of the ecosystem

Let's now consider the biotic or living factors in an ecosystem. Remember, when carrying out fieldwork you must follow the IB ethical practice guidelines and IB animal experimentation policy: that is, animals and the environment should not be harmed during your work.

Methods for estimating abundance of organisms

It is not possible to study every organism in an ecosystem, so limitations are put on how many plants and animals are studied. Trapping methods enable limited samples to be taken.

Trapping methods for organisms that can move around (are motile/mobile) include:

- pitfall traps – beakers or pots buried in the soil which animals walk into and cannot escape from
- nets – sweep, butterfly, seine, and purse
- flight interception traps – fine-meshed nets that intercept the flight of insects – the animals fall into trays where they can be collected (Chapter 3, page 154)
- small mammal traps – often baited, with a door that closes once an animal is inside (e.g. Longworth trap)
- light traps – a UV bulb against a white sheet attracts certain night-flying insects
- Tullgren funnels – paired cloth funnels, with a light source at one end, a sample pot the other, and a wire mesh between:

A light trap attracts nocturnal insects.

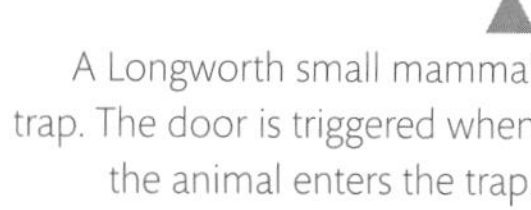

A Longworth small mammal trap. The door is triggered when the animal enters the trap.

invertebrates in soil samples placed on the mesh move away from the heat of the lamp and fall into the collecting bottle at the bottom (Figure 2.57).

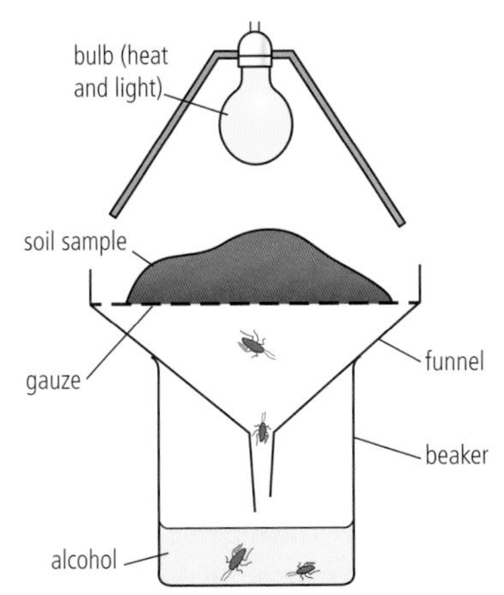

Figure 2.57 A Tullgren funnel

Trapping methods for organisms that cannot move around (are non-motile) or have limited movement):

- quadrats – square frames of different sizes depending on the sample area being studied; frames can be divided into grids of smaller squares to more easily quantify the numbers of organisms present (page 136)
- point frames.

Abundance, as used in ecology, refers to the relative representation of a species in an ecosystem. You can work out the number or abundance of organisms in various ways – either by directly counting the number or percentage cover of organisms in a selected area (for organisms that do not move or are limited in movement), or by indirectly calculating abundance using a formula such as the Lincoln index (for animals that are motile).

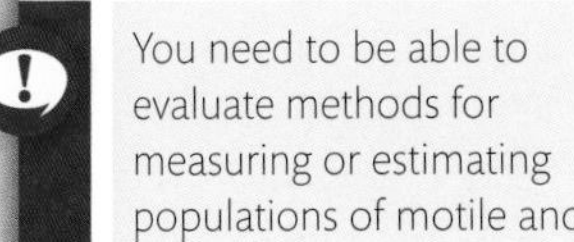

You need to be able to evaluate methods for measuring or estimating populations of motile and non-motile organisms.

The Lincoln index

The Lincoln index is used to estimate the total population size of a motile animal in the study area. In a sample taken using the methods outlined above, it is unlikely that all the animals in a population are sampled, so a mathematical method is used to calculate the total numbers. The Lincoln index is an *indirect* way of estimating the abundance of an animal population because a formula is used to calculate abundance rather than counting the total number of organisms directly.

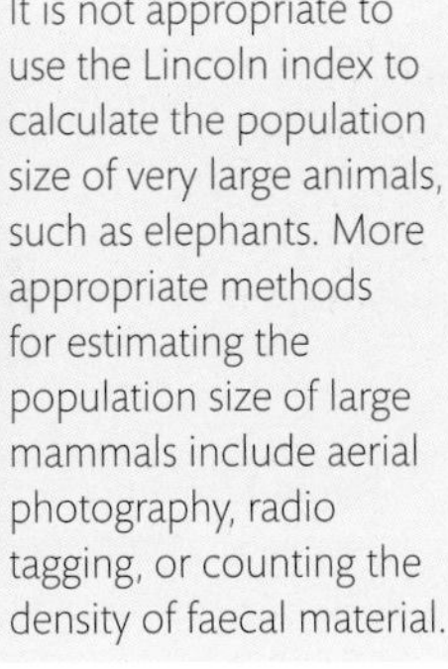

It is not appropriate to use the Lincoln index to calculate the population size of very large animals, such as elephants. More appropriate methods for estimating the population size of large mammals include aerial photography, radio tagging, or counting the density of faecal material.

Using the Lincoln index involves collecting a sample from the population, marking the organisms in some way (paint can be used on insects, or fur clipping on mammals), releasing them back into the wild, then resampling some time later and counting how many marked individuals you find in the second capture. It is essential that marking methods are ethically acceptable (non-harmful) and non-conspicuous (so that the animals are not more easily seen by predators).

Because of the procedures involved, this is called a 'capture–mark–release–recapture' technique. If all of the marked animals are recaptured then the number of marked animals is assumed to be the total population size, whereas if half the marked animals are recaptured then the total population size is assumed to be twice as big as the first sample.

The formula used in calculating population size is:

$$\text{Lincoln index} = \frac{n_1 \times n_2}{n_m}$$

Formulae do not need to be memorized but you should know how to apply them to given data.

Where n_1 is the number caught in the first sample and marked, n_2 is the number caught in the second sample, and n_m is the number caught in the second sample which were previously marked.

There are limitations to the Lincoln index. Animals may move in and out of the sample area, making the capture–mark–release–recapture method less trustworthy and the data invalid. The density of the population in different habitats might vary: there may be many in one area, few in another. The assumption that they are equally spread all

CHALLENGE YOURSELF

ATL Thinking skills

Take a sheet of paper and divide it into 100 squares. Cut these squares out and put them into a tray. Select 20 of these squares and mark them with a cross. Put them back into the tray. Capture another 20 of the pieces of paper. Record how many of these are marked with a cross. Use the Lincoln index to estimate the population size of all pieces of paper. How closely does this agree with the actual number (100)? How could you improve the reliability of the method?

over may not be true. Some individuals may be hidden by vegetation and therefore difficult to find, hence not included in the sample. There may be seasonal variations in animals that affect population size; for example, they may migrate in or out of the study area.

You need to be able to describe and evaluate direct and indirect methods for estimating the abundance of motile organisms.

Direct methods of estimating the abundance of motile animals

Direct methods of estimating abundance include actual counts and sampling. These methods give the relative abundance of different animals in a sample, rather than an estimation of absolute population size (which the Lincoln index does).

Technology is now allowing direct counts of animal populations using aerial photography. Photographs can be taken of animal herds and the number of individuals counted using a computer. There are many ways to sample animal populations, such as those listed on page 132, including pitfall traps (Figure 2.58).

Figure 2.58 A pitfall trap. A plastic pot is buried in the ground. A raincover (e.g. a plastic plate) is placed over the pit to prevent rain from flooding the trap. The raincover can be supported by stones or sticks that elevate it above the pit. Animals fall into the trap and cannot escape.

If the leaf litter on a forest floor is to be sampled, a standardized sample of leaf litter can be put in a tray and a pooter used to suck invertebrates into a small pot. Pooters can also be used to sample insects directly from vegetation. Pooters can be bought or made using a glass jar or plastic pot with tubes or straws (Figure 2.59).

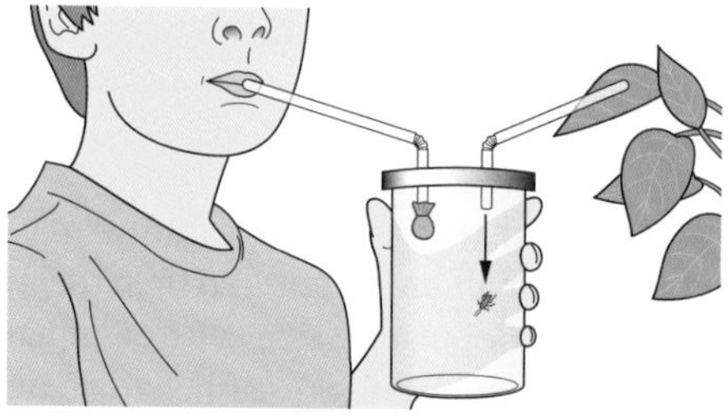

Figure 2.59 A home-made pooter. Plastic straws are attached to a glass jar or pot. One tube is put in the mouth: suction creates a lower pressure in the jar so that small animals are drawn into the jar. A fine mesh is wrapped around the end of the tube so that insects are not ingested when creating the suction.

In order to sample river organisms, the bed of the river is disturbed so that animals found there can be collected. The method involves agitating the riverbed with a boot and collecting disturbed animals downstream in a net. A fixed time is set for this 'kick sampling'. The catch is put in a shallow white tray with at least 2 cm depth of fresh water from the river, and an identification key is used to sort the catch into different groups. Limitations to this method include the difficulty of standardizing the kick-action (different intensities of kicking will disturb different numbers of organisms), and some animals may remain stuck to rocks and so not be sampled.

Sample methods must allow for the collection of data that is scientifically representative and appropriate, and allow the collection of data on all species present. Results can be used to compare ecosystems.

In the early 1980s, Terry Erwin, a scientist at the Smithsonian Institution collected insects from the canopy of tropical forest trees in Panama. He sampled 19 trees and collected 955 species of beetle. Using extrapolation methods, he estimated there could be 30 million species of arthropod worldwide. Although now believed to be an overestimate, this study started the race to calculate the total number of species on Earth before many of them become extinct.

Canopy fogging uses a harmless chemical to knock down insects into collecting trays (usually on the forest floor) where they can be collected. Insects not collected can return to the canopy when they have recovered.

Quadrats

Quadrats are used to limit the sampling area when you want to measure the population size of non-motile organisms (motile ones can move from one quadrat to another and so be sampled more than once thus making results invalid). Quadrats vary in size from 0.25 m square to 1 m square. The size of quadrat should be optimal for the organisms you are studying. To select the correct quadrat size, count the number of different species in several differently sized quadrats. Plot the number of species against quadrat size: the point where the graph levels off, and no further species are added even when the quadrats gets larger, gives you the size of the quadrat you need to use.

If your sample area contains the same habitat throughout, **random sampling** is used. Quadrats should be located at random (use a random number generator). First, you mark out an area of your habitat using two tape measures placed at right angles to each other. Then you use the random numbers to locate positions within the marked-out area. For example, if the grid is 10 m by 10 m, random numbers are generated between 0 and 1000. The random number 596 represents a point 5 m 96 cm along one tape measure. The next random number is the coordinate for the second tape. The point where the coordinates cross is the location for the quadrat.

To explore a random number generator, go to www.pearsonhotlinks.co.uk, enter the book title or ISBN, and click on weblink 2.8.

If your sample area covers habitats very different from each other (e.g. an undisturbed and a disturbed area), you need to use **stratified random sampling**, so you take sets of results from both areas. If the sample area is along an environmental gradient, you should place quadrats at set distances (e.g. every 5 m) along a transect: this is called **systematic sampling**. **Continuous sampling** samples along the whole length of the transect.

Population density is the number of individuals of each species per unit area. It is calculated by dividing the number of organisms sampled by the total area covered by the quadrats, as shown below.

$$\text{population density} = \frac{\text{total number of a species in all quadrats}}{\text{area of one quadrat} \times \text{total number of quadrats}}$$

Plant abundance is best estimated using **percentage cover**. This is an estimate of the area in a given quadrat covered by the organism (usually a plant) in question. This method is not suitable for mobile animals as they may move from the sample area while counting is taking place.

Percentage cover is the percentage of the area within the quadrat covered by one particular species. Percentage cover is worked out for each species present. Dividing the quadrat into a 10 × 10 grid (100 squares) helps to estimate percentage cover (each square is 1 per cent of the total area cover).

The sampling system used depends on the areas being sampled.

- **Random sampling is used if the same habitat is found throughout the area.**
- **Stratified random sampling is used in two areas different in habitat quality.**
- **Systematic sampling is used along a transect where there is an environmental gradient (such as along a succession).**

Percentage frequency is the number of *actual* occurrences divided by the number of *possible* occurrences, expressed as a percentage. For example, if a plant occurs in 7 out of 100 squares in a grid quadrat, its percentage frequency is 7 per cent; or if 8 quadrats out of 10 contain yellow-horned poppy on a transect across a shingle ridge (page 115), their percentage frequency would be 80 per cent. When using whole quadrats to estimate percentage frequency, results depend on the size of the quadrat and so these details need to be included in the conclusion (e.g. yellow-horned poppies occur at a frequency of 80 per cent in a sample of 10 × 1 m^2 quadrats)

The quadrat method is subjective, and different people will end up with different measures. There are many possible sources of error. One species may be covering another and so not be included, and differences between species may be slight, so two or more organisms may be mistakenly identified as the same or different species. It is also difficult to use quadrats for very large or very small plants, or for plants that grow in tufts or colonies. It is possible that plants that appear to be separate are joined by roots: this will affect calculation of population density. It is also difficult to measure the abundance of plants outside their main growing season when plants are largely invisible.

Abundance scales

Another method of estimating the abundance of non-motile organisms is the use of abundance scales: these can be used to estimate the relative abundance of different organisms on, for example, a rocky shore. These are known as DAFOR scales, where each letter indicates a different level of abundance: D = dominant, A = abundant, F = frequent, O = occasional, and R = rare. Quadrats are usually used to define the sample area. Different types of species are put in different categories, for example seaweeds will be in a different group to periwinkles (periwinkles are a type of mollusc). These scales allow for general comparison between different sampling sites. It is a qualitative scale used to judge to abundance of different organisms. Because it is qualitative, it is subjective and so different people may have different judgements of abundance. Also, there are not distinctions between different species in the same category (e.g. all seaweeds will be treated alike, irrespective of size or other differences). The lack of quantitative data makes statistical analysis difficult.

TOK

Applying the rigorous standards used in a physical science investigation would render most environmental studies unworkable. Whether this is acceptable or not is a matter of opinion, although it could be argued that by doing nothing we would miss out on gaining a useful understanding of the environment.

Methods for estimating the biomass of trophic levels

Methods for estimating the biomass and energy of trophic levels in a community include measurement of dry mass, controlled combustion, and extrapolation from samples. Data from these methods can be used to construct ecological pyramids.

You need to be able to evaluate methods for estimating biomass at different trophic levels in an ecosystem.

You have seen how pyramids of biomass can be constructed to show total biomass at each trophic level of a food chain. Rather than weighing the total number of organisms at each level (clearly impractical) an extrapolation method is used: the mass of one organism, or the average mass of a few organisms, is multiplied by the total number of organisms present to estimate total biomass.

Biomass is calculated to indicate the total energy within a living being or trophic level. Biological molecules are held together by bond energy, so the greater the mass of living material, the greater the amount of energy present. Biomass is taken as the mass of an organism minus water content (i.e. dry weight biomass). Water is not included in biomass measurements because the amount varies from organism to organism, it contains no energy and is not organic. Other inorganic material is usually insignificant in terms of mass, so dry weight biomass is a measure of organic content only.

To obtain quantitative samples of biomass, biological material is dried to constant weight. The sample is weighed in a previously weighed container. The specimens are put in a hot oven (not hot enough to burn tissue) – around 80 °C – and left for a specific length of time. The specimen is reweighed and replaced in the oven. This is repeated until a similar mass is obtained on two subsequent weighings (i.e. no further loss in mass is recorded as no further water is present). Biomass is usually stated per unit area (i.e. per metre squared) so that comparisons can be made between trophic levels. Biomass productivity is given as mass per unit area per period of time (usually per year).

To estimate the biomass of a primary producer within a study area, you would collect all the vegetation (including roots, stems, and leaves) within a series of 1 m by 1 m quadrats and then carry out the dry-weight method outlined above. Average biomass can then be calculated.

Dry-weight measurements of quantitative samples can be extrapolated to estimate total biomass.

Once dry biomass has been obtained, combustion of samples under controlled conditions give quantitative data about the amount of energy contained per unit sample (e.g. per gram) in the material. Organic matter can be burned in a calorimeter (Figure 2.60), where the heat released during combustion is measured to determine the energy content. Extrapolation from these samples, by estimating the total biomass of organisms and multiplying this by the energy content per unit mass, can be used to indicate the total energy per trophic level in an ecosystem. From such data, pyramids of productivity can be constructed.

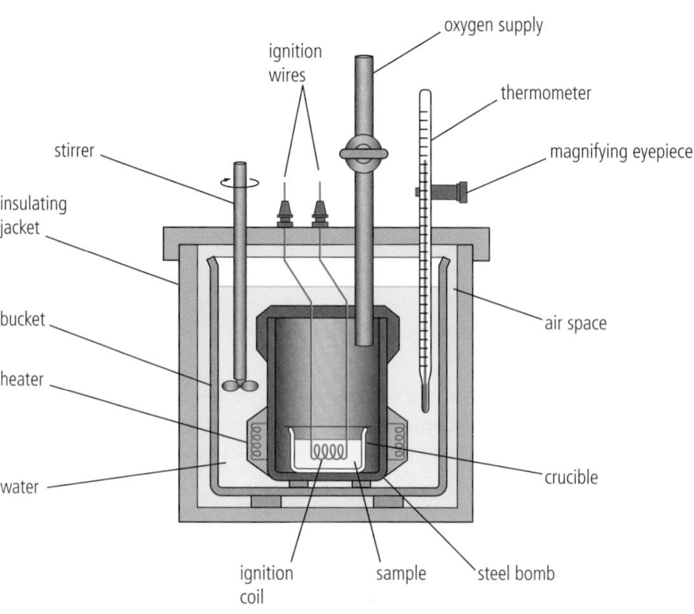

Figure 2.60 A calorimeter – used to calculate the energy content of biomass

One criticism of this method is that it involves killing living organisms. It is also difficult to measure the biomass of very large plants, such as trees. There are further problems in measuring the biomass of roots and underground biomass, as these are difficult to remove from the soil.

TOK

Variables can be measured but not controlled while working in the field. Fluctuations in environmental conditions can cause problems when recording data. Standards for acceptable margins of error are therefore different from laboratory-based experiments. Is this acceptable?

CONCEPTS: Environmental value systems

Ecological sampling can at times involve the killing of wild organisms. For example, to help assess species diversity of poorly understood organisms, it may be necessary to take dead specimens back to the lab for identification; similarly, dead organisms may be needed to assess biomass. An ecocentric worldview, which promotes the preservation of all life, may lead you to question the value of such approaches. Does the end justify the means, and what alternatives (if any) exist?

Species richness and diversity

You need to be able to calculate and interpret data for species richness and diversity.

- **Species richness is the number of species in a community.**
- **Species diversity is the number of species and their relative abundance in a given area or sample.**

Species diversity is considered as a function of two components: the number of different species and the relative numbers of individuals of each species. It is different from **species richness** (the number of species in an area) because the relative abundance of each species is also taken into account.

There are many ways of quantifying diversity. You must be able to calculate diversity using the **Simpson's diversity index** (see also pages 153–154). The index can be used for both animal and plant communities.

The Simpson's index is:

$$D = \frac{N(N-1)}{\Sigma n(n-1)}$$

Where D is the Simpson's diversity index, N is the total number of organisms of all species found and n is the number of individuals of a particular species.

There are many versions of diversity indices but you are only expected to be able to apply and evaluate the result of the Simpson diversity index. You are not expected to memorize the Simpson's diversity formula but must know the meaning of the symbols.

Suppose you want to examine the diversity of beetles within a woodland ecosystem. You could use multiple pitfall traps to establish the number of species and the relative abundance of individuals present. You could then use Simpson's diversity index to quantify the diversity.

Samples must be comprehensive to ensure all species are sampled (Figure 2.61). However, it is always possible that certain habitats have not been sampled and some species missed. For example, canopy fogging (page 135) does not knock down insects living within the bark of the tree, so these species would not be sampled.

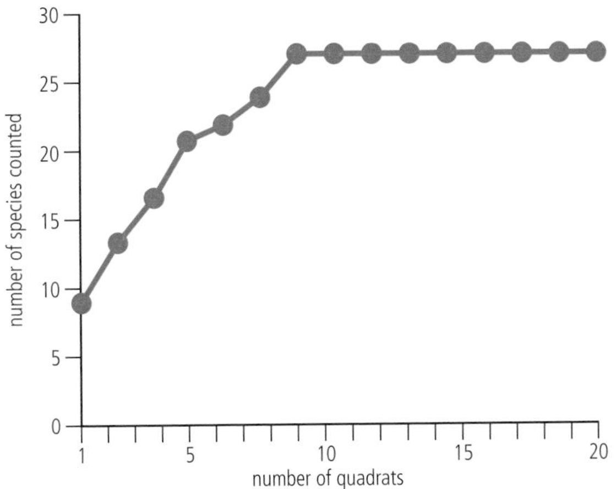

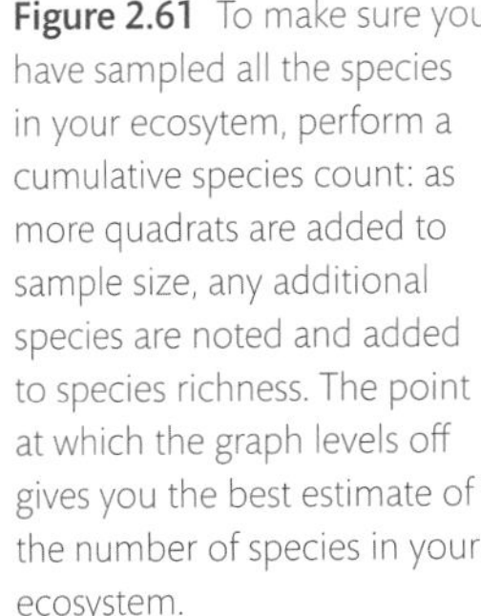

Figure 2.61 To make sure you have sampled all the species in your ecosytem, perform a cumulative species count: as more quadrats are added to sample size, any additional species are noted and added to species richness. The point at which the graph levels off gives you the best estimate of the number of species in your ecosystem.

You need to be able to draw graphs to illustrate species diversity in a community over time or between communities. When plotting changes in diversity over time, values of D would appear on the y-axis, time on the x-axis, and a line graph created. When comparing communities, a bar graph would be plotted showing the values of D for each community.

Measures of diversity are relative, not absolute. They are relative to each other but not to anything else, unlike, say, measures of temperature, where values relate to an absolute scale. Comparisons can be made between communities containing the same type of organisms and in the same ecosystem, but not between different types of community and different ecosystems. Communities with individuals evenly distributed between different species are said to have high 'evenness' and have high diversity. This is because many species can co-exist in the many different niches within a complex ecosystem.

Communities with one dominant species have low diversity, which indicates an ecosystem not able to support as many types of organism. Measures of diversity in communities with few species can be unreliable as relative abundance between species can misrepresent true patterns (by chance, samples may contain unrepresentative numbers of certain species).

Example calculation of Simpson's diversity index

The data from several quadrats in woodland were pooled to obtain Table 2.7.

Table 2.7 Data from woodland

Species	Number, n	$n(n-1)$
woodrush	2	2
holly (seedlings)	8	56
bramble	1	0
Yorkshire fog	1	0
sedge	3	6
total (N)	15	64

To download a Simpson's diversity index calculator, go to www.pearsonhotlinks.co.uk, enter the book title or ISBN, and click on weblink 2.9.

Putting the figures into the formula for Simpson's diversity index:

$N = 15$

$N - 1 = 14$

$N(N - 1) = 15 \times 14 = 210$

$\sum n(n - 1) = 64$

$D = 210/64 = 3.28$

Suppose that the Simpson's diversity index was calculated for a second woodland, where $D = 1.48$. A high value of D suggests a stable and mature site, and a low value of D suggests pollution or agricultural management. The woodland with $D = 3.28$ could be an undisturbed ecosystem and the woodland with $D = 1.48$ could be a disturbed ecosystem. Some ecosystems have naturally low diversity, such as Arctic tundra (page 109), and so reasons for values of D must be attributed based on what is known about the ecosystem being studied. The higher value in the woodland study suggests a more complex ecosystem where many species can coexist. The lower value suggests a simpler ecosystem where fewer species can coexist. The woodland with the higher Simpson's diversity index is an area that would be better for conservation. The woodland with the lower Simpson's Diversity Index is an area that would not be as good for conservation.

Different habitats can be compared using D: the higher the result, the greater the species diversity; a lower value in one habitat may indicate human impact. Low values of D in the Arctic tundra, however, may represent stable and mature sites. This indication of diversity is only useful when comparing two similar habitats or the same habitat over time.

Do not confuse *species richness* with *diversity*. Diversity is the function of two components: the number of different species and the relative numbers of individuals of each species. This is different from species richness, which refers only to the number of species in a sample or area.

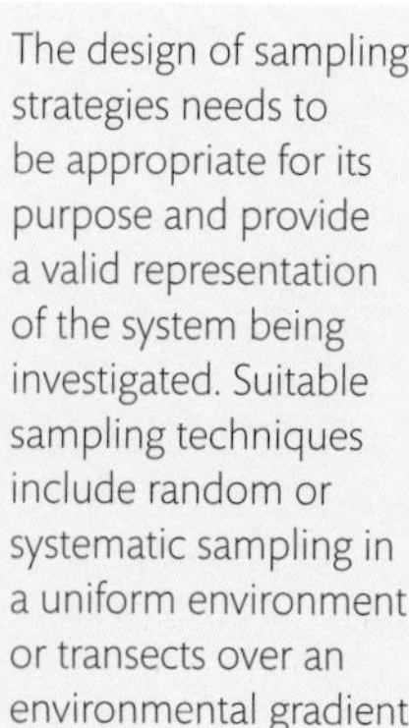

The design of sampling strategies needs to be appropriate for its purpose and provide a valid representation of the system being investigated. Suitable sampling techniques include random or systematic sampling in a uniform environment or transects over an environmental gradient.

Measuring changes in ecosystems

The techniques you have explored in this section can be used to investigate how ecosystems change, either along environmental gradients (such as those found along a succession) or due to human activities. Such investigations make ideal projects for Internal Assessment.

Sampling strategies may be used to measure biotic and abiotic factors and their change in space (along an environmental gradient) or over time (through succession or before and after a human impact (e.g. as part of an EIA). There is an example of the latter in Chapter 1 (pages 46–47).

Measuring change along an environmental gradient

Environmental gradients are changes in environmental factors through space (e.g. decreasing temperature with increasing altitude up a mountain) or where an ecosystem suddenly ends (e.g. at forest edges). In these situations, both biotic and abiotic factors vary with distance and form gradients in which trends can be recorded. The techniques used in sampling such gradients are based on the quadrat method (pages 135–136) and, as such, are more easily done on vegetation and non-motile animals.

Different types of quadrat can be used, depending on the type of organism being studied.

- Frame quadrats are empty frames of known area, such as 1 m^2.
- Grid quadrats are frames divided into 100 small squares with each square representing one per cent. This helps in calculating percentage cover (page 136).
- Point quadrats are made from a frame with 10 holes, which is placed into the ground by a leg. A pin is dropped through each hole in turn and the species touched are recorded. The total number of pins touching each species is converted to percentage frequency data; for example, if a species touched 6 out of the 10 pins it has 60 per cent frequency.

Because environmental variables change along a gradient, random quadrat sampling is not appropriate. All parts of the gradient need to be sampled, so a transect is used. The simplest transect is a **line transect** – a tape measure laid out in the direction of the gradient (e.g. on a beach this would be at 90° to the sea). All organisms touching the tape are recorded. Many line transects need to be taken to obtain valid quantitative data. Larger samples can be taken using a **belt transect**. This is a band of chosen width (usually between 0.5 and 1 m) laid along the gradient (Figure 2.62).

A frame quadrat

A point quadrat

Figure 2.62 Belt transects sample a strip through the sample area. Replication of transects is needed to obtain valid quantitative data.

In both line and belt transects, the whole transect can be sampled (a **continuous transect**) or samples are taken at points of equal distance along the gradient (an **interrupted transect**). If there is no discernible vertical change in the transect, horizontal distances are used (e.g. along a shingle ridge succession), whereas if there is a climb or descent then vertical distances are normally used (e.g. on a rocky shore).

It is important that transects are carried out, as far as possible, at the same time of day, so abiotic variables are comparable. Seasonal fluctuations also mean that samples should be taken either as close together in time as possible or throughout the whole year: datalogging equipment allows the latter to take place, although this may be impractical in school studies.

So that data are reliable and quantitatively valid, transects should be repeated – at least three times is recommended. To avoid bias in placing the transects, a random number generator can be used (page 135). A tape measure is laid at right angles to the environmental gradient: transects can be located at random intervals along the tape or at regular intervals (Figure 2.63).

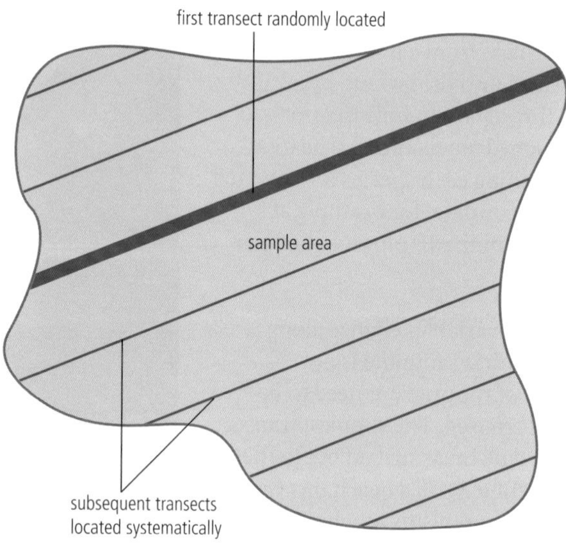

Figure 2.63 All transects can be located randomly or they can be systematically located following the random location of the first. So, for example, subsequent transects might be located every 10 m along a line perpendicular to the ecological gradient.

Measuring changes in an ecosystem due to human activity

Interesting studies can be made using historic maps or GIS (geographic information system) data to track land use change.

Suitable human impacts include toxins from mining activity, landfills, eutrophication (page 255), effluent, oil spills, overexploitation, and change of land use (e.g. deforestation, development, or use for tourism activities).

CONCEPTS: Environmental value systems

Greater awareness of environmental issues has caused EVSs to alter over time. What would not have been seen as a problem in the past (e.g. mining activity) is now understood to produce toxins and to lead to environmental damage. Greater understanding of scientific issues has influenced public perception of human effects on the environment, and has changed worldviews.

Changes in the ecosystem depend on the human activity involved. Methods used for measuring abiotic and biotic components of an ecosystem must be appropriate to the human activity being studied. In your local area there will be locations where you can investigate the effect that human disturbance has had on natural ecosystems. These may be areas of forest that have been harvested for timber, grassland habitats that are regularly trampled by walkers, and so on. There is a variety of methods you can use to study the effect of human activities:

- carry out capture–mark–release–recapture methods on invertebrate species in disturbed and undisturbed sites (pages 133–134)
- measure species diversity using the Simpson's index (page 139)
- use indicator species (Chapter 4), pages 252–255)
- measure variables such as light levels, temperature and wind speed. You could also calculate the average width of tree stems at breast height (DBH), and the degree of canopy openness (the amount of sky can you see through the canopy of the forest), which would give you measures of tree biomass and leaf cover.
- measure soil erosion – in areas with high precipitation this can be simply calculated by measuring the depth of soil remaining under free-standing rocks and stones, where soil around these solid objects has been eroded away
- measure soil variables such as soil structure, nutrient content, pH, compaction levels, and soil moisture (Section 5.1, pages 269–278).

You need to compare measurements taken from the disturbed area with those from undisturbed areas, so that you can work out the magnitude and effect of the disturbance. Where environmental gradients are present, factors should be measured along the full extent of the gradient so that valid comparisons can be made.

Case study

Studying the effects of deforestation

Both pristine and logged forest areas must be studied so comparisons can be made. Stratified random sampling is used in two areas because the pristine and logged forest areas are different in habitat quality. Sampling grids are established in both pristine and logged forest sites. Samples are collected from the grids using random sampling methods. For example, for a grid of 10 m by 10 m, a random number generator could be used to choose random points to sample within the grid. Numbers generated between 0 and 1000 would provide the sample points; for example 580 would represent a point 5 m 80 cm along the bottom of the grid, and 740 a point 7 m 40 cm along the side of the grid (Figure 2.64).

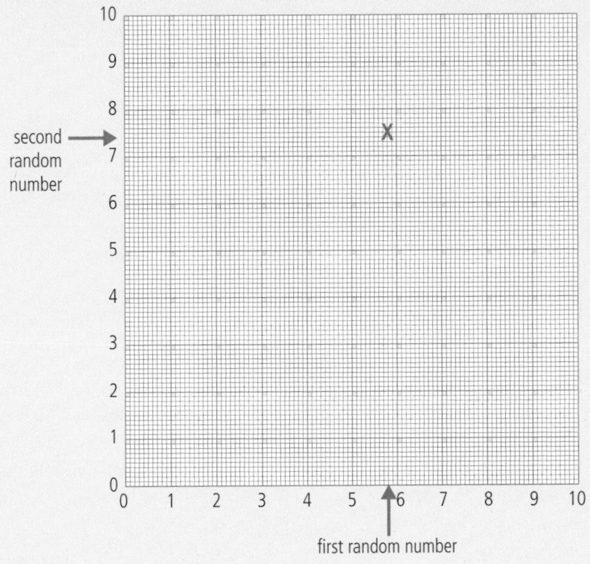

Figure 2.64 Locating a sampling point (X) using random numbers (see text for details)

continued

Abiotic and biotic measurements can be made at each sample point. Abiotic measurements include wind, temperature, and light intensity. Quadrats can be used to sample biotic measurements. Biotic factors include the species of plants and animals present, and the population size of selected indicator species. Motile animals can be sampled using capture–mark–release–recapture methods.

Several samples are taken from each sampling grid. Sampling grids must be repeated in both pristine and logged forest areas so that data are reliable. At least five sampling grids from both pristine and logged forest are recommended. Abiotic and biotic components must be measured over a long period of time to take into account daily and seasonal variations so as to ensure data is valid.

These images show logging and the development of settlements and farming areas.

GIS (geographic information system) data can be used to track changes in ecosystems over time. One example of this is the use of satellite images.

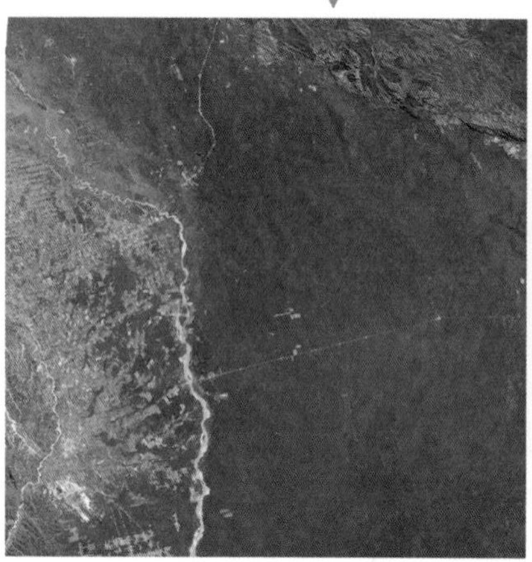

An advantage of satellite images is that the visible nature of the photos is useful for motivating public opinion and action. A disadvantage is that they can be expensive to obtain and may not be available for the area being studied. Another disadvantage is that although some biotic measurements can be taken, such as plant productivity, other biotic and abiotic components cannot be measured such as species diversity and relative humidity. Satellite images are best used in conjunction with ground studies so that the images can be matched with abiotic and biotic data from the ground.

Exercises

1. List as many abiotic factors as you can think of. Say how you would measure each of these factors in an ecological investigation.
2. Evaluate each of the methods you have listed in Exercise 1. What are their limitations, and how may they affect the data you collect?
3. Which methods could you use in (a) marine ecosystems, (b) freshwater ecosystems, and (c) terrestrial ecosystems?
4. Create a key for a selection of objects of your choice. Does your key allow you to accurately identify each object?
5. What ethical considerations must you bear in mind when carrying out capture–mark–release–recapture exercises on wild animals?
6. What is the difference between species diversity and species richness?
7. What does a high value for the Simpson's index tell you about the ecosystem from which the sample is taken? What does a low value tell you?
8. Describe methods for measuring changes along an environmental gradient. Divide these into abiotic and biotic factors. Evaluate each – what are the limitations of each and how will they affect the data you collect?

Big questions

Having read this section, you can now discuss the following big questions:

- What strengths and weaknesses of the systems approach and the use of models have been revealed through this topic?
- How are the issues addressed in this topic of relevance to sustainability or sustainable development?

Points you may want to consider in your discussions:

- Ecological studies allow ecosystems to be monitored, modelled and evaluated over time, with both natural change and human impacts being measured. What are the strengths and weaknesses of models used?
- Carrying out an investigation into the effects of human activities on an ecosystem will indicate the level of damage inflicted on the system. Sustainability is the use and management of resources that allows full natural replacement of the resources exploited and full recovery of the ecosystems affected by their extraction and use: large changes in the ecosystem following disturbance would suggest that the impacts are not sustainable.
- Sustainable development means meeting the needs of the present without compromising the ability of future generations to meet their own needs – by studying the impacts of human activities, using the methods covered in this chapter, you will be able to assess whether development has been sustainable or not.

Practice questions

1 a Deduce, giving a reason, whether the figure below could represent the transfer of energy in a terrestrial ecosystem. [1]

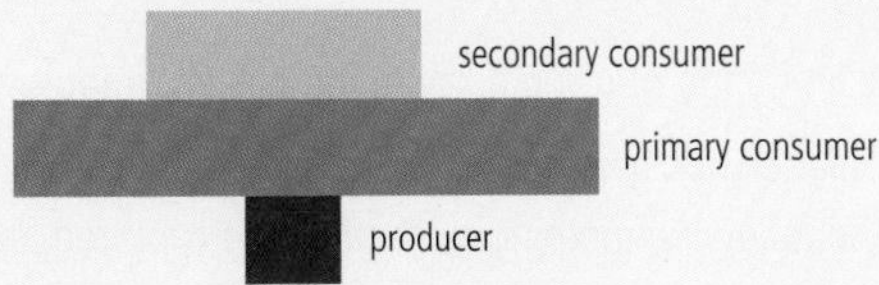

b Define the term *species*. [1]

c The figure below shows the species composition of two areas of forest. There are 100 trees in each area of forest.

	Abundance of organisms	
	Ecosystem A	**Ecosystem B**
White pine	84	50
Red maple	16	50

Simpson's diversity index can be calculated by applying the formula:

$$D = \frac{N(N-1)}{\sum n(n-1)}$$

where: N = total number of organisms of all species and n = number of organisms of a particular species.

The Simpson's diversity index for Ecosystem A is 1.38. Calculate Simpson's diversity index for Ecosystem B. [2]

d The organisms shown below (not drawn to scale) were found in an aquatic ecosystem.

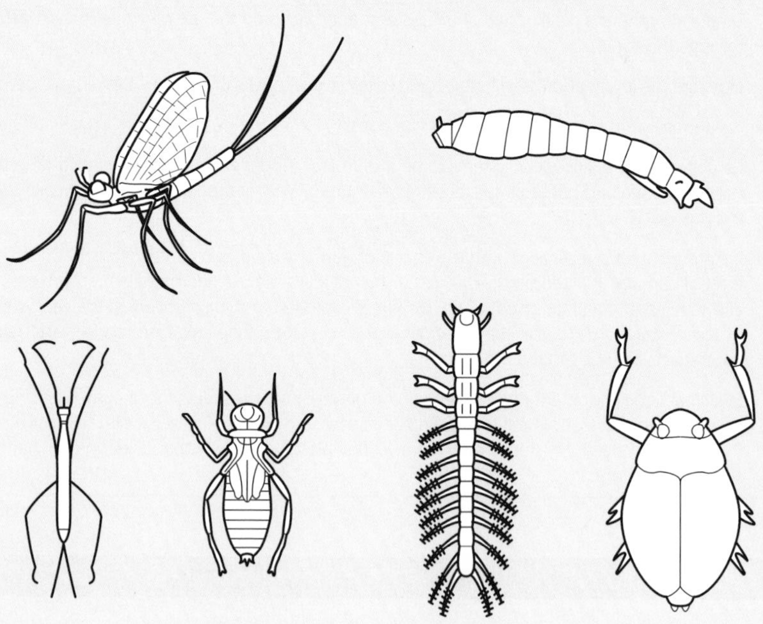

i Suggest two visible characteristics of the organisms shown above which could be used to construct an identification key. [1]

ii Identify one limitation of using a key to identify an organism. [1]

2 To estimate the populations of small mammals in a woodland, ecologists set traps in the area before sunset and the following morning marked all the captured animals before releasing them again.

a State what information the ecologists must record before releasing the animals. [1]

b A week later, the traps are set again as before. State what data must be recorded when the traps are opened and explain how these data may be used to estimate the small mammal populations in the area. [2]

3 The figure opposite shows the flow of energy through a freshwater ecosystem in Florida, USA. The figures are given in kilojoules per square metre per year (kJ m^{-2} yr^{-1}).

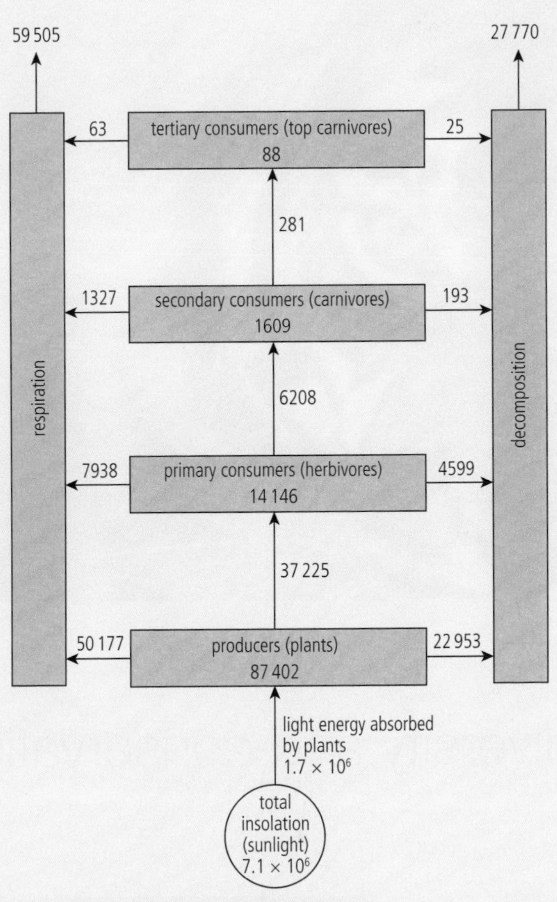

- **a** Define the term *net primary productivity (NPP)*. [1]
- **b** Define the term *gross secondary productivity (GSP)*. [1]
- **c** Calculate the efficiency of conversion of total insolation (sunlight) to NPP in the figure. [1]
- **d** List four possible reasons why not all sunlight emitted by the Sun is used by plants for photosynthesis. [2]
- **e** Explain, giving two reasons, why the net productivity of secondary consumers is much smaller than that of primary consumers. [2]

4 Outline two examples of a transformation of carbon and two examples of a transfer of carbon which occur during the carbon cycle. [4]

5 Explain, with reference to two contrasting biomes, why one biome will be more productive than the other. [5]

6 **a** With reference to examples, distinguish between the terms *succession* and *zonation*. [4]
- **b** With reference to a named example of an ecosystem, explain why the climax community is more diverse and therefore stable, than a community which has been interrupted by human activity. [6]
- **c** Explain why an understanding of how ecosystems work can help people to manage resources effectively. [8]

7 Describe a method for measuring changes in abiotic components in a named ecosystem affected by human activity. [5]

03 Biodiversity and conservation

3.1 An introduction to biodiversity

Opposite: Sumatran tigers are critically endangered. This magnificent animal epitomizes the crisis facing many species on Earth, and the need to conserve biodiversity for future generations.

Significant ideas

Biodiversity can be identified in a variety of forms, including species diversity, habitat diversity, and genetic diversity.

The ability to both understand and quantify biodiversity is important to conservation efforts.

Big questions

As you read this section, consider the following big question:

- How are the issues addressed in this topic of relevance to sustainability or sustainable development?

Knowledge and understanding

- Biodiversity is a broad concept encompassing total diversity which includes diversity of species, habitat diversity, and genetic diversity.
- Species diversity in communities is a product of two variables, the number of species (richness) and their relative proportions (evenness).
- Communities can be described and compared by the use of diversity indices. When comparing communities that are similar, low diversity could be evidence of pollution, eutrophication, or recent colonization of a site. The number of species present in an area is often used to indicate general patterns of biodiversity.
- Habitat diversity refers to the range of different habitats in an ecosystem or biome.
- Genetic diversity refers to the range of genetic material present in a population of a species.
- Quantification of biodiversity is important to conservation efforts so that areas of high biodiversity may be identified, explored, and appropriate conservation put in place where possible.
- The ability to assess changes to biodiversity in a given community over time is important in assessing the impact of human activity in the community.

Biodiversity is a broad concept encompassing total diversity, including species diversity, habitat diversity, and genetic diversity.

What is biodiversity?

The word **biodiversity** is a conflation of 'biological diversity' and was first made popular by ecologist EO Wilson in the 1980s. It is now widely used to represent the variety of life on Earth. *Bio* makes it clear we are interested in the living parts of an ecosystem, and *diversity* is a measure of both the number of species in an area and their relative abundance (Chapter 2, pages 138–140). The term can be used to evaluate both the complexity of an area and its health. Biodiversity can be measured in three different ways: species diversity, habitat diversity, and genetic diversity.

Rainforests have high diversity. They are rich in resources (e.g. food, space) with many different niches available, so many species can co-exist.

Biodiversity refers to the variety of life on Earth. The word was first used by conservation biologists to highlight the threat to species and ecosystems, and is now widely used in international agreements concerning the sustainable use and protection of natural resources.

A species is a group of organisms sharing common characteristics that interbreed and produce fertile offspring.

Species diversity

Species diversity refers to the variety of **species** per unit area; it includes both the number of species present and their relative abundance. The higher the species diversity of a community or ecosystem, the greater the complexity. Areas of high species diversity are also more likely to be undisturbed (e.g. primary rainforest). Species diversity within a community is a component of the broader description of the biodiversity of an entire ecosystem.

Species diversity in communities is a product of two variables, the number of species (richness) and their relative proportions (evenness).

Richness and evenness

As we saw in Chapter 2 (pages 138–140), richness and evenness are components of biodiversity. *Richness* is a term that refers to the number of a species in an area, and *evenness* refers to the relative abundance of each species (Figure 3.1). A community with high evenness is one that has a similar abundance of all species – this implies a complex ecosystem where there are lots of different niches that support a wide range of different species. In contrast, low evenness refers to a community where one or a few species dominate – this suggests lower complexity and a smaller number of potential niches, where a few species can dominate.

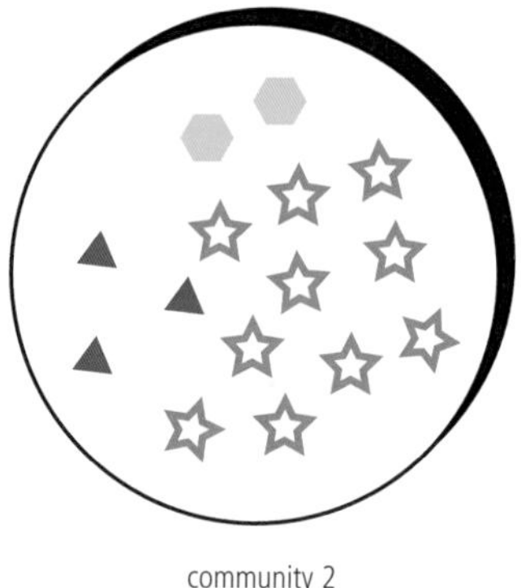

Figure 3.1 Richness and evenness in two different communities. Both communities have the same number of species (species richness) with three species each, but community 1 has greater evenness (all species are equally abundant) than community 2, where one species dominates. Community 1 shows greater species diversity than community 2.

A community is a group of populations living and interacting with each other in a common habitat.

Communities can be described and compared by the use of diversity indices, for example using the Simpson's index (Chapter 2, pages 139–140). When comparing communities that are similar, low diversity could be evidence of pollution, eutrophication or recent colonization of a site. Number of species is often used to assess biodiversity in an area, although measurement of species richness alone can be a misleading measure of disturbance to ecosystems; species diversity measures give more meaningful data (pages 153–154).

The Simpson index (*D*) is a method for measuring diversity (Chapter 2, page 139). Areas with a high *D* value suggest a stable and mature site. A low value of *D* could suggest pollution, recent colonization, or agricultural management.

Measurements of species richness depend on sample size, especially when dealing with small organisms such as insects. Figure 3.2 shows the accumulated species richness of dung beetles (a large group of scarab beetles

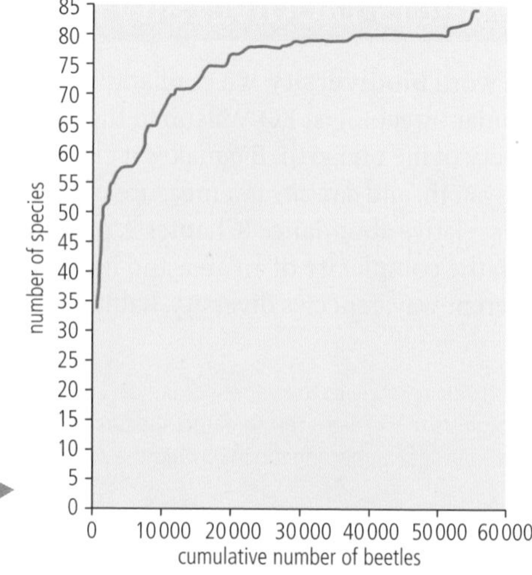

Figure 3.2 Species accumulation graph of beetles collected by pitfall trap in the Bornean lowland rainforest

that feed on faeces, carrion (dead animals), decomposing plant material, as well as other food sources) in a rainforest ecosystem in Borneo. Accurate measurement of species richness was only possible after a large number of beetles had been collected. Clearly, for accurate measures of species richness and, by implication, accurate calculation of diversity indices, appropriate sample sizes are required.

TOK

Does it matter that there are no absolute measurements for diversity indices? Numbers can be used for comparison but on their own mean little. Are there other examples in science of similar relative rather than absolute measurement systems?

You need to be able to comment on the relative values of biodiversity data (e.g. why the value of *D* in one area is higher than that in another). Interpreting diversity is complex, low diversity can be present in natural, ancient and unpolluted sites (e.g. Arctic ecosystems).

Habitat diversity

Habitat diversity is often associated with the variety of ecological niches. For example, a woodland may contain many different habitats (e.g. river, soil, trees) and so have a high habitat diversity, whereas a desert has few (e.g. sand, occasional vegetation) and so has a low habitat diversity.

A habitat is the environment in which a species normally lives.

Habitat diversity refers to the range of different habitats in an ecosystem or biome.

Death Valley desert. Ecosystems such as deserts have low biodiversity as there are fewer opportunities for species to coexist.

Genetic diversity

The term *genetic* refers to **genes**, which are sections of DNA found in the nucleus of all cells. They are essentially the instructions from which a species is produced. **Gene pool** refers to all the different types of gene found within every individual of a species. A large gene pool leads to high **genetic diversity** and a small gene pool to low genetic diversity. Although the term normally refers to the diversity within one species, it can also be used to refer to the diversity of genes in all species within an area.

Genetic diversity refers to the range of genetic material present in a population of a species.

TOK

Early definitions of diversity have become limited as scientific knowledge has increased. Species diversity depends on the correct identification of different organisms and their distribution around the Earth. In the past, this was based on physical characteristics, which we now know can prove unreliable (e.g. two species may look similar but be completely unrelated). Genetic diversity allows for a more accurate way to describe species, although variation within the gene pool of individual species may cause problems (i.e. all species show physical variation in size, colour, and so on – it may be difficult to decide whether individuals are from different species or simply indicate variation within one species).

Species with low genetic diversity, such as cheetahs, are more prone to extinction. This is because if the environment changes, such a species is less likely to have the genes to help it to survive.

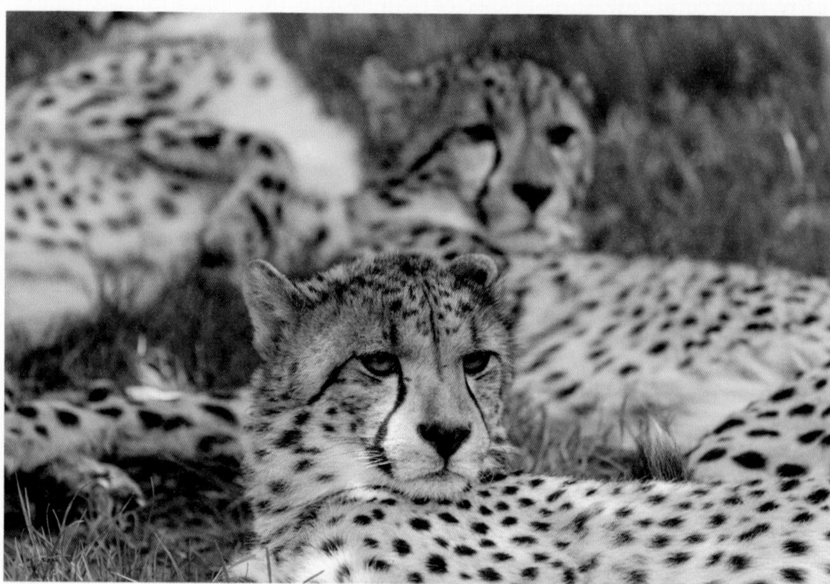

Genetic diversity in cheetahs is low

Overview of biodiversity

The term *biodiversity* is often used as a way of referring to the heterogeneity (variability) of a community, ecosystem or biome, at the species, habitat, or genetic level. The scientific meaning of diversity can become clear from the context in which it is used and may refer to any of the meanings explained above. For the meaning to be obvious, the level should be spelled out by using the correct term (i.e. species diversity, habitat diversity, or genetic diversity).

Conservation of habitat diversity usually leads to the conservation of species and genetic diversity.

CONCEPTS: Biodiversity

Biodiversity is a broad concept encompassing the total diversity of living systems, which includes the diversity of species, habitat diversity, and genetic diversity.

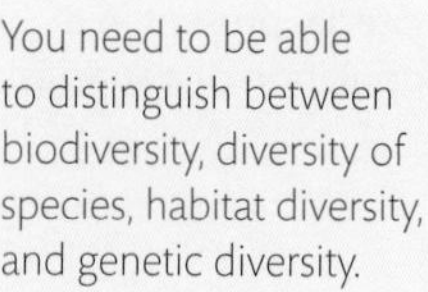

You need to be able to distinguish between biodiversity, diversity of species, habitat diversity, and genetic diversity.

Of the three types of diversity, the increase of habitat diversity is most likely to lead to an increase in the other two. This is because different habitats tend to have different species, and so more habitats will generally have a greater variety of species. Similarly, different species tend to have different genes and so more species will generally include a greater variety of genes. The conservation of habitat diversity will therefore usually lead to the conservation of species and genetic diversity.

Quantification of biodiversity is important to conservation efforts so that areas of high biodiversity may be identified, explored, and appropriate conservation put in place where possible.

Conservation of biodiversity

Conservation means 'keeping what we have'. Conservation aims to protect habitats and ecosystems, and hence species from human-made disturbances, such as deforestation and pollution. Conservation activities aim to slow the rate of extinction caused by the knock-on effects of unsustainable exploitation of natural resources and to maintain biotic interactions between species.

CONCEPTS: Sustainability

Diversity indices can be used to assess whether the impact of human development on ecosystems is sustainable or not.

Ecosystems can be immensely complex systems, as you learned in Chapter 2. How can the effects of human activities be assessed given such complexity? In Chapter 2 you considered ways in which the effects of human disturbance can be measured (pages 142–144). Disturbance, or perturbation, takes ecosystems away from steady-state equilibrium and can lead to new stable states after certain tipping points are reached (Chapter 1, page 33). In Chapter 2, you explored the differences between fundamental and realized niche (page 63): perturbation can simplify ecosystems, or change them so that opportunities for the existence for many species are removed (i.e. ecosystems change so that the realized niche of species no longer exists in the area). Such changes may, for other species, provide an expansion of their usual range because their realized niche spreads into the disturbed area. For example, species found in the canopy and along river banks in a rainforest (rather than the forest interior) may spread into forest that has been logged where new conditions are now found.

▲ Bulldozer making a logging road in rainforest in Sabah, Malaysia. Change in the ecosystem leads to change in the species found there.

CONCEPTS: Equilibrium

Disturbance takes ecosystems away from steady-state equilibrium and can lead to new stable states if certain tipping points are reached.

A key tool used by conservation biologists to assess the effect of disturbance is use of diversity indices, such as the Simpson's index. Quantification of biodiversity in this way is important to conservation efforts so that areas of high biodiversity are identified, explored, and appropriate conservation put in place where possible. Areas that are high in biodiversity are known as hotspots. They contain large numbers of **endemic** species (species not found anywhere else), and so measures of biodiversity are essential in identifying areas that should be protected against damaging human activities. An example of a biological hotspot is Tumbes-Chocó-Magdalena, an area that includes the forests of the South American Pacific coast (from Panama to Peru) and the Galápagos Islands.

Measurements of species richness, on their own, are not sufficient to establish the impacts of human activities. Assessment of species richness varies according to sampling technique. Certain species may be sampled by a given technique but not others; light traps, for example, sample insects drawn to a light bulb, but not the ones that are not. Sample size also affects the assessment of species richness – the bigger the sample, the more species collected. The relative abundance of species in a community must also be taken into account, as you have seen. Care must be taken in giving reasons for differences in species diversity, as measured by the Simpson's index (page 139). For comparisons to be made between different areas using a diversity index, the same sampling method must be used and a similar type of habitat investigated (e.g. forest ecosystems in the same region). Diversity indices also work best when similar groups of organisms are compared (e.g. dung beetle communities

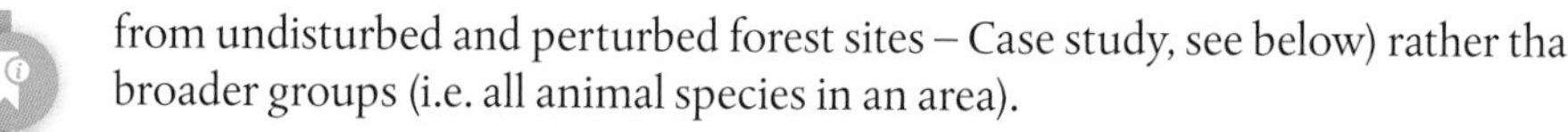

from undisturbed and perturbed forest sites – Case study, see below) rather than broader groups (i.e. all animal species in an area).

Values of *D* are relative to each other and not absolute, unlike measures of, for example, temperature which are on a fixed scale. This means that two different areas can be compared to each other using the index, but a value on its own is not useful. Individual values of *D* give an indication of the composition of the community being investigated (i.e. low values of *D* indicate low evenness, meaning one species may dominate the community) but do not, on their own, help in identifying areas of biodiversity that should be conserved.

The ability to assess changes to biodiversity in a given community over time is important in assessing the impact of human activity in the community.

You need to be able to discuss the usefulness of providing numerical values of species diversity to understanding the nature of biological communities and the conservation of biodiversity.

CONCEPTS: Biodiversity

Species diversity is a measure of the number of species in an area and their relative abundance.

Case study

Species richness and diversity of beetle communities following logging

A flight intercept trap in logged forest. Insects fly into the net and fall into aluminium trays where they are collected.

A study was carried out in the tropical rainforests of Borneo to investigate the effects on dung beetle communities of logging and conversion to plantation. The aim of the investigation was to understand the nature of biological communities in these forest ecosystems and to see the effect of human activities, and to establish the conservation value of these areas. Beetles were collected using a flight interception trap (page 132).

The results are shown in Table 3.1.

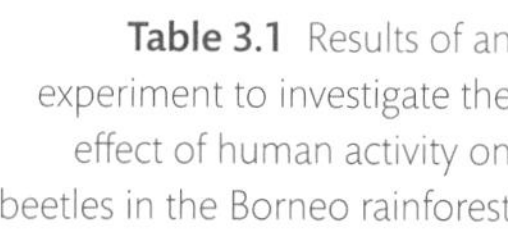

Table 3.1 Results of an experiment to investigate the effect of human activity on beetles in the Borneo rainforest

Trap location	Measurements of species richness and diversity		
	Species richness	Diversity	Evenness
primary forest	36	2.96	0.83
logged forest	42	2.24	0.60
plantation	14	2.05	0.78

Species richness is the number of species, diversity is measured using a diversity index, and evenness is a measure of how evenly (equally) abundance is distributed between species (an evenness value of 1 would indicate that all species are equally abundant). Primary forest is forest that is pristine and has not been affected by human activities.

Species richness is highest in logged forest: this is because disturbed forest contains a mixture of species which are usually separated along environmental gradients and not found in one location in primary forest (e.g. riverine species and those found in the canopy move into logged areas). The species diversity in logged forest is lower than primary forest, indicating a simplified ecosystem where certain species dominate. This is indicated by a low evenness measure. Plantation forest has the lowest species richness and diversity, indicating a loss of primary forest species and a much simpler ecosystem compared to primary rainforest. This study indicates the dangers of only using species richness information to compare different areas: species diversity is a much more robust and accurate method of indicating the health, and therefore conservation value, of ecosystems.

Exercises

1. Define the terms *genetic diversity*, *species diversity*, and *habitat diversity*.
2. Explain how diversity indices can be used to measure the impact of human activities.
3. Discuss the usefulness of providing numerical values of species diversity to understanding the nature of biological communities and the conservation of biodiversity.

Big questions

Having read this section, you can now discuss the following big question:

- How are the issues addressed in this topic of relevance to sustainability or sustainable development?

Points you may want to consider in your discussions.

- What do diversity indices reveal about the state of an ecosystem?
- How can diversity indices be used to measure the impact of human disturbance on an ecosystem and assess whether it is sustainable or not?

3.2 Origins of biodiversity

Significant ideas

Evolution is a gradual change in the genetic character of populations over many generations achieved largely through the mechanism of natural selection.

Environmental change gives new challenges to species, which drives evolution of diversity.

There have been major mass extinction events in the geological past.

Big questions

As you read this section, consider the following big questions:

- How are the issues addressed in this topic of relevance to sustainability or sustainable development?
- In what ways might the solutions explored in this topic alter your predictions for the state of human societies and the biosphere some decades from now?

Knowledge and understanding

- Biodiversity arises from evolutionary processes.
- Biological variation arises randomly and can either be beneficial to, damaging to, or have no impact on the survival of the individual.
- Natural selection occurs through the following mechanism:
  - within a population of one species there is genetic diversity, which is called variation
  - due to natural variation some individuals will be fitter than others
  - fitter individuals have an advantage and will reproduce more successfully
  - the offspring of fitter individuals may inherit the genes that give the advantage.

- This natural selection will contribute to evolution of biodiversity over time.
- Environmental change gives new challenges to the species, those that are suited survive, and those that are not suited will not survive.
- Speciation is the formation of new species when populations of a species become isolated and evolve differently.
- Isolation of populations can be caused by environmental changes forming barriers such as mountain building, changes in rivers, sea level change, climatic change or plate movements. The surface of the Earth is divided into crustal/tectonic plates which have moved throughout geological time. This has led to the creation of both land bridges and physical barriers with evolutionary consequences.
- The distribution of continents has also caused climatic variations and variation in food supply, both contributing to evolution.
- Mass extinctions of the past have been caused by different factors such as tectonic movements, super-volcanic eruption, climatic changes (including drought and ice ages), and meteor impact, which resulted in new directions in evolution and therefore increased biodiversity.

Evolution is the cumulative, gradual change in the genetic composition of a species over many successive generations, ultimately giving rise to species different from the common ancestor.

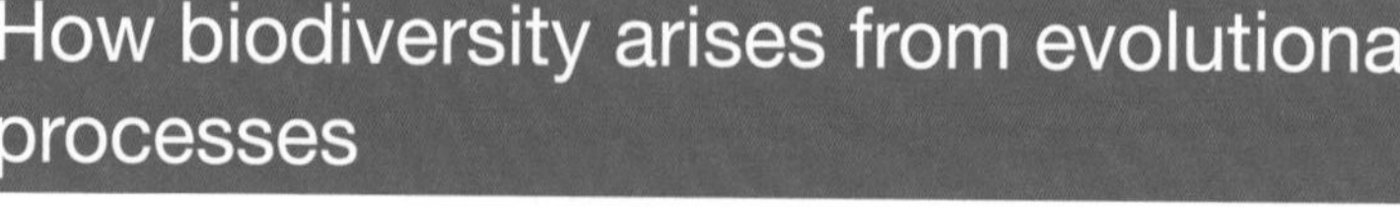

How biodiversity arises from evolutionary processes

Looking at the great diversity of life on Earth, one important question is, how has this biodiversity arisen? The answer lies in the theory of evolution, which describes how species change gradually over many years from ancestral species into entirely new species. A *common ancestor* is the most recent species from which two or more now different species have evolved (humans and chimpanzees, for example, share a common ancestor of some 6 million years ago). Biodiversity arises from evolutionary processes.

Charles Darwin about 20 years after the voyage of HMS *Beagle*. He was in his forties and accumulating evidence in support of his theory of evolution.

Evolution, the development of new species over very long periods of geological time (millions of years), has been accepted by scientists for many years. Evidence is found by examination of the fossil record: older rocks contain fossils of simpler forms of life, more recent rocks contain fossils of more complex life forms. However, the explanation of how evolution actually occurred took longer to work out, and was finally described by Charles Darwin in his book *On the Origin of Species* in 1859. This is one of several theories of evolution but is the only one that is now widely recognized within the scientific community, and has survived the test of time.

Natural selection

Pages from one of Darwin's notebooks, in which he first outlined his ideas on evolution by natural selection

Darwin made a 5-year trip on HMS *Beagle* between 1831 and 1836. The aim of the expedition was to map the coasts and waters of South America and Australia. Darwin was on board as a companion to the captain, but Darwin was also a talented and curious naturalist. During the trip, he was exposed to some of the most diverse ecosystems on Earth (the rainforests of South America), and the Galápagos Islands of the west coast of South America. It was essentially the interrelationship between species and environment in the Galápagos Islands that stimulated Darwin to produce his theory of evolution. Darwin noted that:

- all species tend to over-reproduce
- this leads to competition for limited resources (a 'struggle for existence')
- species show variation (all individuals are not alike, they have subtle differences in appearance or behaviour).

From this Darwin concluded that:

- those best adapted to their surroundings survive
- these can then go on to reproduce.

We now know that variation is caused by genetic diversity: changes in the gene pool of a species arise through mutations (changes in the genetic code) and sexual reproduction. Biological variation arises randomly and can either be beneficial to, damaging to, or have no impact on the survival of the individual. Beneficial change to the gene pool of a species can lead to increased chances of survival and the ability to pass on the same genetic advantage to the next generation. Survival has a genetic basis – nature selects the individuals possessing what it takes to survive. This means successful genes are selected and passed on to the next generation. Over time, a change in the species' gene pool takes place, and such changes ultimately lead to new species. Where changes to the genetic code lead to non-beneficial effects, such as the development of a genetic disease (e.g. cystic fibrosis – a disease that affects the lungs and digestive system), the affected genes can still be passed down through the generations but offer no adaptive advantage. Should such genes be distinctly harmful, individuals with them may die before they can reproduce. Some variation has no effect on the survival of a species (it is said to be neutral).

Darwin called the process **natural selection** because nature does the choosing, as opposed to artificial selection (selective breeding), a common practice in which humans choose animals or plants to breed together based on desirable characteristics. It is selective breeding that has led to all the varieties of domestic and agricultural animals we have today. Over millennia, ,the result of natural selection is not just new varieties but new species. The process of natural selection contributes to the evolution of biodiversity over time.

Darwin collected huge numbers of animals, plants and fossils during his trip on HMS *Beagle*. Many of these are now in the Natural History Museum, London. It was after his return to the UK, and after he had time to examine his specimens, that Darwin began to develop his theory.

He was particularly influenced by specimens of three species of mockingbird from the Galápagos Islands. He noticed that each species was from a different island and each was specifically adapted (in body size and beak shape) to conditions on its island. This led him to start to think that, rather than each species being created separately (as was widely thought at the time), perhaps all were related to a common ancestor from the South American mainland. Moreover, perhaps each had evolved through the process of natural selection to become adapted to different niches on different islands. The mockingbirds are believed to have had a more important role in the development of Darwin's initial ideas than the famous Galápagos finches. In *The Voyage of the Beagle* (1839), Darwin wrote: 'My attention was first thoroughly aroused by comparing together the various specimens ... of the mocking-thrush.'

To learn more about the work of Charles Darwin, go to www.pearsonhotlinks.co.uk, enter the book title or ISBN, and click on weblink 3.1.

Genes are sections of DNA found in the nucleus of all cells. They are essentially the instructions from which a species is produced. Gene pool refers to all the different types of gene found within every individual of a species.

Biogeography is the study of the geographical distribution of species, and explains their current distribution using evolutionary history. Once the historical factors that have been involved in shaping biodiversity are understood, scientists can better predict how biodiversity will respond to our rapidly changing world (e.g. as a result of climate change).

Natural selection occurs through the following mechanism:

- **within a population of one species there is genetic diversity, which is called variation**
- **because of natural variation, some individuals will be fitter than others**
- **fitter individuals have an advantage and will reproduce more successfully**
- **the offspring of fitter individuals inherit the genes that give the advantage; these offspring therefore survive and pass on the the genes to subsequent generations.**

The first publication of the theory of evolution by natural selection was not in 1859 in Darwin's *On the Origin of Species*, but in 1858 in a joint publication by Darwin and Wallace in *Proceedings of the Linnaean Society of London*, following a presentation of their findings at the Society earlier that year.

CONCEPTS: Biodiversity

Biodiversity has occurred through the process of evolution by natural selection.

In other books you will find accounts referring to Darwin's finches. The Galápagos Islands are home to 12 species of finch, each clearly adapted to its specific island's type of vegetation. Although Darwin collected specimens of the finches, he did not label them with the locations where they were found. So, he paid them little attention until he was certain that his three mockingbirds were indeed different species. Fortunately, other finches, which had been collected by members of *HMS Beagle*'s crew, had been labelled with the islands on which they were found. So, the finches were after all able to play a useful back-up role in Darwin's conclusion that new species can develop.

Different species of giant tortoise are found on different islands of the Galápagos, each adapted to local conditions. (a) On islands with tall vegetation, saddle-shaped shell fronts enable the animals to stretch up and reach the plants. (b) Animals with domed-shaped shell fronts are found on islands where vegetation is common on the ground.

In 1858, Charles Darwin unexpectedly received a letter from a young naturalist, Alfred Russel Wallace. Wallace outlined a remarkably similar theory of natural selection to Darwin's own. Wallace had come up with the idea while travelling in South East Asia. The men had developed the same theory independently. Why was this possible? Common experiences seem to have been crucial, and both had read similar books. For two individuals to arrive at one of the most important theories in science independently and at the same time is remarkable.

Portrait of Alfred Russel Wallace at the Natural History Museum, London.

To learn more about Richard Dawkins, an advocate of Darwin's theory, go to www.pearsonhotlinks.co.uk, enter the book title or ISBN, and click on weblink 3.2.

The theory of evolution by natural selection tells us that change in populations is achieved through the process of natural selection. Is there a difference between a convincing theory and a correct one?

CONCEPTS: Environmental value systems

The evidence for Darwin's theory is overwhelming. Despite this, some people (creationists) do not believe it to be correct. These people believe that the Genesis story in the Bible is literally true, with all life on Earth being created within 6 days. Scientific evidence strongly contradicts this version of events. Most religions accept Darwin's theory while maintaining a belief in a creator God. What do you think? Ultimately you must weigh your worldview with the scientific evidence and draw your own conclusions.

The role of isolation in forming new species

Mountain formation leads to evolution and increased biodiversity. Mountains (like the Eiger, Munch, and Jungfrau, in the Swiss Alps) form a physical barrier that isolates populations. The uplift can also create new habitats, with an increase in biodiversity due to populations adapting to new habitats through natural selection.

Natural selection is not, on its own, sufficient to lead to speciation; **isolation** is required. Populations must first become separated (i.e. isolated), one from the other, so that genes cannot be exchanged between them (this is **reproductive isolation**). If the environments of the isolated populations are different, natural selection will work on each population so that, through evolution, new species are formed (i.e. **speciation** occurs).

Speciation is the formation of new species when populations of a species become isolated and evolve differently.

The islands of the Galápagos are quite widely separated and very different from each other (Figure 3.3). This means that animal and plant populations which arrived from mainland South America (ancestral populations) became geographically isolated from each other. For example, an ancestral population of mockingbirds arriving from the mainland would have spread onto several different islands. As local environmental and biological conditions were different on each island, different species evolved to inhabit different ecological niches. The islands are 1060 km from the mainland and the distances between them are sufficiently large to make it difficult for the geographically isolated populations on different islands to interbreed. Thus gene flow (the exchange of genetic material through interbreeding) would be limited.

A species is a group of organisms that interbreed and produce fertile offspring. Sometimes, two species breed together to produce a hybrid, which is a sterile organism (pages 61–62).

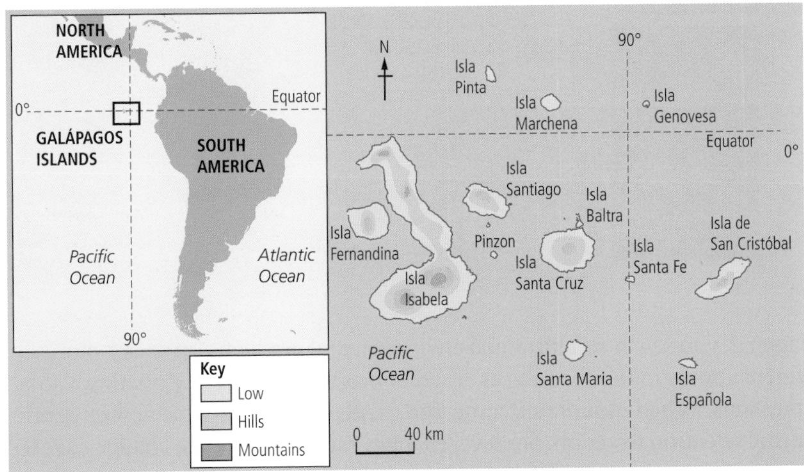

Figure 3.3 The Galápagos Islands

Geographic isolation is essential in the formation of new species. Without it, interbreeding would cause the genes from two populations to continue to mix (Figure 3.4) and characteristics of the ancestral species to remain.

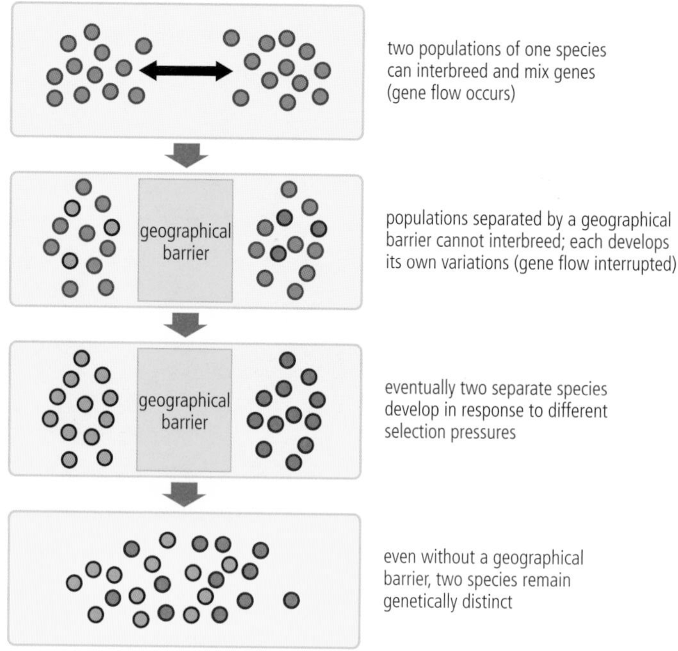
two populations of one species can interbreed and mix genes (gene flow occurs)
geographical barrier
populations separated by a geographical barrier cannot interbreed; each develops its own variations (gene flow interrupted)
geographical barrier
eventually two separate species develop in response to different selection pressures
even without a geographical barrier, two species remain genetically distinct

Figure 3.4 Geographical barriers include mountains, island formation, water (sea, river or lake), or hostile environments.

Geographical isolation is a physical barrier, such as a mountain range, that causes populations to become separated.

Case study

Speciation in spotted owls

Populations of the spotted owl in North America have become geographically separated over time, forming two varieties: the northern spotted owl and the Mexican spotted owl (Figure 3.5). Given enough time and continued isolation, these will eventually be unable to interbreed and produce fertile offspring. They will then be two separate species.

Northern spotted owl *Strix occidentalis caurina*

Mexican spotted owl *Strix occidentalis lucida*

Figure 3.5 The ranges of these two varieties of spotted owl do not overlap and they occupy different niches – geographical isolation means there is little gene flow between the two varieties.

In Chapter 2, you saw how altitudinal environmental gradients on rocky shores lead to different communities forming at different heights (page 113). Zonating also occurs on mountains. When mountains come into existence, they provide new environments for natural selection to act on. Sea level changes caused by climate change have led to higher altitude areas becoming isolated (as sea level rises), or have provided land-bridges for migration of species to new areas (when sea levels drop and once-separated areas of land join up. Environmental change produces new challenges to species: those that are suited survive, and those that are not suited become extinct. The same process takes place whenever environmental change occurs, whether it is by barrier formation (e.g. mountains; sea level change), climatic change, or movement of tectonic plates (page 162).

During the Pleistocene ice ages (which began 2.6 million years ago), a fall in sea levels (due to decrease in temperature and large amounts of water becoming locked up in ice caps and glaciers above sea level) led to a land bridge (Beringia) forming between previously separated Alaska and eastern Siberia (Figure 3.6). It is possible that the earliest human colonizers of the Americas entered from Asia via this route. Between about 17 000 and 25 000 years ago, the islands of South East Asia (Borneo, Java, and Sumatra) were connected to the mainland of Asia forming one land mass, which we call Sundaland (Figure 3.7). Again, this land bridge was caused by a drop in sea level due to climate change. In both cases, as sea levels rose again, the land bridges were lost and areas became isolated once more.

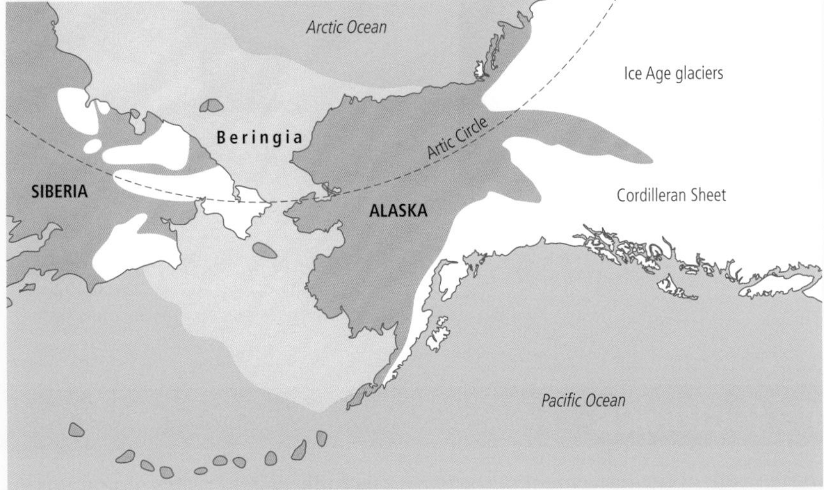

Figure 3.6 A land bridge formed between Siberia and Alaska during the Pleistocene ice age.

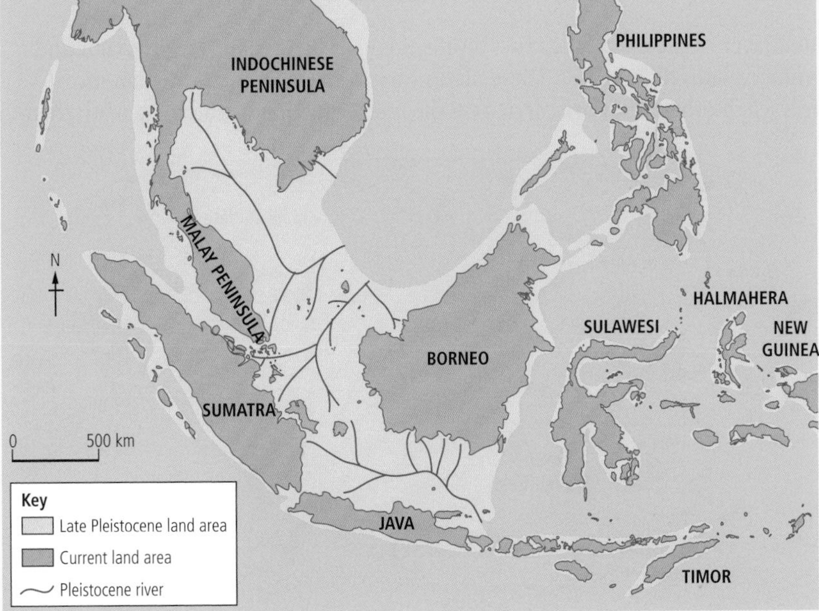

Figure 3.7 Lower sea levels during the late Pleistocene led to mainland Asia joining with the islands of Sumatra, Java, and Borneo.

Isolation of populations can be caused by environmental change such as mountain formation, change in river courses, sea level change, climatic change, or plate movements.

As well as geographical isolation, other isolating mechanisms exist that cause speciation. Speciation can take place in populations that are not separated by geographical barriers, and exist in the same location. For example, behavioural differences that emerge between populations can lead to reproductive isolation. Differences in courtship display in birds of paradise have led to the evolution of numerous species within the same forest: male birds of paradise have bright and colourful feathers which they use to attract females, and different species also have different dancing displays. Changes in the appearance or behaviour of populations may result in males and females of those populations no longer being attracted to each other and therefore not breeding together.

Male Raggiana bird of paradise displaying his plumage to a female. The male bird is brightly coloured but the female is plain. Alfred Russel Wallace was one of the first naturalists to observe these and other species of birds of paradise in the wild.

Ecological differences can also emerge between populations: for example, species may become separated along environmental gradients. In the rainforests of Borneo, a group of dung beetles have become adapted to living in the canopy, where environmental conditions are very different from those on the ground.

Dung beetles feeding on primate dung in the canopy of rainforest in Borneo. These beetles make a ball from the dung: a male and female here can be seen working together. This group is an example of isolation and speciation occurring within the same forest ecosystem as ground-living dung beetles.

Plate tectonics

The surface of the Earth is divided into tectonic plates, which have moved throughout geological time. This has led to the creation of both physical barriers and land bridges with evolutionary consequences.

Let's now consider how movement of the Earth's tectonic plates creates mountains and other phenomena that lead to the isolation of populations and speciation.

The outer layer of the Earth, the crust (lithosphere), is divided into eight major and many minor plates (Figure 3.8). These plates vary in size and shape but can move relative to each other. They are carried on the mantle (asthenosphere) beneath them,

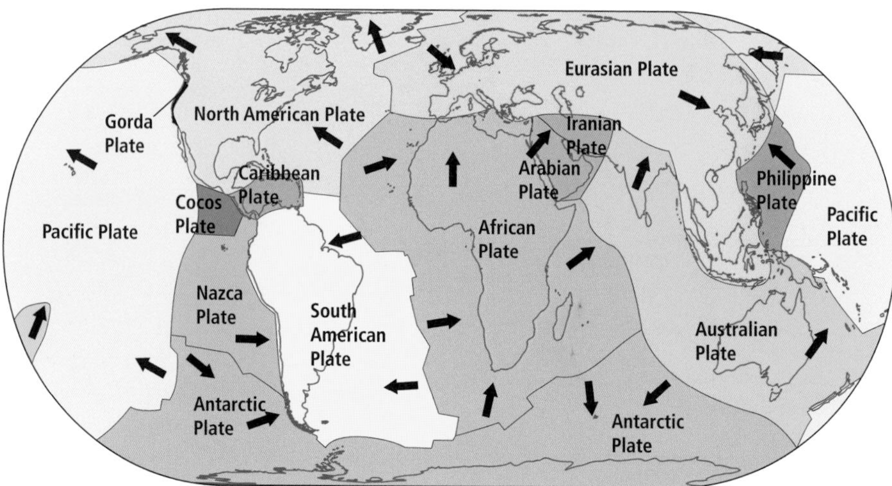

Figure 3.8 Earth's tectonic plates

which can flow like a liquid on geological time scales. The edges of adjacent plates can move parallel to each other, be pushed one under the other, or collide. Earthquakes, volcanoes, and mountain building occur at these boundaries. The movement and forming and reforming of these plates is known as **plate tectonics**.

During the Palaeozoic and Mesozoic eras (about 250 million years ago) all land mass on Earth existed as one supercontinent, Pangaea (Figure 3.9). This name is derived from the Greek for 'entire'. About 175 million years ago, the land mass split into two separate supercontinents, Laurasia and Gondwana. Laurasia contained land that became North America, Eurasia (Europe and Asia) and Greenland, and Gondwana contained the land that became South America, Africa, Australia, Antarctica, and India. The distribution of all extinct and extant (still living) species found in these geographical areas today can be explained in terms of these ancient land masses.

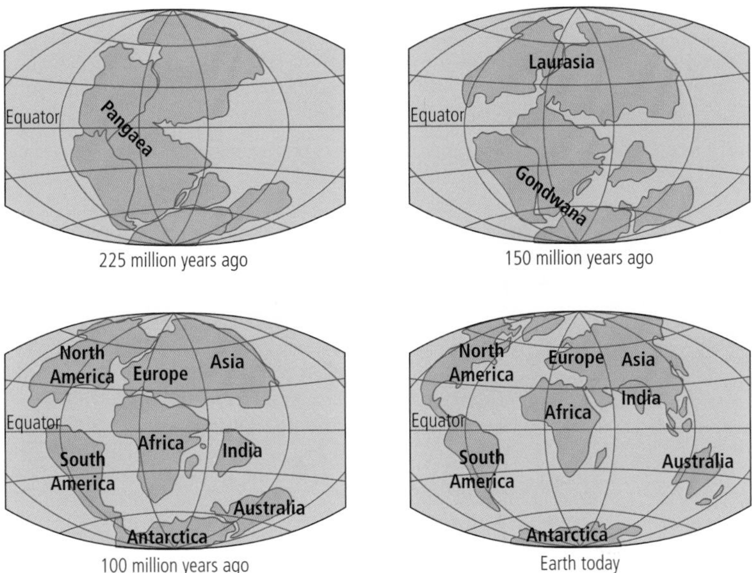

Figure 3.9 Continental drift from 225 million years ago to the present day

Movement of the tectonic plates can produce barriers such as mountain ranges, oceans, and rift valleys that lead to isolation of gene pools and then speciation. Movement apart of the plates can also lead to isolation and the development or preservation of unique species. For example, the separation of Australia led to the preservation of its distinctive flora and fauna (e.g. eucalypts, monotremes, and marsupials such as kangaroos). Similarly, Madagascar is the only place where lemurs are found today.

Formation of land bridges between previously separated plates can provide opportunities for species to spread from one area to another. For example, species from Australia spread onto new islands in Indonesia, and the similarity between caribou and reindeer (in Alaska and Siberia) suggests a common ancestry.

The movement of plates through different climatic zones allows new habitats to present themselves. For example, the northward movement of the Australian plate, and the subsequent drying of much of the continent, has provided changes in the selective forces on species leading to the evolution of drought-tolerant species. The distribution of continents has caused climatic variations and variation in food supply, both contributing to evolution.

Plate movement can generate new and diverse habitats, thus promoting biodiversity (Figures 3.10–3.14).

To learn more about plate tectonics and species diversity, go to www.pearsonhotlinks.co.uk, enter the book title or ISBN, and click on weblink 3.3.

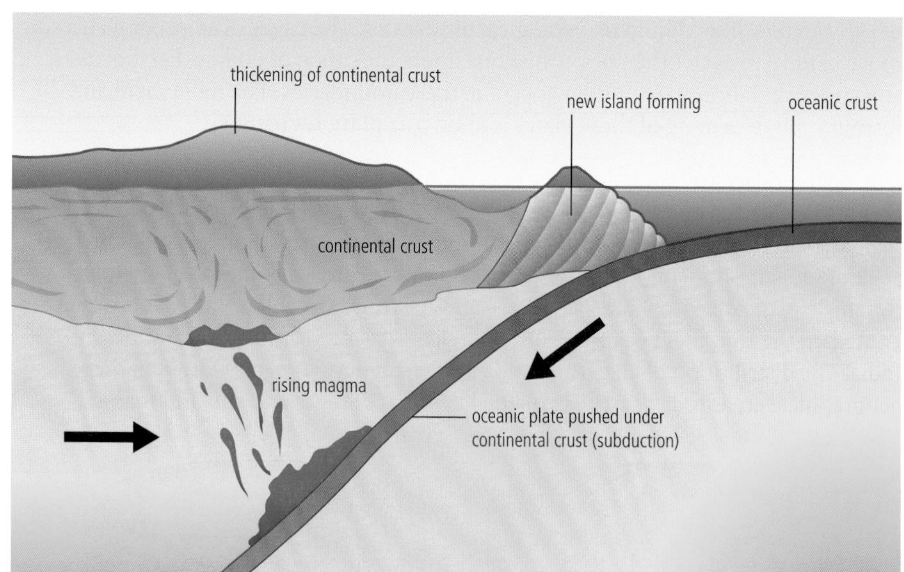

Figure 3.10 Subduction of denser oceanic crust beneath the less dense continental crust. This can lead to new island arcs (e.g. New Zealand, where the Pacific plate is being subducted under the Indian/ Australian plate), and mountain areas where magma rises up from under the subduction area causing volcanic action and thickening of the crust (e.g. the Andes of South America and the Cascade Range of north-western USA).

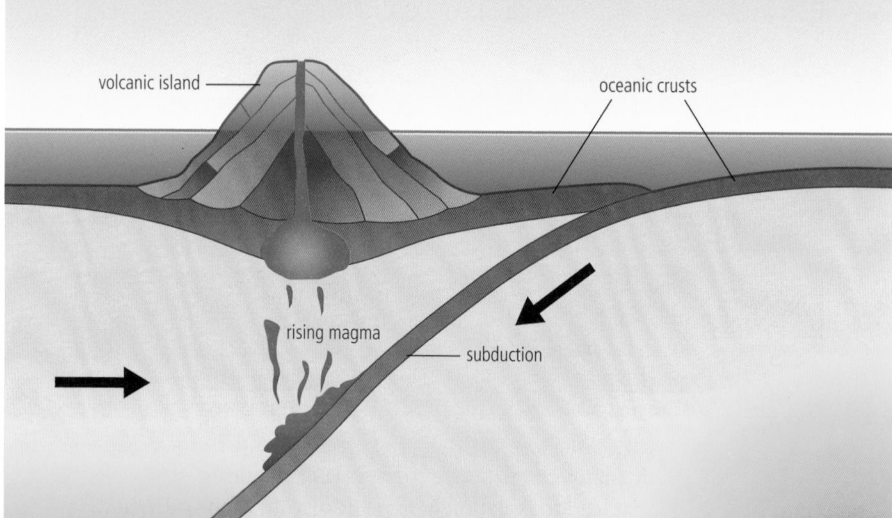

Figure 3.11 Oceanic crust is subducted beneath oceanic crust – as both are the same density, the effect is different from that in Figure 3.10. Resulting volcanic activity from rising magma causes new islands to form, with new habitats providing possibilities for speciation. Japan, the Philippines, the Aleutians of Alaska, and the Leeward Islands of the Caribbean were all created in this way.

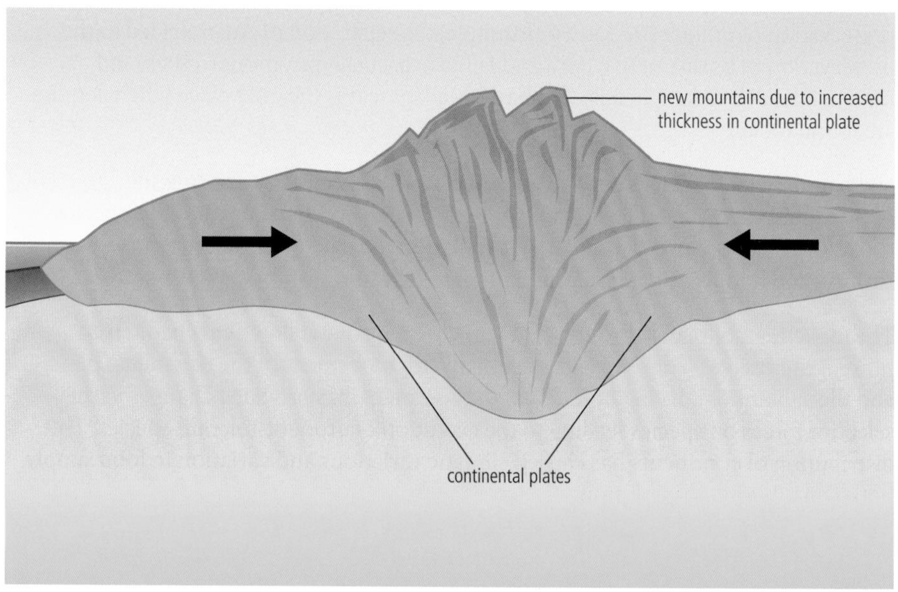

Figure 3.12 Continental plates colliding. This leads to an increase in continental plate thickness and eventually to new mountain ranges (e.g. the Himalayas, where the Indian plate is being pushed against the large Asian plate). Creation of new habitats at different altitudes adds to the biodiversity of the region.

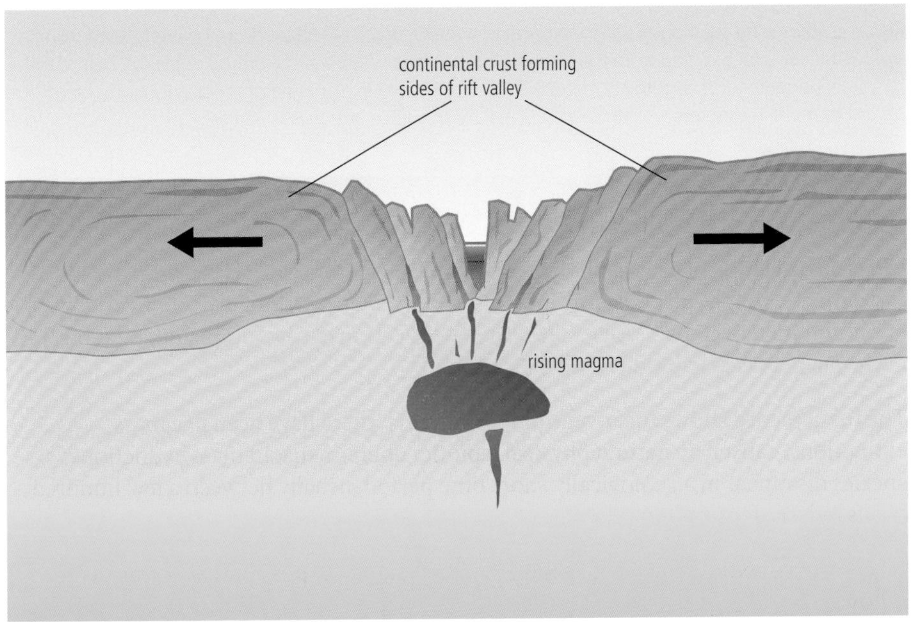

Figure 3.13 Continental plates moving apart cause rift valleys. Deep lakes may form in these valleys (e.g. Lake Tanganyika and Lake Victoria in the East African rift valley and the world's deepest lake, Lake Baikal, in Siberia). Given time, new seas may form – The Red Sea, which separates Africa and Saudi Arabia, is an example. The creation of new aquatic habitats drives speciation in these rift areas. Magma rising from the rift can stick to the separating plates creating new land (e.g. Iceland, Ascension Island, the Azores, and Tristan de Cunha in the Atlantic) again creating new opportunities for species evolution.

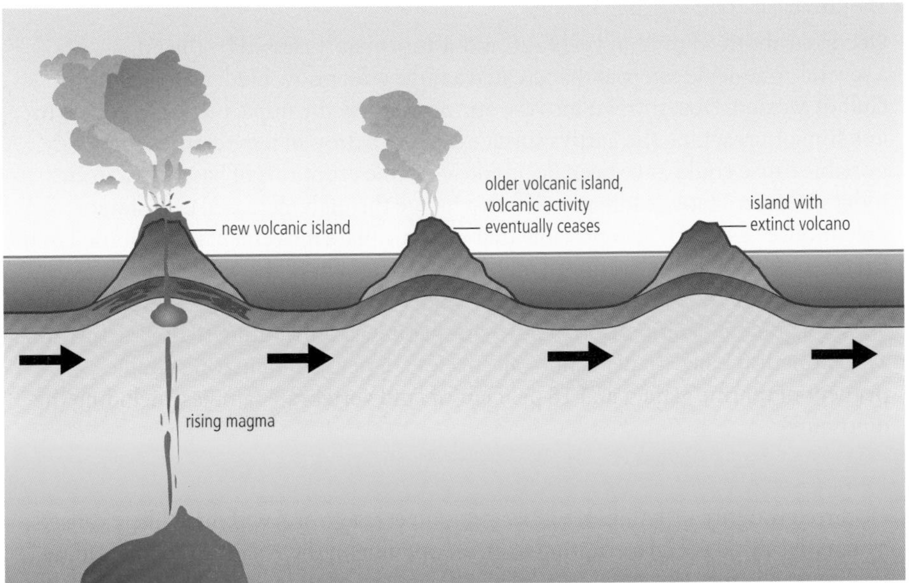

Figure 3.14 In some areas, hot rock rises from deep in the mantle and breaks through the oceanic crust. These hotspots are not caused at plate edges but by movement of plates over areas where magma rises. The hotspots can create chains of islands (e.g. Galápagos Islands and Hawaii). As Darwin found, the creation of volcanic islands and their colonization by animals and plants that become adapted to local conditions can lead to increased regional diversity.

CONCEPTS: Biodiversity

The movement of the Earth's plates has caused isolation of populations, climatic variations, and changes in food supply, contributing to evolution and new biodiversity.

Mount Everest, the world's tallest mountain at nearly 9000 m, has been created over 40 million years by the collision between the Indian and Eurasian plates. The rocks on the summit are 50 million-year-old limestones that formed in the shallow waters of an ocean that once lay between India and Asia. The Himalayas are a physical barrier that has led to the separation of populations and the evolution of new species: this is true of mountain building in general.

You need to be able to explain how plate activity has influenced evolution and biodiversity.

Evidence for the role of plate tectonics in contributing to speciation can be interpreted from the fossil record, and from the current distribution of organisms around the planet. Before the continental drift hypothesis, there was no satisfactory explanation of the distribution of life forms. Why, therefore, didn't scientists establish such a hypothesis earlier?

A mass extinction is a period in which at least 75 per cent of the total number of species on the Earth at the time are wiped out.

Mass extinctions

Mass extinctions of the past have been caused by factors such as tectonic movements, super-volcanic eruptions, climatic changes (including drought and ice ages), and meteor impact. All resulted in new directions in evolution and, therefore, an eventual increased in biodiversity.

The fossil record shows that over millions of years, there have been five mass extinctions, caused by natural physical (abiotic) phenomena. In mass extinctions, species disappear in a geologically short time period, usually between a few hundred thousand to a few million years.

The five mass extinctions in the geological past, and their possible causes, are as follows.

- *The Cretaceous–Tertiary extinction*
  Occurred about 65 million years ago and was probably caused by the impact of a several-mile-wide asteroid that created a huge crater now hidden beneath the Gulf of Mexico. Dust thrown into the atmosphere by the impact would have led to less sunlight reaching the Earth's surface causing a drop in temperature. Another possible cause could have been flood-like volcanic eruptions of basalt lava from India's Deccan Traps, leading to climate change through increased emission of greenhouse gases. It is also possible that tectonic plate movements contributed to the Cretaceous–Tertiary extinction: a major rearrangement of the world's landmasses caused by plate movement would have resulted in climatic changes that could have caused a gradual deterioration of dinosaur habitats, contributing to their extinction. The extinction killed 76 per cent of all species (16 per cent of marine families, 47 per cent of marine genera and 18 per cent of land vertebrate families, including the dinosaurs).
- *The End Triassic extinction*
  Occurred roughly 199 million to 214 million years ago and was most likely caused by massive floods of lava erupting from an opening in the Atlantic Ocean, leading to climate change. The extinction killed 80 per cent of all species (23 per cent of all families and 48 per cent of all genera).
- *The Permian–Triassic extinction*
  Occurred about 251 million years ago, and was the largest of these events. It is suspected to have been caused by a comet or asteroid impact, although direct evidence has not been found. Others believe the cause was flood volcanism (as with the End Triassic extinction) from the Siberian Traps, which destroyed algae and plants, reducing oxygen levels in the sea. Some scientists believe that plate movement may have contributed to the Permian extinction. The joining together of all the land masses to create the supercontinent Pangaea (page 163), which occurred sometime before the Permian extinction, would have led to environmental change on the new land mass, especially in the interior which would have become much drier. The new landmass also decreased the quantity of shallow seas and exposed formerly isolated organisms of the former continents to increased competition. Pangaea's formation would have altered oceanic circulation and atmospheric weather patterns, creating

seasonal monsoons. Pangaea formed millions of years before the Permian extinction, however, and the very gradual changes that are caused by continental drift, are unlikely, on their own, to have led to the simultaneous loss of both terrestrial and oceanic life on the scale seen.

The Permian extinction wiped out 96 per cent of all species (53 per cent of marine families, 84 per cent of marine genera and an estimated 70 per cent of land species such as plants, insects, and vertebrate animals: in total, 57 per cent of all families and 83 per cent of all genera).

- *The Late Devonian extinction*
  Occurred about 364 million years ago, caused by global cooling (followed by global warming), linked to the diversification of land plants (causing less CO_2 in atmosphere and therefore lower levels of greenhouse gases). The extinction killed 75 per cent of all species (19 per cent of all families and 50 per cent of all genera)
- *The Ordovician–Silurian extinction*
  Occurred about 439 million years ago, was caused by a drop in sea levels as glaciers formed, then by rising sea levels as glaciers melted. The extinction killed 86 per cent of all species (27 per cent of all families and 57 per cent of all genera).

The average time between these mass extinctions is around 100 million years. The exception is the gap between the Permian–Triassic and the End Triassic extinctions, which were approximately 50 million years apart.

CONCEPTS: Biodiversity

Mass extinctions have led to initial massive reductions in the Earth's biodiversity. These extinction events have resulted in new directions in evolution and therefore increased biodiversity in the long term.

Although the mass extinction events led to a massive loss of biodiversity, with less than 1 per cent of all species that have ever existed still being alive today, they ultimately led to new biodiversity evolving (Figure 3.15). The large-scale loss of species led to new opportunities for surviving populations, with many groups undergoing adaptive radiation (where an ancestral species evolves to fill different ecological niches, leading to new species).

A 6th mass extinction?

The Earth is believed to be currently undergoing a sixth mass extinction, caused by human activities (biotic factors). If this is the case, it is the first extinction event to have biotic, rather than abiotic causes. The difference between abiotic and biotic factors is important, and represents a significant shift in the cause of extinction.

The sixth extinction can be divided into two discrete phases:

- phase 1 began when the first modern humans began to disperse to different parts of the world about 100 000 years ago
- phase 2 began about 10 000 years ago when humans turned to agriculture.

The development of agriculture and the clearance of native ecosystems accelerated the pace of extinction. Mass extinctions of the past took place over geological time, which allowed time for new species to evolve to fill the gaps left by the extinct species. Current changes to the planet are occurring much faster, over the period of human lifetimes. Over-population, invasive species, and over-exploitation are fuelling the extinction. Pollution and the advent of global warming (Chapters 6 and 7) are also accelerating changes to the planet and increasing extinction rates in species that cannot adapt to the changing conditions or migrate to new areas. Some scientists have predicted that 50 per cent of all species could be extinct by the end of the 21st century.

You need to be able to discuss the causes of mass extinctions.

TOK

We can never know for sure what has caused past extinctions. Scientists can only look at the fossil record and the geology of the Earth and draw conclusions from them. Does this lack of experimental evidence limit the validity of the conclusions drawn?

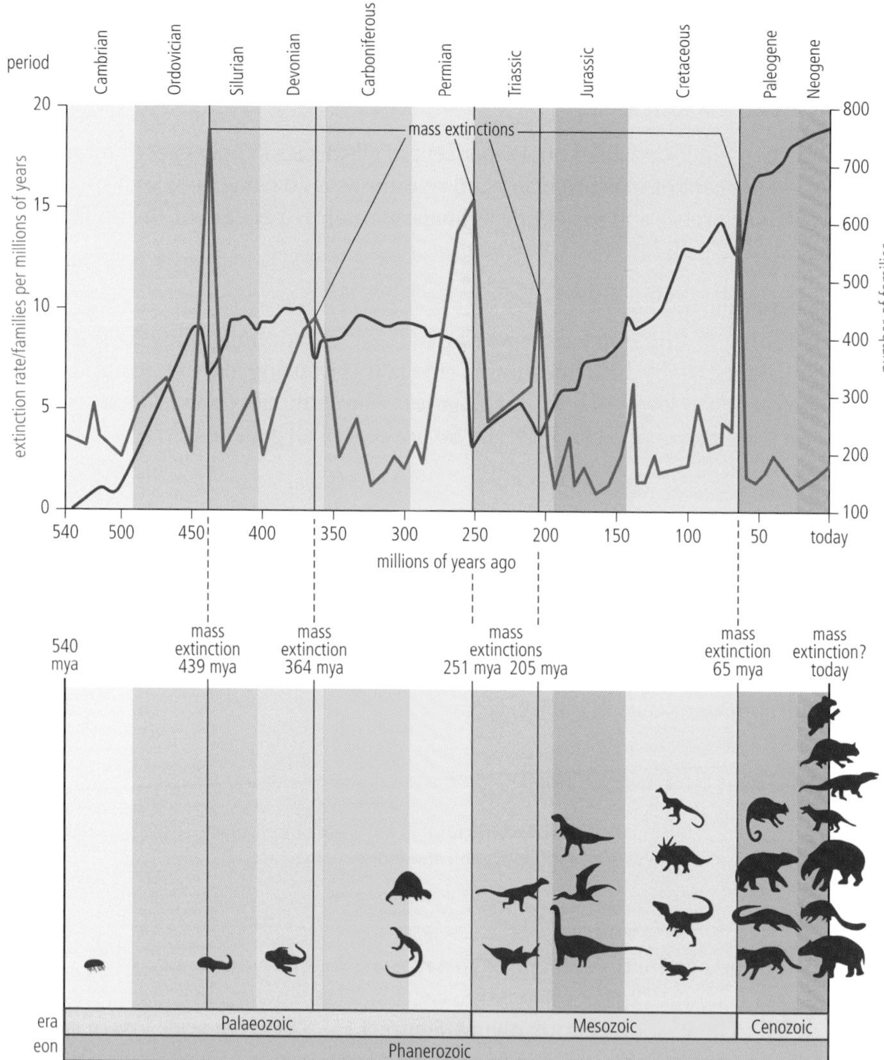

Figure 3.15 The evolution of life – and the mass extinctions that have wiped out 99 per cent of all species that have ever existed on Earth

Exercises

1. Darwin's theory of evolution is explained in terms of natural selection. What is natural selection, and how does it lead to the generation of new species?
2. Isolation mechanisms are essential for the generation of new species. How does the isolation of populations lead to speciation?
3. Name three different ways that the Earth's plates interact. How does each lead to speciation events?
4. Outline how land bridges have contributed to the current distribution of species. Give one example and say how and when land bridges affected species distribution patterns.
5. How many mass extinctions have there been in the past? What was the cause of these extinctions?

Big questions

Having read this section, you can now discuss the following big questions:

- How are the issues addressed in this topic of relevance to sustainability or sustainable development?
- In what ways might the solutions explored in this topic alter your predictions for the state of human societies and the biosphere some decades from now?

Points you may want to consider in your discussions:

- Unsustainable development can lead to species extinction. Given the five mass extinctions of the past, is this something that the human race should be concerned about?
- What effects could species extinctions have on human societies in years to come?

3.3 Threats to biodiversity

Significant idea

Global biodiversity is difficult to quantify but is decreasing rapidly due to human activity. Classification of species conservation status can provide a useful tool in conservation of biodiversity.

Big questions

As you read this section, consider the following big questions:

- To what extent have the solutions emerging from this topic been directed at preventing environmental impacts, limiting the extent of the environmental impacts, or restoring systems in which environmental impacts have already occurred?
- How are the issues addressed in this topic of relevance to sustainability or sustainable development?
- In what ways might the solutions explored in this topic alter your predictions for the state of human societies and the biosphere some decades from now?

Knowledge and understanding

- Estimates of the total number of species on the planet vary considerably. They are based on mathematical models, which are influenced by classification issues and lack of finance for scientific research, so many habitats and groups are significantly under-recorded.
- The current rates of species loss are far greater now than in the recent past, due to increased human influence. The human activities that cause species extinctions include habitat destruction, introducing invasive species, pollution, overharvesting, and hunting.
- The International Union of Conservation of Nature (IUCN) publishes data in the Red List of Threatened Species in several categories. Factors used to determine the conservation status of a species include: population size, degree of specialization, distribution, reproductive potential and behaviour, geographic range and degree of fragmentation, quality of habitat, trophic level, and the probability of extinction.
- Tropical biomes contain some of the most globally biodiverse areas and their unsustainable exploitation results in massive losses in biodiversity and their ability to perform globally important ecological services.
- Most tropical biomes occur in LEDCs and, therefore, there is conflict between exploitation, sustainable development and conservation.

Estimates of the total number of species on the planet vary considerably. They are based on mathematical models, which are influenced by classification issues and lack of finance for scientific research, so many habitats and groups are significantly under-recorded.

How many species are there on Earth?

There are approximately 1.8 million described species stored in the world's museums. The actual number of species on the planet will be much larger than this, although the real figure can only be guessed at currently. It is impossible to get an accurate count on the number of species because the majority of the species that have yet to be discovered and described are very small: insects, and bacteria and other microbes.

Estimates of global species numbers are based on mathematical models that extrapolate from known information. Lack of exploration of the deep sea and rainforest canopies, for example, means that knowledge of the total number of species on Earth is poorly understood, although estimates give an indication of the possible scale. Estimates range from 5 million to 100 million, with the scientific consensus currently being around 9 million species. This estimate is broken down as follows:

- animals: 7.77 million (12 per cent of which are described)
- fungi: 0.61 million (7 per cent of which are described)
- plants: 0.30 million (70 per cent of which are described)
- other species: 0.07 million.

Most described species belong to groups that have been studied extensively in the past – these tend to be the larger organisms (e.g. mammals, birds, flowering plants). Scientists have also focused on what they see as more appealing groups (e.g. those with fur or feathers). Smaller species that are more difficult to identify and study are less well represented, including some of the most species-diverse groups on the planet (insects, spiders, bacteria, fungi, etc.). Funds for taxonomic work (i.e. research into classifying organisms) in natural history museums and universities are generally limited. The lack of finance for scientific research, in terms of collecting specimens from the more inaccessible regions of the Earth and the necessary work needed to identify new species, means that many habitats and groups are significantly under-recorded. Estimations of total species numbers (and current extinction rates) are therefore based on limited data.

To learn more about the diversity of life on Earth through the Natural History Museum website, go to www.pearsonhotlinks.co.uk, enter the book title or ISBN, and click on weblink 3.4.

CONCEPTS: Biodiversity

The total number of species on the planet is unknown. Many areas remain unexplored, and research funding is limited.

Only 1 per cent of described species are vertebrates (Figure 3.16), yet this is the group that conservation initiatives are often focused on.

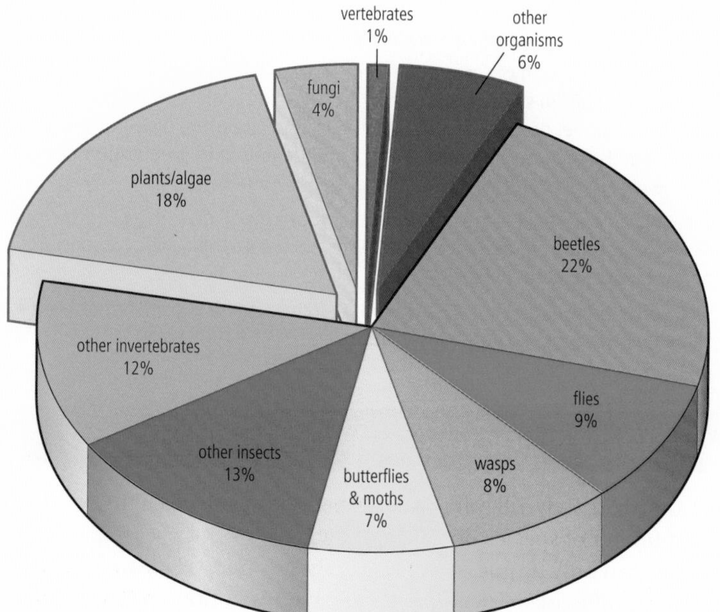

Figure 3.16 Of the total number of described species (about 1.8 million), excluding microbes, over three-quarters are invertebrates. Over half are insects. The most successful group are the beetles, which occupy all ecosystems apart from oceanic ones.

What are the current rates of species loss?

In order to understand how many species are currently going extinct, existing species must be identified and named. Experts who study specific groups of organisms (e.g. moths, beetles, and birds) are found in centres of excellence around the world, as are reference collections. For taxonomy to succeed, scientists from around the world must work together, and major surveys must be carried out using international teams of specialists.

Estimates of extinction rates are varied, but current extinction rates are thought to be between 100–10 000 times greater than background rates. Estimates range from 30 000 to 60 000 species a year.

We know that mass extinction events have happened in the past, but what do we know of current extinction rates? Throughout the history of the Earth, diversity has never remained constant; there have been a number of natural periods of extinction and loss of diversity. More recently, humans have played an increasing role in diversity loss, especially in biodiverse ecosystems such as rainforests and coral reef.

The background (natural) level of extinction known from the fossil record is between 10 and 100 species per year. Human activities have increased this rate. Because the total number of classified species is a small fraction of the estimated total of species, estimates of extinction rates are also varied. Estimates from tropical rainforest suggest the Earth is losing 27 000 species per year from those habitats alone. The rate of extinction differs for different groups of organisms, but examining the figures for one group (mammals) gives an indication of the extent of the problem. Mammal species have an average species lifespan, from origin to extinction, of about 1 million years. There are about 5000 known mammalian species alive at present. The background extinction rate for this group should be approximately one species lost every 200 years. Yet the past 400 years have seen 89 mammalian extinctions, almost 45 times the predicted rate, and another 169 mammal species are listed as critically endangered.

Causes of species loss

Natural causes

Natural hazard events such as volcanoes, drought, ice ages, and meteor impact have led to periods of loss of diversity. The eruption of Krakatau caused a dust plume that reduced sunlight over large areas of the globe, reducing surface temperatures. Changes in the Australian climate through tectonic movement and global warming have caused increased frequency of fires and a general drying of the continent that have led to the prevalence of drought and fire-tolerant species (e.g. *Acacia* and *Eucalyptus*) and the extinction of other species.

Changes in the orbit of the Earth and its tilt, plus tectonic movement, have led to repeated long-term cold periods (Figure 3.17),

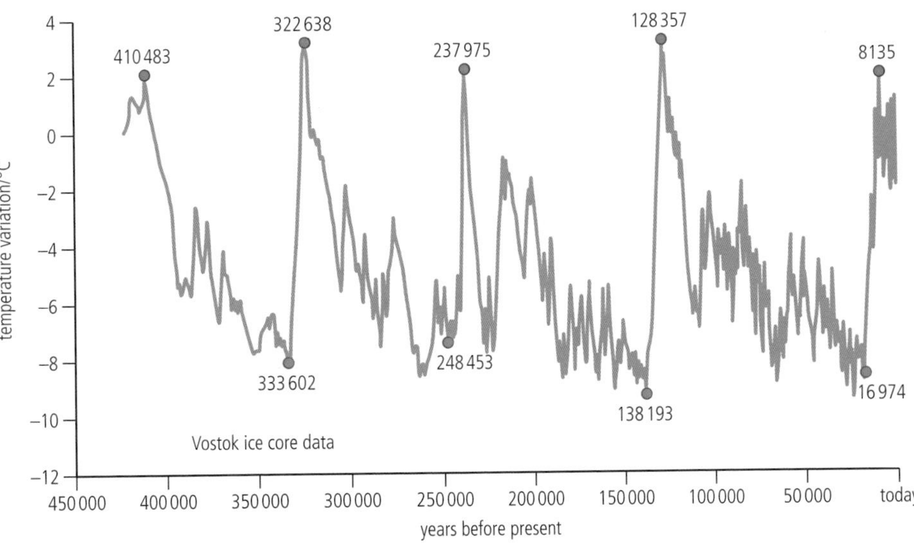

Figure 3.17 Variation in the temperature of the Earth taken using data from ice in the Antarctic. Major ice ages have occurred about every 100 000 years.

which have resulted in the selection of species adapted to the colder conditions and extinction of less-adapted species. One reason for the success of mammals is their ability to generate their own heat and control their temperature, which has enabled them to survive in colder environments and through ice ages.

Human causes

The current rates of species loss are far greater than in the recent past, due to increased human influence. The human activities that cause species extinctions include habitat destruction, introduction of invasive species, pollution, overharvesting, and hunting.

CONCEPTS: Biodiversity

Global biodiversity is difficult to quantify but is decreasing rapidly due to human activity.

Habitat destruction

This includes habitat degradation, fragmentation, and loss. Agricultural practices have led to the destruction of native habitats and replaced them with monocultures (i.e. crops of only one species). Monocultures represent a large loss of diversity compared to the native ecosystems they replace. However, increasing awareness of this has led to the re-establishment of hedgerows and undisturbed corridors that encourage more natural communities to return.

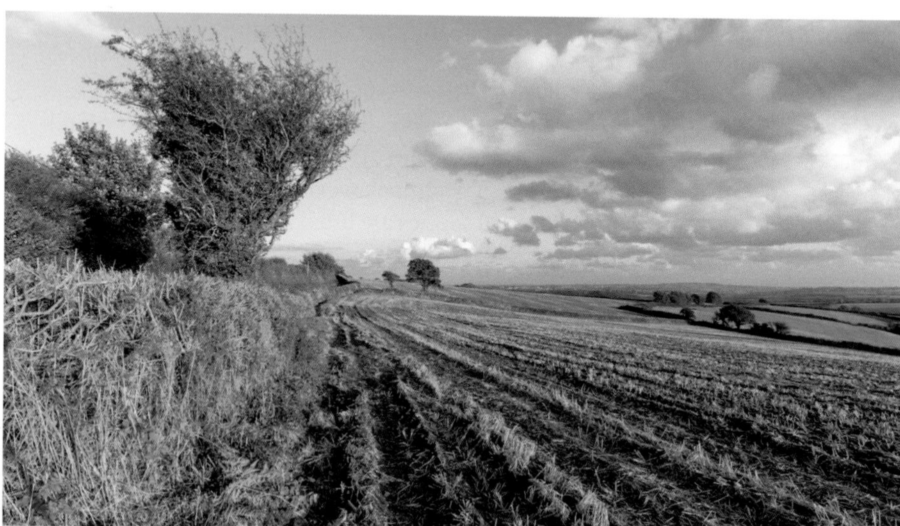

Hedgerows (left of photo) provide habitats for native species. They also act as corridors for the movement of species from one area to another.

Non-specific pesticides used in agriculture can wipe out native as well as imported pest species (i.e. alien species that have been introduced into a country), which again leads to an overall loss of diversity.

Habitats can be lost through mining activities. Mobile phones contain an essential element (tantalum) which is obtained by mining coltan (a metallic ore that contains the elements niobium and tantalum). Coltan is found mainly in the eastern regions of the Democratic Republic of Congo – mining activities in these areas have led to extensive habitat destruction of forests that contain gorillas and other endangered animals.

Natural habitats have also been cleared to make way for plantation crops. Sugar plantations have replaced tropical forest ecosystems, such as mangrove in Australia (page 185), and oil palm plantations throughout South East Asia have led to the widespread loss of tropical forests (page 174).

Introduction of invasive species

Species that are introduced to areas and compete with endemic (native) species: this can lead to the extinction of the native species. The grey squirrel was introduced into the UK from North America. This species competes with native red squirrel and has led to such a reduction in red squirrel numbers that the animal is now rare. Introduced red-clawed signal crayfish (*Pacifastacus leniusculus*), a large, aggressive American species, has wiped out almost 95 per cent of the native UK white-clawed species (*Austropotamobius pallipes*) since its introduction in the late 1970s.

Sea polluted with plastic garbage

Pollution

Pollution includes chemicals, litter, nets, plastic bags, oil spills, and so on. Pollution damages habitats and kills animals and plants, leading to the loss of life and reduction in species' population numbers.

Overharvesting and hunting

Animals are hunted for food, medicines, souvenirs, fashion, and to supply the exotic pet trade. Overharvesting of North Atlantic cod in the 1960s and 1970s led to significant reduction in population number (Chapter 4, page 240).

TOK

Why do people continue to damage the environment even when they know the effects on natural systems? Do people truly understand the consequences of their actions?

Threats to tropical biomes

Tropical biomes contain some of the most globally biodiverse areas and their unsustainable exploitation results in massive losses in biodiversity and their ability to perform globally important ecological services.

A mangrove forest

Tropical biomes include some of the most diverse on Earth, such as tropical rainforests (page 104) and coral reef (page 110). Coastal areas may have areas of mangrove forest that provide natural protection against the sea: mangroves provide a natural filter to sediment run-off from the land and stop erosion into the sea. Many are found within tropical biomes are termed 'biodiversity hotspots' (page 153) as they contain large numbers of species, often endemic to the area (i.e. not found anywhere else).

Tropical rainforests are characterized by long wet seasons and tall trees and plants that grow year-round. These forests presently cover 5.9 per cent of the Earth's land surface (around 1.5 per cent of the entire Earth's surface). In the tropics, the Sun's rays are the most concentrated and shine for nearly the same number of hours every day of the year: this makes the climate warm and stable. About 33 per cent of all rainforest is in the Amazon Basin, 20 per cent is found in Africa and a further 20 per cent in Indonesia (Figure 2.35, page 104). The remainder is scattered around the world. High levels of light and water make rainforests very productive. This explains why they can support such high biomass and wide diversity of life.

As you saw in Chapter 2, rainforests are complex ecosystems with many layers: emergent trees, the canopy, the understory and the ground (page 105). The complex layered structure of rainforests enables them to support many different niches (i.e. many different ways of living). Over 50 per cent of the world's plant species and 42 per cent of all terrestrial vertebrate species are endemic to 34 identified biodiversity hotspots (the majority of which are rainforests). In addition to their biodiversity value, tropical biomes such as rainforest provide many ecosystem functions. For example, they prevent soil erosion and nutrient loss, control the local water cycle (water evaporates from leaves in rainforest and falls locally as rain), act as carbon sinks (locking up in trees and other vegetation carbon that would otherwise be in the atmosphere), and so on. You can remind yourself of ecosystem services on page 43).

Human impact, both direct and indirect, on the world's rainforests is having a major effect on species survival. Uncontrolled hunting (for bush-meat and reasons such as the exotic pet trade) is removing large species and creating an 'empty forest syndrome' – the trees are there but the large species have disappeared. The replacement of natural tropical rainforest by oil palm plantations is replacing a diverse ecosystem with a monoculture ecosystem.

Tropical biomes are under constant threat, with large areas are being lost. An average of 1.5 ha (the size of a football pitch) of tropical rainforest is lost every 4 seconds. Deforestation and forest degradation are driven by external demands for timber, beef, land for crops such as soya and oil palm, and biofuels. Developing 'carbon markets' – which value ecosystems as stores of carbon (in vegetation) – could provide the means to give sufficient monetary value to rainforests to help protect them.

Rainforests have thin, nutrient-poor soils (Chapter 2, page 105). Because there are not many nutrients in the soil, it is difficult for rainforests to re-grow once they have been cleared. Studies in the Brazilian Atlantic forest have shown that parts of the forest can return surprisingly quickly – within 65 years – but for the landscape to truly regain its native identity takes a lot longer – up to 4000 years. Recovery depends on the level of disturbance – a large area of cleared land will take a lot longer to grow back (if at all) than small areas which have been subject to shifting cultivation (Chapter 5, pages 287–288). Forest which has been selectively harvested for timber (only large trees have been removed) can grow back rapidly if not too much timber has been removed. A larger amount of timber removal may mean that the forest never fully recovers because fast-growing, light-loving species (such as vines and creepers) block out the light for slow growers, so the forest remains at a sub-climax level.

By 2020, Indonesia's oil palm plantations are projected to triple in size to 16.5 million hectares. Many conservationists believe that this, in Indonesia and other countries, will lead directly and indirectly to the further clearance of a huge area of rainforest.

Case study

Oil palm and habitat destruction

Oil palm is the second most traded vegetable oil crop after soy. Over 90 per cent of the world's oil palm exports are produced in Malaysia and Indonesia, in areas once covered by rainforest and peat forest. Oil palm is traditionally used in the manufacture of food products, but is now increasingly used as an ingredient in bio-diesel. It is also used as biofuel burned at power stations to produce electricity. This new market has the potential to dramatically increase the global demand for oil palm. In the UK, the conversion of just one oil-fired power station to palm oil could double UK imports. The 6.5 million hectares of oil palm plantation across Sumatra and Borneo is estimated to have caused the destruction of 10 million hectares of rainforest – an increase in demand for palm oil as a biofuel would further increase the threat to natural ecosystems unless checks and balances are put in place.

Land use in tropical areas is a contentious issue. The widespread clearance of natural ecosystems so that land can be made available for plantations leads to biodiversity loss, although the plantations provide valuable financial income (something that the natural ecosystems on their own may not do). Diversification of the local economy into areas such as ecotourism can provide alternative sources of income and take pressure off local habitats, as would the development of conservation areas (page 194).

The recently discovered rainforest tree frog, *Rhacophorus gadingensis*

Some species, such as tree frogs, spend all their time in the rainforest canopy; they never reach the forest floor, so are not commonly seen. *Rhacophorus gadingensis* was recently discovered in a remote forest reserve in the centre of the island of Borneo.

TOK

In order to establish the species that exist in an area, populations must be sampled. When sampling populations of abundant, small, and poorly understood species (e.g. insects), specimens must be returned to natural history museums for identification – animals are killed in the process. Does this raise ethical issues, or does the end justify the means?

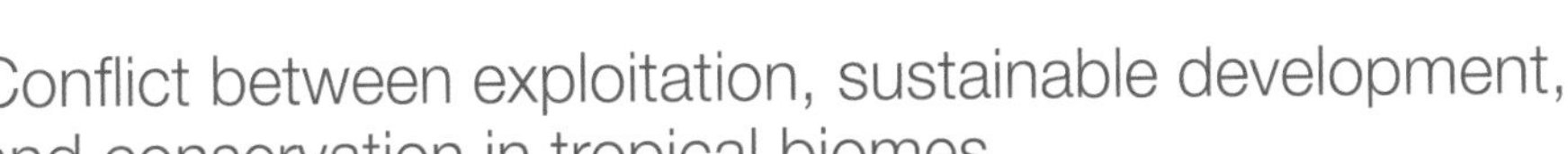

The rate of loss of biodiversity may vary from country to country depending on the ecosystems present, protection policies and monitoring, environmental viewpoints, and the stage of economic development.

Conflict between exploitation, sustainable development, and conservation in tropical biomes

Most tropical biomes occur in LEDCs and therefore there is conflict between exploitation, sustainable development, and conservation.

MEDCs have the luxury of being able to preserve their remaining natural ecosystems as they do not rely on these areas to provide income. In addition, MEDCs cleared the majority of their natural ecosystems (i.e. climax communities) in the past (e.g. in the UK the native forests were cleared to provide land for agriculture and timber to build ships) and so the argument for preserving the remaining diversity is on a different scale to the needs of LEDCs where most tropical biomes are found. For sustainable development to take place in LEDCs, there needs to be a balance between conserving tropical biomes and using the land to provide income for the local economy.

You need to be able to evaluate the impact of human activity on the biodiversity of tropical biomes. You also need to be able to discuss the conflict between exploitation, sustainable development, and conservation in tropical biomes.

One of the traditional incomes from tropical rainforests was timber. At the peak of logging operations in Borneo, for example, trees were removed in large numbers: in terms of volume, up to 100 m^3 of wood per hectare. Conventional logging methods were not selective and caused damage to the remaining forest. More recently, selective logging methods (also known as reduced-impact logging – RIL) have been used. These techniques cause less damage, allow faster regeneration of forest, and preserve forest structure and biodiversity better than conventional methods.

Ecotourism is also a way of providing ongoing income without destroying natural capital (e.g. the Great Barrier Reef, page 184).

LEDCs obviously wish to grow their economies and head towards MEDC status, but the resulting conflict between exploitation, sustainable development and conservation can always be resolved providing there is local support and the political will to protect biodiversity before it is lost forever.

Case study

CAMPFIRE in Zimbabwe

The Communal Areas Management Programme for Indigenous Resources (CAMPFIRE) is a Zimbabwean community-based management programme, which assists rural development and conservation. CAMPFIRE is helping people manage their environment in a sustainable way. Approximately 12 per cent of the natural habitats of Zimbabwe are in **protected areas** and when these were set up, local people were relocated to surrounding areas. When wildlife, such as elephants, leave the parks and enter inhabited areas, conflicts can arise. CAMPFIRE encourages people to see their local wildlife as a resource rather than as a nuisance.

Five main activities help provide extra income to local communities.

- Trophy hunting – professional hunters and safari operators are allowed into the areas; 90 per cent of CAMPFIRE's income is raised this way.
- Sale of wildlife – some areas with high wildlife populations sell animals to national parks or game reserves (e.g. one district raised US$50 000 by selling 10 roan antelope).
- Harvesting natural resources – a number of natural resources such as river-sand and timber are harvested and sold.
- Tourism – income from tourists is now being redirected to local communities; some local people are employed as guides or run local facilities for tourists.
- Selling wildlife meat – some species are abundant (e.g. impala); the National Parks Department supervise killing and selling skins and meat.

The International Union of Conservation of Nature (IUCN) publishes data in the Red List of Threatened Species in several categories.

Determining conservation status

The Red List

For more than four decades, IUCN (page 193) has published documents called the *Red Data Books*. The books assess the **conservation status** of particular species in order to highlight plants and animals threatened with extinction, and to promote their conservation. Known informally as the **Red List**, the books are essentially an inventory of all threatened species. The genetic diversity represented by these plants and animals is an irreplaceable resource which the IUCN is looking to conserve through increased awareness. These species also represent key building blocks of ecosystems, and information on their conservation status provides the basis for making informed decisions about conserving biodiversity from local to global levels.

The purposes of the Red List are:

- to identify species requiring some level of conservation
- to identify species for which there is concern about their conservation status
- to catalogue plants and animals facing a high risk of global extinction
- to raise awareness of animals and plants that face a higher risk of global extinction than others and require conservation efforts.

Factors used to determine a species' Red List conservation status

Various factors are used to determine the conservation status of a species, and a sliding scale operates (from severe threat to low risk). The range of factors used to determine conservation status includes the following.

- *Population size*
  Smaller populations are more likely to go extinct. Species with small populations also tend to have low genetic diversity – inability to adapt to changing conditions can prove fatal. Many of the large cat species are in this category (e.g. cheetah, snow leopard, and tiger).
- *Trophic level*
  Top predators are sensitive to any disturbance in the food chain and any reduction in numbers of species at lower trophic levels can have disastrous consequence (e.g. snow leopard, Chapter 2 page 85). Also, because of the '10 per cent rule' of energy loss through ecosystems (Chapter 2, pages 89–90), large fierce animals tend to be rare and are therefore particularly sensitive to hunters and reductions in population size.
- *Reduction in population size*
  A reduction in population size may indicate that a species is under threat. For example, numbers of European eel (*Anguilla anguilla*) are at their lowest levels ever in most of its range and it continues to decline.
- *Degree of specialization*
  Many species have a specific diet or habitat requirements: if their specific resource or habitat is put under threat, so are they. Some animals can only live on certain tree species, such as the palila bird (a Hawaiian honeycreeper), which is dependent on the mamane tree (*Sophora chrysophylla*) for its food and is losing habitat as the mamane tree is cut down. Other examples include the giant panda (dependent on bamboo) and the koala (dependent on a particular eucalypt).
- *Geographic range*
  Species that occupy a restricted habitat are likely to be wiped out. For example, the slender-billed grackle (*Cassidix palustris*), a bird which once occupied a single marsh near Mexico City, was driven to extinction when a reduction in the water table drained the marsh.
- *Distribution*
  Species that live in a small area are under greater threat from extinction than those that are distributed more widely. Loss of the area they live in will lead to loss of the species. Golden lion tamarin monkeys (*Leontopithecus rosalia*) are only found in one small area of southern Brazil, and are therefore especially prone to extinction. Any change in the habitat of a species with a limited area of occupancy (e.g. deforestation of the Mata Atlântica), or a small decrease in population size, could lead to their extinction.
- *Reproductive potential and behaviour*
  Animals that live a long time and have long gestation times, for example elephants and rhinos, have low rates of reproduction, and can take many years to recover from any reduction in population number. This makes them vulnerable to extinction. If a change in habitat or the introduction of a predator occurs, the population drops and there are too few reproductive adults to support and maintain the population.

Factors used to determine the conservation status of a species include: population size, degree of specialization, distribution, reproductive potential and behaviour, geographic range and degree of fragmentation, quality of habitat, trophic level, and the probability of extinction.

Because they are slow-reproducing, any loss in numbers means a fast decline. The Steller's sea cow was heavily hunted and unable to replace its numbers fast enough. Orang-utans have one of the slowest reproductive rates of all mammal species: they give birth to a single offspring only once every 6 to 8 years; with such a low reproductive rate, even a small decrease in numbers can lead to extinction.

- *Degree of fragmentation*
  Species in fragmented habitats may not be able to maintain large enough population sizes. The Sumatran rhinoceros (*Dicerorhinus sumatrensis*) lives in tropical rainforest in South East Asia. Fragmentation of the forest through deforestation and conversion to plantation forest, has led to a reduction in habitat area for this species.
- *Quality of habitat*
  Species that live in habitats that are poorer in quality are less likely to survive than species in habitats that are better in quality. For example, the fishing cat (*Prionailurus viverrinus*) is found in South East Asian wetland areas where it is a skilful swimmer: drainage of wetlands where it lives for agriculture has led to a reduction in habitat quality.
- *Probability of extinction*
  Even without human intervention, many species are likely to go extinct and so are of especial need of conservation efforts.

Irrespective of human interference, any animal or plant which is rare, has a restricted distribution, has a highly specialized habitat or niche, or a low reproductive potential, or is at the top of the food chain, is prone to extinction.

Puya raimondii, also known as 'Queen of the Andes', is a spectacular high-Andean plant found from Peru to Bolivia. Reasons for being on the Red List: isolated and very small population size; seeds only once in 80 years before dying; climate change may be limiting its ability to flower.

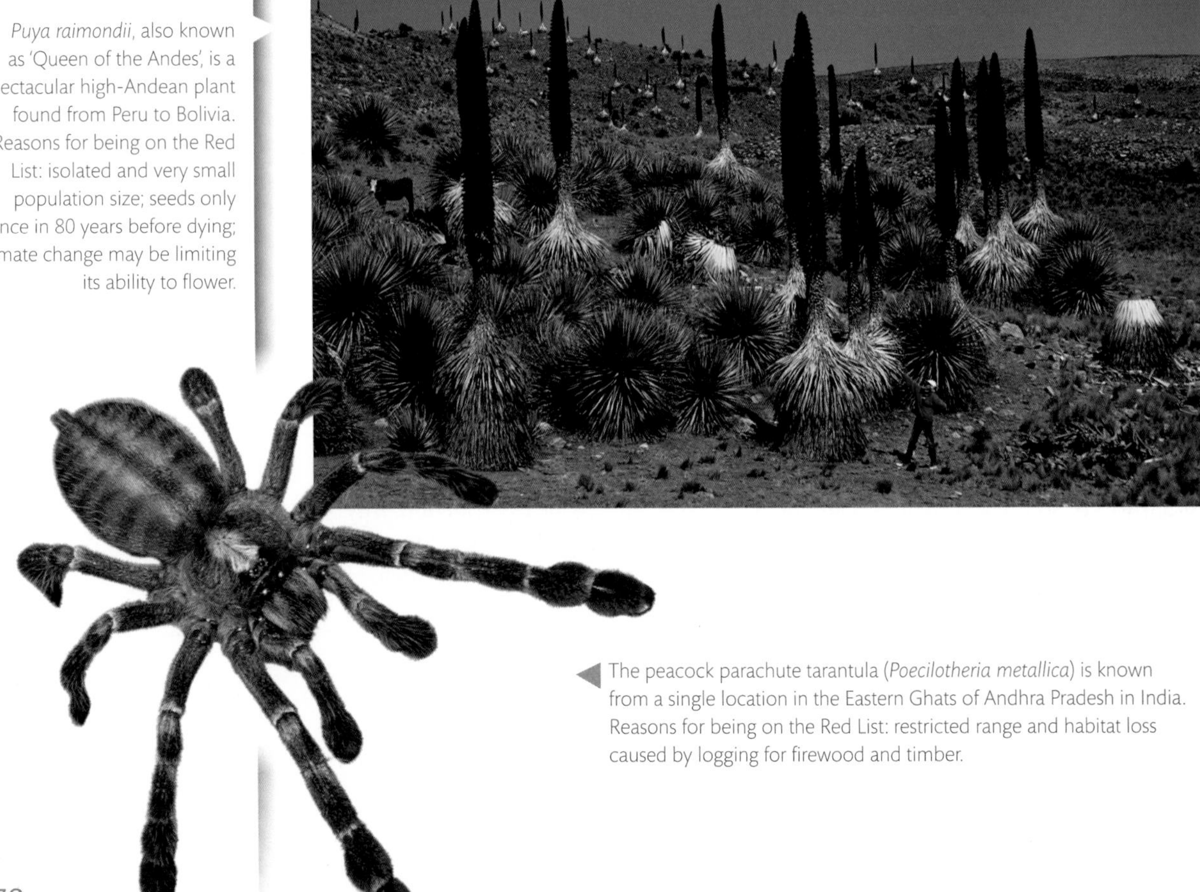

The peacock parachute tarantula (*Poecilotheria metallica*) is known from a single location in the Eastern Ghats of Andhra Pradesh in India. Reasons for being on the Red List: restricted range and habitat loss caused by logging for firewood and timber.

This topic raises some engaging issues of debate concerning the moral justification for exploiting species and the moral imperative for conserving them. Think carefully about the following questions (there are no correct answers).

- Do some organisms have more of a right to conservation than others? How can this be justified?
- Do pandas have a greater right to conservation than lichens?
- Do 'pests' or pathogenic organisms have a right to be conserved?
- To what extent are these arguments based on emotion and to what extent on reason? And how does this affect their validity?

Extinct, critical, and back from the brink

You need to be able to discuss the case histories of three different species: one that has become extinct due to human activity, another that is **critically endangered**, and a third species whose conservation status has been improved by intervention. In each case, the ecological, socio-political or economic pressures that are impacting on the species should be explored. The species' ecological roles and the possible consequences of their disappearance should be considered.

Case study

Extinct: Falkland Islands wolf

The Falkland Islands wolf

Description

The Falkland Islands wolf was the only native land mammal of the Falkland Islands. The islands were first sighted in 1692. In 1833, Charles Darwin visited the islands and described the wolf as 'common and tame'.

The genus name, *Dusicyon*, means 'foolish dog' in Greek (Dusi = foolish, cyon = dog).

Ecological role

The Falkland Islands wolf is said to have lived in burrows. As there were no native rodents on the islands (the usual wolf prey), it is probable that its diet consisted of ground-nesting birds (such as geese and penguins), grubs, insects, and some seashore scavenging.

Pressures

The many settlers of the Islands (mainly the Scottish inhabitants, but also the French and some English) considered the Falkland Islands wolf a threat to their sheep. A huge-scale operation of poisoning and shooting the wolf began with the aim of leading it to extinction. The operation was successful very rapidly, assisted by the lack of forests and the tameness of the animal (due to the absence of predators, the animal trusted humans who would lure it with a piece of meat and then kill it).

Consequences of disappearance

The Falkland Islands wolf was not particularly threatening nor was it a significant predator, although the removal of a top predator would have had an impact on the rest of the food chain (e.g. increase in population of its prey).

Case study

Critically endangered: Iberian lynx

The Iberian lynx

Description

The Iberian lynx (*Lynx pardinus*) is also known as the Spanish lynx and is native to the Iberian Peninsula. It has distinctive, leopard-like spots with a coat that is often light grey or various shades of light brownish-yellow. It is smaller than its northern relatives such as the Eurasian lynx, and so typically hunts smaller animals, usually no larger than hares. It also differs in habitat choice, inhabiting open scrub whereas the Eurasian lynx inhabits forests.

Ecological role

The Iberian lynx is a specialized feeder, and rabbits account for 80–100 per cent of its diet. Lynx often kill other carnivore species, including those regarded as pests by humans, such as feral cats and foxes, but do not eat them.

Pressures

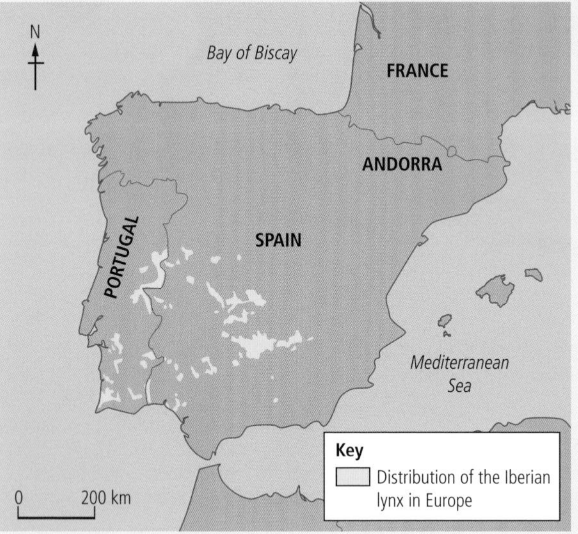

Figure 3.18 The present day distribution of the Iberian lynx in Europe

The lynx's highly specialized diet makes it a naturally vulnerable species and the rapid decline in rabbit populations since the 1950s has had a direct impact on lynx numbers. The Iberian lynx occurs only in isolated locations of Spain and possibly Portugal (Figure 3.18). Habitat destruction, deterioration, and alteration have impacted negatively on the lynx for centuries. The Iberian lynx were protected from hunting in the early 1970s, since when hunting has declined. Some lynxes are still shot and killed in traps and snares set for smaller predators, particularly on commercial hunting and shooting estates.

Methods of restoring population

The Iberian lynx is fully protected under national law in Spain and Portugal, and public awareness and education programmes have helped to change attitudes towards the animal, particularly among private landowners. Two international seminars have been held (2002 and 2004) to establish a coordinated strategy to save the Iberian lynx from extinction. A captive breeding programme has been started in Spain. In Portugal, the National Action Plan foresees a reintroduction programme. The construction of facilities for breeding and reintroduction has been prepared. Further protection stems from the fact that one lynx's endemic areas has been turned into the Doñana National Park.

Case study

Improved by intervention: American bald eagle

The American bald eagle

Description

The bald eagle (*Haliaeetus leucocephalus*), also known as the American eagle, was officially declared the National Emblem of the United States in 1782. It was selected by the Founding Fathers of the USA because it is a species unique to North America. It has since become the living symbol of the USA's spirit and freedom.

Bald eagles are one of the largest birds in North America with a wing span of 6–8 feet. Females tend to be larger than males. They live for up to 40 years in the wild, and longer in captivity. Bald eagles are monogamous and have one life partner.

Ecological role

Bald eagles live near large bodies of open water such as lakes, marshes, seacoasts, and rivers. They nest and roost in tall trees. The eagles live in every US state except Hawaii. They use a specific territory for nesting, winter feeding, or a year-round residence. Their natural domain is from Alaska to California, and from Maine to Florida. Bald eagles that live in the northern USA and Canada migrate to the warmer southern areas during the winter to obtain easier access to food. Some bald eagles that live in the southern states migrate slightly north during the hot summer months. They feed primarily on fish, but also eat small animals (ducks, coots, muskrats, turtles, rabbits, snakes, etc.) and occasional carrion.

Pressures

Bald eagle population numbers have been estimated to have been 300 000 to 500 000 birds in the early 1700s. Their population fell to fewer than 10 000 nesting pairs by the 1950s, and to fewer than 500 pairs by the early 1960s. This population decline was caused by the mass shooting of eagles, the use of pesticides on crops, the destruction of habitat, and the contamination of waterways and food sources by a wide range of poisons and pollutants. For many years, the use of DDT pesticide on crops caused thinning of eagle egg shells, which often broke during incubation.

Methods of restoring population

The use of DDT pesticide was outlawed in the USA in 1972 and in Canada in 1973. This action contributed greatly to the return of the bald eagle.

The bald eagle was listed as 'endangered' in most of the USA from 1967 to 1995. The number of nesting pairs of bald eagles in 48 of the states increased from fewer than 500 in the early 1960s to over 10 000 in 2007. That was enough to remove them from the list of threatened species on 28 June 2007.

Since de-listing, the primary law protecting bald eagles has shifted from the Endangered Species Act to the Bald and Golden Eagle Act. Although bald eagles have made an encouraging comeback throughout the USA since the early 1960s, they continue to be face hazards that must be closely monitored and controlled. Even though it is illegal, bald eagles are still harassed, injured, and killed by guns, traps, power lines, windmills, poisons, contaminants, and destruction of habitat.

CONCEPTS: Strategy

Carefully planned strategies are needed to improve the conservation status of critically endangered species: these strategies need to address the ecological, socio-political or socio-economic pressures that are impacting on the species.

The golden toad

A species first discovered in 1966 was recorded as extinct by the IUCN in 2004. The golden toad (*Incilius periglenes*) was a small, shiny, bright toad that was once common in a small region of high-altitude, cloud-covered tropical forests, about 30 km^2 in area, above the city of Monteverde in Costa Rica. The last recorded sighting of the toad was in 1989. Possible reasons for its extinction include a restricted range, global warming, airborne pollution, increase in UV radiation, fungus or parasites, or lowered pH levels.

CHALLENGE YOURSELF

 Research skills

Research and summarize the case history of three different species: one that is extinct, one that is critically endangered, and a third species whose conservation status has been improved by intervention. For each, list the ecological, socio-political, and economic pressures that are involved, and outline the possible consequences of their disappearance on the ecosystem.

Threats to an area of biological significance

You need to be able to describe the threats to biodiversity from human activity in a given natural area of biological significance or conservation area.

The Great Barrier Reef Marine Park is 345 000 km^2: larger than the entire area of the UK and Ireland combined. The reef is the world's biggest single structure made by living organisms and is large enough to be seen from space. The Great Barrier Reef is an important part of the Aboriginal Australian culture and spirituality. It is also a very popular destination for tourists, especially in the Cairns region, where it is economically significant. Fishing also occurs in the region, generating AU$1 billion per year.

Case study

The Great Barrier Reef

The Great Barrier Reef

Coral reef, like rainforest, is amazingly diverse (and for similar reasons – such as its location, complexity, and high productivity). The Great Barrier Reef stretches 2300 km along the Queensland coastline of northern Australia. It is home to 1500 species of fish, 359 types of hard coral, a third of the world's soft corals, 6 of the world's 7 species of threatened marine turtle and more than 30 species of marine mammals including vulnerable dugongs (sea cows). In addition, there are 5000 to 8000 molluscs and thousands of different sponges, worms and crustaceans, 800 species of echinoderms (starfish, sea urchins) and 215 bird species, of which 29 are seabirds (e.g. reef herons, ospreys, pelicans, frigate birds, and shearwaters).

There are many and varied threats to this ecosystem.

Human threats

Ecological, socio-political, and economic pressures are causing the degradation of the coral reef, and as a consequence are threatening the biodiversity of the area. Tourism is now a major contributor to the local economy, but tourism can have negative impacts: coral is very fragile and is easily damaged by divers' fins and anchors. Although it is illegal to take pieces of coral from the country of origin, tourism inevitably leads to coral being damaged as tourists break bits off for souvenirs. As the sea is rich in fish, over-fishing can disrupt the balance of species in the food chain and there may also be accidental damage from anchors and pollution from boats. Seafloor trawling for prawns is still permitted in over half of the marine park, resulting in the unintentional capture of other species and also the destruction of the seafloor.

Land use in Australia has shifted from low-level subsistence agriculture to large-scale farming. Queensland has extensive sugar plantations where once forests stood. The plantations need heavy input of fertilizers and pesticides, so now run-off from the soils into the sea has caused inorganic nitrogen pollution to increase by 3000 per cent. Combined with sewage and pollution from coastal settlements such as Cairns, this means there are excessive nutrients in the water and algal blooms occur.

In addition, sedimentation (leading to mud pollution) has increased by 800 per cent due to deforestation of mangroves to make space for tourist developments, housing, and farming. Traditionally, coastal wetland ecosystems provided a natural filter to sediment run-off. Extensive mangrove forests along the coast chiefly fulfilled this function, but clearance has caused serious mud pollution issues. Mud pollution makes the water cloudy and reduces coral reef productivity thus disrupting the interdependence of the coral ecosystem with sea-grass beds and mangrove ecosystems.

Socially, there is pressure to raise important revenues for the country through agriculture, which is backed-up politically at the national level. Increasing awareness of the effect of this agriculture on the environment is causing people to rethink their priorities.

Global warming (Chapter 6, pages 313–314) is also affecting the reef. Increases in sea temperature have caused two mass coral bleaching events (plant and algal life on the reef dies, so the reef loses colour) in 1998 and 2002. Bleaching was more severe in 2002, when aerial surveys showed that almost 60 per cent of reefs were bleached to some degree. Increases in sea level and changes to sea temperatures may have a permanent effect on the Great Barrier Reef causing loss in biodiversity and ecological value of the area. In addition, climate change may be causing some fish species to move away from the reef to seek waters which have their preferred temperature. This leads to increased mortality in seabirds that prey on the fish.

The available habitat for sea turtles (e.g. coral reef and seagrass beds) are being damaged by sedimentation, nutrient run-off, tourist development, destructive fishing techniques, and climate change, causing reduction in population numbers.

Natural threats

All the human impacts have knock-on effects and thereby make the coral even more vulnerable to natural threats such as disease and natural predators. One such predator is the crown-of-thorns starfish which preys on the coral polyps (Figure 2.7, page 68) that form the coral reef. The starfish climbs onto the reef and extrudes its stomach over the coral, releasing digestive enzymes that digest the polyps so they can be absorbed. One adult crown-of-thorns starfish can destroy 6 m of coral in a year. Outbreaks of these starfish are thought to be natural, but the frequency and size of outbreaks has increased due to human activity. Reduction in water quality enables the starfish larvae to thrive, and unintentional over-fishing of natural predators (e.g. the giant triton, a large aquatic snail) is believed to have caused an increase in starfish numbers.

Crown-of-thorns starfish is one of the threats to the Great Barrier Reef.

continued

CHALLENGE YOURSELF

Research skills

Research and describe a local example of a natural area of biological significance that is threatened by human activities. List the ecological, socio-political, and economic pressures that caused or are causing the degradation of the area, and outline the possible impacts on biodiversity.

Structural damage to coral can be caused by storms and cyclones, which are becoming intensified and more frequent due to climate change. Another key atmospheric effect, linked to changes in seawater temperature, is El Niño. In this regular event, fluctuations in the surface waters of the tropical eastern Pacific Ocean lead to increases in sea temperature across the east-central and eastern Pacific Ocean area, including Australian waters. Increased sea temperature, as we have already seen, can lead to coral bleaching – this has knock-on effects on the fish species that depend on the reef for food and protection, and for nurseries for their young.

Consequences

Coral reefs are able to withstand some threats, but the current combined effect of human and natural processes can lead to irreversible damage to the reef, and the species that depend on it. In turn, these effects can lead to the breakdown of the reef ecosystem. When a 'critical threshold' is reached, the problems may well become irreversible and the ecosystem will not recover even if the threats stop. Loss of biodiversity and the valuable role that the ecosystem provides (e.g. in conjunction with mangroves and sea-grass beds as a line of coastal defence against erosion and sediment run-off) will inevitably lead to a reduction in its value as an economic resource.

A tourist watches a green turtle on the Great Barrier Reef.

The United Nations Educational, Scientific and Cultural Organization (UNESCO) encourages the protection and preservation of cultural and natural heritage sites considered to be of outstanding value to humanity. There are 679 cultural and 174 natural World Heritage sites so far listed, including the Great Barrier Reef, Yosemite National Park and the Galápagos Islands.

Exercises

1. List five factors that lead to the loss of diversity. How does each result in biodiversity loss?
2. Why is rainforest vulnerable to disturbance?
3. Evaluate the impact of human activity on the biodiversity of tropical biomes.
4. Discuss the conflict between exploitation, sustainable development, and conservation in tropical biomes.
5. What factors are used to determine a species' Red List status? List five.
6. Which types of species are common in the Red List, and which are less common? What implication does this have for the conservation of biodiversity?

Big questions

Having read this section, you can now discuss the following big questions:

- To what extent have the solutions emerging from this topic been directed at preventing environmental impacts, limiting the extent of the environmental impacts, or restoring systems in which environmental impacts have already occurred?
- How are the issues addressed in this topic of relevance to sustainability or sustainable development?
- In what ways might the solutions explored in this topic alter your predictions for the state of human societies and the biosphere some decades from now?

Points you may want to consider in your discussions:

- What indicators can be taken to suggest that a species is at threat from extinction?
- How can the population of a species facing extinction be restored?
- What threats do biologically significant areas face and how can the extent of the environmental impacts be limited?
- What issues arise when attempts are made to balance conservation with economic development? What conflicts exist between exploitation, sustainable development, and conservation in tropical biomes?

3.4 Conservation of biodiversity

Significant ideas

- The impact of losing biodiversity drives conservation efforts.
- The variety of arguments given for the conservation of biodiversity depend on environmental value systems.
- There are various approaches to the conservation of biodiversity, with associated strengths and limitations.

Big questions

As you read this section, consider the following big questions:

- To what extent have the solutions emerging from this topic been directed at preventing environmental impacts, limiting the extent of the environmental impacts, or restoring systems in which environmental impacts have already occurred?
- What value systems can you identify at play in the causes and approaches to resolving the issues addressed in this topic?
- How does your own value system compare with others you have encountered in the context of issues raised in this topic?
- In what ways might the solutions explored in this topic alter your predictions for the state of human societies and the biosphere some decades from now?

Knowledge and understanding

- Arguments about species and habitat preservation can be based on aesthetic, ecological, economic, ethical, and social justifications.
- International, governmental and non-governmental organizations (NGOs) are involved in conserving and restoring ecosystems and biodiversity, with varying levels of effectiveness due to their use of media, speed of response, diplomatic constraints, financial resources and political influence.
- Recent international conventions on biodiversity work to create collaboration between nations for biodiversity conservation.
- Conservation approaches include habitat conservation, species-based conservation and a mixed approach.
- Criteria for consideration when designing protected areas include: size, shape, edge effects, corridors, and proximity to potential human influence.

- Alternative approaches to the development of protected areas are species-based conservation strategies that include:
  - the Convention on International Trade in Endangered Species (**CITES**)
  - captive breeding and reintroduction programmes, and zoos
  - selection of charismatic species to help protect others in an area (flagship species)
  - selection of keystone species to protect the integrity of the food web.
- Community support, adequate funding and proper research influence the success of conservation efforts.
- The location of a conservation area in a country is a significant factor in the success of the conservation effort. Surrounding land use for the conservation area and distance from urban centres are important factors for consideration in conservation area design.

Arguments about species and habitat preservation can be based on aesthetic, ecological, economic, ethical, and social justifications.

Arguments for preserving biodiversity

The value of biodiversity can be difficult to quantify. Goods harvested from an ecosystem are easier to evaluate than indirect values such as the aesthetic or cultural aspects of an ecosystem. For example, it is easy to value rainforest in terms of amount of timber present because this has direct monetary value. But intact rainforests also provide valuable ecosystem services for the local, national and global communities (Figure 3.19). Rainforests are vital to the hydrologic (water) cycle, stabilize some of the world's most fragile soils by preventing soil erosion, and are responsible for regulating temperature and weather patterns in the areas surrounding the forest. In addition, they sequester (isolate) and store huge amounts of carbon from the atmosphere. They cool and clean the world's atmosphere. They are a huge source of the world's biodiversity, and they provide fresh water (the Amazon provides 20 per cent of the world's fresh water).

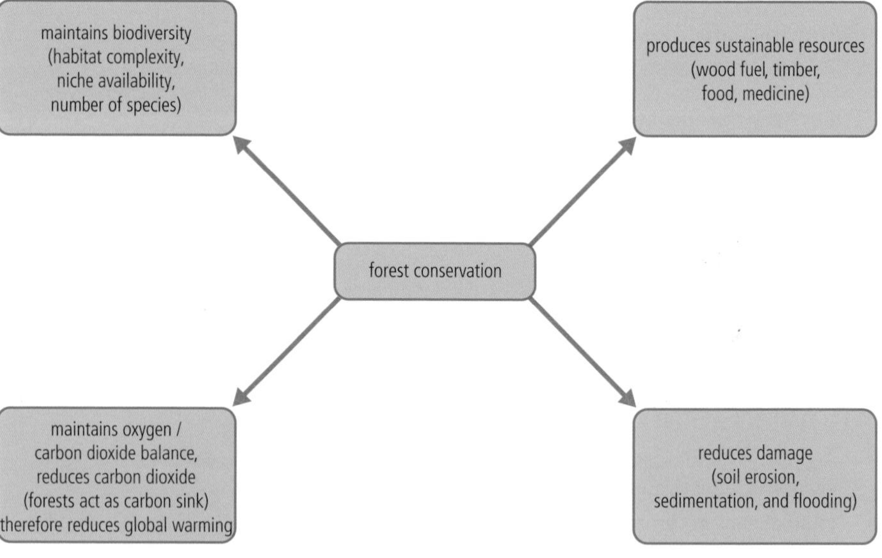

Figure 3.19 The biological significance of a forest

CONCEPTS: Biodiversity

Tropical rainforests should be conserved for a variety of reasons:

- they have an economic value to humans
- they contain food, medicines, and materials for human use
- rainforest has an intrinsic value (page 17)
- they provide life-support function (e.g. water cycles, carbon sink, oxygen provider)

- they contain high biodiversity
- they have aesthetic value
- the tourism function can bring income
- they provide a home to indigenous people
- regeneration rate is slow
- they provide spiritual, cultural, or religious value to local communities
- the current human population has a duty to protect rainforests for future generations.

The Iban are an indigenous people of Sarawak (Malaysia), Brunei, and western Kalimantan (Indonesian Borneo). Traditionally, they live in communal longhouses, hunting and fishing in rainforest areas, and growing crops using shifting cultivation. Pulp and paper companies have cleared Iban land and planted acacia trees, and other areas have been cleared for oil palm. The Iban have appealed against the loss of their traditional lands, although these rights have so far been denied as courts decided that land-ownership based on continuous occupation should 'not be extended to areas where the natives used to roam to forage for their food and building materials in accordance with their tradition'.

Iban woman with baby near her home

Most of these benefits are difficult to give monetary value to: every person on the planet benefits from these services, but none of us pay for them. Intact rainforests are aesthetically pleasing and this makes people want to visit them, which gives rainforest value from an ecotourism point of view. As rainforests contain such a high percentage of the existing global biodiversity, it can also be argued that we have an ethical responsibility to conserve them.

The value of ecosystems depends on cultural background as well as economic status. The value of a rainforest to someone who lives in and relies on it for their livelihood is very different from an outsider who does not have these concerns.

To learn about Survival International (an organization that supports indigenous people's rights), go to www.pearsonhotlinks.co.uk, enter the book title or ISBN, and click on weblink 3.5.

To learn about the Forests NOW campaign (which aims to raise awareness of the need to protect forests in order to prevent climate change), click on weblink 3.6.

Forest people are found in rainforests in Brazil, Colombia, Ecuador, Paraguay, Canada, Peru, Argentina, Botswana, Kenya, Ethiopia, Sudan, Central Africa, Australia, Indonesia, the Philippines, India, Bangladesh, Russia, Malaysia, and Sri Lanka. The majority are under threat from logging and rainforest loss, for example the Awá tribe in Brazil. The Awá's territory has been invaded and destroyed. Cattle ranchers illegally occupy Awá land and, in another part of their territory, groups of heavily armed loggers have destroyed much of the forest.

There are many arguments for preserving species and habitats, as we have seen above. These arguments can be divided into five groups.

- *Aesthetic reasons*

  Species and habitats are pleasant to look at and provide beauty and inspiration.

- *Ecological reasons*

  Rare habitats should be conserved as they may contain endemic species that require specific habitats. In addition, ecosystems with high levels of biodiversity are generally more stable and more likely to survive into the future. Healthy ecosystems are also more likely to provide ecosystem services such as pollination and flood prevention. Species should be preserved because if they disappear, they could have effects on the rest of the food chain and ecosystem.

Ecological reasons are concerned with ecosystems and their functioning.

Economic arguments for preservation often involve valuation of ecotourism, the genetic resource, and commercial considerations of the natural capital.

- *Economic reasons*
  Species and habitats provide financial income. Species should be preserved to maintain genetic diversity, so that genetic resources will be available in the future. For example, genetic diversity will allow crops to be improved in the future. Other reasons for preserving biodiversity are that commercial resources (e.g. new medicines) are still waiting to be discovered. The rosy periwinkle, a plant endemic to Madagascar, is used in cancer treatment. Ecotourism is successful when habitats high in biodiversity are preserved because they attract people to visit.

Ethical arguments are very broad, and can include the intrinsic value of the species or the utilitarian (i.e. the usefulness of species).

- *Ethical reasons*
  Everyone has a responsibility to protect resources for future generations. Ethical reasons also include the idea that every species has a right to survive.

TOK

What is a valuable species to one culture may not be so to another. For example, in South East Asia, elephants are valued by tourists, but to locals they are pests that eat crops and destroy their forest plantations.

- *Social reasons*
  Many natural ecosystems around the world provide places to live for indigenous peoples. Loss of these areas would mean loss of these peoples' homes, source of livelihood, and culture. In addition, many areas of great biodiversity provide an income for local people, such as tourism and wildlife protection. These areas therefore support social cohesion and cultural services.

It is easier to give commercial value to resources such as timber, medicine, and food. It is more difficult to give value to ecosystem services, cultural services, and ethical and aesthetic factors, although this does not mean that these are not equally valid reasons for preserving biodiversity.

CONCEPTS: Environmental value systems

Nomadic Penan hunting with blowpipe

The Penan of Borneo are nomadic hunter-gatherers who have historically relied on the rainforests for their survival. They have a comprehensive knowledge of the forest and are highly skilled in surviving there (e.g. a poison-headed dart from a blowpipe can strike an animal 40 m high in the upper canopy.

Forest peoples' views of rainforest differ from the views of people from developed countries. To forest people, the forest is their home, from which they derive food, medicine, and their cultural values. Economically developed countries see the rainforest as an opportunity to exploit natural resources and use land for new settlements. To forest people, losing the forest is losing their home, their source of food, and the destruction of their culture which has developed through generations of forest living.

Conservation organizations

International, governmental, and non-governmental organizations (NGOs) are involved in conserving and restoring ecosystems and biodiversity, with varying levels of effectiveness due to their use of media, speed of response, diplomatic constraints, financial resources, and political influence.

It is often difficult to make your voice heard by those who influence global policies (e.g. national governments). Combined voices are more effective and conservation organizations that work at both local and global levels are good at campaigning on key environmental issues such as climate change and the preservation of biodiversity.

Non-governmental organizations (NGOs) are not run by, funded by, or influenced by governments of any country (e.g. Greenpeace and the World Wide Fund for Nature, WWF). **Intergovernmental Organizations (IGOs)** are bodies established through international agreements to protect the environment and bring together governments to work together on an international scale (e.g. the European Environment Agency (EEA), United Nations Environment Programme (UNEP), and IUCN).

Each type of organization has its own strengths and weaknesses (Table 3.2). IGOs tend to be more conservative (i.e. have a more conventional approach to conservation and are not likely to be controversial), whereas NGOs tend to be more radical (and often have to be to get their message across and to be heard). NGOs also tend to be field based, gathering information to back up their arguments, whereas IGOs tend to gather information from scientific research which they pay for.

Table 3.2 Differences between IGOs and NGOs

	IGO (e.g. UNEP)	NGO (e.g. Greenpeace)
use of media	• works with media so communicates its policies and decisions effectively to the public	• may gain media coverage through variety of protests (e.g. protest on frontlines) • often run campaigns focused on large charismatic species such as whales/seals/pandas • sometimes access to mass media is hindered, especially in non-democratic countries • public protests put pressure on governments
speed of response	• slow to respond – agreements require consensus from members • can be bureaucratic and take time to act • directed by governments, so sometimes may be against public opinion	• fast to respond – usually its members already have reached consensus (or they wouldn't have joined in the first place)
political pressures	• decisions can be politically (and economically) driven rather than by best conservation strategy	• can be idealistic, and driven by best conservation strategy • focus on the environment • often hold the high moral ground over other organizations, although may be extreme in actions or views
public image	• organized as businesses with concrete allocation of duties • cultivate a measured image based on a scientific and business-like approach	• can be confrontational and have a radical approach to an environmental issue like biodiversity
legislation	• enforce decisions via laws (may be authoritarian sometimes)	• serve as watchdogs (suing government agencies or businesses who violate environmental law) • rely on public pressure rather than legal power to influence governments as they have no power to enforce laws
agenda	• provide guidelines and implement international treaties	• use public pressure to influence national governments • lobby governments over policies and legislation • buy and manage land to protect habitats, wildlife, etc.
funding	• fund environmental projects with monies coming from national budget • usually manage publicly owned lands	• fund environmental projects with monies coming from private donations
extent of geographical influence	• have influence both locally and globally	• focus more on local and/or national information, aiming at education – produce learning materials and opportunities for schools and public

Both IGOs and NGOs are trying to promote conservation of habitats, ecosystems, and biodiversity. Other similarities between the two organizations include the following.

- *Use of media*
  Both provide environmental information to the public on global trends, publishing official scientific documents and technical reports gathering data from a variety of sources.
- *Public image*
  Both lead and encourage partnership between nations and organizations to conserve and restore ecosystems and biodiversity.
- *Legislation*
  Both seek to ensure that decisions are applied.
- *Agenda*
  Both collaborate in global, transnational scientific research projects.

  Both provide forums for discussion.
- *Geographical influence*
  IGOs monitor regional and global trends; NGOs also monitor species and conservation areas at a variety of levels, from local to global.

If you are asked to compare the roles of an intergovernmental organization (IGO) and a named non-governmental organization you need to refer to named examples (e.g. IGO – UNEP; NGO – WWF). Answers should include similarities and differences between the two organizations.

Recent international conventions on biodiversity aim to create collaboration between nations for biodiversity conservation.

International conventions on biodiversity

CONCEPTS: Strategy

International conventions provide governments with strategies for conserving biodiversity.

In Chapter 1, you learned how international conferences have led to international conventions on biodiversity (e.g. the Earth Summit in 1992 led to the Convention on Biological Diversity, page 8).

The IUCN (aka World Conservation Union) was founded in 1948. It is concerned with the importance of conservation of resources for sustainable economic development. You have already seen how the IUCN plays a role in species conservation via the Red List (page 176); the IUCN has also helped establish international conventions to help protect biodiversity.

In 1980, the IUCN established the World Conservation Strategy (WCS) along with UNEP and WWF. The WCS outlined a series of global priorities for action and recommended that each country prepare its own national strategy as a developing plan that would take into account the conservation of natural resources for long-term human welfare. The strategy also drew attention to a fundamental issue: the importance of making the users of natural resources become their guardians. It stressed that without the support and understanding of the local community, whose lives are most closely dependent on the careful management of natural resources, the strategies cannot succeed.

The WCS consisted of three factors:

- maintaining essential life support systems (climate, water cycle, soils) and ecological processes
- preserving genetic diversity
- using species and ecosystems in a sustainable way.

The WCS focused on specific factors for preserving biodiversity. It chose these issues because they are the ones that people with different environmental viewpoints are most likely to agree on. Ethical and aesthetic arguments are more difficult to define and vary between different communities. The arguments used by the WCS are also more scientifically verifiable than ethical or aesthetic arguments. Most nations place more value on scientific validity than other arguments.

History of the IUCN

- *1948*
  Foundation of the organization, named International Union for the Protection of Nature (IUPN).
- *1949*
  Main focus on protecting habitats and species from the exploitative tendencies of humans.
- *1956*
  IUPN seen as too preservationist; changed its name to the International Union for the Conservation of Nature and Natural Resources.
- *1961*
  Lack of funds led to the establishment of an independent fund-raising organization, WWF, to raise funds and support IUCN.
- *1966*
  Species Survival Commission published Red Data Lists to provide detailed information on status, distribution, breeding rate, causes of decline, and proposed protective measures for all endangered species.
- *1967*
  UN List of National Parks and Equivalent Reserves produced (gives definitions and classification of types of protected areas; regularly updated and revised).
- *1973*
  Convention on the International Trade of Endangered Species of Wild Fauna and Flora (CITES) established.
- *1980*
  World Conservation Strategy (WCS) published.
- *1991*
  Update of the WCS *Caring for the Earth: A Strategy for Sustainable Living* launched in 65 countries. Stated the benefits of sustainable use of natural resources, and the benefits of sharing resources more equally among the world population.
- *1992*
  Global Biodiversity Strategy. The aim of the strategy was to aid countries to integrate biodiversity into their national planning. Three main objectives:
  - conservation of biological variation
  - sustainable use of its components
  - equitable sharing of the benefits arising out of the utilization of genetic resources.

Global and local approaches to environmental problem solving

Some environmental problems are global, so it makes sense that international cooperation is used in addressing them. For example, global warming will have far-reaching global impacts so a united response to monitoring and mitigation is more likely to be effective. International agreements can help to motivate governments to take action and honour their commitments (e.g. to cut carbon dioxide emissions – such action was taken to establish the Kyoto Protocol, page 390). As an international organization, UNEP has the resources to mobilize and coordinate action (e.g. environmental research) when individual nations, especially LEDCs, might not have access to funds or expertise. When problems cross borders (e.g. smuggling endangered species), international cooperation is vital (e.g. CITES, pages 199–201). On the other hand, problems are often local, so local people should be involved in providing a local solution. This is recognized by the WCS. The motivation for addressing problems often starts at local level, when individuals feel passionately about an issue. Issues such as recycling and landfill are local ones, so a global strategy would be cumbersome, bureaucratic, and inappropriate.

Global summits and the conventions that come out of them have shaped attitudes towards sustainability. The UN Conference on Human Environment (Stockholm, 1972) was the first meeting of the international community to consider global environment and development needs (Chapter 1, pages 7–8). Summits play a pivotal role in setting targets and shaping action at both an international and local level. As you saw in Chapter 1, the UN Rio Earth summit led to Agenda 21 and the Rio Declaration (page 8). The 2000 UN Millennium Summit agreed a set of Millennium Development Goals (MDGs) (Chapter 8, page 410). The subsequent World Summit in New York, USA, recommended that each country developed its own strategy for fulfilling the MDGs. However, should countries break these agreements, there is little the international community can do about it, unless they are legally binding. Even when conventions do not achieve their initial goals, they may act as a catalyst in changing the attitudes of governments, organizations, and individuals.

You need to be able to evaluate different approaches to protecting biodiversity.

In situ vs. *ex situ* conservation

Conservation approaches include habitat conservation, **species-based conservation** and a mixed approach. ***In situ* conservation** is the conservation of species in their natural habitat. This means that endangered species, for example, are conserved in their native habitat. Not only are the endangered animals protected, but also the habitat and ecosystem in which they live, leading to the preservation of many other species. *In situ* conservation works within the boundaries of conservation areas or nature reserves.

***Ex situ* conservation** is the preservation of species outside their natural habitats. This usually takes place in botanic gardens and zoos, which carry out captive breeding and reintroduction programmes. The species-based approach to conservation is an approach that focuses on specific individual species (usually animals) that are vulnerable. The aim is to attract interest in their conservation and therefore funding and public pressure for conservation.

CONCEPTS: Strategy

There are different strategies for conservation: *in situ* conservation preserves biodiversity in natural habitats (e.g. protected areas, safari parks); *ex situ* conservation preserves biodiversity outside natural habitats (e.g. zoos).

You need to be able to explain the criteria used to design and manage protected areas.

Designing protected areas

Most countries have large areas of land that have been cleared of native habitat for development purposes (e.g. cities). The remaining areas of native habitat can be made into protected areas. Protected areas are often isolated and in danger of becoming islands within areas of disturbance, such as cleared land. When protected areas become islands, they lose some of their diversity due to increased **edge effects** (the impact of changed environmental conditions at the edge of the reserves) and localized extinctions.

'Island biogeography' theory was developed in the 1960s by Robert MacArthur and Edward Wilson. They showed that smaller conservation areas contain comparatively fewer species and lower diversity than larger areas. Ever since, reserve designers have been using these ideas to ensure maximum preservation of species within conservation areas. Size, shape, edge effects, and whether or not reserves are linked by corridors, are all taken into account when designing conservation areas (Figure 3.20).

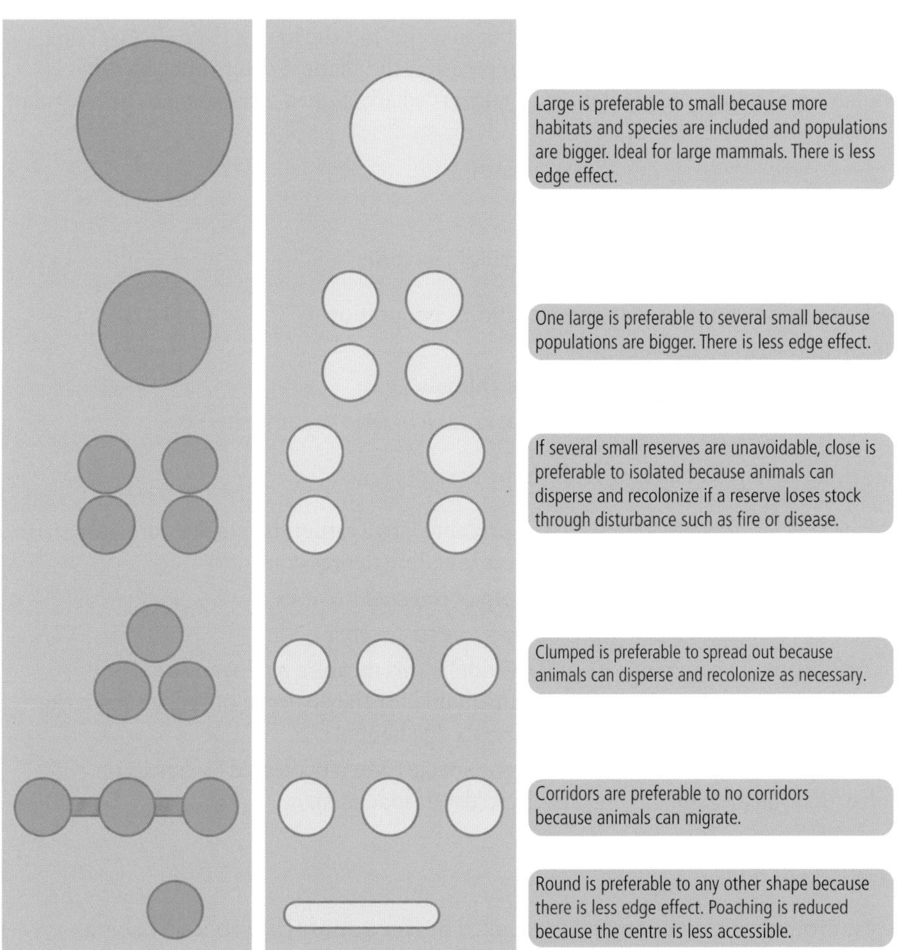

Large is preferable to small because more habitats and species are included and populations are bigger. Ideal for large mammals. There is less edge effect.
One large is preferable to several small because populations are bigger. There is less edge effect.
If several small reserves are unavoidable, close is preferable to isolated because animals can disperse and recolonize if a reserve loses stock through disturbance such as fire or disease.
Clumped is preferable to spread out because animals can disperse and recolonize as necessary.
Corridors are preferable to no corridors because animals can migrate.
Round is preferable to any other shape because there is less edge effect. Poaching is reduced because the centre is less accessible.

Figure 3.20 The shape, size and connectivity of reserves are important in the design of protected areas.

Area

One of the great debates in reserve design is known as SLOSS (single large or several small): is it better to have one large reserve (e.g. 10 000 ha) or several smaller ones (four at 2500 ha)? Much depends on location of the habitats – if habitats to be preserved are not all found reasonably close together, then several small reserves may be necessary. But overall, bigger is better because one large area can support more species than several smaller areas (they have more habitats and can support more top carnivores). The best indicator of species survival and success of the reserve's size is the population size of individual species. In an ideal situation, several large reserves would allow the protected habitats to be replicated thus guarding against the possible effects of fire or a disease which could lead to the extinction of species contained within the affected reserve.

Edge effects

At the edge of a protected area, there is a change in abiotic factors (e.g. more wind, or warmer and less humid conditions compared to the interior of the reserve). Edge effects attract species that are not found deeper in the reserve, and may also attract exotic species from outside the reserve, leading to competition with forest species and overall reduction in diversity.

Shape

The best shape for a reserve is a circle because this has the lowest edge effects. Long thin reserves have large edge effects. In practice, the shape is determined by what is available and where the habitats to be conserved are located. Parks tend to be irregular shapes.

Corridors

The benefits of linking reserves by corridors include:

- allowing gene flow by immigration and/or emigration
- allowing seasonal movements
- reducing collisions between cars and animals
- having fewer or no roads as these can act as a barrier to some species.

The disadvantages of linking reserves by corridors include:

- some species may breed outside the protected area rather than in it leading to reduction in numbers (this is called 'outbreeding depression')
- invasion of exotic pests or disease from connected reserves
- poachers can easily move from one reserve to another
- corridors may be narrow (30–200 m wide) – this means a big increase in edge conditions rendering the corridors unsuitable for the dispersal of species from the centre of reserves, which normally avoid edge habitat
- corridors may become barriers to some species when protected by fences or obstructions (designed, for example, to deter poachers).

Criteria for consideration when designing protected areas include: size, shape, edge effects, corridors, and proximity to potential human influence. Well-designed protected areas:

- **are large because this promotes large population sizes and high biodiversity; enables protection of large vertebrates/top carnivores; reduces perimeter relative to area, so edge effects and disturbance are minimized**
- **are unfragmented and connected to other reserves (by corridors) to allow movement and migration between reserves**
- **do not have roads that can act as barriers to migration and increase disturbance and edge effects.**

Buffer zone

Areas around conservation areas are called **buffer zones**. They contain habitats and may be either managed or undisturbed. These areas minimize disturbance from outside influences such as people, agriculture or invasion by diseases or pests. For example, a nearby large town or extensive disturbance (e.g. logging) can directly impact a protected area if it is not surrounded by an area that buffers (protects) it from effects of the disturbance. Most successful protected areas are surrounded by buffer zones.

CONCEPTS: Strategy

Protected areas need to be designed following strategies that enable the maximum amount of biodiversity to be conserved.

Evaluating the success of a protected area

You need to be able to evaluate the success of a given protected area. A specific example of a protected area and the success it has achieved should be studied.

Case study

Danum Valley Conservation Area (DVCA), Malaysian Borneo

Danum Valley Field Centre, Malaysia. Research at the centre focuses on local primary forest ecology as well as the effect of logging on rainforest structure and communities.

The Danum Valley Conservation Area (DVCA) is a protected area located in the Malaysian state of Sabah on the island of Borneo, at latitude 5° North. The DVCA and surrounding areas is a model of how effective conservation can be matched with local economic needs.

Granting protected status to a species or ecosystem is no guarantee of protection without community support, adequate funding and proper research. In north-eastern Borneo, the third largest island in the world, a large area of commercial forest is owned by the Sabah Foundation (also known as Yayasan Sabah). The Yayasan Sabah Forest Management Area (YSFMA) is an extensive area of commercial hardwood forest containing within it protected areas of undisturbed forest, areas that are being rehabilitated with 'enrichment planting' (adding seedlings to heavily disturbed logged forest), and areas of commercial softwood forestry. Research of the primary rainforest within the DVCA has established the biological importance of the native forest and acted as a focus for conservation in the region. DVCA covers 43 800 ha (Figure 3.21), comprising almost entirely lowland dipterocarp forest (dipterocarps are valuable hardwood trees). The DVCA is the largest expanse of pristine forest of this type remaining in Sabah.

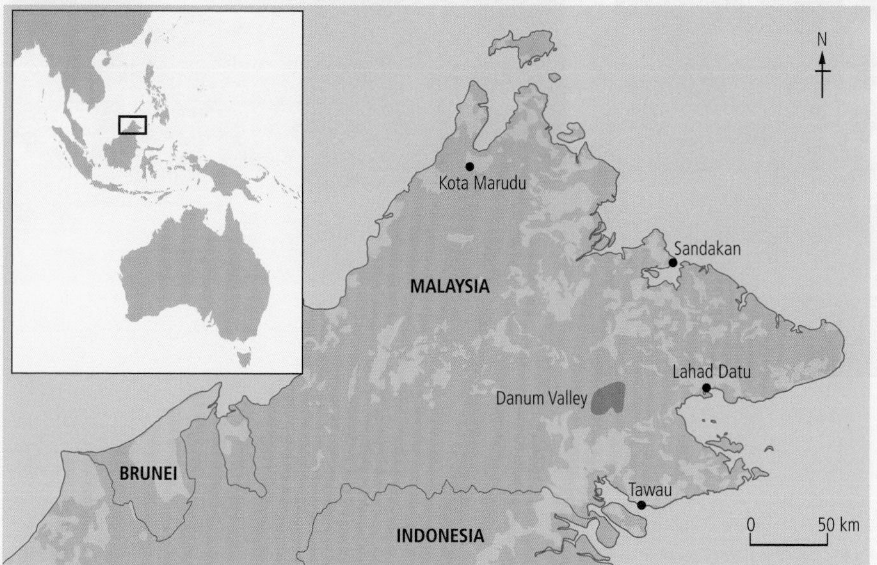

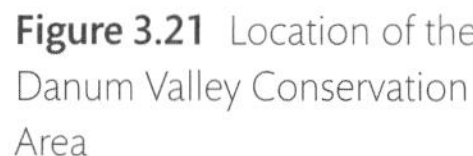

Figure 3.21 Location of the Danum Valley Conservation Area

Until the late 1980s, the area was under threat from commercial logging. The establishment of a long-term research programme between Yayasan Sabah and the Royal Society in the UK (the oldest scientific body in the world) has created local awareness of the conservation value of the area and provided important scientific information about the forest and what happens to it when it is disturbed through logging. Danum Valley is controlled by a management committee representing all the relevant local institutions – wildlife, forestry, and commercial sectors are all represented.

continued

Two other conservation areas, the Maliau Basin and Imbak Canyon, are linked by commercial forest corridors. To the east of DVCA is the 30 000 ha Innoprise-FACE Foundation Rainforest Rehabilitation Project (INFAPRO), one of the largest forest rehabilitation projects in South East Asia, which is replanting areas of heavily disturbed logged forest. The Innoprise–IKEA project (INIKEA) to the west of DVCA, is a similar rehabilitation project (Figure 3.22).

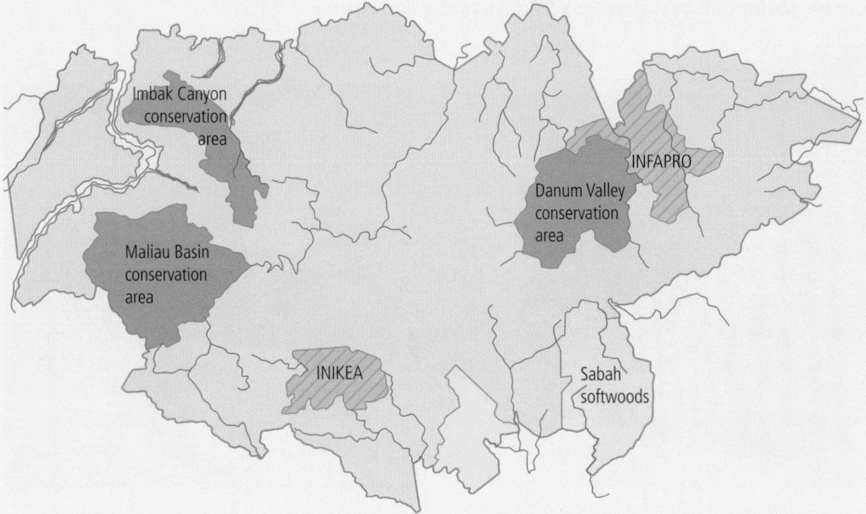

Figure 3.22 Location of conservation areas, rehabilitation projects and commercial softwood forestry within YSFMA. The combined network of different types of forest has enabled effective conservation of animals and plants important to the region.

Because all areas of conservation and replantation are within the larger commercial forest, the value of the whole area is greatly enhanced. Movement of animals between forest areas is enabled and allows the continued survival of some important and endangered Borneo animals such as the Sumatran rhino, the orang-utan and the Borneo elephant.

Orang-utans are found on the islands of Borneo and Sumatra. They are high-profile animals and are used to promote the conservation of rainforest.

The Borneo Rainforest Lodge – an ecotourism destination at the edge of the DVCA

In the late 1990s, a hotel was established on the north-eastern edge of the DVCA. It has established flourishing ecotourism in the area and exposed the unique forest to a wider range of visitors than was previously possible. As well as raising revenue for the local area, it has raised the international profile of the area as an important centre for conservation and research.

Such projects require significant funding which has come from Yayasan Sabah (a state foundation funded by the Sabah Government and Federal Government of Malaysia) and companies such as Malaysia's Petra Foundation, Shell, BP, the Royal Society, and others. The now high international profile of the Danum Valley, and key research over a long period of time (the programme is now the longest running in South East Asia), have helped to establish the area as one of the most important conservation areas in the region, if not the world.

Community support

The Danum Valley Field Centre is managed and maintained by a large staff of local people. Many are from the nearest town (Lahad Datu) or from east-coast kampongs (villages) such as Kampong Kinabatangan.

The field centre and surrounding conservation area provides opportunities for employment, education, and training. Support from the local community in running the various facilities on site (e.g. field centre office, accommodation, research support, and education centre) and in local towns, and much interest from nature groups in schools, have been important to the success of the project.

As well as the strengths outlined above, Danum Valley does have some limitations.

- Oil palm plantations are being grown near to the northern border of the DVCA. This could affect the ecotourism potential of Danum Valley as tourists do not want to see agricultural areas so close to a protected area. The presence of people so near to the conservation area may also lead to increased poaching activity or illegal logging activity.
- The funding that supports the DVCA has been raised by logging and conversion of land once covered by rainforest to forest plantation. Some conservationists may see a conflict between the activities that have provided revenue for the DVCA and the aims of a protected area.
- The DVCA and surrounding area is currently designated a conservation area, but a change of leadership within those involved with the DVCA could see this designation changed. The establishment of the DVCA as a World Heritage Site would give international protection to the DVCA and ensure its long-lasting protection.

Overall, however, the impacts of the DVCA have been overwhelmingly positive. In June 2013, the Sabah State Assembly reclassified several forests as protected areas in the YSFMA, creating an unbroken stretch of continuous unbroken forest, including Maliau Basin, Imbak Canyon, and Danum Valley. This created the single largest protected area in Malaysia, covering nearly 500 000 ha (about five times the size of Penang Island).

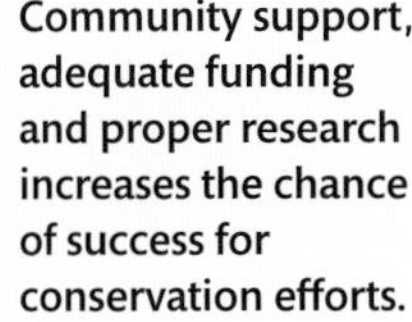

Community support, adequate funding and proper research increases the chance of success for conservation efforts.

The DVCA contains more than 120 mammal species including 10 species of primate. The DVCA and surrounding forest is an important reservation for orang-utan. These forests are particularly rich in other large mammals including the Asian elephant, Malayan sun bear, clouded leopard, bearded pig, and several species of deer. The area also provides one of the last refuges in Sabah for the critically endangered Sumatran rhino. Over 340 species of bird have been recorded at Danum, including the argus pheasant, Bulwer's pheasant, and seven species of pitta bird. Higher plants include more than 1300 species in 562 genera of 139 families, representing 15 per cent of the species recorded for Sabah.

A Sumatran rhino

The location of a conservation area in a country is a significant factor in the success of the conservation effort. Use of surrounding land and distance from urban centres are important factors for consideration in conservation area design.

Species-based conservation strategies

Alternative approaches to the development of protected areas are species-based conservation strategies that include:

- **CITES**
- **captive breeding and reintroduction programmes, and zoos**
- **selection of charismatic species to help protect others in an area (flagship species)**
- **selection of keystone species to protect the integrity of the food web.**

The Convention on International Trade in Endangered Species (CITES)

CITES was established in 1973 and celebrated its 40th anniversary in 2013. It is an international agreement aimed at regulating trade in endangered species of both plants and animals. This trade is worth billions of dollars every year and involves hundreds of millions of plant and animal specimens. Trade in animal and plant specimens (whole organisms, alive or dead, or their parts and derivatives), as well as factors such as habitat loss, can seriously reduce their wild populations and bring some species close to extinction. CITES' aim is to ensure that international trade in specimens of wild animals and plants does not threaten the survival of the species in the wild. CITES gives varying degrees of protection to 35 000 species of animals and plants. Species under threat from extinction are protected under 'Appendix I' of CITES. Commercial trade in wild-caught specimens of these species is illegal (permitted only in exceptional licensed circumstances). Many wildlife species in trade are not endangered: these are listed under 'Appendix II'. CITES aims to ensure that trade of Appendix II species remains sustainable and does not endanger wild populations, so as to safeguard these species for the future. Countries who sign up are agreeing to monitor trade in threatened species and their products that are exported and imported. Illegal imports and exports can result in seizures, fines and imprisonment, which discourages illegal trade.

How CITES works

Membership of CITES is voluntary. Each member country agrees to adopt legislation to implement CITES at the national level. All import, export, re-export, or

Case study

The effect of reclassifying African elephants from Appendix I to Appendix II

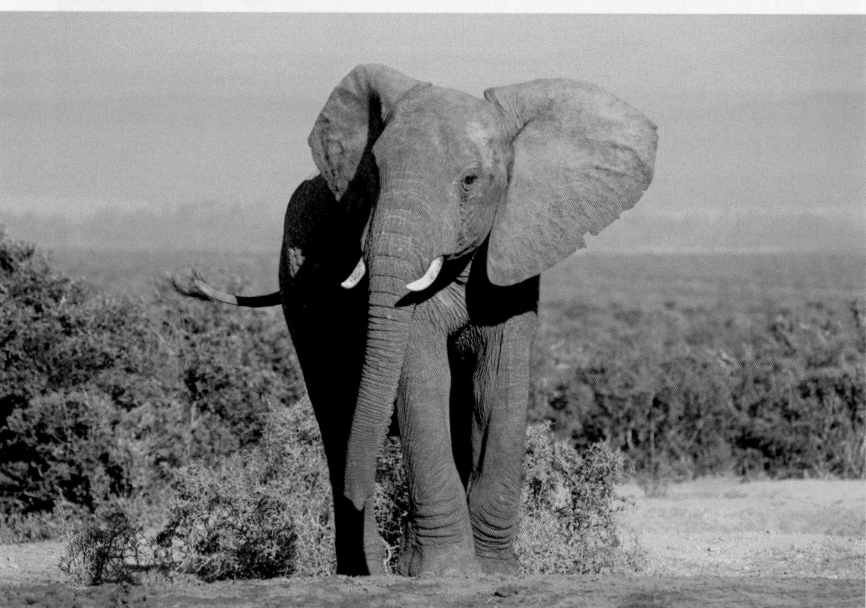

An African elephant

One of the biggest threats to elephant populations is the ivory trade, as the animals are poached for their ivory tusks. Other threats to wild elephants include habitat destruction and conflicts with local people. African elephants were listed on Appendix I of CITES in 1990. Appendix I prohibits the trade of wild-caught specimens completely (so as to protect plants and animals under considerable threat of extinction), whereas under Appendix II specimens can be exported but with trade restricted by a tightly controlled permitting process (i.e. classification is extended to species that are not necessarily threatened but could easily become so). As elephant populations grew in Zimbabwe, Botswana, and Namibia, in 1997 the classification of African elephants in these countries was changed to Appendix II (i.e. they were down-listed). The down-listing of African elephants in these countries resulted in a single shipment of stockpiled ivory, estimated to be ca. 50 000 kg, to Japan in 1999. African elephants were down-listed to Appendix II in South Africa in 2000. Delisting may have led to increased ivory poaching and a decline in many wild elephant populations African elephants provide an example of the effect of reclassification on wild populations.

introduction of specimens or parts and derivatives of any species covered by CITES, has to be authorized through a licensing system and permits must be obtained. The scheme has its limitations: it is voluntary and countries can 'enter reservations' on specific species when they join or when Appendices are amended, and penalties may not necessarily match the seriousness of the crime or be sufficient deterrent to wildlife smugglers (particularly given the large amounts of money that can be earned by poachers). In addition, unlike other international agreements such as the Montreal Protocol, CITES lacks its own financial mechanism for implementation at the national level and member states must contribute their own resources. However, taken overall, CITES has been responsible for ensuring that the international trade in wild animals and plants remains sustainable (Appendix II species), and for protecting species at risk of extinction (Appendix I).

CITES in numbers:

- 180: the number of countries ('contracted parties') signed up to the agreement
- 5500: the number of listed animals
- 29 500: the number of listed plants
- 35 000: the number of listed species.

Captive breeding and reintroduction programmes, and zoos

Zoos have become increasingly focused on conservation and many now lead the way in the preservation of species threatened with extinction. In prioritizing species for conservation, zoos have to answer many crucial questions.

How to select what to conserve?

- What is the level of threat? It is better to conserve endangered animals than ones that are not endangered.
- What to focus on? Different zoos have different expertise and areas of influence; they focus on their particular strengths.
- Can the zoo afford to financially support the project in the long term?
- Should species that are threatened for natural reasons (natural ecology) be conserved, such as those threatened by natural predation?
- What is the economic status of the country concerned? Zoos are more likely to support *in situ* conservation in LEDCs than MEDCs (which can help themselves).

In situ or *ex situ* conservation?

- How big is the animal? Smaller ones are easier to keep in zoos.
- Species facing habitat loss need to be conserved *ex situ* (e.g. Livingstone fruit bat – where 90 per cent of the habitat was lost due to cyclone damage).
- Animals threatened by diseases need to be kept *ex situ* (e.g. amphibian species are currently under threat globally from a fungus which is wiping them out in the wild, and so are being kept in quarantine in zoos).
- Decisions on which projects to undertake, will be influenced by staff expertise and whether or not the zoo vet has the knowledge to look after the species.
- If local people are willing to help, *in situ* conservation may be appropriate. If there are local political problems, *ex situ* may be preferred.
- Zoos often use species that are attractive to the public (e.g. lemurs and meerkats) to bring in visitors to provide funds for conservation. *Ex situ* conservation is therefore often used, even if the species is not especially threatened.

Is intervention helping?

Research to see if intervention is helping can be carried out by studying whether or not numbers are improving in the wild. Local expertise can assess whether the conservation effort is effective (e.g. in February 2015, a giant panda census was carried out in China, indicating that populations had grown by 268 to a total of 1864 since the last survey in 2003).

How are zoo populations managed?

When keeping animals in zoos, the welfare of the species must be taken into account. Behavioural studies can indicate whether or not animals are under stress. These studies may look at male and female social interactions, and how the animal uses their enclosures. The zoo would also consider whether the five freedoms are being met. The five freedoms were established in 1965 and were important in establishing modern zoo standards.

1 Freedom from thirst, hunger, and malnutrition through ready access to fresh water and a diet to maintain full health and vigour.
2 Freedom from thermal and physical discomfort by providing an appropriate shelter and a comfortable resting area.
3 Freedom from pain, injury, and disease by prevention or rapid diagnosis and treatment.
4 Freedom to express normal behaviour by providing sufficient space, proper facilities, and company of the animal's own kind.
5 Freedom from fear and distress by ensuring conditions and treatment avoid mental suffering.

How are breeding programmes managed?

For effective conservation and re-establishment of species in the wild, breeding programmes can be used. To be effective, details of the species' natural breeding behaviour must be known.

- Is it acceptable to choose the mate? Do you allow mate choice?
- The zoo may want to look at genes and the genetic compatibility of mates so as to avoid inbreeding. Leaving it to chance may lead to an animal choosing an unsuitable partner.
- Stud books can be used to establish genetic compatibility.
- Is artificial insemination a possibility? This will get round the problem of shipping in a mate.
- Birth control may be needed as the zoo may not want to have animals breeding (if zoo capacity is full).
- Keeper intervention may be needed – females sometimes reject young.
- Latest knowledge of reproductive biology and genetics is needed. Research is used (e.g. DNA testing by establishing parentage within a population).

Correct enclosure design and enrichment schemes mean that a species is more likely to breed.

Strengths and weaknesses of zoos

Among the strengths of zoos is their role in educating the public about the need for conservation. They also provide a way for people to empathize with wildlife. Although captivity is not the best solution, it acts as a good substitute and zoos can use breeding programmes to increase the population sizes of endangered animals while ensuring genetic diversity (i.e. by genetic monitoring). Well-managed zoos provide a proper diet and enough space while keeping species in a controlled environment which protects individual organisms. They offer a temporary safe haven while efforts are made to preserve habitats, so that species can be reintroduced later.

Weaknesses of zoos include the following. Some animals may have problems of re-adapting to wild, and captive animals released into the wild may become easy prey for predators. Not all species breed easily in captivity (e.g. it has proved extremely difficult to breed giant pandas). People may get used to seeing species in zoos and assume it is normal. Habitats in zoos are very different from natural habitats, especially for animals that have complex interactions with their environment such as orang-utans. There are ethical issues around caged animals, and some people object to animals being kept in captivity for profit. The best solution for endangered animals lies in the protection of their habitats.

The golden lion tamarin is one of the great success stories of zoo conservation. This small primate has been saved from extinction through captive breeding programmes.

Coordination of efforts between zoos helps in the effective conservation of species. The European Association for Zoos and Aquaria (EAZA) works out where specific zoos can help in specific areas. They have a number of Regional Collection Plans (RCPs). One of the RCPs is for the Callitrichid group of monkeys. The golden lion tamarin is a member of this group and has been brought back from the brink of extinction.

Flagship species

Flagship species are 'charismatic' species selected to appeal to the public and thereby help to protect other species in an area (e.g. the giant panda, meerkats, gorillas). By focusing on high-profile, iconic species there is a greater chance that conservation issues will catch the public attention, both nationally and internationally, and raise the necessary money for conservation initiatives. The advantages of this approach are twofold: money can be raised for the conservation of other species that may be equally endangered but are less appealing, and by preserving the habitat of the high-profile animal, other organisms in the habitat are also be preserved. Disadvantages of the approach include the favouring of

Giant panda eating bamboo, Chengdu, China

charismatic species (including those that may not be endangered in the wild) at the expense of less publically attractive species (even though they may be more critically endangered). Another disadvantage is that while species are preserved in zoos, their native habitat may be destroyed (as has happened with the giant panda).

How do we justify the species we choose to protect? Is there a focus on animals we find attractive (the ones with fur and feathers) and is there a natural bias within the system? Do tigers have a greater right to exist than endangered and endemic species of rat?

Keystone species

An agouti feeding on a Brazil nut in a forest

Keystone species are species that are vital for the continuing function of the ecosystem: without them the ecosystem may collapse. For example, the agouti of tropical South and Central America, which feeds on the nuts of the Brazil nut tree.

The Brazil nut tree (*Bertholletia excelsa*) is a hardwood species that is found from eastern Peru, eastern Colombia and eastern Bolivia through Venezuela and northern Brazil. They are the tallest trees in the Amazon (they grow up to 50 m). The agouti is a large forest rodent, and the only animal with teeth strong enough to open the Brazil nut tree's tough seed pods. The agouti buries many of the seeds around the forest floor so it has access to food when the Brazil nuts are less abundant. Some of these seeds germinate and grow into adult plants. Without the agouti, the Brazil nut tree would not be able to distribute its seeds and the species would eventually die out. Without the Brazil nut tree, other animals and plants that depend on it would be affected; for example, harpy eagles use the trees for nesting sites. Brazil nuts are one of the most valuable non-timber products found in the Amazon: they are a protein-rich food source, and their extracted oils are a popular ingredient in many cosmetic products. The sale of Brazil nuts provides an important source of income for many local communities.

A keystone is the central stone at the top of an arch. It supports the whole arch and ultimately the building it is part of.

TOK

There are various approaches to the conservation of biodiversity. How can we know when we should act on what we know?

Given the complexity of ecosystems, keystone species may be difficult to identify. In addition, many keystone species may be species that are as yet unidentified. By conserving whole ecosystems (i.e. establishing protected areas), rather than attempting to conserve individual species, the complex interrelationships that exist there will be preserved, including the keystone species.

Comparing different approaches to conservation

Table 3.3 summarizes advantages and disadvantages of some of the *in situ* and *ex situ* conservation strategies explored in this chapter.

Table 3.3 Evaluating habitat conservation (protected areas) and species-based conservation (CITES and zoos)

	Advantage	Disadvantage
protected areas	• can conserve whole ecosystems • allows research and education • preserves many habitats and species • prevents hunting and other disturbance from humans • visiting an intact ecosystem enables it to be studied to increase understanding of its functions • preservation of diversity more likely with a holistic approach as diversity can be species/habitat/ genetic • many species have not been discovered yet but are protected	• requires sufficient funding and protection to ensure area is not disturbed • difficult to establish in first place due to political issues/economic interests • areas can become islands and therefore may lose biodiversity due to size, shape, edge effects, etc. • may be subject to outside forces that are difficult to control
CITES	• can protect many species • signed by many countries • treaty works across borders • CITES is legally binding on the parties and so they must implement the convention	• difficult to enforce • implementation varies from country to country • it does not take the place of national legislation and countries must make their own laws to ensure that CITES is applied at the national level
zoos	• allows controlled breeding and maintenance of genetic diversity • allows research • allows for education • effective protection for individuals and species • education/empathy	• have historically preferred popular animals not necessarily those most at risk • problem of reintroducing zoo animals to wild • *ex situ* conservation and so do not preserve native habitat of animals

The main strengths of species-based conservation are that it attracts attention and therefore funding for conservation, and successfully preserves vulnerable species in zoos, botanic gardens, and seed-banks (i.e. preserve genetic diversity for future restocking of habitats). The main limitation of this approach is that if the ecosystem is not treated as a holistic unit, and habitats are not directly preserved, it will be difficult to ultimately preserve species.

The main strength of protected areas is that they protect the whole ecosystem and the complex interrelationships that exist there, so long-term survival of species is more likely. They also allow research to take place on intact ecosystems, greatly adding to our understanding of the factors that support biodiversity. Ecotourism raises awareness and profits are recycled back into biodiversity programmes. However, they do require considerable funding and protection to ensure the areas are not disturbed. Limitations may come from the fact they may become islands and may therefore lose biodiversity through their size, increased edge effects, or reduced gene flow between populations.

A mixed approach

Combining both *in situ* (e.g. protected areas) and *ex situ* (e.g. zoos and captive breeding) methods can be the best solution for species conservation in many instances. A good example of this is giant panda conservation. You have already seen how these animals can act as flagship species (page 203), and they were listed Appendix I by CITES in March 1984. Other species-based approaches include breeding programmes in zoos. Chengdu Zoo began breeding giant pandas in 1953, and Beijing Zoo in 1963. From 1963 to the present time, the giant panda has been bred in 53 zoos and nature reserves within China and internationally. Beijing Zoo has an impressive giant panda house, and has established a successful breeding programme.

Giant pandas enjoy a high profile within Chinese culture. Billboard showing pandas at Beijing zoo.

To learn more about the Chengdu Panda Base, go to www.pearsonhotlinks.co.uk, enter the book title or ISBN, and click on weblink 3.7.

When asked to evaluate different conservation strategies do not simply say 'raises awareness' unless this statement is directly linked to action which enhances diversity (e.g. education of public leads to increased donations to conservation organizations leading to improved biodiversity protection).

Raising giant pandas in captivity has three main difficulties: getting the female to come into heat (become reproductively receptive), conducting artificial insemination (introducing sperm into the female), and raising the cubs. In 1963, Beijing Zoo had the first success in artificially breeding giant pandas, and in 1978 the same zoo was the first to successfully carry out artificial insemination. In 1992, Beijing Zoo succeeded in raising a panda cub that had been artificially bred.

In situ conservation of giant pandas has involved the establishment of protected areas. The first five nature reserves for giant pandas in China were established in 1963, of which four are in Sichuan province. The giant panda nature reserves have expanded from the initial 5 to 56 in 2008.

The Chengdu Research Base of Giant Panda Breeding (also known as the Chengdu Panda Base) is involved in both *in situ* and *ex situ* conservation, with an emphasis on wildlife research, captive breeding, conservation education, and educational tourism. As well as breeding pandas, the Chengdu Panda Base covers an area of about 200 ha, with habitat that also contains red pandas and other endangered species.

Exercises

1. Outline arguments about species and habitat preservation.
2. Draw up a table contrasting governmental organizations and NGOs in terms of use of the media, speed of response, diplomatic constraints, and political influence.
3. State the criteria used to design protected areas. Your answer should address size, shape, edge effects, corridors, and proximity to other reserves.
4. What makes a protected area a success? List at least five essential factors that are required.
5. Evaluate the success of a named protected area.
6. Evaluate different approaches to protecting biodiversity, including habitat conservation, species-based conservation, and a mixed approach.

Big questions

Having read this section, you can now discuss the following big questions:

- To what extent have the solutions emerging from this topic been directed at preventing environmental impacts, limiting the extent of the environmental impacts, or restoring systems in which environmental impacts have already occurred?
- What value systems can you identify at play in the causes and approaches to resolving the issues addressed in this topic?
- How does your own value system compare with others you have encountered in the context of issues raised in this topic?
- In what ways might the solutions explored in this topic alter your predictions for the state of human societies and the biosphere some decades from now?

Points you may want to consider in your discussions:

- How do different conservation measures (e.g. *in situ* and *ex situ*) prevent environmental impacts, limit the extent of the environmental impacts, or restore systems in which environmental impacts have already occurred?
- How would a technocentric view of biodiversity differ from an ecocentric one? How do different EVSs affect approaches to conservation?
- If you are from a MEDC, how would your EVS differ from that of someone from a LEDC, or from someone who relies on the preservation of natural ecosystems for survival?
- Do you think that the conservation measures being taken today will be sufficient to preserve the Earth's biodiversity for the future?

Practice questions

1 a Distinguish between *biodiversity*, *species diversity*, *habitat diversity* and *genetic diversity*. [2]

b Explain how species diversity for an area may be calculated. [4]

c Outline the reasons why tropical biomes should be conserved. [5]

2 The map in the figure below shows plate movements and three biodiversity hotspots in Asia and Australasia. Hotspots are regions with especially high biodiversity.

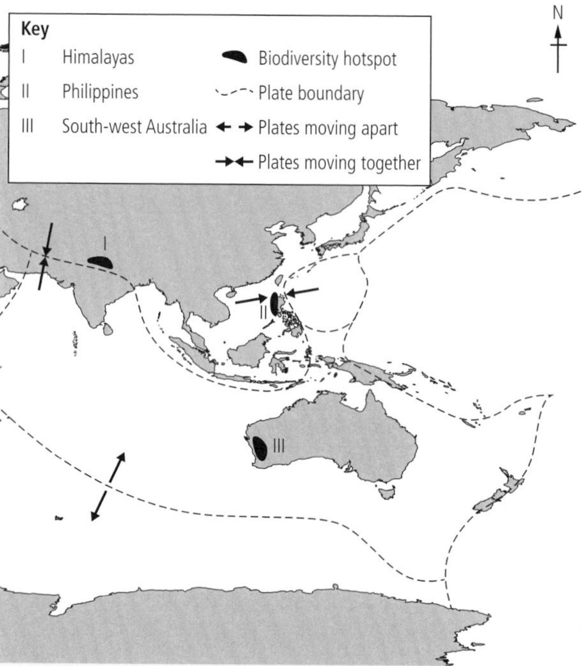

a Explain how the plate movements shown in the figure may have contributed to the biodiversity of the hotspot regions. [4]

b Below is a photograph of a clouded leopard (*Neofelis nebulosa*), one of the Himalayan species that is listed as 'vulnerable' on the Red List.

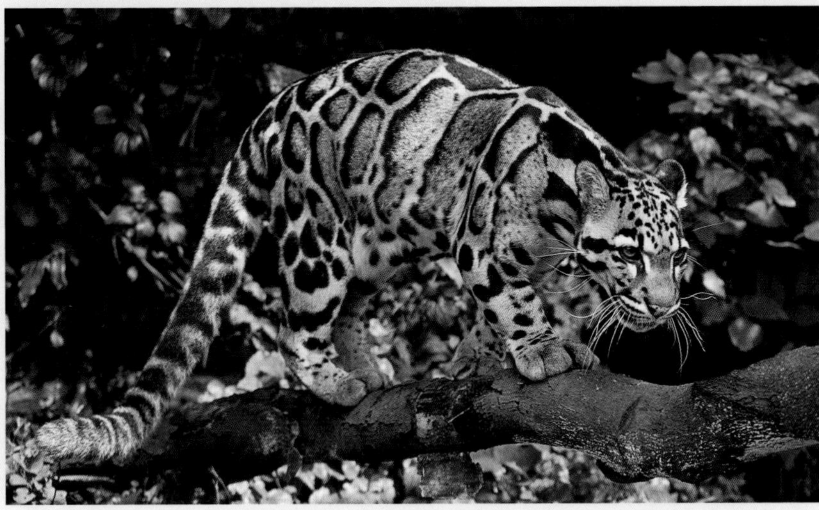

i Outline four factors that are used to determine the conservation status of an organism on the Red List. [2]

ii With reference to the case history of a named critically endangered species or endangered species, describe the human factors that have led to its conservation status. [2]

3 a The World Wide Fund for Nature (WWF) estimates that there are now more tigers in captivity than in the wild. Evaluate the use of zoos for the preservation of the tiger population. [3]

b Suggest two criteria that should be used to design a protected area for tigers. [2]

4 a Population dynamics can be defined as, 'the study of changes and the reasons for changes in population size'. Discuss why an understanding of the population dynamics of an endangered species is essential to the efforts for its conservation. [8]

b Outline two reasons for the extinction of a named species and suggest how intervention measures can improve the conservation status of a species. [8]

c Justify which criteria you think should be used to judge the success of a conservation area. Evaluate the success of a named protected area using the criteria you have identified. [7]

04 Water, aquatic food production systems, and societies

Opposite: Volunteers try to clear water that is full of discarded plastic bottles and other garbage, blocking the Vacha Dam near the town of Krichim, Bulgaria

4.1 Introduction to water systems

Significant ideas

The hydrological cycle is a system of stores and flows that can be easily disrupted by human activities.

The ocean circulatory system influences global climates by transporting water and energy around the Earth.

Big questions

As you read this section, consider the following big questions:

- What strengths and weaknesses of the systems approach and the use of models have been revealed through this topic?
- To what extent have the solutions emerging from this topic been directed at *preventing* environmental impacts, *limiting* the extent of the environmental impacts, or *restoring* systems in which environmental impacts have already occurred?
- How are the issues addressed in this topic of relevance to sustainability or sustainable development?
- In what ways might the solutions explored in this topic alter your predictions for the state of human societies and the biosphere some decades from now?

Knowledge and understanding

- Solar radiation drives the hydrological cycle.
- Only a small fraction (approximately 2.6 per cent by volume) of the Earth's water storages are fresh water.
- Storages in the hydrological cycle include the atmosphere, organisms, soil, and various water bodies such as oceans, groundwater (aquifers), lakes, rivers, glaciers, and ice caps.
- Flows in the hydrological cycle include evapotranspiration, sublimation, evaporation, condensation, advection (wind-blown movement), precipitation, melting, freezing, flooding, surface run-off, infiltration, percolation and stream-flow/currents.
- Human impacts such as agriculture, deforestation and urbanization have a significant impact on surface run-off and infiltration.
- Ocean circulation systems are driven by differences in temperature and salinity that affect water density. The resulting differences in water density drive the ocean conveyor belt which distributes heat around the world, so affecting climate.

The hydrological cycle

Solar radiation drives the hydrological cycle.

SYSTEMS APPROACH

The global hydrological cycle refers to the movement of water between atmosphere, lithosphere, biosphere, and pedosphere (Figure 4.1). At a global scale, it can be thought of as a closed system with no losses. In contrast, at a local scale, the cycle generally has a single input – *precipitation* (PPT) – and two major losses (outputs) – *evapotranspiration* (EVT) and *run-off*. Exotic rivers are an exception – they bring water into a region from a different climate zone (e.g. the Nile flowing through the Sahara desert brings water from the Ethiopian Highlands).

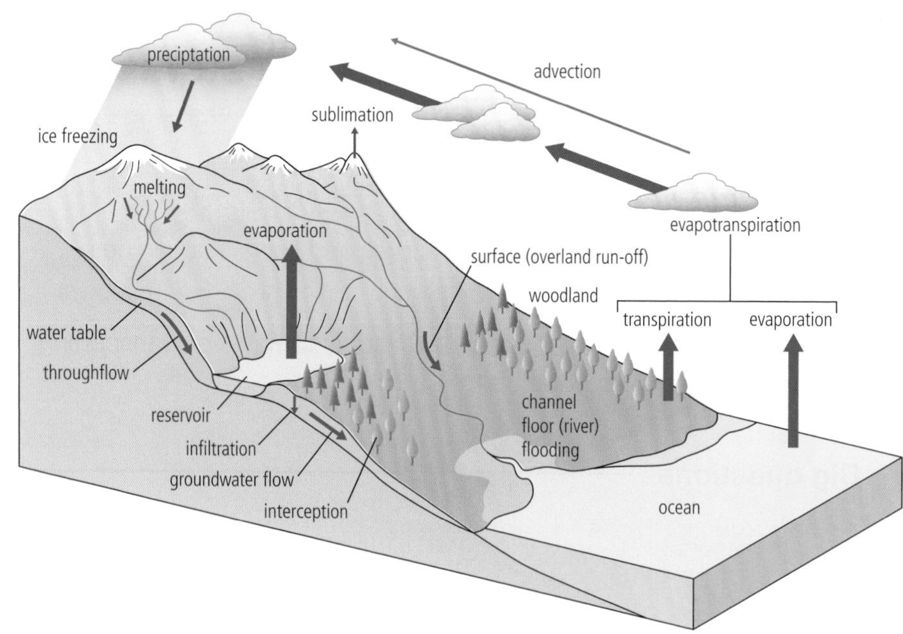

Figure 4.1 The global hydrological cycle

TOK

The global hydrological cycle is complex. To enable our understanding of how the cycle works, scientists use a systems model. This is a simplified structuring of reality which shows just a few parts of the system. Many parts of the cycle cannot be observed. Hence, scientists use the drainage basin hydrological cycle (Figure 4.2). The drainage basin is taken as the unit of study rather than the global system, as it is easier to observe and measure more parts of the cycle. The basin cycle is an open system: the main input is precipitation which is regulated by various means of storage.

Water can be stored at a number of places within the cycle. These stores include organisms, depressions on the Earth's surface, soil moisture, groundwater and water bodies such as rivers and lakes, and bodies of ice such as glaciers and ice caps. The global hydrological cycle also includes stores in the oceans and the atmosphere.

Solar radiation drives the hydrological cycle. This is because the main source of energy available to the Earth is the Sun. In some places, there are important local sources of heat; for example, geothermal heat in Iceland and human-related (anthropogenic) sources in large-scale urban-industrial zones. However, solar heating is the main cause of variations in the hydrological cycle, as well as the main cause of global temperature patterns and global wind patterns.

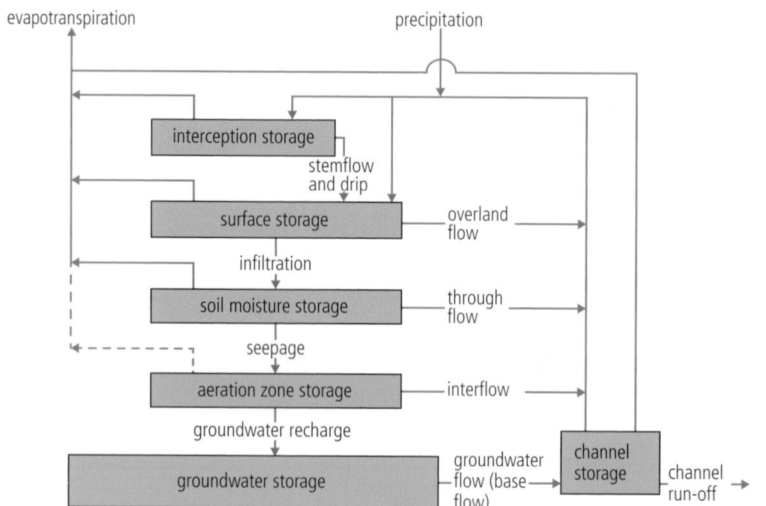

Figure 4.2 Systems diagram of the hydrological cycle

Global water stores

Only a small fraction (approximately 2.5 per cent by volume) of the Earth's water storages is fresh water.

Table 4.1 summarizes the Earth's water stores. Of the fresh water, around 70 per cent is in the form of ice caps and glaciers, around 30 per cent is groundwater. The

Reservoir		Thousands of cubic kilometres	% of total
ocean		1350000	97.403
atmosphere		13.0	0.00094
land		35977.8	2.596
of which	in rivers	1.7	0.00012
	in freshwater lakes	100.0	0.0072
	in inland seas	105.0	0.0076
	in soil water	70.0	0.0051
	in groundwater	8200.0	0.592
	in ice caps/glaciers	27500.0	1.984
	in biota	1.1	0.00088

Table 4.1 Global water reservoirs

water in lakes, rivers and biota, soil water, and atmospheric water vapour is a very tiny percentage of the whole. Fresh water on the surface of the Earth to which we have direct access (lakes and rivers) is around 0.3 per cent of the total. Atmospheric water vapour contains around 0.001 per cent of the Earth's total water volume. Taken together, all the forms in which the Earth's water can exist are called the hydrosphere.

Storages in the hydrological cycle include organisms, soil water and the atmosphere, and various water bodies such as oceans, groundwater (aquifers), lakes, rivers, glaciers, and ice caps.

The different forms of water in the Earth's water budget are fully recycled during the hydrological cycle but at very different rates. The time for a water molecule enter and leave a part of the system (i.e. the time taken for water to completely replace itself in part of the system) is called the **turnover time.** Turnover time varies enormously between different parts of the system (Table 4.2).

The degree to which water can be seen as a renewable or non-renewable resource depends on where it is found in the hydrological cycle. Renewable water resources

Water location	Turnover time
polar ice caps	10000 years
ice in the permafrost	10000 years
oceans	2500 years
groundwater	1500 years
mountain glaciers	1500 years
large lakes	17 years
bogs	5 years
upper soil moisture	1 year
atmospheric moisture	12 days
rivers	16 days
biological water	a few hours

Table 4.2 Turnover time for different parts of the hydrosphere

are waters that are recycled yearly or more frequently in the Earth's water turnover processes. Thus, groundwater is a non-renewable source of water as turnover time is very long. An aquifer is an underground formation of permeable rock or loose material which stores groundwater. Aquifers can produce useful quantities of water when tapped by wells. Aquifers come in all sizes, from small (a few hectares in area) to very large (covering thousands of square kilometres). They may be only a few metres thick, or they may measure hundreds of metres from top to bottom. Unsustainable use of aquifers results in depleting the storage and has unfavourable consequences: it depletes the natural resource and disturbs the natural equilibrium established over centuries. Restoration requires tens to hundreds of years.

Flows

Flows in the hydrological cycle include evapotranspiration (EVT), sublimation, evaporation, condensation, advection (wind-blown movement), precipitation, melting, freezing, flooding, surface run-off, infiltration, percolation, and stream-flow/ currents.

The hydrological cycle comprises evaporation from oceans, water vapour, condensation, precipitation, run-off, groundwater and EVT (Table 4.3). If 100 units represents global precipitation (on average 860 mm per year), 77 per cent falls over the oceans and 23 per cent on land. A total of 84 units enter the atmosphere by evaporation via the oceans, thus there is a horizontal transfer of 7 units from the land to the sea. Of precipitation over the land, 16 units are evaporated or transpired and 7 units run off to the oceans. There may be some time lag between precipitation and eventual run-off. About 98 per cent of all free water on the globe is stored in the oceans.

Precipitation includes all forms of rainfall, snow, frost, hail and dew. It is the conversion and transfer of moisture in the atmosphere to the land. *Interception* refers to water that is caught and stored by vegetation. It has three main components:

- *interception loss* – water which is retained by plant surfaces and which is later evaporated away or absorbed by the plant
- *throughfall* – water which either falls through gaps in the vegetation or which drops from leaves, twigs or stems
- *stemflow* – water which trickles along twigs and branches and finally down the main trunk.

Table 4.3 Global water exchanges

Annual exchange		Thousands of cubic kilometres	
Evaporation		496.0	
of which	from oceans		425.0
	from land		71.0
Precipitation		496.0	
of which	to oceans		385.0
	to land		111.0
Run-off to oceans		41.5	
of which	from rivers		27.0
	from groundwater		12.0
	from glacial meltwater		2.5

Interception loss varies with different types of vegetation. Interception is less from grasses than from deciduous woodland owing to the smaller surface area of the grass shoots. From agricultural crops, and from cereals in particular, interception increases with crop density. Coniferous trees intercept more than deciduous trees in winter, but the reverse is true in summer.

Evaporation is the process by which a liquid or a solid is changed into a gas. It is the conversion of solid and liquid precipitation (snow, ice, and water) to water vapour in the atmosphere. It is most important from oceans and seas. Evaporation increases under warm, dry, and windy conditions and decreases under cold, calm conditions. Evaporation losses will be greater in arid and semi-arid climates than they will be in polar regions. Factors affecting evaporation include meteorological factors such as temperature, humidity, and windspeed. Of these, temperature is the most important factor. Other factors include the amount of water available, vegetation cover, and colour of the surface (*albedo* or reflectivity of the surface).

Transpiration is the process by which water vapour escapes from living plants, mainly from the leaves, and enters the atmosphere. The combined effects of evaporation and transpiration are normally referred to as evapotranspiration (EVT). EVT represents the most important aspect of water loss, accounting for the loss of nearly 100 per cent of the annual precipitation in arid areas and 75 per cent in humid areas. Only over ice and snow fields, bare rock slopes, desert areas, water surfaces, and bare soil will purely evaporative losses occur.

Infiltration is the process by which water soaks into or is absorbed by the soil. The *infiltration capacity* is the maximum rate at which rain can be absorbed by a soil in a given condition. Infiltration capacity decreases with time through a period of rainfall until a more or less constant value is reached. Infiltration rates of 0–4 mm h^{-1} are common on clays whereas 3–12 mm h^{-1} are common on sands. Vegetation also increases infiltration. This is because it intercepts some rainfall and slows down the speed at which it arrives at the surface. For example, on bare soils where rainsplash impact occurs, infiltration rates may reach 10 mm h^{-1}. On similar soils covered by vegetation rates of between 50 and 100 mm h^{-1} have been recorded. Infiltrated water is chemically rich as it picks up minerals and organic acids from vegetation and soil.

Infiltration is inversely related to overland run-off and is influenced by a variety of factors such as duration of rainfall, antecedent soil moisture (i.e. pre-existing levels of soil moisture), soil porosity, vegetation cover, raindrop size and slope angle. In contrast *overland flow* (surface run-off) is water that flows over the land's surface. It occurs in two main circumstances:

- when precipitation exceeds the infiltration rate
- when the soil is saturated (all the pore spaces are filled with water).

In areas of high precipitation intensity and low infiltration capacity, overland run-off is common. This is clearly seen in semi-arid areas and in cultivated fields. By contrast, where precipitation intensity is low and infiltration is high, most overland flow occurs close to streams and river channels.

Condensation is the process by which vapour passes into a liquid form. It occurs when air is cooled to its dew point or becomes saturated by evaporation into it. Further cooling leads to condensation on surfaces to form water droplets or frost.

Sublimation refers to the conversion of a solid into a vapour with no intermediate liquid state. Under conditions of low humidity, snow can be evaporated directly into water vapour without entering the liquid water state. Sublimation is also used to describe the direct deposition of water vapour onto ice.

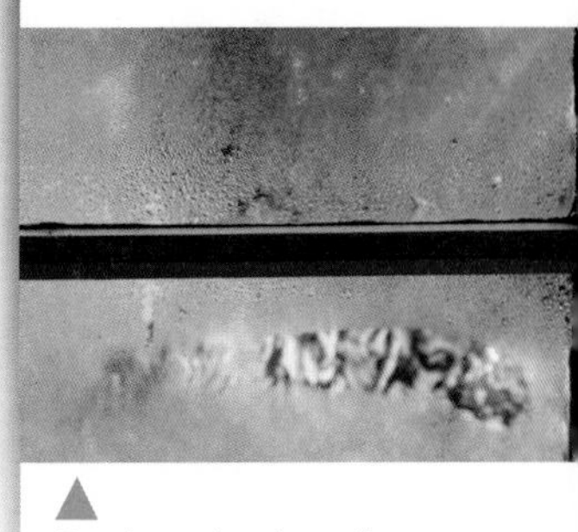

Condensation is easily observable on a window.

Advection is the horizontal transfer of energy or matter. It refers particularly to the movement of air in the atmosphere which results in the redistribution of such elements as warm or cold air, moisture and pollutants.

Freezing refers to the change of liquid water into a solid ice, once temperatures fall below 0 °C. *Melting* is the change from a solid ice to a liquid water when the air temperature rises above 0 °C.

Stream-flow or *currents* refers to the movement of water is channels, such as streams and rivers. The water may enter the stream as direct channel precipitation (it falls on the channel), or it may reach the channel by surface run-off, groundwater flow (baseflow), or throughflow (water flowing through the soil).

You should be able to construct and analyse a hydrological cycle diagram.

Flooding refers to the covering (inundation) of normally dry land by water. It occurs when the river channel is unable to contain the amount of water added to it. Flooding may occur for a variety of reasons (e.g. heavy rainfall, prolonged rain, snowmelt, tidal surges, dam failure). Human activities in the drainage basin may intensify flood conditions and increase flood frequency.

Human impacts such as agriculture, deforestation, and urbanization have a significant impact on surface run-off and infiltration.

Human influences on the hydrological cycle

Human modifications are made at every scale. Good examples include large-scale changes of channel flow, irrigation and drainage, and abstraction of groundwater and surface water for domestic and industrial use.

Eutrophication and dead zones are discussed on pages 255–61 and 262, respectively.

The impact of agriculture on water systems

Irrigation

Irrigation is the addition of water to areas where there is insufficient for adequate crop growth. Water can be taken from surface stores, such as lakes, dams, reservoirs and rivers, or from groundwater. Types of irrigation range from total flooding, as in the case of paddy fields, to spray and drip irrigation, where precise amounts are measured out to each individual plant (Figure 4.3).

Drip irrigation.

Centre pivot irrigation.

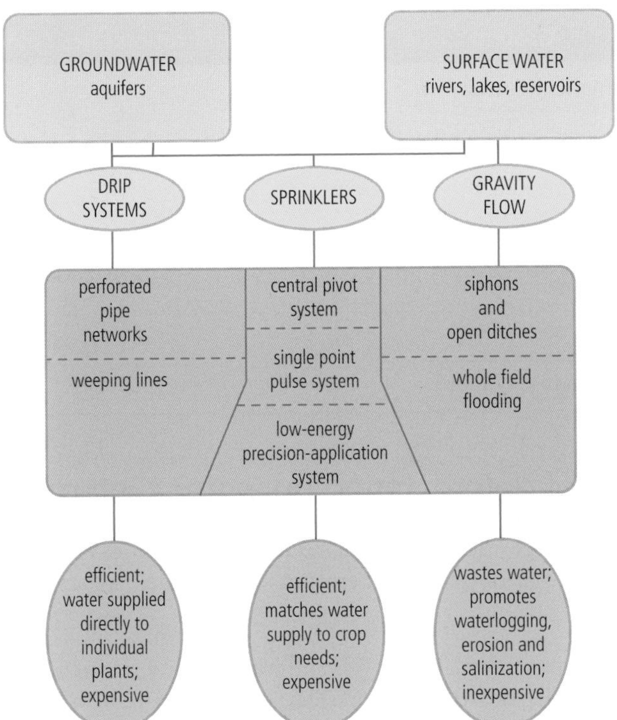

Figure 4.3 Types of irrigation

Irrigation occurs in MEDCs and LEDCs. For example, large parts of the USA and Australia are irrigated. In Texas, irrigation has lowered the water table by as much as 50 m. By contrast, in the Indus Plain in Pakistan, irrigation has raised the water table by as much as 6 m in the last 100 years and caused widespread *salinization*. This occurs when groundwater levels are close to the surface. Capillary forces bring water to the surface where it may be evaporated leaving behind any soluble salts that it is carrying. Some irrigation, especially for growing rice crops in paddies, requires huge amounts of water. As water evaporates in the hot sun, the salinity levels of the remaining water increase. This could lead to the promotion of salt-tolerant organisms.

Irrigation can reduce the Earth's albedo (reflectivity) by as much as 10 per cent. This is because a reflective sandy surface may be replaced by one with dark green crops. Irrigation can also cause changes in precipitation. Large-scale irrigation in semi-arid areas, such as the High Plains of Texas, have been linked with increased rainfall, hail storms, and tornadoes. Under natural conditions semi-arid areas have sparse vegetation and dry soils in summer. However, when irrigated, these areas have moist soils in summer and a complete vegetation cover. EVT rates increase and result in increases in the amount of summer rainfall, as has been seen in Kansas, Nebraska, Colorado, and Texas. In addition, hail storms and tornadoes are more common over irrigated areas compared with non-irrigated areas.

Farming can also have a major impact on interception and infiltration. Interception is determined by vegetation type and density. In farmland areas, cereals intercept less than broad leaf crops. Row crops leave a lot of soil bare. Infiltration is up to five times greater under forests compared with grassland. This is because the forest channels water down tree trunks and roots.

Land use practices are also important (Table 4.4). Grazing leads to a decline in infiltration due to compaction of the soil. Ploughing increases infiltration because it loosens soils. Waterlogging and salinization are common if there is poor drainage.

Table 4.4 Influence of ground cover on infiltration rates

Ground cover	Infiltration rates / mm h^{-1}
old permanent pasture	57
permanent pasture: moderately grazed	19
permanent pasture: heavily grazed	13
strip-cropped	10
weeds or grain	9
clean tilled	7
bare, crusted ground	6

The impact of deforestation on water systems

Deforestation has a large impact on water systems: after deforestation, flood levels in rivers increase. Changes in run-off and erosion following deforestation are shown in Table 4.5.

Following forest regeneration, flood levels and water quality return to pre-removal levels. But this may take decades to occur. The return to pre-removal levels after regeneration include:

- higher interception rates of mature forests
- decreased overland run-off beneath a mature forest
- higher infiltration rates beneath forests
- deeper soils beneath a cover of trees.

The replacement of natural vegetation by crops needs to be carefully managed. The use of shade trees and cover crops is a useful way of reducing soil erosion following deforestation. Grazing tends to increase overland run-off because of surface compaction and vegetation removal.

Deforestation is also linked with increases in the sediment and chemical loads of streams as nitrates are lost from soil and erosion occurs (Table 4.6). In an extreme case of deforestation in the north-east of the USA, sediment loads increased 15-fold and nitrate loads increased by almost 50 times! Other examples are not so extreme - much depends on how the forest is managed. If deforestation is only partial, there is less sediment load. If replanting takes place quickly the effects of deforestation are reduced.

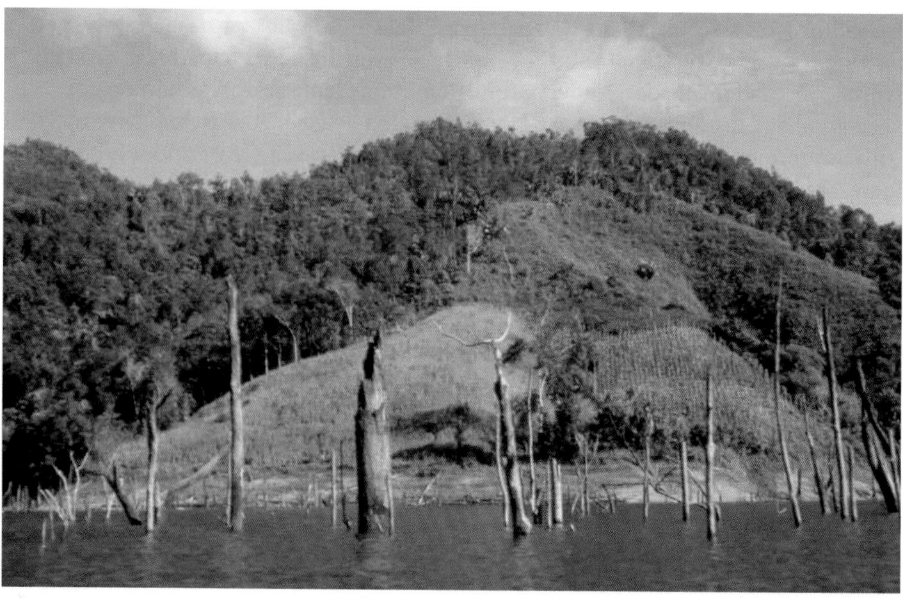

Deforestation, agriculture and flooding at Batang Ai, Sarawak, Malaysia

Table 4.5 Changes in run-off and erosion following deforestation

Locality	Average annual rainfall / mm	Slope / %	Annual run-off / %			Erosion / t ha^{-1} yr^{-1}		
			A	B	C	A	B	C
Ougadougou (Burkina Faso)	850	0.5	2.5	2–32	40–60	0.1	0.6–0.8	10–20
Sefa (Senegal)	1300	1.2	1.0	21.2	39.5	0.2	7.3	21.3
Bouake (Ivory Coast)	1200	4.0	0.3	0.1–26	15–30	0.1	1–26	18–30
Abidjan (Ivory Coast)	2100	7.0	0.4	0.5–20	38	0.03	0.1–90	108–170
Mbapwa (Tanzania)	approx. 570	6.0	0.4	26.0	50.4	0	78	146

A = forest or ungrazed thicket; B = crop; C = barren soil

Table 4.6 Changes in nitrate–nitrogen levels after deforestation

Site	Nature of disturbance	Nitrate–nitrogen loss / kg ha^{-1} yr^{-1} Control	Nitrate–nitrogen loss / kg ha^{-1} yr^{-1} Disturbed
Hubbard Brook (New Hampshire)	clear-cutting without vegetation removal, herbicide inhibition of re-growth	2.0	97
Gale River (New Hampshire)	commercial clear-cutting	2.0	38
Fernow (West Virginia)	commercial clear-cutting	0.6	3.0
Coweeta (North Carolina)	complex	0.05	7.3*
HJ Andrews forest (Oregon)	clear-cutting with slash burning	0.08	0.26
Alsea River (Oregon)	clear-cutting with slash burning	3.9	15.4
	Mean	1.44	26.83

* Second year of recovery after a long-term disturbance: all other values are for first year of recovery

Indeed, young plants that are growing rapidly take up large amounts of water and nutrients from the soil, thereby reducing the rate of overland run-off and the chemical load of streams. Much depends on the type of vegetation, its relative density, size and rates of growth.

Deforestation can also have an important effect on local climate. The removal of trees leads to an increase in light intensity, temperature, wind speed, and moisture at ground level. This has a number of consequences, including:

- organic matter is decomposed at a faster rate
- raindrop impact increases
- EVT rates decrease
- overland run-off increases.

The reduced forest traps less rain; the litter layer is reduced and this, in turn, intercepts less rainfall; the proportion of bare ground increases, and raindrop impact compacts the soil. As forests are cut down, more light gets through to the ground so new vegetation can grow there. This encourages grazing animals which eat the buds of growing trees. Consequently, vegetation that grows from the base (e.g. grasses) is favoured over vegetation that grows from buds (e.g. trees). In addition, the grazers compact the soil and increase its density. This leads to decreasing infiltration, and increased overland flow, which increases soil erosion.

Thinner soils store and transport less moisture. Consequently, there is increased surface run-off and sediment discharge. Moreover, the removal of some trees may lead to a reduction in the amount of groundwater. This happens because the thin soils of the cut forest are more exposed to direct sunlight and lose more moisture through evaporation. Soils under a complete forest canopy are shaded, thus evaporation losses are less, and they provide more water to groundwater stores.

Deforestation is, therefore, associated with reduced infiltration rates, reduced soil water storage, and increased rates of surface run-off and soil erosion (Table 4.7). There are also changes in stream morphology (shape and size), increases in the mean annual flood, and an increase in the frequency of landslides. This is because on a forested slope, tree roots bind the soil, whereas on deforested slopes there is less anchorage of soil and an increase in landslides.

As a result of the intense surface run-off and soil erosion, rivers have a higher *flood peak* and a shorter time lag. However, in the dry season, river levels are lower, and rivers have greater turbidity (murkiness due to more sediment) as they carry more silt and clay in suspension.

Other changes relate to *climate*. As deforestation progresses, there is a reduction in the amount of water that is transpired from the vegetation, hence the recycling of water slows down. EVT rates from savannah grasslands are estimated to be about a third of that of tropical rainforest. Thus, mean annual rainfall is reduced, and the seasonality of rainfall increases.

Table 4.7 Soil erosion and deforestation in the Himalayas

Rainfall intensity / mm hr^{-1}	Average soil losses / kg ha^{-1} by percentage of area forested			
	20–30%	40–50%	60–70%	80–90%
0–9	6.1	4.0	2.9	2.6
10–19	19.1	19.2	9.8	10.6
>20	43.6	25.2	28.1	16.9

Urban development in Seoul, South Korea – notice the relative lack of vegetation and the large amount of impermeable surfaces.

The effect of urbanization on water systems

There are many changes to the water cycle that occur in urban areas (Table 4.8). The changes depend, in part, on the size of the urban area and the nature of land use. Due to the increase in impermeable surfaces (Table 4.9), there is more overland run-off. In most cities, due to the many storm sewers and drainage channels, water is diverted into underground channels very quickly. Due to the relative lack of vegetation in some parts, temperatures become quite high, increasing evaporation. Flash floods may occur owing to rapid run-off, little absorption, and a lack of storage.

Table 4.8 The potential impact of urbanization on water systems

The result of urbanization	Potential hydrological response
removal of trees and vegetation	• decreased EVT and interception
initial construction of houses, streets, and culverts	• decreased infiltration and lowered groundwater table • increased storm flows and decreased base flows during dry periods • increased stream sedimentation
development of residential, commercial and industrial areas	• greatly increased volume of run-off and flood damage potential
construction of storm drains and channel improvements	• local relief from flooding • concentration of floodwaters may increase flood problems downstream
drainage density	• basins with a high drainage density (e.g. urban basins with a network of sewers and drains) respond very quickly • networks with a low drainage density have a very long time lag
land use	• land uses which create impermeable surfaces, or reduce vegetation cover, decrease interception and increase overland flow
porosity and impermeability of 'artificial surface' rocks and soils	• urban areas contain large areas of impermeable surfaces which cause more water to flow overland; this causes greater peak flows • rocks such as chalk and gravel are permeable and allow water to infiltrate and percolate; this reduces peak flow and increases the time lag • sandy soils allow water to infiltrate, whereas clay is much more impermeable and causes water to pass overland

Type of surface	Impermeability / %
water-tight roof surfaces	70–95
asphalt paving in good order	85–90
stone, brick and wooden block pavements: with tightly cemented joints with open or uncertain joints	 75–85 50–70
inferior block pavements with open joints	40–50
tarmacadam roads and paths	25–60
gravel roads and paths	15–30
unpaved surfaces, railway yards, vacant land	10–30
parks, gardens, lawns, meadows – depending on the surface slope and character of the sub-soil	5–25

Table 4.9 Impermeability of urban surfaces

! You should be able to discuss the human impact on the hydrological cycle.

CHALLENGE YOURSELF

Thinking skills 

To what extent can the hydrological cycle be considered an open or closed system?

Ocean circulation

The distribution of the oceans and ocean currents

The oceans cover approximately 70 per cent of the Earth's surface, and are of great importance to humans. Particularly important is through the atmosphere–ocean link, by which oceans regulate climatic conditions. Warm ocean currents move

Ocean circulation systems are driven by differences in temperature and salinity that affect water density. The resulting difference in water density drives the ocean conveyor belt which distributes heat around the world, so affecting climate.

Figure 4.4 The world's main ocean currents

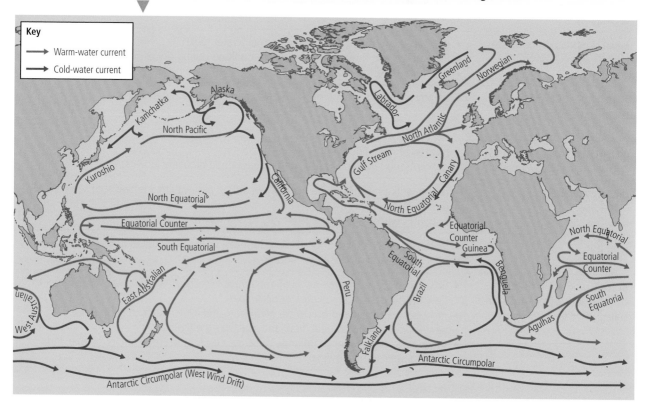

water away from the equator towards the poles, whereas cold ocean currents move water away from the cold regions towards the equator (Figure 4.4). The warm Gulf Stream, for instance, transports 55 million cubic metres per second from the Gulf of Mexico towards north-west Europe. Without it, the temperate lands of north-west Europe would be more like the sub-Arctic. The cold Peru current brings nutrient-rich waters dragged to the surface by offshore winds. In addition, there is the Great Ocean Conveyor Belt (page 224). This deep, global-scale circulation of the ocean's waters effectively transfers heat from the tropics to colder regions.

Salinity

Oceanic water varies in salinity (Figure 4.5). Average salinity is about 35 parts per thousand (ppt). Concentrations of salt are higher in warm seas, because of the high evaporation rates of the water. In tropical seas, salinity decreases sharply with depth. The run-off from most rivers is quickly mixed with ocean water by the currents, and has little effect on reducing salinity. However, a large river such as the Amazon in South America may result in the ocean having little or no salt content for over a kilometre or more out to sea.

The freezing and thawing of ice also affects salinity. The thawing of large icebergs (made of frozen fresh water and lacking any salt) decreases salinity, while freezing of seawater increases the salinity temporarily. Salinity levels increase with depth.

The predominant mineral ions in seawater are chloride (54.3 per cent) and sodium (30.2 per cent), which combine to form salt. Other important minerals in the sea include magnesium and sulfate ions.

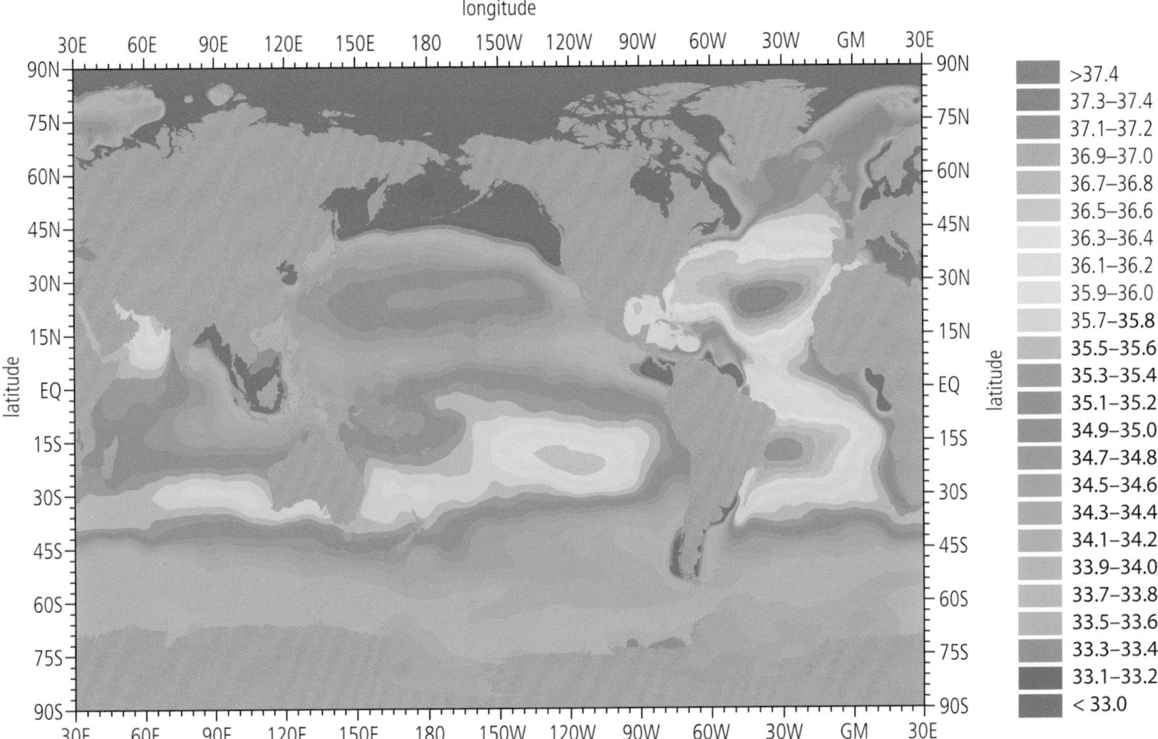

Figure 4.5 Global variations in oceanic salinity

Temperature

Temperature varies considerably at the surface of the ocean, but there is little variation at depth (Figure 4.6). In tropical and subtropical areas, sea surface temperatures in excess of 25 °C are caused by insolation. From about 300 to 1000 m, the temperature declines steeply to about 8–10 °C. Below 1000 m, the temperature decreases to a more uniform 2 °C in the ocean depths.

The temperature profile is similar in the mid-latitudes (40–50° N and S), although there are clear seasonal variations. Summer temperatures may reach 17 °C, whereas winter sea temperatures are closer to 10 °C. There is a more gradual decrease in temperature with depth (thermocline).

Density

Temperature, salinity, and pressure affect the density of seawater. Large water masses of different densities are important in the layering of the ocean water (denser water sinks). As temperature increases, water becomes less dense. As salinity increases, water becomes more dense. As pressure increases, water becomes more dense. A cold, highly saline, deep mass of water is very dense, whereas a warm, less saline, surface water mass is less dense. When large water masses with different densities meet, the denser water mass slips under the less dense mass. These responses to density are the reason for some of the deep ocean circulation patterns.

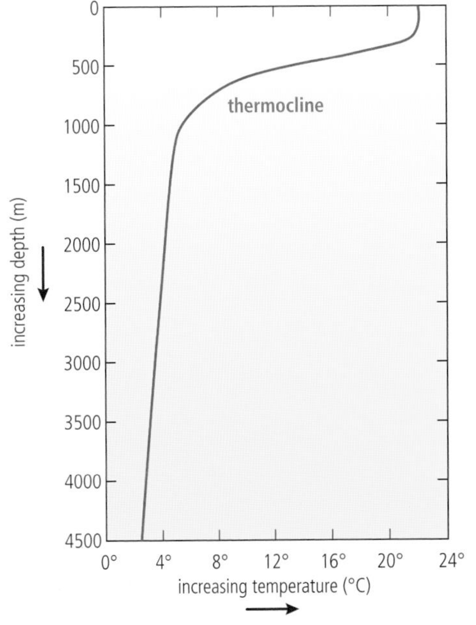

Figure 4.6 Ocean temperature and depth

The Great Ocean Conveyor Belt

The oceanic conveyor belt is a global thermohaline circulation, driven by the formation and sinking of deep water and responsible for the large flow of upper ocean water (Figure 4.7). In addition to the transfer of energy by wind and the transfer of energy by ocean currents, there is also a transfer of energy by deep-sea currents. In polar regions, cold, salty water sinks to the depths and makes its way towards the equator. It then spreads into the deep basins of the Atlantic, the Pacific, and the Indian Oceans. Surface currents bring warm water to the North Atlantic from the Indian and Pacific Oceans. These waters give up their heat to cold winds which blow from Canada across the North Atlantic. This water then sinks and starts the reverse convection of the deep ocean current. The amount of heat given up is about a third of the energy received from the Sun.

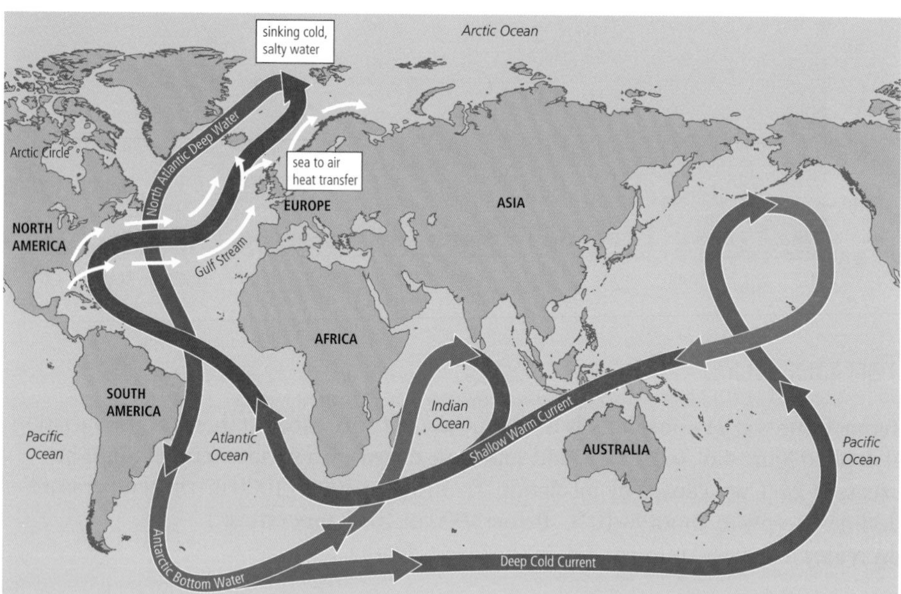

Figure 4.7 The Great Ocean Conveyor Belt

Because the conveyor operates in this way, the North Atlantic is warmer than the North Pacific, so there is proportionally more evaporation there. The water left behind by evaporation is saltier and therefore much denser, which causes it to sink. Eventually the water is transported into the Pacific where it mixes with warmer water and its density is reduced.

Specific heat capacity

The specific heat capacity is the amount of energy it takes to raise the temperature of a body. It takes more energy to heat up water than it does to heat land. However, it takes longer for water to lose heat. This is why the land is hotter than the sea by day, but colder than the sea by night. Places close to the sea are cool by day, but mild by night. With increasing distance from the sea this effect is reduced.

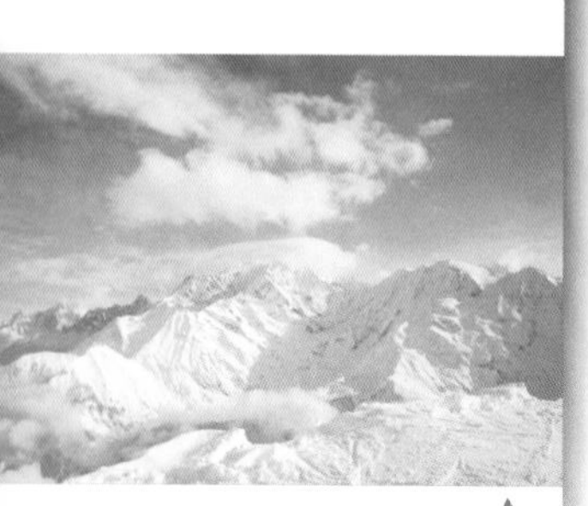

Mountainous scene, the Alps

Exercises

1. Define the term *hydrological cycle*.
2. Study the photograph opposite of a mountainous scene from the European Alps. Identify two visible stores of fresh water.

3. Study the photograph opposite of part of the River Thames, UK. Identify three stores in the hydrological cycle that are visible in the photograph.
4. Use the data in Table 4.1 to construct a pie chart to show the main components of the global hydrological cycle.
5. Define the terms *transpiration* and *sublimation.*
6. Outline the factors that increase the rate of evaporation.
7. Construct a flow diagram to show the main characteristics of the global water exchanges (Table 4.3)
8. Study Figure 4.3. Identify the least efficient method of irrigation.
9. Which two forms of irrigation are most efficient?
10. Explain why drip irrigation systems are more efficient than whole field flooding (gravity flow) types of irrigation.
11. Study Table 4.4.
    a. Define the term *infiltration.*
    b. Suggest why pastoral farming (pasture) allows more infiltration than arable farming (cropped or grain).
    c. Suggest why farmed land has a higher infiltration than bare earth.
12. Using Table 4.5 describe how erosion varies with annual run-off. How do rates of erosion and run-off vary with average annual rainfall?
13. Compare annual run-off under forest or ungrazed thicket, crop, and barren soil.
14. Table 4.6 shows changes in nitrate-nitrogen levels after deforestation. How do nitrate-nitrogen losses differ between disturbed (deforested) plots and control plots? Use evidence to support your answer.
15. Table 4.7 shows data for soil loss, rainfall intensity and percentage forest cover in part of the Himalayas.
    a. Describe how average soil loss varies with (i) forest cover and (ii) rainfall intensity.
    b. Suggest reasons for your answers to part A.
16. Study Table 4.8. In what ways does the drainage of cities differ from natural drainage systems? In what ways does this influence the hydrological cycle within urban areas?
17. Study Table 4.9. In which parts of a city would you expect there to be:
    a. most impermeable surfaces
    b. least impermeable surfaces?
18. Explain briefly how different land uses may influence the hydrological cycle within urban areas.
19. Describe the global variations in oceanic salinity as shown in Figure 4.5.
20. Briefly explain the operation of the Great Ocean Conveyor Belt (Figure 4.7).

River Thames, UK

Big questions

Having read this section, you can now discuss the following big questions:

- What strengths and weaknesses of the systems approach and the use of models have been revealed through this topic?
- To what extent have the solutions emerging from this topic been directed at *preventing* environmental impacts, *limiting* the extent of the environmental impacts, or *restoring* systems in which environmental impacts have already occurred?
- How are the issues addressed in this topic of relevance to sustainability or sustainable development?
- In what ways might the solutions explored in this topic alter your predictions for the state of human societies and the biosphere some decades from now?

Points you may want to consider in your discussions:

- Many hydrological cycles cross international boundaries. How does this affect the management of water?
- Identify the solutions to the impacts of agriculture, deforestation, and urbanization on the hydrological cycle.
- Can agriculture, deforestation, and urbanization allow for the natural functioning of the hydrological cycle?
- In what ways may population growth and human activities have an impact on the hydrological cycles of the future?

4.2 Access to fresh water

Significant ideas

The supplies of freshwater resources are inequitably available and unevenly distributed and this can lead to conflict and concerns over water security.

Freshwater resources can be sustainably managed using a variety of different approaches.

Big questions

As you read this section, consider the following big questions:

- What strengths and weaknesses of the systems approach and the use of models have been revealed through this topic?
- To what extent have the solutions emerging from this topic been directed at *preventing* environmental impacts, *limiting* the extent of the environmental impacts, or *restoring* systems in which environmental impacts have already occurred?
- How are the issues addressed in this topic of relevance to sustainability or sustainable development?
- In what ways might the solutions explored in this topic alter your predictions for the state of human societies and the biosphere some decades from now?

Knowledge and understanding

- Access to an adequate supply of fresh water varies widely.
- Climate change may disrupt rainfall patterns and further affect this access.
- As population, irrigation, and industrialization increase, the demand for fresh water increases.
- Freshwater supplies may become limited through contamination and unsustainable abstraction.
- Water supplies can be enhanced through reservoirs, redistribution, desalination, artificial recharge of aquifers, and rainwater harvesting schemes. Water conservation (including grey-water recycling) can help to reduce demand but often requires a change in attitude by water users.
- The scarcity of water resources can lead to conflict between human populations particularly where sources are shared.

Access to fresh water

Access to an adequate supply of fresh water varies widely.

Human populations require water for home use (drinking, washing, and cooking), agriculture (irrigation and livestock), industry (manufacturing and mining), and hydroelectric power. Given the scarcity of freshwater resources, the pressure put on them is great and likely to increase in the future in parts of the world (Figure 4.8). Without sustainable use it is likely that humans will face many problems. Already there are a billion people who live without clean drinking water, and 2.6 billion who lack adequate sanitation.

The world's available freshwater supply is not distributed evenly around the globe, either seasonally, or from year to year. About three-quarters of annual rainfall occurs in areas containing less than a third of the world's population, whereas two-thirds of the world's population live in the areas receiving only a quarter of the world's annual rainfall. For instance, about 20 per cent of the global average run-off each year is accounted for by the Amazon Basin, a vast region with fewer than 10 million people. Similarly, the Congo Basin accounts for about 30 per cent of the Africa's annual run-off, but has less than 10 per cent of its population.

African women using a river for washing clothes

As Figure 4.8 suggests, the availability of fresh water is likely to become more stressed in the future. This may be the result, in part, of climate change, whereby rising temperatures lead to melting glaciers and increased evaporation. Unequal access to water may cause a conflict between those who have an abundance of water and those who do not (case study, page 228).

Every year, more people die from poor quality water than from all forms of violence, including war.

Climate change may disrupt rainfall patterns and further affect access to fresh water.

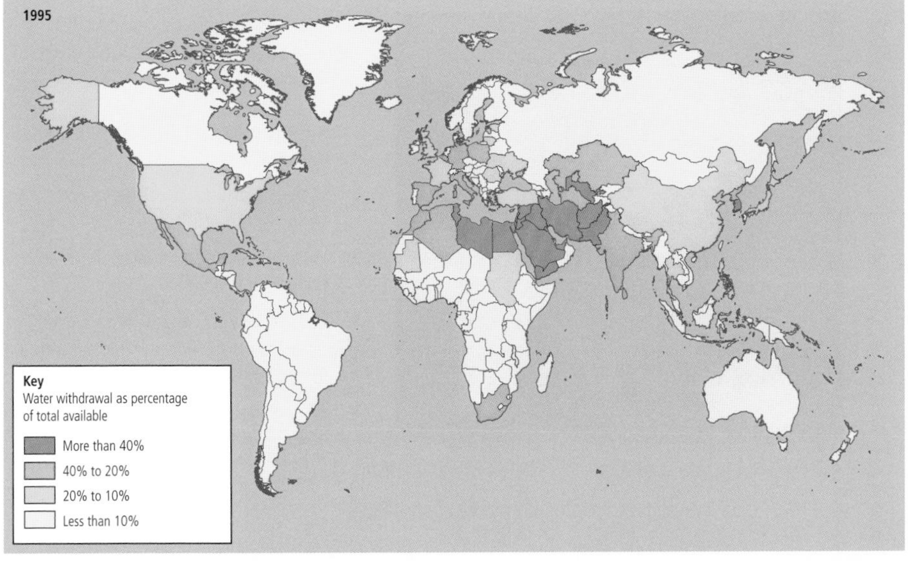

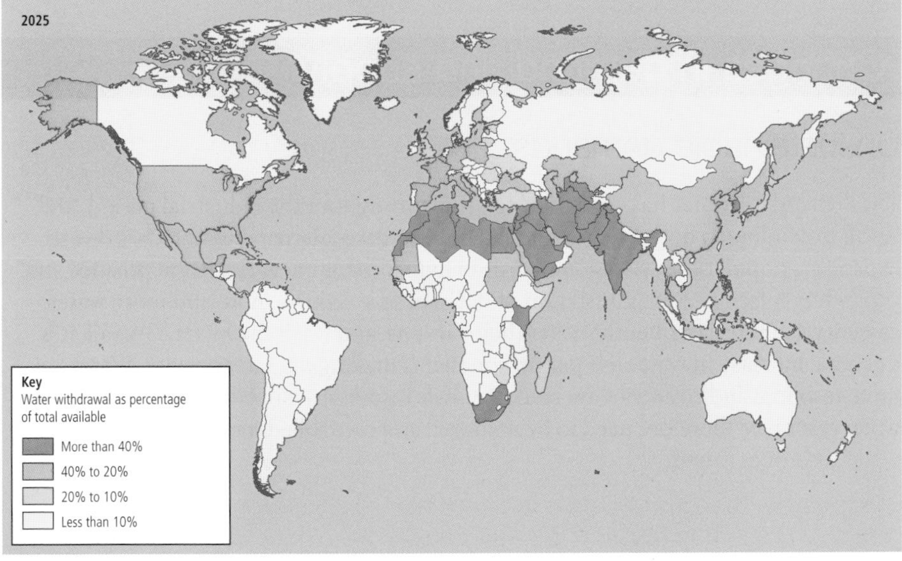

Figure 4.8 Water stress, 1995 and 2025

Case study

The price of water for the world's poor

In some cases, the poor actually pay more for their water than the rich. For example, in Port-au-Prince, Haiti, surveys have shown that households connected to the water system typically paid around US$1.00 per cubic metre, while unconnected customers forced to purchase water from mobile vendors paid from US$5.50 to a staggering US$16.50 per cubic metre.

Urban residents in the United States typically pay US$0.40–0.80 per cubic metre for municipal water of excellent quality. In Lima, Peru, poor families on the edge of the city pay vendors roughly US$3.00 per cubic metre, 20 times the price for families connected to the city system.

Residents of Jakarta, Indonesia, purchase water at US$0.09–0.50 per cubic metre from the municipal water company, US$1.80 from tanker trucks, and US$1.50–2.50 from private vendors – as much as 50 times more than residents connected to the city system. Jakarta's water supply and disposal systems were designed for 500 000 people, but today are struggling with more than 15 million.

The city suffers continuous water shortages, and less than 25 per cent of the population has direct access to water supply systems. The water level in what was previously an artesian aquifer is now generally below sea level – in some places 30 m below. Saltwater intrusion and pollution have largely ruined this as a source of drinking water.

TOK

A child collecting water from a well

Aid agencies frequently make use of emotive advertisements around water and food security issues.

Look at the photograph opposite and consider these questions.

1 Why is a child collecting water? Why not an adult?
2 Why is there such a large 'hole' in the ground?
3 Is this likely to be a rural or urban environment? What is the evidence for your answer?
4 Why is water scarce in this environment?
5 What impact could water scarcity have on children's education?
6 How could their lives be affected by water shortages?
7 To what extent can emotion be used to manipulate knowledge and actions?

As population, irrigation, and industrialization increase, the demand for fresh water increases.

Changes in demand and supply

Unsustainable demands

The demand for water has continued to grow throughout the industrial period, and is still expanding in both MEDCs and LEDCs. Increased demand in LEDCs is due to expanding populations, rising standards of living, changing agricultural practice and expanding (often heavy) industry. In MEDCs, people require more and more water as they wash more frequently, water their gardens, and wash their cars. Overall, it is a general increase in water use per person that is making demand heavier. Water is a finite resource and countries are reaching their resource availability limits – existing water resources therefore need to be managed and controlled more carefully, and new water resources found.

Water scarcity may become more widespread and have an increasing impact in the future – declining soil moisture has a very important impact on plant productivity.

As world population and industrial output have increased, the use of water has accelerated, and this is projected to continue. By 2025, global availability of fresh water may drop to an estimated 5100 m^3 per person per year, a decrease of 25 per cent on the 2000 figure. Rapid urbanization results in increasing numbers of people living in urban shanty towns where it is extremely difficult to provide an adequate supply of clean water or sanitation.

Freshwater supplies may become limited through contamination and unsustainable extraction.

Irrigation, industrialization, and population increase all make demands on the supply of fresh water. Global warming may disrupt rainfall patterns and water supplies. The hydrological cycle supplies humans with fresh water but we are withdrawing water from underground aquifers and degrading it with wastes at a greater rate than it can be replenished.

As water quality declines in some regions, more than half of native freshwater fish species and nearly one-third of the amphibians are at risk of extinction.

CONCEPTS: Environmental value systems

Technocentrists would argue that solutions can be found to sustain human populations and overcome unsustainable use of water resources. The cases outlined in this section demonstrate the serious situations many areas of the world face regarding their water resources and that how, even with technological progress, water supply remains of critical concern.

While some uses of river resources can be unsustainable (e.g. the siltation caused by dams), rivers can generally be replenished over a short period of time. Unsustainable use of fresh water largely concerns the overuse of aquifers. These non-renewable sources of water cannot be replenished at a fast-enough rates to make current usage sustainable. The USA is one of the world's largest agricultural producers. In certain areas, irrigation has been depleting groundwater resources beyond natural recharge rates for several years. For example, the High Plains (Ogallala) aquifer irrigates more than 20 per cent of USA cropland It is close to depletion in parts of Kansas because the water level has fallen so much (Figure 4.9). In some regions, water depletion now poses a serious threat to the sustainability of the agricultural and rural economy.

Water supplies can be enhanced through reservoirs, redistribution, desalination, artificial recharge of aquifers, and rainwater harvesting schemes. Water conservation (including grey-water recycling) can help to reduce demand but often requires a change in attitude by the water users.

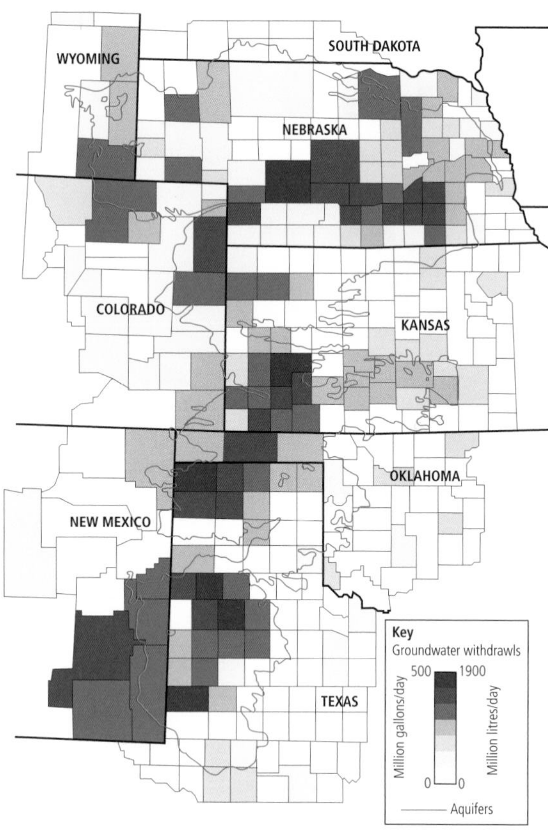

Figure 4.9 Impact of water use on High Plains aquifer

Projections over the next decade suggest that demand for water from irrigators will continue to rise, notably in countries where irrigated farming provides the major share of agricultural production (e.g. Australia, Mexico, Spain, and the USA). Groundwater pumping in Saudi Arabia exceeds replenishment by five times. This will lead to stiffer competition for water among other users (e.g. domestic use). Pressure on irrigated farming in many drier and semi-arid areas is being caused by the growing incidence and severity of droughts over the past decade, perhaps related to the impact of climate change. Demand for water will increase as a result, in part, of more people and greater demand for more food. Moreover, some of the water is being contaminated through excessive use of fertilizers and chemical waste dumping. Consequently, there will be further increased pressure of water resources.

Sustainably managing water resources

Water resources can be managed sustainably if individuals and communities make changes locally *and* this is supported by national government. Water usage needs to be coordinated within natural processes, and management strategy should ensure that non-renewable sources of fresh water (e.g. aquifers) are not used at an unsustainable rate. Use can be reduced by self-imposed restraint; for example, using water only when it is essential, not causing waste, and reusing supplies such as bath water. Education campaigns can increase local awareness of issues and encourage water conservation. There are many opportunities to increase fresh water supplies:

- retain water in reservoirs for use in dry seasons
- redistribute water from wetter areas to drier areas (e.g. from southern China to northern China)
- desalinate sea water (but this is expensive)
- water conservation (e.g. recycle grey-water – water that has already been used so is not fit for drinking but could be used for other purposes).

You must be able to evaluate the strategies that can be used to meet increasing demand for fresh water. You must give their strengths and their weaknesses

Water harvesting in Antigua – water flows over the concrete surface and is directed into a collection tank.

Water harvesting refers to making use of available water before it drains away or is evaporated. Water can be harvested in many ways. The main ones are:

- extraction from rivers and lakes (e.g. by primitive forms of irrigation such as the shaduf and Archimedes screw) – aided by gravity
- trapping behind dams and banks (bunds)
- pumping from aquifers (water-bearing rocks)
- desalinating saltwater to produce fresh water.

These can be achieved with either high technology or low technology methods. Efficient use or storage of water can also be achieved in many ways, for example:

- irrigation of individual plants rather than of whole fields
- covering expanses of water with plastic or chemicals to reduce evaporation
- storage of water underground in gravel-filled reservoirs (to reduce evaporation losses).

Sustainable use of water in cities and populated areas could be achieved by:

- making new buildings more water-efficient (e.g. recycling rainwater for sanitation and showers)
- offsetting new demand by fitting homes and other buildings with more water-efficient devices and appliances (e.g. dishwashers and toilets)
- expanding metering to encourage households to use water more efficiently.

In rural areas, solutions for sustainable water use could include selecting drought-resistant crops to reduce the need for irrigation (which uses up fresh water – much of it wasted through evaporation – and can cause soil degradation). Contamination of water supplies through fertilizers and pesticides can be addressed by reducing their use: organic fertilizers cause less pollution and biocontrol (i.e. natural predators of pests) can be used to reduce crop pests. Industries can be forced to remove pollutants from their wastewater through legislation.

The response of individuals and governments to make their use of fresh water more sustainable depend on the level of development of their country. Competing demands on fresh water vary between countries. Domestic water consumption is the minority water use in all countries, so the biggest impacts in terms of sustainable water use will be within the agricultural sector in LEDCs, and within the industrial sector in MEDCs (Figure 4.10).

Figure 4.10 Water use in LEDCs and MEDCs

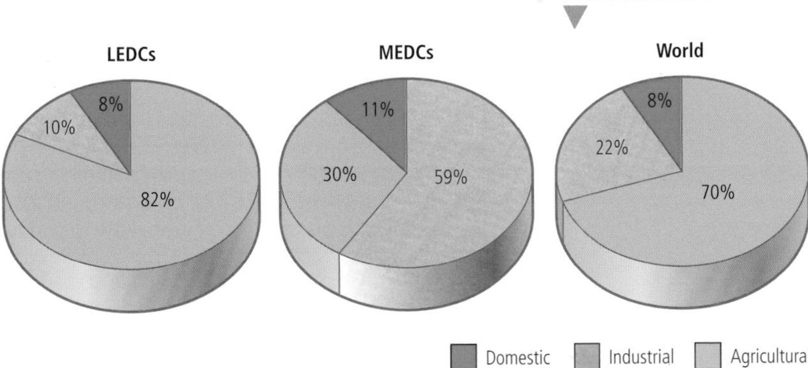

CONCEPTS: Environmental value systems

Environmental value systems determine water usage: ecocentrist managers plan to use these resources sustainably (i.e. by not diminishing them to such a degree as to make them non-replenishable). How would this differ from a technocentrist approach?

There is a long list of measures that can increase agricultural water productivity (Table 4.10). Drip irrigation is the method with the most untapped potential for farmers. Drip irrigation is a system of plastic tubes installed at or just below the soil surface to deliver water to individual plants. The water, which can be enhanced with fertilizer, is delivered to the roots of plants, so that there is very little lost to evaporation. Drip irrigation can achieve as high as 95 per cent efficiency compared with 50–70 per cent for conventional flood systems. In surveys across the USA, Spain, Jordan, Israel, and India, drip irrigation has been shown to cut water use by between 30 per cent and 70 per cent and to increase crop yields by 20–90 per cent, even leading to a doubling of productivity. Nevertheless, drip irrigation accounts for only 1 per cent of all irrigated land worldwide. Low-cost irrigation methods are summarized in Table 4.11.

Table 4.10 Options for improving irrigation water productivity

Category	Measures
technical	• land levelling to apply water more uniformly • efficient sprinklers to apply water more uniformly • drip irrigation to cut evaporation and other water losses and to increase crop yields
managerial	• applying water when most crucial to a crop's yield • better maintenance of canals and equipment
institutional	• establishing water user organizations for better involvement of farmers and collection of fees • reducing irrigation subsidies and/or introducing conservation-orientated pricing
agronomic	• selecting crop varieties with high yields per litre of transpired water • better matching of crops to climate conditions and the quality of water available • selecting drought-tolerant crops where water is scarce or unreliable

Table 4.11 Low-cost irrigation methods for small farmers

Technology or method	General conditions where appropriate	Examples
cultivating wetlands, delta lands, valley bottoms	seasonally waterlogged floodplains or wetlands	• Niger and Senegal river valleys • fdambos of Zambia and Zimbabwe
Archimedean screw, shaduf or beam and bucket, hand pump	very small (less than 0.5 ha) farm plots underlain by shallow groundwater	• eastern India • Bangladesh • parts of South East Asia
Persian wheel, bullocks and other animal-powered pumps, low-cost mechanical pumps	similar to those above, but where the average size of farm plots is roughly 0.5–2.0 ha	• as above, plus: • parts of North Africa and Near East
various forms of low-cost micro-irrigation (including bucket kits, drip systems, micro-sprinklers)	areas with perennial but scarce water supply; hilly, sloping or terraced farmlands	• north-west, central and southern India • Nepal • central Asia, China, near East
tanks, check dams, terracing	semi-arid and/or drought-prone areas	• much of semi-arid South Asia

Groundwater pollution from fertilizer run-off is causing depletion of the stock of fresh water. Over a fifth of groundwater monitoring sites in agricultural areas of Denmark, the Netherlands, and the USA have recorded nitrate levels that exceed drinking water standards: a particular concern where groundwater provides the main source of drinking water for both people and livestock. The situation is likely to deteriorate as phosphates (also widely used in agriculture) can take many years to seep into the groundwater from the soil.

Over-exploitation of water resources by agriculture has damaged some aquatic ecosystems, and has harmed recreational and commercial fishing.

Conflict arising from shared water resources

As populations grow, greater demands are made on water resources. Water resources are now becoming a limiting factor in many societies, and the availability of water for drinking, industry and agriculture needs to be considered. Many societies are now dependant primarily on groundwater, which is non-renewable. As societies develop, water needs increase. The increased demand for fresh water can lead to inequity of usage and political consequences. When water supplies fail, populations will be forced to take drastic steps, such as mass migration. Water shortages may also lead to civil unrest and wars.

The scarcity of water resources can lead to conflict between human populations particularly where sources are shared.

You should be able to discuss, with reference to a case study, how shared freshwater resources have given rise to international conflict.

Case study

Water and conflict in the Nile Basin

Table 4.12 Characteristics of countries in the Nile Basin

Country	Population in millions	Growth rate / %	PPP / US$	Water footprint / $m^3\ yr^{-1}$
Burundi	10	3.3	600	719
D R Congo	77	2.5	400	552
Egypt	86	1.8	6600	1341
Eritrea	6	2.3	1200	1089
Ethiopia	96	2.9	1300	1167
Kenya	45	2.1	1800	1101
Rwanda	12	2.6	1500	821
Sudan	35	1.8	2600	1736
South Sudan	11	4.1	1400	1736*
Tanzania	49	2.8	1700	1026
Uganda	36	3.2	1500	1079

*No separate reading for South Sudan

The Nile has three main tributaries - the White Nile, the Blue Nile, and the Atbara. The 11 countries in the Nile Basin depend heavily on the river. It is the only major renewable source of water in the region, hence it is vital for water and food security. The Nile basin countries have a total population of over 450 million people. Over 200 million people rely directly on the Nile for their food and water security. The population is expected to double within 25 years - putting immense pressure on the river for water for agriculture, industry and domestic uses.

The Nile's origin is outside the borders of Egypt, but this did not prevent Egypt from getting the lion's share of its waters. A 1929 treaty between Egypt and Britain's East African colonies (Burundi, Kenya, Rwanda, Tanzania, and Uganda) awarded 57 per cent of the waters to Egypt while also requiring other nations to clear with Cairo any major water project on the river. In 1959, Egypt and Sudan signed the Nile Water Agreement in which Egypt was allocated three-quarters of the total water volume (55.5 billion cubic metres) and Sudan one quarter (15.5 billion cubic metres). These two signatories, allocating virtually all of the Nile waters, did not consult Ethiopia, the main source of the river. After the 1959 accord, both Egypt and Sudan built mega-dams to exploit the water for irrigation.

The upstream Nile Basin countries (Figure 4.11) — Burundi, Ethiopia, Kenya, Rwanda, Tanzania, and Uganda — initiated negotiations in 1999 to find an equitable and reasonable way to share the Nile waters. The decade-long negotiations resulted in the 2010 Cooperative Framework Agreement, known as the Entebbe Agreement. The landmark accord, signed by the six upstream countries, was rejected outright by both Egypt and Sudan.

The independence gained by South Sudan in 2011 changed the geopolitical balance of the Nile Basin. It joined the Nile Basin Initiative in 2012. South Sudan controls some 28 per cent of the Nile's flow.

Upstream countries have managed to gain a greater share of the Nile's water resources in recent years. The Nile River Co-operative Framework came into force as international law in 2011 (the Entebbe Agreement). The Entebbe Agreement allows the countries of the Upper Nile Basin to build dams and undertake other water development projects. Current signatories include Ethiopia, Rwanda, Uganda, Kenya, Tanzania, and Burundi.

continued

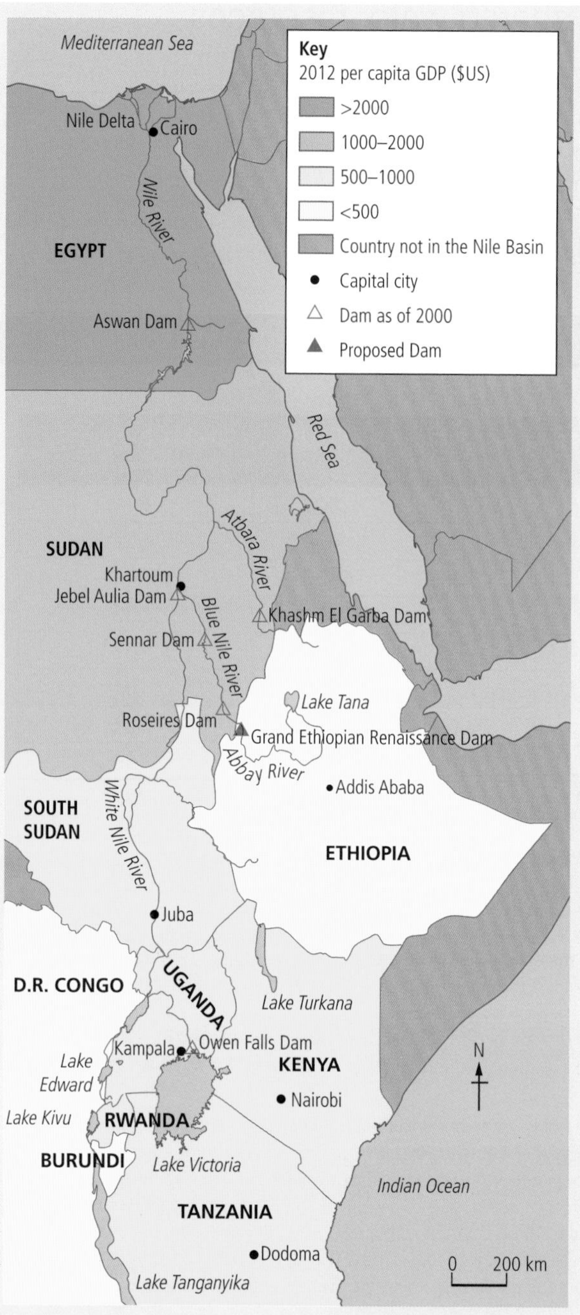

Figure 4.11 The Nile Basin

Egypt's 'historic rights' to the Nile have led to over-dependency on the river – the Nile accounts for 97 per cent of Egypt's water needs. Few countries are as dependent on a single river as Egypt. Water shortages and the limited amount of arable land in Egypt have led to a heavy reliance on food imports to feed the Egyptian population. Although the agricultural sector uses 80 per cent of Egypt's water, 60 per cent of food has to be imported. Egypt is one of the world's leading grain importers. This makes it extremely vulnerable to global food price increases and shortages in supply.

Until recently, Ethiopia had not bothered to make use of its many rivers. Political instability and poverty in the Nile Basin countries limited their ability to move toward socio-economic development of the Nile. However, in 2011, the country announced plans for the construction of its Great Ethiopian Renaissance Dam (GERD), designed to generate a staggering 6000 megawatts of electricity. The dam is situated on the vast Blue Nile gorge, where the land is unsuitable for agriculture. In 2014, Ethiopia turned down Egypt's demand that it suspend construction of the US$4.2 billion mega-dam on the Nile.

GERD will be the largest dam in Africa. The construction has triggered many protests, especially from Egypt. There is some concern that it could result in the evaporation of 3 billion cubic metres of Nile water (Egypt's Aswan Dam is responsible for the evaporation of 12 billion cubic metres of Nile water annually). Together these dams could lead to much-reduced flows downstream.

The Nile is also threatened by many environmental pressures – climate change, salinization, pollution, land degradation, reduced river flow, and increased likelihood of drought and floods.

Large-scale water developments carry a dual threat – conflict with neighbouring countries, and internal conflict due to the displacement of many communities. Egypt views the building of GERD as a threat to its national security. In the past, Egypt has tended to use military threats in Nile disputes but is unlikely to be able to follow such threats through. Greater co-operation with other Nile Basin countries is likely to be the most feasible way forward for Egypt.

A UN-backed plan suggests using the Nubian Sandstone Aquifer, below the eastern part of the Sahara Desert, for water. Egypt, Sudan, Chad, and Libya have agreed the plan. The UN suggests there could be 400 years worth of water available through this aquifer.

Exercises

1. Compare the areas that are predicted to have water stress in 2025 (Figure 4.8) with those that experienced water stress in 1999.
2. Explain why poor people often pay more for their water than rich people.
3. Evaluate the use of drip irrigation.
4. For urban areas, examine the range of ways in which water use can be made more sustainable. What is the percentage of water that is used for agriculture:
   - **a.** globally?
   - **b.** in LEDCs?
   - **c.** in MEDCs?
5. Explain how one technical, one managerial, one institutional and one agronomic method may help improve irrigation water proficiency.
6. Outline the range of low cost irrigation methods available to small farmers

   Look at the Case study: 'Water and conflict in the Nile Basin', and answer the following questions.
   - **a.** Which three countries have the largest populations in the Nile Basin?
   - **b.** Which of the countries in Table 4.12 has the highest population growth rate?
   - **c.** Comment on the variations in water footprint between the countries. NB Ignore the data for South Sudan.
   - **d.** Plot a scattergraph to show the relationship between PPP (US$) and water footprint. Comment on the result you have produced.
   - **e.** What are the implications of population growth and poverty in the Nile Basin for the future use of water resources.
   - **f.** Suggest reasons why conflict may develop between different countries in the Nile Basin over the use of water.

Big questions

Having read this section, you can now discuss the following big questions:

- What strengths and weaknesses of the systems approach and the use of models have been revealed through this topic?
- To what extent have the solutions emerging from this topic been directed at *preventing* environmental impacts, *limiting* the extent of the environmental impacts, or *restoring* systems in which environmental impacts have already occurred?
- How are the issues addressed in this topic of relevance to sustainability or sustainable development?
- In what ways might the solutions explored in this topic alter your predictions for the state of human societies and the biosphere some decades from now?

Points you may want to consider in your discussions:

- How does a systems approach help in our understanding of unequal access to water resources?
- To what extent are there solutions for increasing greater access to freshwater resources?
- Outline the opportunities and barriers to managing freshwater resources sustainably.
- Suggest how, and why, access to freshwater resources is likely to change in the future.

4.3 Aquatic food production systems

Significant ideas

Aquatic systems provide a source of food production.

Unsustainable use of aquatic ecosystems can lead to environmental degradation and collapse of wild fisheries.

Aquaculture provides potential for increased food production.

Big questions

As you read this section, consider the following big questions:

- What strengths and weaknesses of the systems approach and the use of models have been revealed through this topic?
- To what extent have the solutions emerging from this topic been directed at *preventing* environmental impacts, *limiting* the extent of the environmental impacts, or *restoring* systems in which environmental impacts have already occurred?
- How are the issues addressed in this topic of relevance to sustainability or sustainable development?
- In what ways might the solutions explored in this topic alter your predictions for the state of human societies and the biosphere some decades from now?

Knowledge and understanding

- Demand for aquatic food resources continues to increase as human population grows and diet changes.
- Photosynthesis by phytoplankton supports a highly diverse range of food webs.
- Aquatic (freshwater and marine) flora and fauna are harvested by humans.
- The highest rates of productivity are found near the coast or in shallow seas where upwellings and nutrient enrichment of surface waters occur.
- Harvesting some species can be controversial (e.g. seals and whales). Ethical issues arise over biorights, rights of indigenous cultures, and international conservation legislation.
- Developments in fishing equipment and changes to fishing methods have led to dwindling fish stocks and damage to habitats.
- Unsustainable exploitation of aquatic systems can be mitigated at a variety of levels (international, national, local, and individual) through policy, legislation and changes in consumer behaviour.
- Aquaculture has grown to provide additional food resources and support economic development and is expected to continue to rise.
- Issues around aquaculture include loss of habitats, pollution (from feed, antifouling agents, and antibiotics and other medicines added to the fish pens), spread of diseases and escaped species (some involving genetically modified organisms).

Aquatic systems provide a source of food production.

Aquatic food production systems

Demand for aquatic food resources continues to increase as the human population grows and as more people become wealthy, more health conscious, and change their diet to include more meat, fish, and dairy products.

World fisheries

World fisheries and aquaculture produced almost 150 million tonnes of fish in 2010, valued at over US$215 billion (Figure 4.12). Over 125 million tonnes were used as food for people. The world's fish food supply has grown dramatically since 1961, with an average growth rate of 3.2 per cent per year compared with a growth rate of 1.7 per cent per year for the world's population (Figure 4.13). World food fish supply increased from an average of 9.9 kg *per capita* in the 1960s to 18.4 kg in 2009. Fish consumption was lowest in Africa, while Asia accounted for two-thirds of total consumption. Consumption in Asia reached 85.4 million tonnes, of which 42.8 million tonnes was consumed outside China.

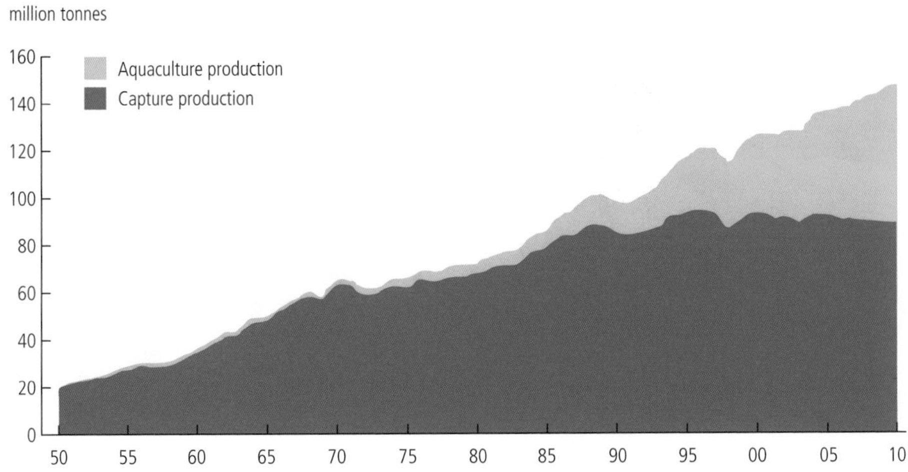

Figure 4.12 World capture and aquaculture production

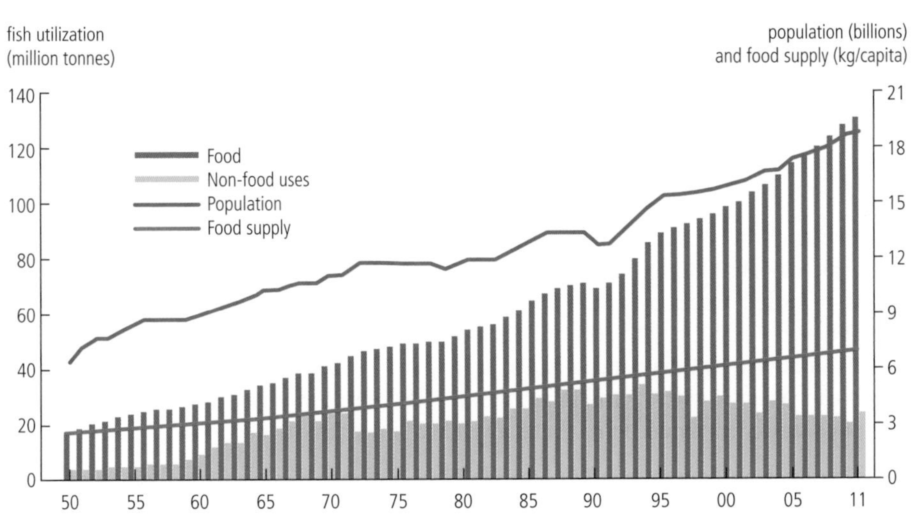

Figure 4.13 World food utilization and supply

Although annual *per capita* consumption of products has grown steadily in LEDCs, it is still considerably lower than in MEDCs, although the gap is narrowing.

China has been responsible for much of the increase in world *per capita* fish supply because of a substantial increase in aquaculture. China's share in world fish production grew from 7 per cent in 1961 to 35 per cent in 2010.

In 2009, fish accounted for 16.6 per cent of the world population's intake of animal protein. Globally, fish provides about 3.0 billion people with almost 20 per cent of their intake of animal protein.

Overall, global capture fisheries production remains relatively stable at about 90 million tonnes. Between 2004 and 2010, catches of all marine species except the Peruvian anchovy (*Engraulius ringens*, anchoveta) ranged between 72.1 million and 73.3 million tonnes. However, anchoveta catches in the south-east Pacific decreased from 10.7 million tonnes in 2004 to 4.2 million tonnes in 2010. This was due, in part, to fishing closures to protect the high number of juveniles present as a consequence of the La Niña event (intensification of 'normal' weather conditions).

To learn more about El Niño and La Niña, go to www.pearsonhotlinks.co.uk, enter the book title or ISBN, and click on weblink 4.1.

The Tokyo fish market – unsustainable demand for fish (shown here are tuna)

The north-west Pacific is still by far the most productive fishing area. Catches peaked in the north-west Atlantic, north-east Atlantic, and north-east Pacific temperate fishing areas many years ago, and total production has declined continuously from the early 2000s, but in 2010 this trend was reversed. Total catches grew in the west and east Indian Ocean and in the west central Pacific.

Fish stocks

Photosynthesis by phytoplankton supports a highly diverse range of food webs.

Production of the world's marine fisheries increased from 16.8 million tonnes in 1950 to a peak of 86.4 million tonnes in 1996, and then stabilized at about 80 million tonnes. Global recorded production was 77.4 million tonnes in 2010. The north-west Pacific had the highest production with 20.9 million tonnes (27 per cent of the global marine catch) in 2010, followed by the west central Pacific with 11.7 million tonnes (15 per cent), the north-east Atlantic with 8.7 million tonnes (11 per cent). After 1990, the number of over-exploited stocks continued to increase, albeit at a slower rate (Figure 4.14).

Food production systems can be compared and contrasted in terms of their trophic levels and efficiency of energy conversion (Figure 4.15). You have already seen that the second law of thermodynamics means that energy conversion through food chains is inefficient, and energy is lost by respiration and waste production at each level within a food web.

In terrestrial systems, most food is harvested from relatively low trophic levels (producer and herbivores). Systems that produce crops (arable) are more energy efficient that those that produce livestock. This is because in the former, producers are at the start of the food chain and contain a greater proportion of the Sun's energy than subsequent trophic levels.

In aquatic systems, perhaps largely due to human tastes, most food is harvested from higher trophic levels where the total storages are much smaller. This is less energy

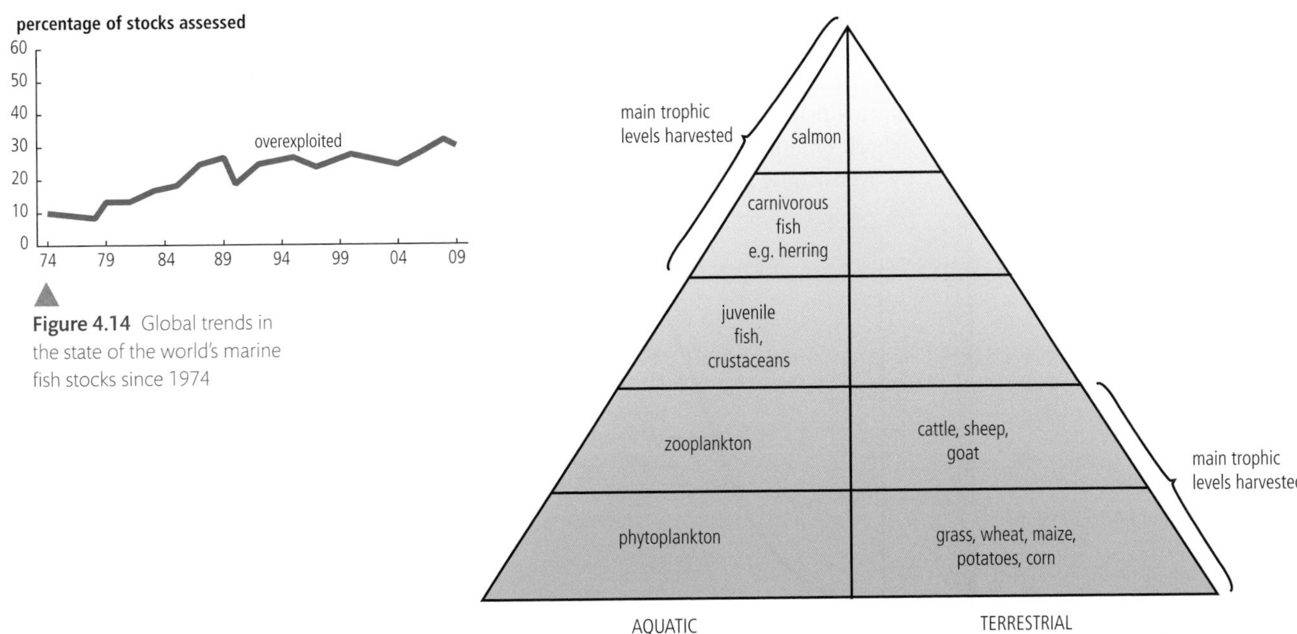

Figure 4.14 Global trends in the state of the world's marine fish stocks since 1974

Figure 4.15 Aquatic and terrestrial food production systems

efficient than crop production (i.e. crops capture energy directly from the primary source; fish are several steps away from primary production). Although energy conversions along an aquatic food chain may be more efficient than in a terrestrial chain, the initial fixing of available solar energy by aquatic primary producers tends to be less efficient due to the absorption and reflection of light by water.

In the oceans, photosynthesis by phytoplankton supports a highly diverse range of food webs. Some ecosystems are more productive than others – coral reefs and those with an abundant supply of light and nutrients tend to support a greater range of food webs.

The highest rates of productivity are found near the coast in shallow waters where upwelling currents cause nutrient enrichment of surface waters (e.g. off the coast of Peru). However, during El Niño episodes, the nutrient enrichment occurs further from the coast, beyond the reach of small boats.

Sustainability

Sustainable yield (SY) is calculated as the rate of increase in natural capital (i.e. natural income) that can be exploited without depleting the original stock or its potential for replenishment. Exploitation must not decrease long-term productivity. So, the annual sustainable yield for a given crop may be estimated as the annual gain in biomass or energy through growth and recruitment (in-migration of species).

SY is the amount of increase per unit time (i.e. the rate of increase).

Where

t = the time of the original natural capital

$t + 1$ = the time of the original capital plus yield,

SY = (total biomass at $t + 1$) – (total biomass at t)

or

SY = (total energy at $t + 1$) – (total energy at t)

The relationship can be simplified as:

SY = (annual growth and recruitment) – (annual death and emigration)

Maximum sustainable yield (MSY) is the largest yield (or harvest) that can be taken from the stock of a species over an indefinite period. MSY aims to maintain the population size at the point of maximum growth rate by harvesting the individuals that would normally be added to the population, allowing the population to continue to be productive indefinitely (Figure 4.16). MSY is the point where the highest rate of capture fisheries can occur (this is often difficult to determine). It is used extensively by fisheries management (Figure 4.17). Populations of cod have been particularly affected by over-fishing in the North Atlantic (Figure 4.18).

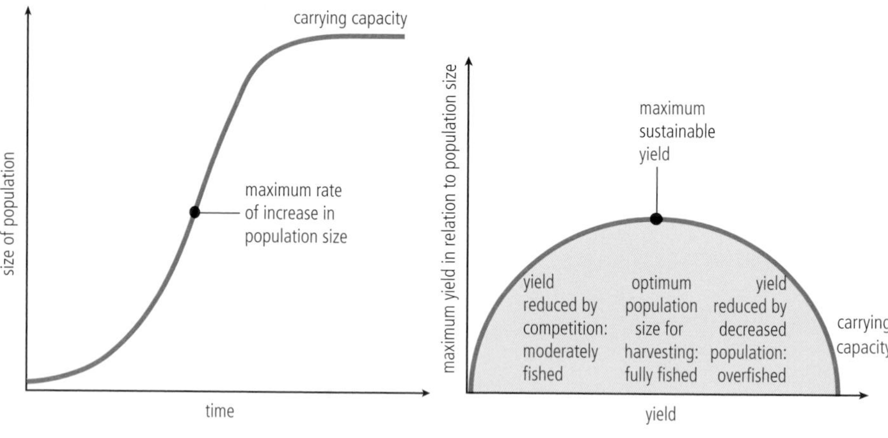

Figure 4.16 Maximum sustainable yield occurs at maximum. rate of increase in population. Near the carrying capacity, the yield reduces.

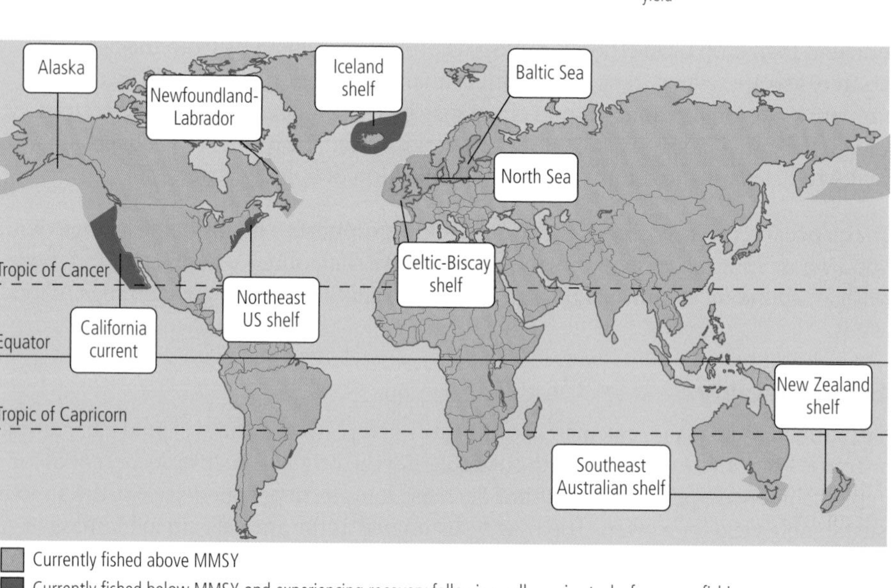

Figure 4.17 Multi-species maximum sustainable yield (MMSY): MSY values for whole fish communities worldwide

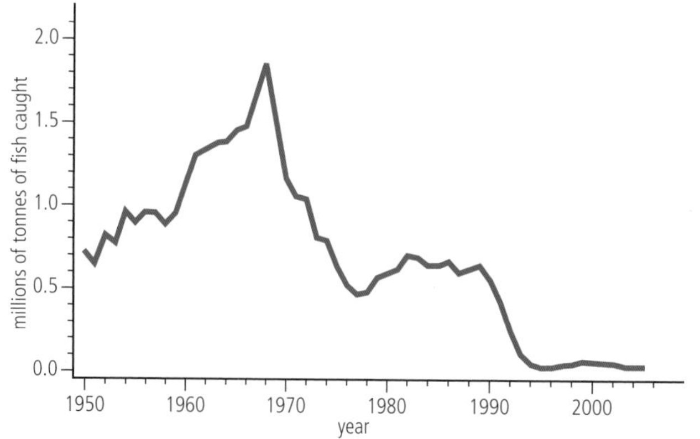

Figure 4.18 North-west Atlantic cod populations, off the coasts of the USA and Canada, were particularly affected by over-fishing. By 1992, cod populations were only 1 per cent of their 1960 levels.

Most of the stocks of the top 10 species, which account in total for about 30 per cent of world marine capture fisheries production, are fully exploited and, therefore, have no potential for increases in production. The two main stocks of anchoveta in the south-east Pacific, Alaska pollock in the north Pacific, and blue whiting in the Atlantic are fully exploited. Among the seven principal tuna species, one-third were estimated to be over-exploited.

CONCEPTS: Sustainability

The declining global marine catch over the last few years together with the increased percentage of over-exploited fish stocks and the decreased proportion of non-fully exploited species suggests that the state of world marine fisheries is worsening and has had a negative impact on fishery production. Over-exploitation not only causes negative ecological consequences, but it also reduces fish production, which further leads to negative social and economic consequences.

The Johannesburg Plan of Implementation that resulted from the World Summit on Sustainable Development, demands that all over-exploited stocks be restored to the level that can produce maximum sustainable yield by 2015, a target that seems unlikely to be met.

Global overfishing

Modern fishing ships are able to stay at sea for months, while freezing and storing their catch. Consequently, they can fish further out at sea than was previously possible, often in international waters. In smaller-scale (but often more damaging) fishing, dynamite and cyanide are still used in some areas to stun reef fish to bring them to the surface. The explosions and poisons also kill the reefs. Fuelled in part by subsidies, many fishing fleets have grown beyond the point of sustainability. This overcapacity has led to overfishing and, indirectly, to issues such as Somali piracy. Each year the UN FAO State of World Fisheries and Aquaculture (SOFIA) report examines global fisheries production and shows that overfished fisheries gradually increased up to a peak of 32 per cent of global fisheries in 2010.

Developments in fishing equipment and changes to fishing methods, coupled with over capacity in fleets and inadequate international management, have lead to dwindling fish stocks and damage to habitats.

The solutions to such a complex issue are not simple. It is an irony that the technological developments that can lead to overfishing can also contribute to a solution. One example is in northern European mackerel and herring fleets: sonar techniques used to find fish can now distinguish between shoals of the two species. This allows boats to target only herring when they are in season, avoiding the mixed shoals or wrong fish. This selectivity contributed to the herring fishery's Marine Stewardship Council certification as a sustainable and well-managed fishery.

While the statistics often paint a bleak picture, there are examples of success: fisheries that have recovered, stocks that are improving and environmental impacts that have moved to a sustainable basis. There is also a growing body of opinion that suggests the tide is changing on unsustainable fishing practices. In Europe, an EU-wide ban on discarding is being implemented, largely driven by consumer and political pressure to end the wasteful practice of discarding over-quota, undersize and low-value fish at sea. In future, all catches will have to be landed, increasing the pressure for more selective fishing.

In the developing world, governments are partnering with international NGOs to drive change through Fisheries Improvement Projects (FIPs). In the SOFIA reports, the overfished fisheries have gradually decreased, dropping to 28.8 per cent by the 2014 report. At the same time, the percentage of fisheries fully exploited (defined as fished at Maximum Sustainable Yield) has gradually increased from 47 per cent in 2002 to 61.3 per cent in 2014, while under-exploited fisheries have dropped from 25 per cent to 9.9 per cent over the same period.

Nearly 70 per cent of the world's stocks are in need of management. Cod stocks in the North Sea are less than 10 per cent of 1970 levels. Fishing boats from the EU now regularly fish in other parts of the world, such as off the coastlines of Africa and South America, to make up for the shortage of fish in EU waters. More than half the fish consumed in Europe is now imported.

The closure of Grand Banks

Once a fish stock is over-fished to the point of collapse, it is very difficult for it to recover. The Grand Banks off Newfoundland were once the world's richest fishery. In 1992, the area had to be closed to allow stocks to recover. It was expected to be closed for 3 years, but fish numbers, especially cod, have not yet recovered and the areas is still closed. The cod's niche in the ecosystem has been taken by other species, such as shrimp and langoustines. (Langoustine stocks were previously kept low as young langoustine were predated by mature cod. Now the situation is reversed and young codlings and eggs are predated by mature langoustine.)

Unsustainable exploitation of aquatic systems can be mitigated at a variety of levels (international, national, local, and individual) through policy, legislation, and changes in consumer behaviour.

Managing fisheries

Many argue that measures such as quotas, bans, and the closing of fishing areas still fail to address the real problems of the fishing industry: too many fishermen are chasing too few fish and too many immature fish are being caught. For the fisheries to be protected and for the industry to be competitive on a world scale, the number of boats and the number of people employed in fishing must be reduced. At the same time, the efficiencies which come from improved technology must be embraced.

A World Bank and FAO report in 2008 showed that up to US$50 billion per year is lost in poor management, inefficiency, and overfishing in world fisheries. The report puts the total loss over the last 30 years at US$2.2 trillion. The industry's fishing capacity continues to increase. The number of vessels is increasing slowly. However, each boat has greater capacity due to improved technology. Because of the increase in capacity, much of the investment in new technology is wasted. The amount of fish caught at sea has barely changed in the last decade. Fish stocks are depleted, so the effort to catch the ones remaining is greater.

Figure 4.19 Logo of the Marine Stewardship Council

Consumer behaviour is changing in some societies. Awareness of biorights and the problem of over-fishing has led some people to demand that the fish they consume are taken from sustainable sources. The Marine Stewardship Council identifies and labels sustainable fisheries. Its website states that by 'choosing MSC labelled seafood, you reward fisheries that are committed to sustainable fishing practices' (Figure 4.19). MSC-labelled seafood is now available in 140 countries around the world and the certified fisheries represent 10 per cent of global wild-capture seafood. Researchers have identified nearly a thousand improvements made by MSC-certified fisheries in the past 10 years, improving the sustainability of the fisheries to global best-practice levels of environmental performance.

A number of publications and films have promoted sustainable fishing practices or highlighted unsustainable ones. These aim to further change consumer behaviour and to put pressure on governments and the fishing industry to stop unsustainable fishing.

Strategies for the European fishing industry

The table below suggests some possible strategies for the future, but there are clearly no simple solutions to the problems associated with such a politically, economically and environmentally sensitive industry.

Table 4.13 Possible strategies to manage fisheries in the European Union

Action	Type of measure	Objectives
Conservation of resources		
technical measures	• small meshed nets • minimum landing sizes • boxes	• protect juveniles and encourage breeding • discourage marketing of illegal catches
restrict catches	• total allowable catches (TACS) and quotas	• match supply to demand • plan quota uptake throughout the season • protect sensitive stocks
limit number of vessels	• fishing permits (tradable within and between countries)	• restrict numbers of EU and other countries' vessels fishing in EU waters
surveillance	• check landings by EU and other vessels (log books, computer/satellite surveillance)	• apply penalties to overfishing and illegal landings
structural	• structural aid to the fleet	• finance investment in fleet modernization (although commissioning of new vessels must be closely controlled) • provide reimbursement for scrapping, transfer, and conversion
reduction in employment leading to an increase in productivity	• inclusion of zones dependent on fishing in European Regional Development Fund	• facilitate restructuring of the industry • finance alternative local development initiatives to encourage voluntary/early retirement schemes
Markets		
tariff policy	• minimum import prices • restrictions on imports	• ensure EU preference (although still bound under WTO)
Other measures		
restrict number of vessels	• fishing licenses	• large license fees would discourage small, inefficient boats
increase the accountability of fishermen	• rights to fisheries	• where fish stay put (e.g. shellfish) sections of the seabed can be auctioned off

A fish farm

Fish farming (aquaculture)

Fish farming was first introduced when over-fishing of wild Atlantic salmon in the north Atlantic and Baltic seas caused their populations to crash. Open-net fish farming was introduced in Norway in the 1960s. Since then, the industry has expanded to Scotland, Ireland, Canada, the USA, and Chile, and is dominated by a handful of multinational corporations. Aquaculture involves raising fish commercially, usually for food. (In contrast, a fish hatchery releases juvenile fish into the wild for recreational fishing or to supplement a species' natural numbers.) The most important fish species raised by fish farms are salmon, carp, tilapia, catfish, and cod. Salmon make up 85 per cent of the total sale of Norwegian fish farming, but most global aquaculture production now uses non-carnivorous fish species, such as tilapia and catfish. Technological costs are high, and include using drugs, such as antibiotics, to keep fish healthy, and steroids to improve growth. Breeding programmes are also expensive. Outputs are high per hectare and per farmer, and efficiency is also high.

Issues around aquaculture include loss of habitats, pollution (feed, antifouling agents, and antibiotics and other medicines added to the fish pens), spread of diseases, use of wild-capture fish to make fish-meal, and escaped species.

There are many issues related to aquaculture. One is the loss of natural habitats. In some cases, aquaculture may occur in sea lochs. In other cases, it occurs in fish pens off the coastline. In both cases, natural ecosystems are transformed into commercial operations, in which human activities interfere with natural processes, such as migration, predator–prey relationships, removal of competing species and vegetation, and so on.

Environmental effects can be damaging, especially with salmon farming. Salmon are carnivores and need to be fed pellets made from other fish. It is possible that farmed salmon actually represent a net loss of protein in the global food supply, as it takes 1–5 kg of wild fish to grow 1 kg of salmon. Fish like herring, mackerel, sardines and anchovies are used to produce the feed for farmed salmon, so the production of salmon requires the capture of other fish species on a global scale. Other environmental costs include the sea lice and disease that spread from farmed fish into wild stocks, and pollution (created by uneaten food, faeces, and chemicals) contaminating surrounding waters. Accidental escape of genetically modified fish can affect local wild fish gene pools when the escaped fish interbreed with wild populations. This reduces the wild fish genetic diversity, and potentially introduces non-natural genetic variation. In some parts of the world, escapes from farmed salmon threaten native wild fish, as it may be an alien species. For example, the British Columbia salmon farming industry has inadvertently introduced a non-native species – Atlantic salmon – into the Pacific ocean.

Despite the disadvantages, the positive environmental benefits of not removing fish from wild stocks, but growing them in farms, are great. Wild populations are allowed to breed and maintain stocks, while the farmed variety provides food.

Global aquaculture

Between 1980 and 2010, world food fish production by aquaculture grew by, on average, 8.8 per cent per annum. World aquaculture production in 2010 was about 60 million tonnes (Figure 4.20), worth US$125 billion.

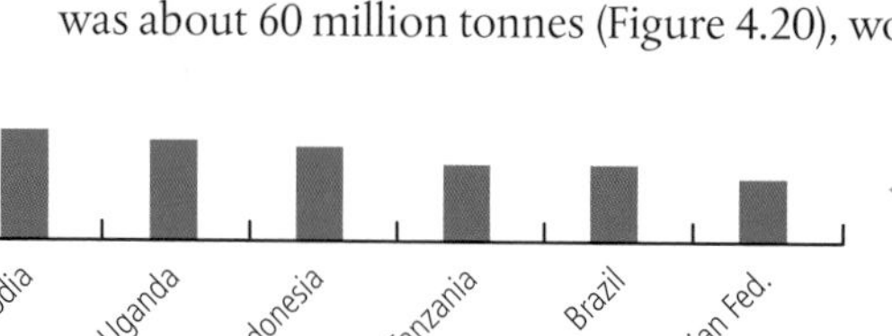

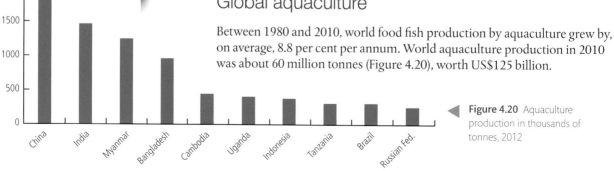

Figure 4.20 Aquaculture production in thousands of tonnes, 2012

Aquaculture production is vulnerable to adverse impacts of disease and environmental conditions. Disease outbreaks in recent years have affected farmed Atlantic salmon in Chile, oysters in Europe, and marine shrimp farming in several countries in Asia, South America, and Africa.

The global distribution of aquaculture remains imbalanced. In 2010, the top 10 producing countries accounted for 87.6 per cent by quantity and 81.9 per cent by value of the world's farmed food fish. Asia accounted for 89 per cent of world aquaculture production by volume in 2010. In Asia, the share of freshwater aquaculture has been gradually increasing, up to 65.6 per cent in 2010 from around 60 per cent in the 1990s.

LEDCs, mostly in sub-Saharan Africa and in Asia, remain minor in terms of their share of world aquaculture production.

Case study

Rice-fish farming in Thailand

Rice-fish farming in Thailand. Once the newly planted rice is established, fish are released into the flooded fields from holding pens.

Cultivating rice and fish together has been a tradition for over 2000 years in South East Asia. This **polyculture** system (paddy rice field stocked with fish) was gradually abandoned due to population pressure and decreasing stocks of wild fish. The fall in fish stocks was due to the toxic effects of the pesticides and herbicides used in high-yield rice monoculture. However, this farming method experienced a revival in the early 1990s, as concerns over the widespread use of pesticides emerged. Implementation is relatively inexpensive and low risk.

The system requires farmers to dig small ponds or trenches in low-lying areas of rice, which become refuges for fish during rice planting and harvesting, or when water is scarce. The excavated soil is used to raise banks around the field to grow other crops on (e.g. vegetables and fruit trees). Once the paddy fields are flooded, young fish (fingerlings) are introduced to the trenches: carp, tilapia, catfish, or other species. After 3 weeks, when the rice is well established, the fish are let into the rice fields. They obtain their food from the fields, but carnivorous species can be fed if necessary. The fish contribute to a decrease of disease and pest incidence in the rice, and rice yields are higher. Because rice productivity increases, farmers do no need to use fertilizers (the fish produce faeces and excreta which naturally fertilize the soil). Rice-fish culture may increase rice yields by up to 10 per cent, and increase income by 50–100 per cent over rice alone, while providing farmers with an important source of protein.

The process counters the decrease in available wild fish in many countries. The most common and widespread fish species used in rice-fish farming are the common carp (*Cyprinus carpio*) and the Nile tilapia (*Oreochromis niloticus*). Both feed on the vegetation and plankton available and do not attack

continued

CHALLENGE YOURSELF

Communication skills

Organize a class debate. Choose one of these topics.

- To what extent do the disadvantages of aquaculture outweigh the advantages?
- Can sustainable fishing ever be achievable?

other fish, so they are preferred species in the culture systems. Other species such as catfish (*Clarias* spp.) are well adapted to the swamp-like conditions of rice fields, but are carnivorous and will feed on other introduced fish.

This food production system is an example of intensive subsistence farming. The cost of feeding the fish is low but demands on labour are high. Technology is low. Other inputs include water for irrigation, and the cost of the breeding stock. The outputs are high per hectare but low per farmer, but overall efficiency is high, and yields of rice improve significantly with this system compared to monoculture rice agriculture. Environmental impacts are low, but include change in the nutrient balance, and the introduction of alien species, which may have impacts for local biodiversity - both plant and animal.

There are several cultural issues regarding the system: other sources of animal protein may be preferred (e.g. poultry, beef, and pork), and the commonly cultured fish species (e.g. tilapia and carp) are not highly valued by people who have access to marine species and wild species. Fish predators such as snakes can lower the fish yield. The system may only be appropriate if there is a reliable water supply, a source of young fish, and fields located close to the family house so they can be monitored.

Harvesting some species can be controversial (e.g. seals and whales). Ethical issues arise over biorights, rights of indigenous cultures, and international conservation legislation.

Whaling – contrasting views

Harvesting some species (such as seals, sharks, and whales) is controversial. There are ethical issues regarding biorights – the rights of an endangered species, a unique species, or landscape to remain unmolested. The rights of indigenous culture and international conservation legislation also should be considered

In the late 1930s, more than 50 000 whales were killed annually. The International Whaling Commission (IWC) was set up to decide on hunting quotas based on the findings of its Scientific Committee. In 1982, the IWC voted to establish a ban on commercial whaling, which took effect in 1986. Japan now wants to lift the ban on stocks that have recovered. However, anti-whaling countries and environmental groups believe that whale species remain vulnerable and that whaling is both immoral and unsustainable.

According to the IWC, indigenous subsistence whaling occurs in Greenland (fin, bowhead, humpback, and minke whales), in Siberia, (gray and bowhead whales), St Vincent and The Grenadines (Bequia, humpback whales) and North America (bowhead whales; Washington State, gray whales).

Inuit and whaling

North American whaling is carried out by small numbers of the Inuit population. The only species hunted is the bowhead whale. Whaling is a central part of Inuit culture and provides a vital source of protein in their diet (see also TOK chapter, page 454).

Indigenous Inuit catching a whale

The 10 000 Inuit in Alaska are allowed to kill a total of up to 336 bowhead whales between 2013 and 2018, with no more than 67 whales in any one year. This represents about half of the meat in the Inuit diet. Scientific research suggests that the bowhead whales are not an endangered species, and their hunt is sustainable. But conservationists take a very different view and state that whales have biorights and should not be killed, especially in a way that causes them great pain and suffering.

The Inuit hunt a variety of animals for meat – including caribou, walrus, seal, and geese – depending on the season and the migratory movements of the species.

In Greenland, Inuit whalers catch around 175 whales per year, making them the third largest hunt in the world after Japan and Norway, which annually averaged around 730 and 590 whales, respectively from 1998 to 2007. The IWC allows the more densely populated west coast to take over 90 per cent of the catch. In a typical year around 150 minke and 10 fin whales are taken from west coast waters and around 10 minkes are from east coast waters.

Japan and whaling

Japan was, for many years, the greater hunter of whales. It reluctantly stopped commercial hunting in 1986. However, it continued to hunt whales for 'scientific research' to establish the size and dynamics of whale populations.

Japan clashed repeatedly with Australia and other western countries, which strongly oppose whaling on conservation grounds. Australia took a case to the UN's International Court of Justice (ICJ) and argued that Japan's scientific research whaling programme was simply commercial whaling in disguise. Japan argued that the suit brought by Australia was an attempt to impose its cultural norms, and furthermore, that minke whales and a number of other species are plentiful and that its whaling activities are sustainable.

In 2014 the ICJ ruled that the Japanese government must halt its whaling programme in the Antarctic. The ICJ believed that the programme was not for scientific research as claimed by Tokyo. It claimed Japan had caught some 3600 minke whales since its current programme began in 2005, but the scientific output was limited. Japan agreed to abide by the ruling but added it 'regrets and is deeply disappointed by the decision'.

There are other threats to whales apart from whaling. These include collision with ships, chemical pollution, habitat degradation, noise pollution, and by-catch (the unintentional capture of whales in fish nets). To protect whales, the Southern Ocean around Antarctica was declared a whale sanctuary in 1994.

Exercises

1. Describe the trend in world fish capture and aquaculture between 1950 and 2010, as shown in Figure 4.12.
2. Compare the trends in population growth, food supply, food, and non-food uses as shown in Figure 4.13.
3. Comment on the changes in world fish stocks, as shown in Figure 4.14.
4. Examine the importance of global fisheries.
5. What is meant by the term 'predatory fish'?
6. Visit the hotlink opposite and answer the following questions.
   a. When was the Common Fisheries Policy of the European Union introduced?
   b. What happened to North Sea cod stocks in 2008?
   c. What are the arguments for and against a common fisheries policy?
   d. Define the term *aquaculture*.
7. Figure 4.20 shows global aquaculture production for 2012.
   a. What was the level of aquaculture production in (i) China and (ii) India?
   b. What proportion of the production shown in Figure 4.20 is from Asian countries?
   c. Suggest two or more reasons why the five largest aquaculture producers are from Asia.
8. To what extent was whaling part of (a) Inuit culture and (b) Japanese culture.
9. Define the term *biorights*. Should all species of whales have biorights? Give reasons to support your answer.

continued

To learn more about the Common Fisheries Policy of the European Union, go to www.pearsonhotlinks.co.uk, enter the book title or ISBN, and click on weblink 4.2.

To learn more about the status of whales, go to www.pearsonhotlinks.co.uk, enter the book title or ISBN, and click on weblink 4.3.

To learn more about the number of whales in the wild, go to www.pearsonhotlinks.co.uk, enter the book title or ISBN, and click on weblink 4.4.

10. To what extent can whaling ever be sustainable?

11. Visit the hotlinks opposite and answer the following questions.

- **a.** Briefly describe the status of whales.
- **b.** How many whales are there in the wild?

Big questions

Having read this section, you can now discuss the following big questions:

- What strengths and weaknesses of the systems approach and the use of models have been revealed through this topic?
- To what extent have the solutions emerging from this topic been directed at *preventing* environmental impacts, *limiting* the extent of the environmental impacts, or *restoring* systems in which environmental impacts have already occurred?
- How are the issues addressed in this topic of relevance to sustainability or sustainable development?
- In what ways might the solutions explored in this topic alter your predictions for the state of human societies and the biosphere some decades from now?

Points you may want to consider in your discussions:

- How far does a systems approach help our understanding of aquatic food production systems?
- Compare and contrast the environmental impact of capture fisheries and aquaculture.
- To what extent can fisheries be managed sustainably?
- Outline the likely pressures on, and potential solutions for the world fisheries in decades to come.

4.4 Water pollution

Significant idea

Water pollution, both groundwater and surface water, is a major global problem whose effects influence human and other biological systems.

Big questions

As you read this section, consider the following big questions:

- What strengths and weaknesses of the systems approach and the use of models have been revealed through this topic?
- To what extent have the solutions emerging from this topic been directed at *preventing* environmental impacts, *limiting* the extent of the environmental impacts, or *restoring* systems in which environmental impacts have already occurred?
- How are the issues addressed in this topic of relevance to sustainability or sustainable development?
- In what ways might the solutions explored in this topic alter your predictions for the state of human societies and the biosphere some decades from now?

Knowledge and understanding

- There are a variety of freshwater and marine pollution sources.
- Types of aquatic pollutant include floating debris, organic material, inorganic plant nutrients (nitrates and phosphates), toxic metals, synthetic compounds, suspended solids, hot water, oil, radioactive pollution, pathogens, light, noise, and biological entities (invasive species).

- A wide range of parameters can be used to directly test the quality of aquatic systems (e.g. pH, temperature, suspended solids/turbidity, metals, nitrates, and phosphates).
- Biodegradation of organic material uses oxygen and can lead to anoxic conditions and subsequent anaerobic decomposition, which in turn leads to formation of methane, hydrogen sulphide, and ammonia (toxic gases).
- Biochemical oxygen demand (BOD) is a measure of the amount of dissolved oxygen required to break down the organic material in a given volume of water through aerobic biological activity. BOD is used to indirectly measure the amount of organic matter in a sample.
- Some species can be indicative of polluted waters and are used as indicator species.
- A biotic index indirectly measures pollution by assaying the impact on species in the community according to their tolerance, diversity, and relative abundance.
- Eutrophication can occur when lakes, estuaries and coastal waters receive inputs of nutrients (nitrates and phosphates) which result in an excess growth of plants and phytoplankton.
- Dead zones in both oceans and fresh water can occur when there is not enough oxygen to support aquatic life.
- Water pollution management strategies include:
  - reducing human activities producing pollutants (e.g. alternatives to current fertilizers and detergents)
  - reducing release of pollution into the environment (e.g. treatment of wastewater to remove nitrates and phosphates)
  - removing pollutants from the environment and restoring ecosystems (e.g. removal of mud from eutrophic lakes and reintroducing plant and fish species).

Water pollution

There are a variety of freshwater and marine pollution sources.

Freshwater and marine pollution sources include run-off, sewage, industrial discharge, solid domestic waste, transport, recreation and tourism, and energy waste. Sources of marine pollution include rivers, pipelines, atmosphere, oil spills, deliberate and accidental discharges from ships, sewage from cruise ships, aquaculture farms, power stations, and industry.

Types of aquatic pollutant include floating debris, organic material, inorganic plant nutrients (nitrates and phosphates), toxic metals, synthetic compounds, suspended solids, hot water, oil, radioactive pollution, pathogens, light, noise, and biological entities (invasive species).

Storm water that washes off the roads and roofs can be a worse source of pollutants than sewage. Such water may contain high levels of heavy metals, volatile solids, and organic chemicals. Studies of water quality during in floods of the Silk Stream in London recorded that between 20–40 per cent of storm water sediments were organic and mostly biodegradable. Highway run-off has 5–6 times the concentration of heavy metals as roof run-off. Annual run-off from 1 km of a single carriageway of the M1 (a highway in the UK) included 1.5 tonnes of suspended sediment, 4 kg of lead, 126 kg of oil and 18 g of hazardous polynuclear aromatic hydrocarbons.

Water quality

Standard water quality tests on drinking water, rivers, and other sites can be performed with portable equipment that enables detection of nitrate, nitrite, free chlorine, chloride, fluoride, hardness, and heavy metals such as lead. Water-quality tests on rivers include **biochemical oxygen demand** (BOD), chemical oxygen demand (COD), turbidity, ammonia, and dissolved oxygen.

There are two main ways of measuring water quality, by direct and indirect measures. Direct measures take samples of the water and measure the concentrations of different

A wide range of parameters can be used to directly test the quality of aquatic systems, including pH, temperature, suspended solids (turbidity), metals, nitrates, and phosphates.

A report in 2014 confirmed that nearly 60 per cent of China's underground water was polluted. The country's land and resources ministry found that of 4778 testing spots in 203 cities, 44 per cent had 'relatively poor' underground water quality; the groundwater in another 15.7 per cent tested as 'very poor'. Water quality improved year on year at 647 spots, and worsened in 754 spots.

Water of very poor quality cannot be used as source of drinking water. The Chinese government is only now beginning to address the noxious environmental effects of its long-held growth-at-all-costs development model. While authorities have become more transparent about air quality data since 2013, information about water and soil pollution in many places remains relatively well guarded.

In 2013, about a third of China's water resources were groundwater based, and only 3 per cent of the country's urban groundwater could be classified as 'clean'. About 70 per cent of groundwater in the north China plain – 400 000 km^2 of some of the world's most densely populated land – was considered unfit for human consumption.

Few Chinese urban dwellers consider tap water safe to drink – most either boil their water or buy it bottled. In 2014, a chemical spill poisoned the water supply of Lanzhou (a city of 2 million people in north-west China) with the carcinogen benzene.

While Beijing's noxious smog has become internationally infamous, drought and water pollution may pose even greater threats to the city. Beijing's annual *per capita* water availability is about 120 cubic metres, about a fifth of the UN's cut-off line for 'absolute scarcity'.

chemicals that it contains. If the chemicals are dangerous or the concentrations are too great, the water is polluted. Measurements like this are known as chemical indicators of water quality (Table 4.14). Indirect methods involve examining the fish, insects, and other invertebrates that live in the water. If many different types of creature can live in a river, the water quality is likely to be very good; if the river supports no fish life at all, the quality is obviously much poorer. Measurements like this are called biological indicators of water quality.

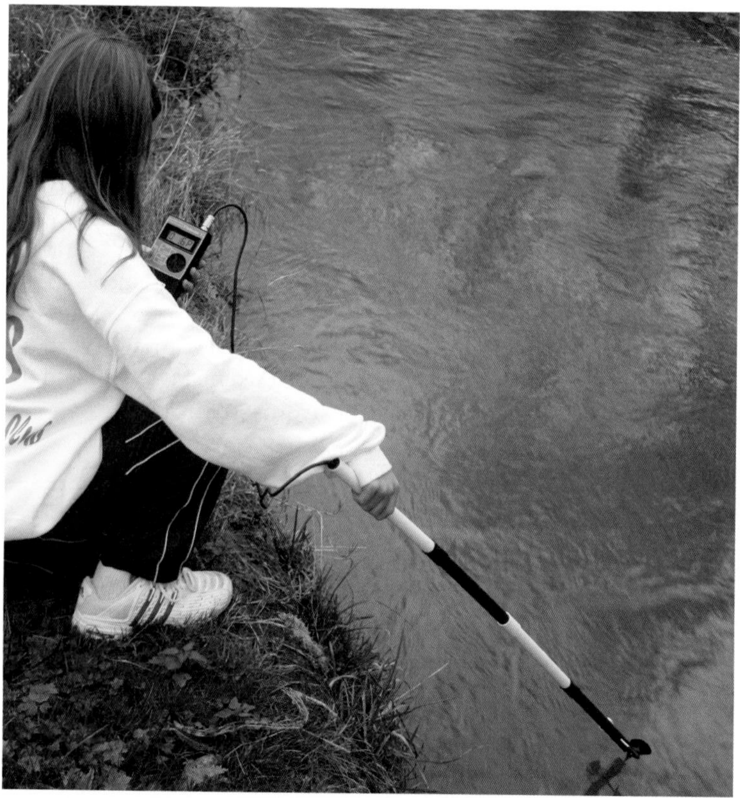

Student measuring water quality

Table 4.14 Chemical indicators of water quality

Indicator	Method	What the results show
dissolved oxygen	test kit, meter, or sensor; oxygen usually measured as percentage saturation; follow instructions to measure oxygen saturation	• 75% oxygen saturation = healthy, clean water • 10–50% oxygen saturation = polluted water • <10% oxygen saturation = raw sewage
pH	indicator paper dipped in sample and compared to a colour chart; record the pH number (e.g. pH = 8)	• pH1–6 = water is acidic • pH7 = water is neutral • pH8–14 = water is alkaline
phosphate	test kit; phosphate is measured in mg dm^{-3}; follow instructions to measure phosphate	• <5 mg dm^{-3} = clean water • 15–20 mg dm^{-3} = polluted water
nitrate	test kit; nitrate is measured in mg dm^{-3}; follow instructions to measure nitrate	• 4–5 mg dm^{-3} = clean water • 6–15 mg dm^{-3} = polluted water
salt (as chloride)	test kit or meter or sensor; the amount of chloride (salinity) is measured in mg dm^{-3}; follow instructions to measure salinity	• 20 000 mg dm^{-3} = seawater • 100–20 000 mg dm^{-3} = tidal or brackish water
ammonia	test kit; ammonia is measured in mg dm^{-3}; follow instructions to measure ammonia	• 0.05–1.00 mg dm^{-3} = clean water • >1–10 mg dm^{-3} = polluted water • 40 mg dm^{-3} = raw sewage

mg dm^{-3} is the symbol for milligrams per decimetre cubed (i.e. milligrams per litre); 1 dm^3 = 1 litre

You should be able to analyse water pollution data.

Biochemical oxygen demand

Biochemical oxygen demand (BOD) is a measure of the amount of dissolved oxygen required to break down the organic material in a given volume of water through aerobic biological activity. BOD is used to indirectly measure the amount of organic matter in a sample.

Aerobic organisms use oxygen in respiration. The more organisms there are at a particular site (e.g. in a river) and the faster their rate of respiration, the more oxygen they will use. So, the biochemical oxygen demand (BOD) at any particular point in the river is determined by the number of aerobic organisms at that point and their rate of respiration.

BOD can indicate whether or not a particular part of a river is polluted with organic matter (e.g. sewage, silage). This is because the presence of an organic pollutant stimulates an increase in the population of organisms that feed on and break down the pollutant. In doing so, they respire and use up a lot of oxygen. This could eventually lead to a lack of oxygen and subsequent anaerobic decomposition which then leads to formation of methane, hydrogen sulfide, and ammonia (toxic gases).

BOD is measured in the following way.

1. Take a sample of water of measured volume.
2. Measure the oxygen level.
3. Place the sample in a dark place at 20 °C in an air tight container for 5 days (lack of light prevents photosynthesis which would release oxygen and give an artificially low BOD).
4. After 5 days, re-measure the oxygen level.
5. BOD is the difference between the two measurements.

Certain species are tolerant of organic pollution and the low oxygen levels associated with it. They are found in high population densities where an organic pollution incident occurs. Other species cannot tolerate low oxygen levels and, if organic pollution enters the river where they live, they move away or die (Figure 4.21). We can use these groups as indicator species (i.e. indicators of organic pollution).

Figure 4.21 Indicator species

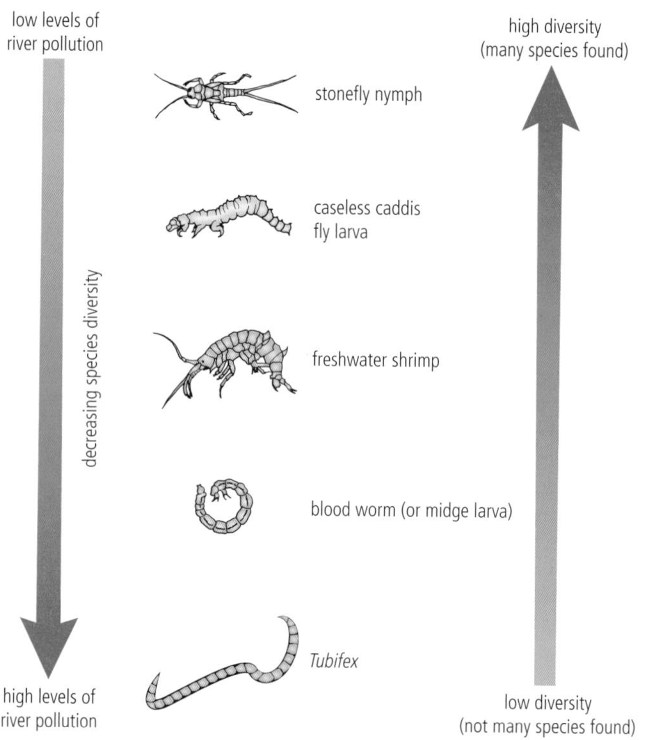

Organic pollutants in water can be more dangerous in summer. This is because the solubility of oxygen decreases as the water temperature increases. So on warm days there is less available oxygen in the water. Aquatic invertebrates and fish can do little to regulate their body temperature; as water temperature increases, so does their internal temperature and so does their rate of respiration. This means they need more oxygen – but the amount of oxygen dissolved in the water is going down. If warm organic pollutants are released into rivers, the effect can be devastating.

Figure 4.22 illustrates how *Tubifex* worms and stonefly nymphs can be used as indicator species. The *Tubifex* worms feed on and tunnel into effluent; their populations increase rapidly immediately downstream of any effluent entry. A high population of these organisms in any river could indicate that organic pollution has recently occurred. In contrast, the population of stonefly nymphs crashes as soon as effluent enters their habitat. They need clean water and, at the point of effluent entry, either die or move away. Thus, the absence of stonefly nymphs in a particular river might indicate organic pollution has occurred and large populations might indicate clean, unpolluted water.

You should be able to evaluate the uses of indicator species and biotic indices in measuring aquatic pollution.

It is often faster and cheaper to measure organisms such as *Tubifex* or stonefly nymphs than it is to try to measure the concentration of specific pollutants – this is one of the advantages of using indicator species. Table 4.15 shows how the types of fauna present in a waterway relate to levels of BOD.

1. Organic pollution enters the river.

2. Population of *Tubifex* increases. These organisms feed on and begin to break down the organic matter. BOD increases.

3. The amount of organic matter has fallen by dispersal and because it has been broken down. BOD is decreasing.

4. Pollutant now much reduced. BOD just above normal.

5. BOD back at normal point – all organic matter broken down.

river

direction of flow

oxygen

BOD

number of individuals

stonefly nymphs

Tubifex

distance downstream

effluent enters river

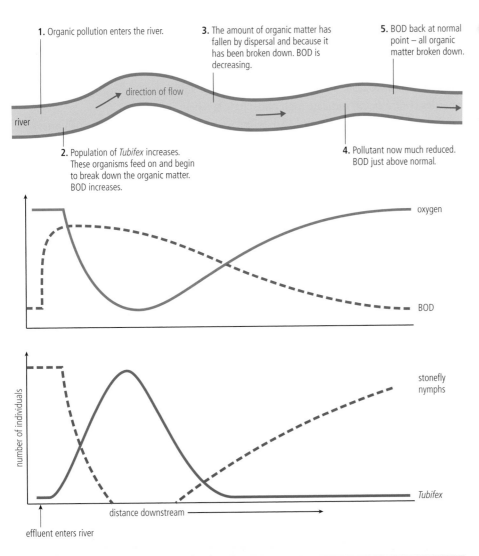

Figure 4.22 Using indicator species to estimate river pollution

Table 4.15 BOD and the type of fauna present

Class of water	Fauna present	BOD / mg O_2 absorbed per dm^3 of water at 20 °C in 5 days	Water suitable for
I	salmon, trout, grayling, stonefly and mayfly nymphs, caddis larvae, *Gammarus*	0–3	• domestic supply
II	trout rarely dominant; chub, dace, caddis larvae, *Gammarus*	4–10 (increases in summer in times of low flow)	• agriculture • industrial processes
III	roach, gudgeon, *Asellus*, mayfly nymphs and caddis larvae rarely	11–15	• irrigation
IV	red chironomid larvae (bloodworms), *Tubifex*	16–30 (completely deoxygenated from time to time)	• unsuitable for amenity use
V	barren / sewage fungus and small *Tubifex*	>30	• none

Indicator species are certain species of fauna whose presence can be indicative of the level of pollution.

An indicator species is one whose presence, absence or abundance can be used as an indicator of pollution. It doesn't have to be water pollution – some species can be used to indicate air pollution, soil nutrient levels and abiotic water characteristics:

- lichen *(Usnea alliculata)* indicates very low levels of sulfur dioxide in air
- nettles (*Ullica dioica*) indicate high phosphate levels in soil
- red alga (*Corauina officinalis*) indicate saline rock pools (absent from brackish ones).

A biotic index indirectly measures pollution by assaying the impact on species within the community according to their tolerance, diversity, and relative abundance.

Trent Biotic Index

The **Trent Biotic Index** is based on the disappearance of indicator species as the level of organic pollution increases in a river. This occurs because the species are unable to tolerate changes in their environment such as decreased oxygen levels or lower light levels. Those species best able to tolerate the existing conditions become abundant – which can lead to a change in diversity. In extreme environments (e.g. a highly polluted river) diversity is low, although numbers of individuals of pollution-tolerant species may be high. Diversity decreases as pollution increases.

The Trent Biotic Index has a maximum value of 10. The indices are in the form of marks out of 10 and give a sensitive assessment of pollution level: 10 indicates clean water and zero indicates highly polluted water.

Here is how it works.

1. Sort your sample, separating the animals according to group (taxonomic Order).
2. Count the number of groups.
3. Note which indicator species are present, starting from the top of the list in Table 4.16.

Table 4.16 Indicator species for the Trent Biotic Index

		Total number of groups present				
		0–1	2–5	6–10	11–15	16+
Indicator present	Number of species	Trent Biotic Index				
stonefly nymphs (Plecoptera)	>1	–	7	8	9	10
	1	–	6	7	8	9
mayfly nymphs (Ephemeroptera)	>1	–	6	7	8	9
	1	–	5	6	7	8
cassias fly larvae (Trichoptera)	>1	–	5	6	7	8
	1	4	4	5	6	7
Gammarus	all above absent	3	4	5	6	7
shrimps, crustaceans (*Asellus*)	all above absent	2	3	4	5	6
Tubifex / chironomid larvae	all above absent	1	2	3	4	–
all above absent	organisms not requiring dissolved oxygen may be present	0	1	2	–	–

To learn more about the Trent Biotic Index for Chesapeake Bay (2006–08), go to www.pearsonhotlinks.co.uk, enter the book title or ISBN, and click on weblink 4.5.

4. Take the highest indicator species on the list and read across the row, stopping at the column with the appropriate number of groups for your sample.

So, if your highest indicator animal belongs to the Trichoptera, you have more than one species and a total of 7 groups, the Trent Biotic Index for your sample is 6.

Eutrophication

Eutrophication refers to the nutrient enrichment of streams, ponds, and groundwater. It is caused when increased levels of nitrogen or phosphorus are carried into water bodies. It can cause algal blooms, oxygen starvation and, eventually, the decline of biodiversity in aquatic ecosystems.

Eutrophication can occur when lakes, estuaries and coastal waters receive inputs of nutrients (nitrates and phosphates), which result in an excess growth of plants and phytoplankton.

You should be able to explain the process and impacts of eutrophication.

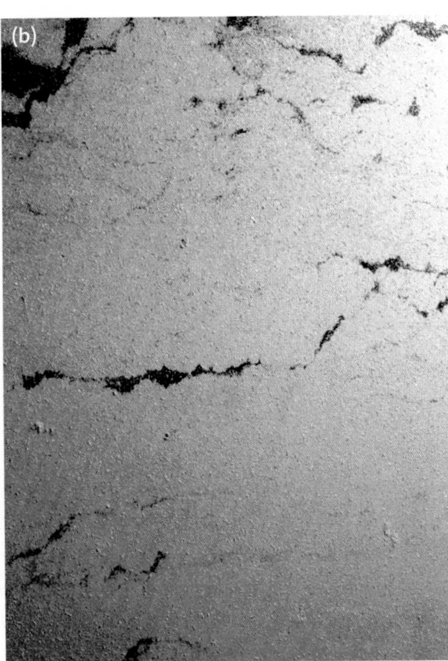

(a) Overgrowth of algae due to eutrophication, Cambridgeshire, UK. (b) Close-up of surface algal bloom due to eutrophication.

CONCEPTS: Equilibrium

In eutrophication, increased amounts of nitrogen and/or phosphorus are carried in streams, lakes, and groundwater causing nutrient enrichment. This leads to an increase in algal blooms as plants respond to the increased nutrient availability. As the algae, die back and decompose, further nutrients are released into the water. This is an example of positive feedback. However, the increase in algae and plankton shade the water below, cutting off the light supply for submerged plants. The prolific growth of algae and cyanobacteria, especially in autumn as a result of increased levels of nutrients in the water and higher temperatures, results in **anoxia** (oxygen starvation in the water). The increased plant biomass and decomposition lead to a build up of dead organic matter and to changes in species composition.

Some of these changes are the direct result of eutrophication (e.g. stimulation of algal growth in water bodies), while others are indirect (e.g. changes in the diversity of fish species due to reduced oxygen concentration). Eutrophication is very much a dynamic system – as levels of nitrates and phosphorus in streams and groundwater change, there is a corresponding change in species composition.

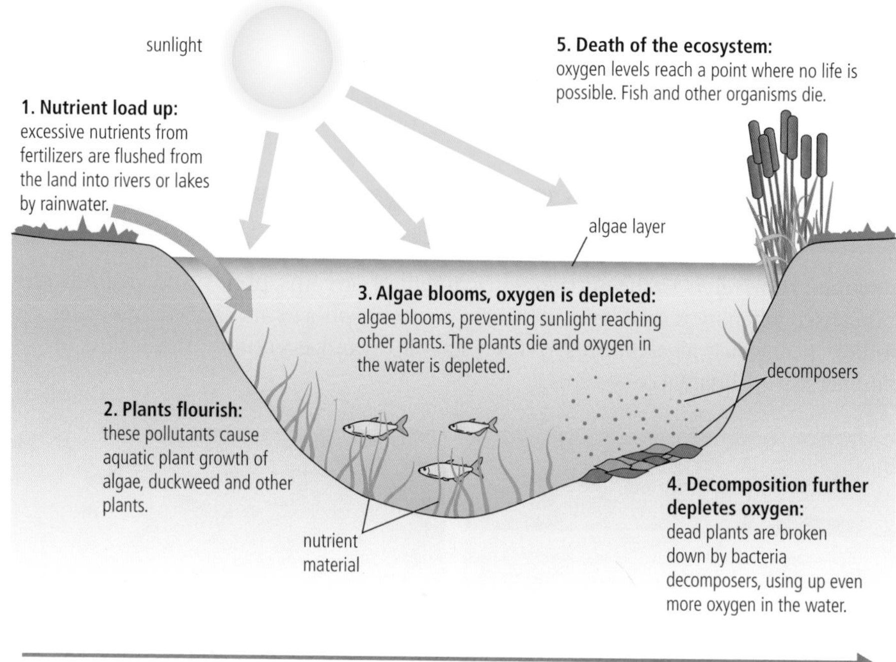

sunlight
1. Nutrient load up:
excessive nutrients from fertilizers are flushed from the land into rivers or lakes by rainwater.
2. Plants flourish:
these pollutants cause aquatic plant growth of algae, duckweed and other plants.
3. Algae blooms, oxygen is depleted:
algae blooms, preventing sunlight reaching other plants. The plants die and oxygen in the water is depleted.
4. Decomposition further depletes oxygen:
dead plants are broken down by bacteria decomposers, using up even more oxygen in the water.
5. Death of the ecosystem:
oxygen levels reach a point where no life is possible. Fish and other organisms die.
algae layer
decomposers
nutrient material
time

Figure 4.23 The process of eutrophication

A number of changes may occur as a result of eutrophication (Figure 4.23).

- Turbidity (murkiness) increases and reduces the amount of light reaching submerged plants.
- Rate of deposition of sediment increases because of increased vegetation cover. This reduces the speed of water and decreases the lifespan of lakes.
- Net primary productivity is usually higher compared with unpolluted water and may be seen by extensive algal or bacterial blooms.
- Dissolved oxygen in water decreases, as organisms decomposing the increased biomass respire and consume oxygen.
- Diversity of primary producers changes and finally decreases; the dominant species change. Initially, the number of primary producers increases and may become more diverse. However, as eutrophication proceeds, early algal blooms give way to cyanobacteria.
- Fish populations are adversely affected by reduced oxygen availability, and the fish community becomes dominated by surface-dwelling coarse fish, such as pike and perch. Other species migrate away from the area, if they can.

In freshwater aquatic systems, a major effect of eutrophication is the loss of the submerged *macrophytes* (aquatic plants). Macrophytes are thought to disappear because they lose their energy supply (sunlight penetrating the water). Sunlight is intercepted by the increased biomass of phytoplankton exploiting the high nutrient conditions. In principle, the submerged macrophytes could also benefit from increased nutrient availability, but they have no opportunity to do so because they are shaded by the free-floating microscopic organisms.

Natural eutrophication

The process of primary succession (Chapter 2, pages 114–119) is associated with gradual eutrophication as nutrients are trapped and stored by vegetation, both as living tissue and organic matter in soil or lake sediments. Nutrient enrichment occurs through addition of sediment, rainfall and the decay of organic matter and waste products. Starting from an oligotrophic (nutrient-poor) state with low productivity, a typical temperate lake increases in productivity fairly quickly as nutrients accumulate.

Anthropogenic eutrophication

Human activities worldwide have caused the nitrogen and phosphorus content of many rivers to double and, in some countries, local increases of up to 50 times have been recorded.

Phosphorus

Phosphorus is a rare element in the Earth's crust. Unlike nitrogen, there is no reservoir of gaseous phosphorus compounds available in the atmosphere. In natural systems, phosphorus is more likely to be a growth-limiting nutrient than nitrogen.

Domestic detergents are a major source of phosphates in sewage effluents. Estimates of the relative contribution of domestic detergents to phosphorus build-up in Britain's watercourses vary between 20 per cent and 60 per cent. As phosphorus increases in a freshwater ecosystem, the amount of plankton increases and the number of freshwater plants decreases.

Nitrogen

Nearly 80 per cent of the atmosphere is nitrogen. In addition, air pollution has increased rates of nitrogen deposition. The main anthropogenic source is a mix of nitrogen oxides (NO_x), mainly nitrogen monoxide (NO), released during the combustion of fossil fuels in vehicles and power plants. Despite its abundance, nitrogen is more likely to be the limiting nutrient in terrestrial ecosystems (as opposed to aquatic ones), where soils can typically retain phosphorus while nitrogen is leached away.

Nutrients applied to farmland through fertilizers may spread to the wider environment by:

- drainage water percolating through the soil, leaching soluble plant nutrients
- washing of excreta, applied to the land as fertilizer, into watercourses
- erosion of surface soils or the movement of fine soil particles into subsoil drainage systems.

In Europe, large quantities of slurry from intensively reared and housed livestock is spread on the fields. Animal excreta are very rich in both nitrogen and phosphorus and, therefore, their application to land can contribute to problems from polluted run-off.

Evaluating the impact of eutrophication

There are three main reasons why the high concentrations of nitrogen in rivers and groundwater are a problem. First, nitrogen compounds can cause undesirable effects in the aquatic ecosystems, especially excessive growth of algae. Second, the loss of fertilizer is an economic loss to the farmer. Third, high nitrate concentrations in drinking water may affect human health, and have been linked to increased rates of stomach cancer.

Algae and cyanobacteria are tiny organisms occurring in fresh water and saltwater. Algae belong to the eukaryotes – single-celled or multicellular organisms whose cells contain a nucleus. The cyanobacteria belong to the prokaryotes – single celled organisms without a membrane-bound nucleus. The cyanobacteria used to be called blue–green algae (a term you may still come across) but they have been reclassified as bacteria. The first members of the cyanobacteria to be discovered were indeed blue–green in colour, but since then new members of the group have been found that are not this distinctive colour.

The mining of phosphate-rich rocks has increased the mobilization of phosphorus. A total of 12×10^{12} g yr $^{-1}$ are mined from rock deposits. This is six times the rate at which phosphorus is locked up in ocean sediments from which the rocks are formed.

About three-quarters of the world's production of phosphorus comes from the USA, China, Morocco, and Russia.

Case study

Eutrophication of Lake Erie

Aerial view of lake 227 in 1994. The green colour is caused by cyanobacteria stimulated by the experimental addition of phosphorus for the 26th consecutive year. Lake 305 in the background is unfertilized.

Natural eutrophication normally takes thousands of years to progress. In contrast, anthropogenic or cultural eutrophication is very rapid. During the 1960s, Lake Erie (on the USA–Canada border) was experiencing rapid anthropogenic eutrophication and was the subject of much concern and research.

Eutrophication of Lake Erie caused algal and cyanobacterial blooms, which caused changes in water quality. The increase in cyanobacteria at the expense of water plants led to a decline in biodiversity. With fewer types of primary producer, there were fewer types of consumer, and so the overall ecosystem biodiversity decreased. Cyanobacteria are unpalatable to zooplankton, thus their expansion proceeds rapidly. The cyanobacterial blooms led to oxygen depletion and the death of fish. In addition, algal and bacterial species can cause the death of fish by clogging their gills and causing asphyxiation. Many indigenous fish disappeared and were replaced by species that could tolerate the eutrophic conditions. Low oxygen levels caused by the respiration of the increased lake phytomass killed invertebrates and fish. The death of macrophytes on the lake floor increased the build up of dead organic matter in the thickening lake sediments. Rotting bacterial masses covered beaches and shorelines.

Researchers at the University of Manitoba set up the Experimental Lakes Area (ELA) in 1968 to investigate the causes and impacts of eutrophication in Lake Erie. Between June 1969 and May 1976, it was the main focus of experimental studies at the ELA.

Aerial view lake 226 in August 1973. The green colour is due to cyanobacteria growing on phosphorus added to the lake on the nearside of the dividing curtain.

Over a number of years, seven different lakes (ELA lakes 227, 304, 302, 261, 226, 303, and 230) were treated in different ways. Lakes 227 and 226 were especially important in showing the effect of phosphorus in eutrophication. Studies of gas exchange and internal mixing in lake 227 during the early 1970s clearly demonstrated that algae in lakes were able to obtain sufficient carbon dioxide, via diffusion from the atmosphere to the lake water, to support eutrophic blooms.
The blue-green algae (now called cyanobacteria) were found to be able to fix nitrogen that had diffused naturally into the lake from the air, making nitrogen available for supporting growth.

ELA lake 226 was the site of a very successful experiment. The lake was divided into two relatively equal parts using a plastic divider curtain. Carbon and nitrogen were added to one half of the lake, while carbon, nitrogen and phosphorus were added to the other half of the lake. For 8 years, the side receiving phosphorus developed eutrophic cyanobacterial blooms, while the side receiving only carbon and nitrogen did not. The experiment suggested that in this case phosphorus was the key nutrient. A multibillion dollar phosphate control programme was soon instituted within the St Lawrence Great Lakes Basin. Legislation to control phosphates in sewage, and to remove phosphates from laundry detergents, was part of this programme.

By the mid-1970s, North American interest in eutrophication had declined. Nevertheless, the nutrient-pollution problem remains the number one water-quality problem worldwide.

Algae may be a nuisance but they do not produce substances toxic to humans or animals. Cyanobacteria, on the other hand, produce substances that are extremely toxic causing serious illness and death if ingested. This is why cyanobacteria are a very worrying problem in water sources or reservoirs used for leisure facilities.

Use of nitrogen fertilizers has increased by 600 per cent in the last 50 years and up to 30 per cent of nitrogen used in agriculture ends up in our fresh water.

Loss to farmers

Eutrophication can result in an economic loss for farmers. Farmers are keen to use NPK (nitrogen, phosphorus, and potassium) fertilizers because these products increase crop growth, improve farmers' income and may help increase crop self-sufficiency in a country. However, the removal of these nutrients from the soil reduces these benefits. Arable soils often contain much inorganic nitrogen: some is from fertilizer unused by the previous crop but most is from the decomposition of organic matter caused by autumn ploughing – ploughing releases vast quantities of nitrogen. However, unless a new crop is planted quickly, much of this is lost by leaching. Another influence is climate – there is normally more decomposition in the autumn when warm soils get wet. In still-growing grass pasture, the nitrate is absorbed but when fields are bare soil, the nitrate is prone to leaching. This problem is especially severe where a wet autumn follows a dry summer. Much soil organic matter may be decomposed and leached at such a time.

Health concerns

The concern for health relates to increased rates of stomach cancer (caused by nitrates in the digestive tract) and to blue baby syndrome (methaemoglobinaemia), caused by insufficient oxygen in the mother's blood for the developing baby. However, critics argue that the case against nitrates is not clear – stomach cancer could be caused by a variety of factors and the number of cases of blue baby syndrome is statistically small. However, in parts of Nigeria, where nitrate concentrations have exceeded 90 mg dm^{-3}, the death rate from gastric cancer is abnormally high.

Case study

Eutrophication in England and Wales

The amount of nitrates in tap water is a matter of general concern. The pattern of nitrates in rivers and groundwater shows marked regional and temporal variations. In the UK, it is concentrated towards the arable areas of the east, and concentrations are increasing. In England and Wales, over 35 per cent of the population derive their water from the aquifers of lowland England and over 5 million people live in areas where there is too much nitrate in the water. The problem is that nitrates applied on the surface slowly make their way down to the groundwater zone - this may take up to 40 years. Thus, increasing levels of nitrate in drinking water will continue to be a problem well into the 21st century. The cost of cleaning nitrate-rich groundwater is estimated at between £50 million and £300 million a year.

Case study

Eutrophication in Kunming City, China

Dianchi Lake, near Kunming City in the Yannan Province of China, has huge problems with eutrophication. Untreated sewage has been drained into the lake since before the 1980s. Cyanobacteria (*Microcystis* spp.) have killed over 90 per cent of native water weed, fish, and molluscs, so destroying the fish industry. The lake is largely green slime but because water supplies have run short, lake water from Dianchi Lake has been used since 1992 to supply Kunming's 1.2 million residents.

The city opened its first sewage treatment plant in 1993, but this copes with only 10 per cent of the city's sewage. Billions of dollars have been spent since the 1980s in attempts to clean up the lake, but with no real success.

There are three main ways of dealing with eutrophication:

- altering human activity
- regulating and reducing pollutants at the point of emission
- clean-up and restoration of polluted water.

Management strategies for eutrophication

Water pollution management strategies include:

- reducing human activities that produce pollutants (for example, using alternatives to current fertilizers and detergents)
- reducing release of pollution into the environment (for example, treatment of waste water to remove nitrates and phosphates)
- removing pollutants from the environment and restoring ecosystems (for example, removal of mud from eutrophic lakes and reintroduction of plant and fish species).

Altering human activities

Public campaigns in Australia have encouraged people to:

- use zero- or low-phosphate detergents
- wash only full loads in washing machines
- wash vehicles on porous surfaces away from drains or gutters
- reduce use of fertilizers on lawns and gardens
- compost garden and food waste
- collect and bury pet faeces.

Possible measures to reduce nitrate loss (based on the mid-latitude northern hemisphere) include the following.

- Avoid using nitrogen fertilizers during the wet season when soils are wet and fertilizer is most likely to be washed through the soil.
- Give preference to autumn-sown crops – their roots conserve nitrogen in the soil and use up nitrogen left from the previous year.
- Sow autumn-sown crops as early as possible and maintain crop cover through autumn and winter to conserve nitrogen.
- Do not apply nitrogen when the field is by a stream or lake.
- Do not apply nitrogen just before heavy rain is forecast (assuming that forecasts are accurate).
- Use less nitrogen if the previous year was dry because less will have been lost. This is difficult to assess precisely.

Regulating and reducing the nutrient source

Prevention of eutrophication at source has the following advantages (compared with treating its effects or reversing the process).

- Technical feasibility – in some situations, prevention at source may be achieved by diverting a polluted watercourse away from the sensitive ecosystem, while removal of nutrients from a system by techniques such as mud-pumping is more of a technical challenge.
- Cost – nutrient stripping at source using a precipitant is relatively cheap and simple to implement. Biomass stripping of affected water is labour-intensive and therefore expensive.
- Products – restored wetlands may be managed to provide economic products such as fuel, compost or thatching material more easily than trying to use the biomass stripped from a less managed system.

Phosphate stripping

Up to 45 per cent of total phosphorus loading to fresh water in the UK comes from sewage works. This input can be reduced by 90 per cent or more by carrying out phosphate stripping. The effluent is run into a tank and dosed with a precipitant, which combines with phosphate in solution to create a solid, which then settles out and can be removed.

CONCEPTS: Environmental value systems

Different users and organizations view eutrophication in different ways – farmers claim to need to use fertilizers to improve food supply; chemical companies argue they produce fertilizers to meet demand from farmers; water companies seek money from the government and the consumer to make eutrophic water safe to drink; the consumers see rising water bills and potential health impacts of eutrophication.

Clean-up strategies

Once nutrients are in an ecosystem, it is much harder and more expensive to remove them than it would have been to tackle the eutrophication at source.

The main clean-up methods available are:

- precipitation (e.g. treatment with a solution of aluminium or ferrous salt to precipitate phosphates)
- removal of nutrient-enriched sediments; for example, by mud pumping
- removal of biomass (e.g. harvesting of common reed) and using it for thatching or fuel.

Temporary removal of fish can allow primary consumer species to recover and control algal growth. Once water quality has improved, fish can be re-introduced.

Managing eutrophication using barley bales. The bales of barley straw are just visible (brown) beneath the water surface at the right-hand edge of the lake.

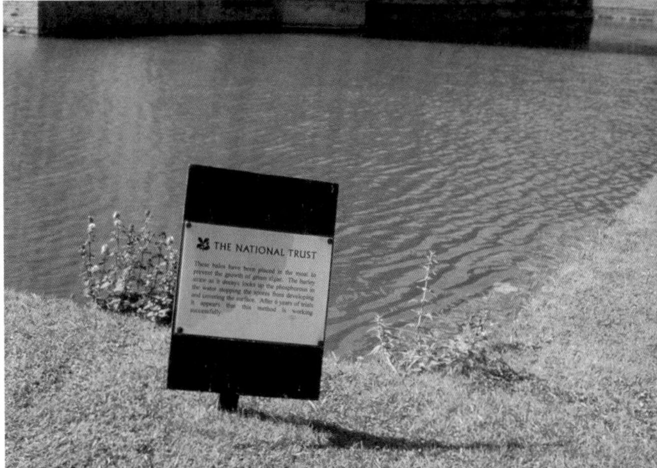

Mechanical removal of plants from aquatic systems is a common method for mitigating the effects of eutrophication. Efforts may be focused on removal of unwanted aquatic plants (e.g. water hyacinth) that tend to colonize eutrophic water. Each tonne of wet biomass harvested removes about 3 kg of nitrogen and 0.2 kg of phosphorus from the system. Alternatively, plants may be introduced deliberately to mop-up excess nutrients.

You should be able to evaluate pollution management strategies with respect to water pollution.

Case study

Effluent diversion at Lake Washington, USA

In some circumstances, it may be possible to divert sewage effluent away from a water body. This was achieved at Lake Washington, near Seattle, USA. In 1955, Lake Washington was affected by cyanobacteria. The lake was receiving sewage effluent from about 70 000 people. The sewerage system was redesigned to divert effluent away from the lake to the nearby sea inlet of Puget Sound.

CHALLENGE YOURSELF

Thinking skills 

'It is easier and more cost-effective to control the causes of eutrophication rather than to deal with the symptoms (results) of eutrophication.' Critically examine this statement.

To learn more about experiments related to eutrophication, go to www.pearsonhotlinks.co.uk, enter the book title or ISBN, and click on weblink 4.6.

Dead zones in both oceans and fresh water can occur when there is not enough oxygen to support marine life.

Dead zones and red tides

Dead zones, red tides and their associated plagues of jellyfish seem to have occurred naturally for centuries, but their appearance is becoming increasingly frequent. Red tides, for example, regularly form off the Cape Coast of South Africa, fed by nutrients brought up from the deep, and off Kerguelen Island in the Southern Ocean. Nowadays, though, most are associated with a combination of phenomena including overfishing, warmer waters, and the washing into the sea of farm fertilizers and sewage.

Most of the larger fish in shallow coastal waters have already been caught. As the larger species disappear, so the smaller ones thrive. These smaller organisms are also stimulated by nitrogen and phosphorus nutrients running off the land. The result is an explosion of growth among phytoplankton and other algae, some of which die, sink to the bottom and decompose, combining with dissolved oxygen as they rot. Warmer conditions, and sometimes the loss of mangroves and marshes, which once acted as filters, encourage the growth of bacteria in these oxygen-depleted waters. A situation develops where there is not enough oxygen to support marine life. (A similar effect occurs with eutrophication in freshwater rivers.)

The result may be a sludge-like soup, apparently lifeless – hence the name dead zones – but in fact teeming with simple, and often toxic, organisms. These may be primitive bacteria whose close relations are known to have thrived billions of years ago. Or they may be algae which colour the sea green or red-brown. In such places, red tides tend to form, some producing toxins that get into the food chain through shellfish, and rise up to kill bigger fish (if there are any left), birds, and even seals and manatees.

Red tides and similar blights do not necessarily last long, nor do they cover much of the surface of the sea. But they are increasing in both size and number: dead zones have now been reported in more than 400 areas. And increasingly they affect not only estuaries and inlets, but also continental seas such as the Baltic, the Kattegat, the Black and East China Seas and the Gulf of Mexico. All of these are traditional fishing grounds.

The winners in these newly polluted, over-exploited, oxygen-starved seas are simple, primitive forms of life, whereas the losers are the species that have taken millenia to develop.

The impacts of waste on the marine environment

Over 80 per cent of marine pollution comes from land-based activities. When waste is dumped, it is often close to the coast and very concentrated. The most toxic waste material dumped into the ocean includes dredged material, industrial waste, sewage sludge, and radioactive waste. Dredging contributes about 80 per cent of all waste dumped into the ocean. Rivers, canals, and harbours are dredged to remove silt and sand build up or to establish new waterways. About 20–22 per cent of dredged material is dumped into the ocean. About 10 per cent of all dredged material is polluted with heavy metals such as cadmium, mercury, and chromium, hydrocarbons such as heavy oils, nutrients including phosphorus and nitrogen, and organochlorines from pesticides. When dredged material is dumped into the ocean, fisheries suffer adverse affects, such as unsuccessful spawning in herring and lobster populations.

Over 60 million litres of oil run off America's roads and, via rivers and drains, find their way into the oceans each year. Through sewage and medical waste, antibiotics and hormones enter the systems of seabirds and marine mammals. Mercury and other metals turn up in tuna, orange roughy, seals, polar bears, and other long-lived animals. In the 1970s, 17 million tonnes of industrial waste were legally dumped into the ocean.

In the 1980s, 8 million tonnes were dumped, including acids, alkaline waste, scrap metals, waste from fish processing, flue desulfurization sludge, and coal ash. The peak of sewage dumping was 18 million tonnes in 1980, a number that fell to 12 million tonnes in the 1990s. Alternatives to ocean dumping include recycling, producing less wasteful products, saving energy and changing the dangerous material into more benign waste.

Oil pollution

All over the world, oil spills regularly contaminate coasts. Oil exploration is a major activity in such regions as the Gulf of Mexico, the South China Sea and the North Sea. The threats vary. For example, there is evidence of widespread toxic effects on benthic (deep-sea) communities on the floor of the North Sea in the vicinity of the 500+ oil production platforms in British and Norwegian waters.

Meanwhile, oil exploration in the deep waters of the North Atlantic, north-west of Scotland, threatens endangered deep-sea corals. There is evidence, too, that acoustic prospecting for hydrocarbons in these waters may deter or disorientate some marine mammals.

Shipping is a huge cause of pollution. Ships burn bunker oil, the dirtiest of fuels, so more carbon dioxide is released and more particulate matter, which may be responsible for about 60 000 deaths each year from chest and lung diseases, including cancer. Most of these occur near coastlines in Europe, and East and South Asia. Some action is being taken. Oil spills should become rarer after 2010, when all single-hulled ships were banned.

Efforts are also being made to prevent the spread of invasive species through the taking on and discharging of ships' ballast water. And a UN convention may soon ban the use of tributyltin, a highly toxic chemical added to the paint used on almost all ships' hulls, in order to kill algae and barnacles.

Case study

Deepwater Horizon oil spill

The Deepwater Horizon oil spill is the largest in US history. In April 2010, an explosion ripped through the Deepwater Horizon oil rig in the Gulf of Mexico, 80 km off the coast. Two days later the rig sank, with oil pouring into the sea at a rate up to 62 000 barrels a day. The oil threatened wildlife along the US coast as well as livelihoods dependent on tourism and fishing. Over 160 km of coastline were affected, including oyster beds and shrimp farms.

The extent of the environmental impact is likely to be severe and last a long time. A state of emergency was declared in Louisiana. The cost to BP, who operated the rig, may reach US$20 billion. BP's attempts to plug the oil leak were eventually successful. Dispersants were used to break up the oil slick but BP was ordered by the US government to limit their use, as they could cause even more damage to marine life in the Gulf of Mexico. By the time the well was capped (in July 2010), about 4.9 million barrels of crude oil had been released into the sea.

Radioactive waste

Radioactive effluent also makes its way into the oceans. Between 1958 and 1992, the Arctic Ocean was used by the Soviet Union, or its Russian successor, as the resting place for 18 unwanted nuclear reactors, several still containing their nuclear fuel. Radioactive waste is also dumped in the oceans, and usually comes from the nuclear power process, medical and research use of radioisotopes, and industrial uses. Nuclear waste usually remains radioactive for decades. Following the explosion at the Daichi nuclear power in Japan in March 2011, radioactive material was carried by air and water across the Pacific towards North America. It reached Vancouver Island, Canada, in February 2015.

Plastic

More alarming still is the plague of plastic. In 2006, the UN Environment Programme reckoned that every square kilometre of sea held nearly 18 000 pieces of floating plastic. Much of it was and still is in the central Pacific, where scientists believe as much as 100 million tonnes of plastic waste are suspended in two separate gyres (large rotating ocean currents) of garbage in the Great Pacific Garbage Patch. To read more about plastic pollution in the Great Pacific Garbage Patch, see pages 428–429.

Dead albatross with plastic debris (in its stomach).

Exercises

1. Define the term *biochemical oxygen demand* (*BOD*) and explain how this indirect method is used to assess pollution levels in water.
2. Describe and explain an indirect method of measuring pollution levels using a biotic index.
3. Figure 4.22 (page 253) shows changes in characteristics of a stream below an outlet of pollution.
   a. Describe the relative changes in *Tubifex* and stonefly nymphs along the course of the river.
   b. Suggest reasons for these changes.
   c. Compare and contrast the presence and abundance of stonefly nymphs and *Tubifex* worms.
   d. Explain the variations in BOD and dissolved oxygen.
4. Outline the processes of eutrophication.
5. Evaluate the impacts of eutrophication.
6. Describe and evaluate pollution management strategies with respect to eutrophication.
7. Outline the effects of eutrophication on natural and human environments.

8. The data below show the main sources of waste entering the sea.

run-off and land-based discharge	44 per cent
atmosphere	33 per cent
maritime transportation	12 per cent
dumping	10 per cent
offshore production	1 per cent

a. Choose a suitable method to present this data.

b. Comment on the results you produce.

Big questions

Having read this section, you can now discuss the following big questions:

- What strengths and weaknesses of the systems approach and the use of models have been revealed through this topic?
- To what extent have the solutions emerging from this topic been directed at *preventing* environmental impacts, *limiting* the extent of the environmental impacts, or *restoring* systems in which environmental impacts have already occurred?
- How are the issues addressed in this topic of relevance to sustainability or sustainable development?
- In what ways might the solutions explored in this topic alter your predictions for the state of human societies and the biosphere some decades from now?

Points you may want to consider in your discussions:

- To what extent can water pollution be considered as a system?
- Are the existing solutions to pollution likely to cope with current levels of water pollution?
- Which is the lesser evil – less food production or eutrophication? How are they linked?
- How is water pollution likely to change in the next decades? Give reasons for your answer.

Practice questions

1 Study the model of the hydrological cycle below.

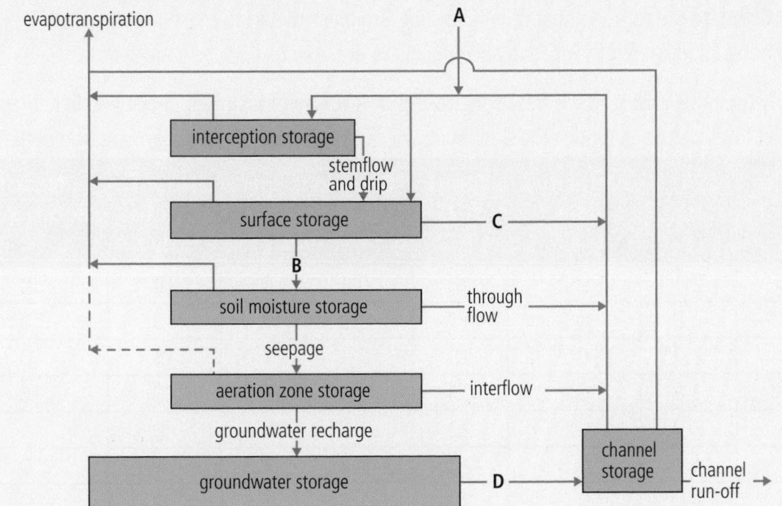

a Identify the flows A, B, C, and D. [2]

b Explain how the hydrological cycle may be changed in urban areas. [4]

c How can the use of groundwater be sustainable? [3]

2 The graph below shows Japan's whale catch between 1985 and 2010.

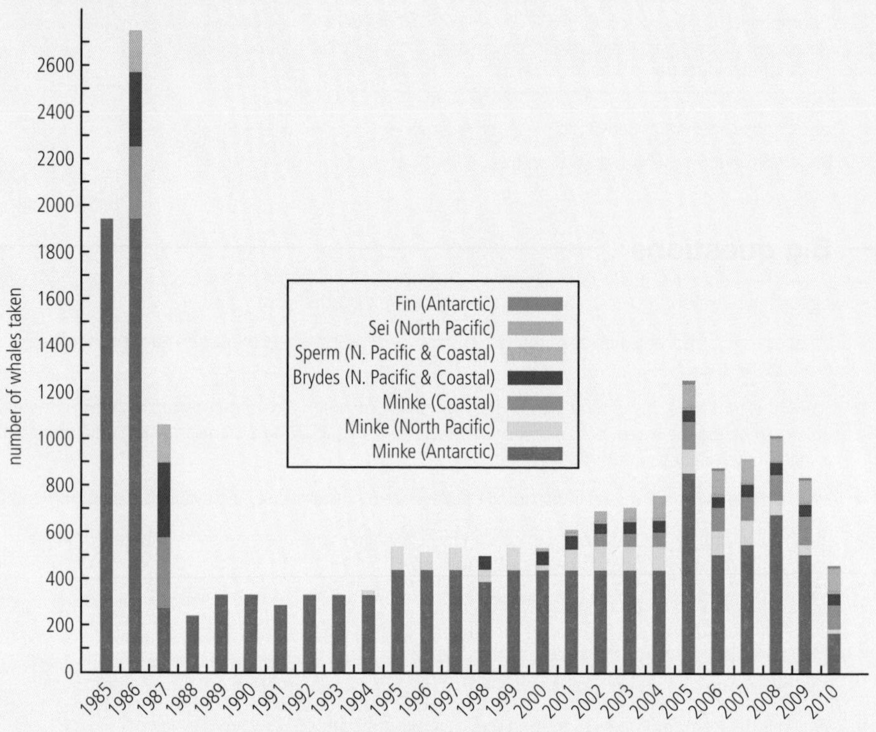

a Describe the trend in the number of whales caught by Japan between 1985 and 2010. [3]

b What is the main species of whale caught, and from where is it mainly taken? [2]

c Suggest why the number of whales killed fell so much in 1987, but then began to rise again. [2]

d Compare the total number of whales that Japan are taking to those that the Inuit populations of North America and Greenland are taking. [2]

3 Study the table below which shows the results of a survey of a stream above and below an outlet from a sewage works. The figure below is a sketch map of the stream and the outlet.

Site	CSA* / m^2	Velocity / $m\,sec^{-1}$	Temp / °C	Oxygen / %	pH	No. of caddis fly	No. of bloodworms
1	2.1	0.2	18	0.1	6.0	12	0
2	2.3	0.2	17	0.2	6.0	15	0
3	2.2	0.3	18	0.1	7.0	11	0
4	3.8	0.3	23	0.3	6.5	0	16
5	3.9	0.6	22	1.8	7.0	0	1
6	4.1	0.8	22	1.7	7.5	1	0
7	3.9	0.7	20	1.6	6.5	2	0
8	4.0	0.7	22	1.5	7.0	7	0

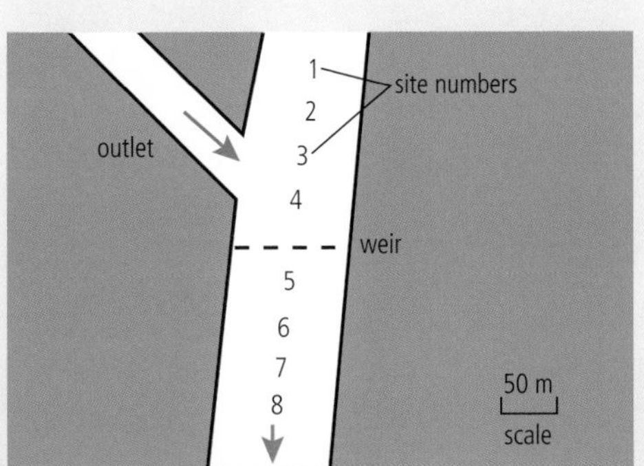

a Define the terms *water quality*, *pollution*, and *discharge*. [3]

b Plot the results for variations in oxygen content along the course of the stream. How does the oxygen content change at sites 4 and 5? Explain why. [6]

c What is the trend in temperature levels between site 1 and site 8? [3]

d i What does pH measure? [1]

ii How do you account for the relatively small linear variations in the stream's pH? [3]

4 Study the figure below which shows sources of cultural eutrophication.

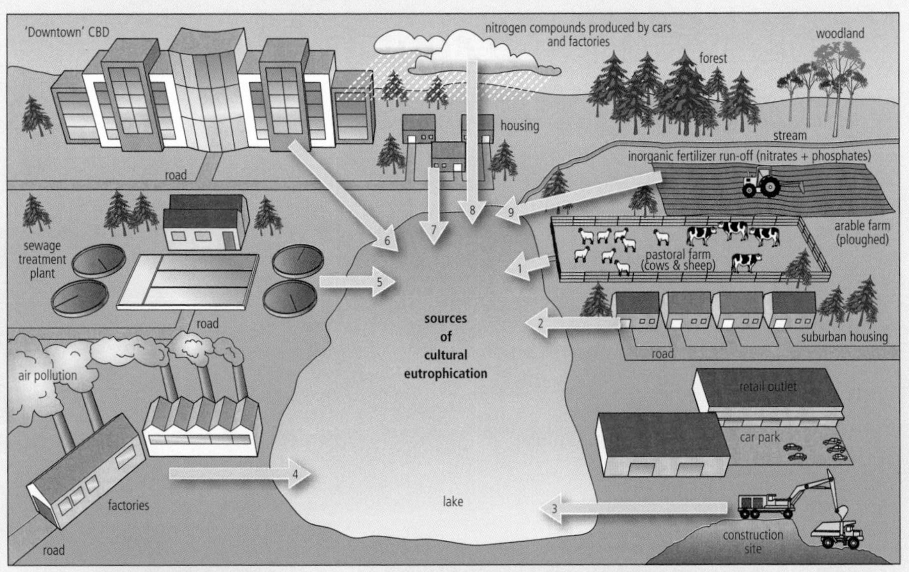

a Explain what is meant by the term *cultural eutrophication*? [1]

b Suggest two ways in which urban areas may contribute to eutrophication. [2]

c What are the natural sources of nutrients as suggested by the figure above? [1]

d Briefly explain the process of eutrophication. [5]

05 Soil systems, terrestrial food production systems, and societies

5.1 Introduction to soil systems

Opposite: A wide variety of foods are produced – and sold – as seen at this market in Vijaupur, India

Significant ideas

The soil system is a dynamic ecosystem that has inputs, outputs, storages, and flows.

The quality of soil influences the primary productivity of an area.

Fertile soil is a non-renewable resource.

Big questions

As you read this section, consider the following big questions:

- What strengths and weaknesses of the systems approach and the use of models have been revealed through this topic?
- To what extent have the solutions emerging from this topic been directed at *preventing* environmental impacts, *limiting* the extent of the environmental impacts, or *restoring* systems in which environmental impacts have already occurred?
- What value systems can you identify at play in the causes and approaches to resolving the issues addressed in this topic?
- In what ways might the solutions explored in this topic alter your predictions for the state of human societies and the biosphere some decades from now?

Knowledge and understanding:

- The soil system may be illustrated by a soil profile that has a layered structure (horizons).
- Soil system storages include organic matter, organisms, nutrients, minerals, air, and water.
- Transfers of material within the soil including biological mixing and leaching (minerals dissolved in water moving through soil) contribute to the organization of the soil.
- There are inputs of organic material including leaf litter and inorganic matter from parent material, precipitation, and energy. Outputs include uptake by plants and soil erosion.
- Transformations include decomposition, weathering, and nutrient cycling.
- The structure and properties of sand, clay, and loam soils differ in many ways, including: mineral and nutrient content, drainage, water-holding capacity, air spaces, biota, and potential to hold organic matter. Each of these variables is linked to the ability of the soil to promote primary productivity.
- A soil texture triangle illustrates the differences in composition of soils.

Soil profiles

A **soil profile** is a vertical section through a soil, and is divided into horizons (distinguishable layers) as shown in Table 5.1. These layers have distinct physical and chemical characteristics, although the boundaries between horizons may be blurred by earthworm activity.

The soil system may be illustrated by a soil profile that has a layered structure (horizons).

Table 5.1 Soil horizons

	Horizons	Sub-horizons/variations within the horizon
	O organic horizon	l undecomposed litter f partly decomposed (fermenting) litter h well-decomposed humus
	A mixed mineral–organic horizon	h humus p ploughed, as in a field or a garden g gleyed or waterlogged
	E eluvial or leached horizon	a strongly leached, ash coloured horizon, as in a podzol b weakly bleached, light brown horizon, as in a brown earth
	B illuvial or deposited horizon	Fe iron deposited t clay deposited h humus deposited
	C bedrock or parent material	r rock u unconsolidated materials

To learn more about the International Year of Soil, go to www.pearsonhotlinks.co.uk, enter the book title or ISBN, and click on weblink 5.1.

You do not need to be able to identify different types of soil, but you do need to know about the processes that operate in soils, the effect of water movement, and the effect of soil organisms on soil development.

Very acid soils may have very distinct soil horizons due to the lack of earthworms to mix the horizons.

TOK

The soil system may be represented by a soil profile. Since a model is strictly speaking not real, how can it lead to knowledge?

Soils are complex features. Simplified soil profiles allow us to study soils in a way that uncovers some of the main processes and features of soils.

Soil system storages include organic matter, organisms, nutrients, minerals, air, and water.

The top layer of vegetation is referred to as the organic (O) horizon. Beneath this is the mixed mineral–organic layer (A horizon). It is generally a dark colour due to the presence of organic matter. An Ap horizon is one that has been mixed by ploughing.

The E horizon is the eluvial or leached horizon found in some soils. Leaching removes material from the horizon. Consequently, the layer is much lighter in colour. Where leaching is intense, an ash-coloured Ea horizon is formed. By contrast, in a brown earth, where leaching is less intense, a light brown Eb horizon is found. The B horizon is the deposited or illuvial horizon. It contains material that has been moved from the E horizon, such as iron (Fe), humus (h), and clay (t).

At the base of the profile is the parent material or bedrock. Sometimes labels are given to distinguish rock (r) from unconsolidated loose deposits (u).

A soil showing clear horizon development

Soil systems

SYSTEMS APPROACH

Soils are a major component of the world's ecosystems (Figure 5.1). They form at the interface of the Earth's atmosphere, lithosphere (rocks), **biosphere** (living matter) and hydrosphere (water). Soils form the outermost layer of the Earth's surface, and comprise weathered bedrock (regolith), organic matter (both dead and alive), air, and water.

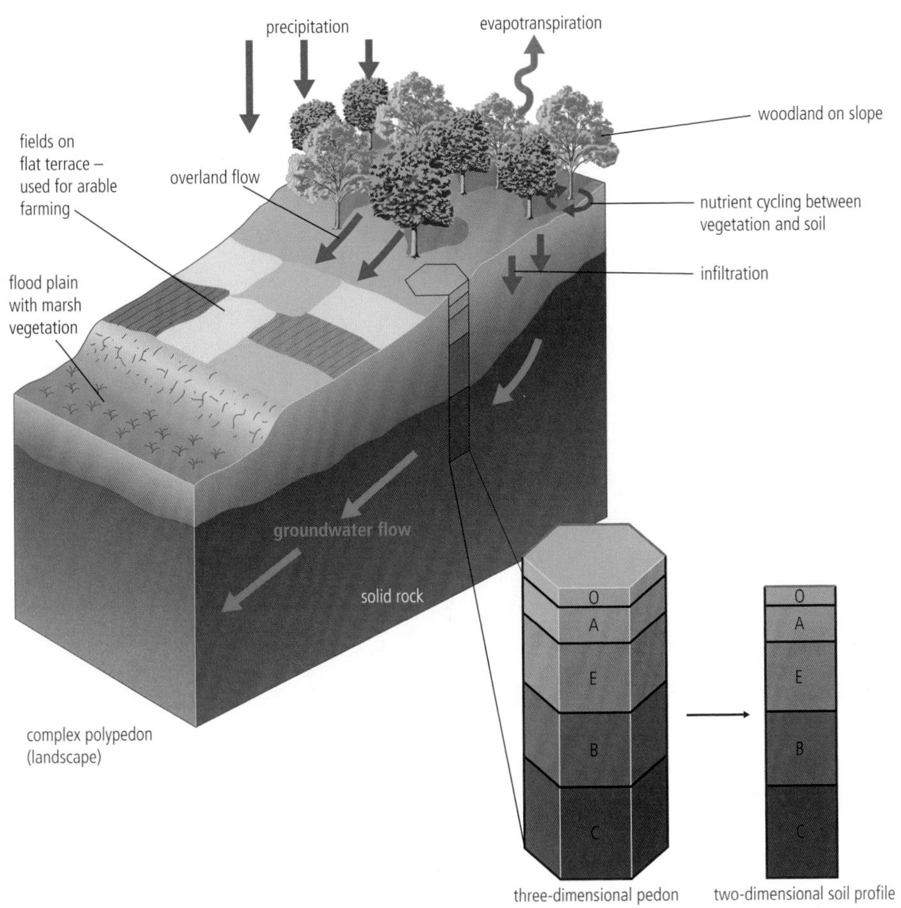

Figure 5.1 Soil systems in the environment

Soils perform a number of vital functions for humans.

- Soils are the medium for plant growth – most foodstuffs for humans are grown in soil.
- Soils contain an important store of relatively accessible fresh water – approximately 70 000 km^3 or 0.005 per cent of the global freshwater total.
- Soils filter materials added to the soil thereby maintaining water quality.
- Some recycling of nutrients takes place in the soil through the breakdown of dead organic matter.
- Soil acts as a habitat for billions of microorganisms as well as for some larger animals.
- Soils provide raw materials in the form of peat, clays, sands, gravels, and minerals.

Soils support our planet's biodiversity.

To learn more about the importance of soils, go to www.pearsonhotlinks.co.uk, enter the book title or ISBN, and click on weblink 5.2.

SYSTEMS APPROACH

Soil system storages include organic matter, organisms, nutrients, minerals, air, and water. Thus, soil has matter in all three states:

- organic and inorganic matter form the solid state
- soil water (from precipitation, groundwater, and seepage) forms the liquid state
- soil atmosphere forms the gaseous state.

Soils are a vital resource for humans but they take a long time to develop. Fertile soil is considered to be a non-renewable resource because of the current rate of resource use compared to the length of time required for resource replacement.

Peat cutting; peat is a store of energy that can be burned to provide heat.

Soil is a non-renewable resource, its preservation is essential for food security and our sustainable future.

Soil-forming processes

Soil-forming processes involve:

- gains and losses of material to and from the profile
- movement of water between the horizons
- chemical transformations within each horizon.

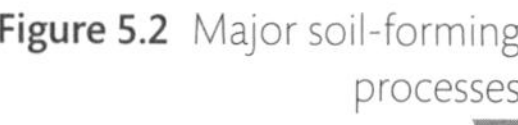

Figure 5.2 Major soil-forming processes

Therefore, soils must be considered as open systems in a steady-state equilibrium, varying constantly as the factors and processes that influence them change.

The principal soil-forming processes include weathering, transfer of materials, organic changes, and waterlogging (Figure 5.2). The weathering of bedrock gives the soil its C horizon and also its initial bases and nutrients (fertility), structure, and texture (drainage).

Gains and losses of material

Transfers of materials within the soil contribute to the organization of the soil. There are inputs of organic material including leaf litter and inorganic matter from parent material, precipitation, and energy. Outputs include uptake by plants and soil erosion.

Inputs include organic material (e.g. leaf litter) and inorganic matter (from parent material), precipitation and energy. Outputs include energy, uptake by plants, and soil erosion.

Movement of water

Translocation includes many processes, mostly by water and mostly downwards. *Leaching* refers to the downward movement of soluble material.

Soils store and filter water, improving our resilience to floods and droughts.

In arid and semi-arid environments, evapotranspiration (EVT) is greater than precipitation, so the movement of soil solution is upwards through the soil. Water is drawn to the drying surface by capillary action and leaching is generally ineffective apart from during occasional storms. Calcium carbonates and other solutes remain in the soil. This process is known as *calcification*. In grasslands, calcification is enhanced because grasses require calcium; they draw it up from the lower layers and return it to the upper layers when they die down. In extreme cases where EVT is intense, sodium or calcium may form a crust on the surface. This may be toxic to plant growth. Excessive sodium concentrations may occur due to capillary rise of water from a water table that is saline and close to the surface – as in the case of the Punjab irrigation scheme in India. Such a process is known as salinization or alkalization.

Transfers of materials include biological mixing and leaching (i.e. minerals dissolved in water moving through soil), which contribute to the organization of the soil.

Chemical transformations

Decomposition

Decomposition changes leaf litter into humus. Organic changes occur mostly at or near the surface. Plant litter is *decomposed* (humified) into a dark mass. It is also degraded gradually by **decomposers** and detritivores such as fungi, algae, small insects, bacteria, and worms. Under very wet conditions, humification forms peat. Over a long period of time, humus decomposes due to *mineralization*, which releases nitrogenous compounds.

Transformations include decomposition, weathering and nutrient cycling.

Weathering

Weathering is the decomposition and disintegration of rocks *in situ*. Decomposition refers to *chemical* weathering and creates altered rock substances, such as kaolinite (china clay) from granite. By contrast, disintegration or *mechanical* weathering produces smaller, angular fragments of the same rock, such as scree. *Biological* weathering has been identified as a process in which plants and animals chemically alter rocks and physically break rocks through their growth and movement. Biological weathering is not a separate type of weathering, but a form of disintegration and decomposition. It is important to note that these processes are *interrelated* rather than operating in isolation.

Soils in tropical areas such as Borneo may be extremely deep, due to the warm, wet year-round climate. In contrast, soils in cold areas such as Iceland may be very thin or non-existent, due to the lack of chemical weathering.

Weathering helps break down rock and forms **regolith**. With the addition of plants and animals, air and water, regolith helps form soil.

Nutrient cycling

A nutrient cycle involves interaction between soil, plants, animals, and the atmosphere, and many food chains. There is great variety between the cycles. Nutrient cycles can be sedimentary, in which the source of the nutrient is from rocks – or they can be atmospheric, as in the case of the nitrogen cycle. Generally, gaseous cycles are more complete than sedimentary ones as the latter are more susceptible to disturbance, especially by human activity.

Soils help to combat and adapt to climate change by playing a key role in the carbon cycle.

Significant differences exist in arable (potential to promote primary productivity) soil availability around the world. Soil processes vary between humid (wet) and arid (dry) areas. These differences have socio-political, economic, and ecological influences.

Nutrients are circulated and reused frequently. All natural elements are capable of being absorbed by plants, either as gases or soluble salts. Only oxygen, carbon, hydrogen and nitrogen are needed in large quantities. These are known as *macronutrients*. The rest are *trace elements* or micronutrients, such as magnesium, sulfur, and phosphorus. These are needed only in small doses. Nutrients are taken in by plants and built into new organic matter. When animals eat the plants, they take up the nutrients. The nutrients eventually return to the soil when the plants and animals die and are broken down by decomposers.

Nutrient cycles can be shown by means of simplified diagrams (Gersmehl's nutrient cycles) which indicate the stores of nutrients as well as the transfers (see Figure 2.30, page 95).

TOK

Is soil a renewable resource or a non-renewable resource?

How does the length of time that it takes a soil to form affect its renewable or non-renewable status?

Soil structures and properties

Soils provide plants with a number of benefits. These include:

- anchorage for roots
- a supply of water
- a supply of oxygen
- a supply of mineral nutrients (e.g. nitrogen)
- protection against adverse changes of temperature and pH.

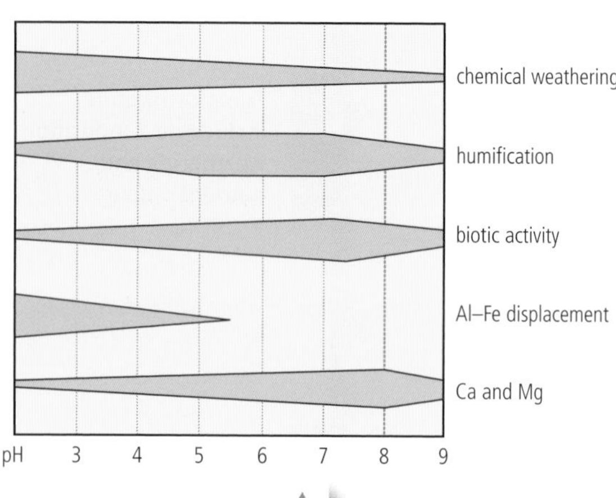

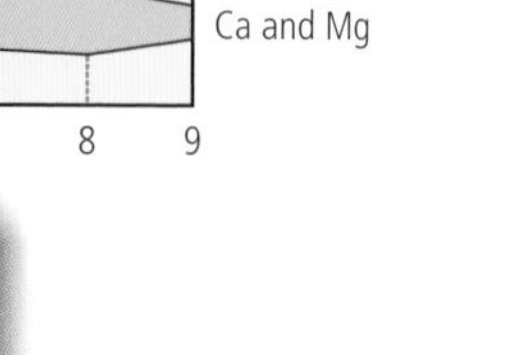

Figure 5.3 Soil pH, characteristics, and processes

Nevertheless, there are several soil conditions that restrict root growth. These are a mix of physical and chemical conditions. Physical conditions include:

- mechanical barriers (usually associated with a high bulk density, as occurs in compacted soils)
- absence of cracks
- shortage of oxygen due to waterlogging
- dryness
- temperatures that are too high or too low.

Chemical conditions include:

- high aluminium concentration, usually associated with low pH (Figure 5.3)
- low nutrient supply
- phytotoxic chemicals in anaerobic soil (e.g. trace metals or salinity associated with insecticides or herbicides).

A soil texture triangle illustrates the differences in composition of soils.

Soil texture

Soil structure refers to the shape and arrangement of individual soil particles (called peds). The ideal soil for cultivation is a *loam* in which there is a balance between water-holding ability and freely draining, aerated conditions. This balance is influenced by a number of factors, especially soil texture. Soil texture refers to the proportion of differently sized materials – usually sand, silt and clay – present in a soil. A loam is a well-balanced soil with significant proportions of sand, silt, and clay.

You can check whether you have interpreted a triangular graph correctly or not, as the percentages added up should total 100 per cent.

Triangular graphs (Figure 5.4) are used to show data that can be divided into three parts, such as sand, silt, and clay for soil. The data must be in the form of a percentage, and the percentage must add up to 100 per cent. The main advantage of triangular graphs are that:

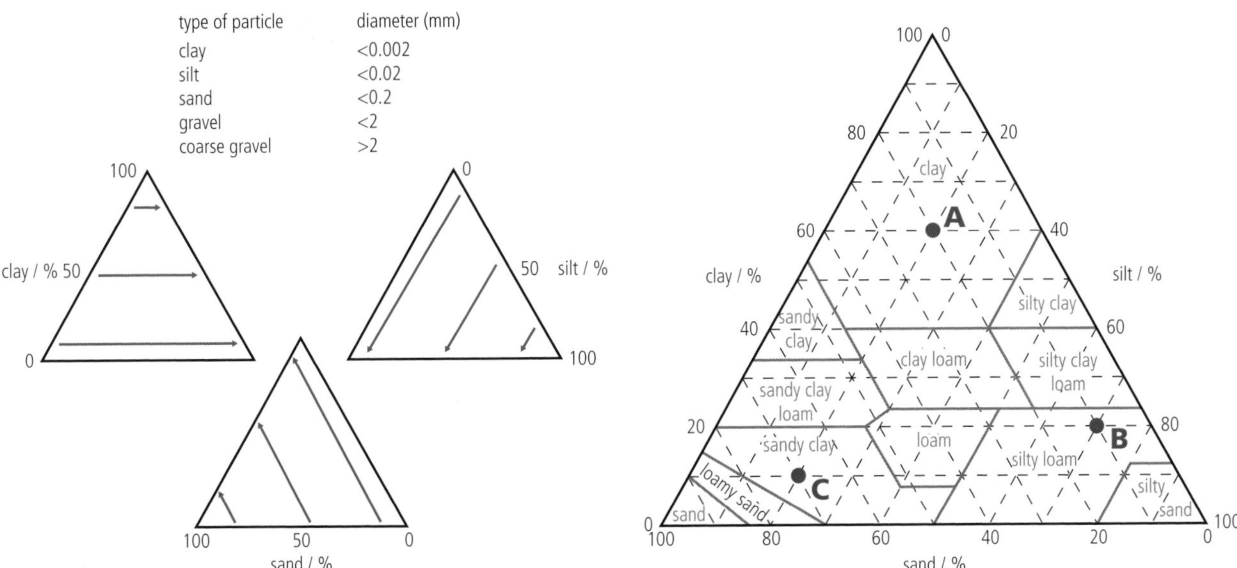

- a large number of data can be shown on one graph
- groupings are easily recognizable (e.g. loams)
- dominant characteristics can be shown
- classifications can be drawn up.

Triangular graphs can be difficult to interpret and it is easy to get confused.

Figure 5.4 Triangular graph showing soil textural groups. (points A, B, and C refer to a question on page 277.)

Soil types

The agricultural potential of a soil depends on:

- the porosity and permeability of the soil
- the surface area of the soil peds.

Drainage and water holding capacity

The pore spaces in soil determine the rate at which water drains through a soil. The surface area of the peds determines the amount of water and nutrients in solution that can be retained against the force of gravity. The terms 'light', 'medium', and 'heavy' are used refer to the workability of a soil.

For example, a heavy clay soil can hold twice as much water as a light soil. Light soils (over 80 per cent sand) are coarse textured and are easily drained of water and nutrients. But they warm up more quickly than heavy clay soils and so allow early growth in spring (this is useful, for example, for root crops such as potatoes). Heavy soils contain more than 25 per cent clay and are fine-textured. Many of their pores are <0.001 mm and the very large chemically active surface area means that these soils are water and nutrient retentive. Clay absorbs water, so that the soil swells when wet and shrinks when dry.

Silt particles are larger than clay and smaller than sand. Hence, silt drains faster than clay but more slowly than sand. It does not retain water or chemicals well, so its potential for farming is reduced.

Table 5.2 summarizes the implications of water retention in various soils.

To watch a selection of soil-related animations, go to www.pearsonhotlinks.co.uk, enter the book title or ISBN, and click on weblink 5.3.

The structure and properties of sand, clay and loam soils differ in many ways. These include mineral and nutrient content, drainage, water-holding capacity, air spaces, biota, and the potential to hold organic matter. Each of these variables is linked to the ability of the soil to promote primary productivity.

Table 5.2 Soil properties and texture

Soil	Water infiltration rate	Water-holding capacity	Nutrient status	Aeration	Ease of working
clay	poor	good	good	poor	poor
silt	medium	poor–medium	medium	poor–medium	medium
sand	good	poor	poor	good	good
loam	medium	medium	medium	medium	medium

Soils formed on chalk or limestone may consist of just an O horizon and a C horizon, as the weathering of the bedrock produces dissolved substances. These are carried away by the percolating water leaving just degraded organic matter and rock impurities.

Air spaces

The ideal soil structure is a crumb structure in which peds are small. The soil structural condition can also be measured by its porosity – this determines its air capacity and water availability (Table 5.3).

Table 5.3 Structural quality of top soil

Structural quality	Air capacity (%)	Available water (%)
very good	>15	>20
good	11–15	16–20
moderate	5–10	10–15
poor	< 5	< 10

Primary productivity

Different soil types have different levels of primary productivity. These can be summarized as follows:

- sandy soil – low primary productivity due to poor water-holding capacity and low nutrient status
- clay soil – quite low primary productivity due to poor aeration and poor water infiltration
- loam soil – high primary productivity due to medium infiltration rate, water-holding capacity, nutrient status, aeration, and ease of working.

Soils were first cultivated in Mesopotamia, between the Tigris and Euphrates rivers, about 10 000 years ago.

For optimum structure, a variety of pore sizes is required to allow root penetration, free drainage and water storage. This is because pore spaces of over 0.1 mm allow root growth, oxygen diffusion and water movement, whereas pore spaces below 0.05 mm help store water.

The workability of a soil depends on the amount of clay present. As Table 5.4 shows, the force needed to pull a plough increases with clay content.

Table 5.4 Force needed to pull a plough and clay content of soil

Clay content (%)	23.6	30.0	31.1	34.3
Force needed to pull a plough (kg)	580	635	680	703

Primary productivity of soil depends on:

- mineral content
- drainage
- water-holding capacity
- air spaces
- biota
- potential to hold organic matter.

Healthy soils are the basis for healthy food production.

Soils are the foundation for vegetation, which is cultivated or managed for feed, fibre, fuel, and medicinal products.

Suitability of soils for food production

The main limiting factor for light soils is drought during the growing season because these soils have a poor nutrient- and water-holding capacity.

Heavy soils in which the clay content is over 28 per cent are the most difficult for arable cultivation. They are highly water retentive, have low permeability and field drainage is slow. Drying out is slow. Heavy soils can become waterlogged when wet or hard when too dry. The number of days in which they can be ploughed is small in comparison with other soils (Table 5.5).

The water and nutrient retention by humus is considerably greater than that in clay.

Table 5.5 Influence of rainfall and texture on the average number of days in which soils can be worked

	Number of days in which soil can be worked								
Month	*February*			*March*			*April*		
Soil type	*Light*	*Medium*	*Heavy*	*Light*	*Medium*	*Heavy*	*Light*	*Medium*	*Heavy*
wetter than average	3	2	0	16	14	9	21	19	16
average rainfall	8	5	3	25	24	20	26	23	16
drier than average	11	9	8	29	29	27	28	26	25

Exercises

1. Study Figure 5.3 (page 274) which shows soil pH, characteristics, and processes.
   a. Describe how chemical weathering varies with pH.
   b. At which level of pH are aluminium and iron (Al and Fe) displaced?
   c. Between which levels of pH is biotic activity increased?
   d. At which pH is calcium and magnesium (Ca and Mg) most common?
   e. Between which pH levels is humification increased?
2. Study Figure 5.4 (page 275).
   a. Identify the composition of the soils marked A, B, and C.
   b. Name a soil that has
      i. 40% clay, 25% silt, and 35% sand
      ii. 20% clay, 30% silt, and 50% sand
      iii. 35% clay, 35% silt, and 30% sand.
   c. Outline the advantages of using triangular graphs.
   d. Briefly explain two or more problems with using triangular graphs.
3. Study Table 5.2 Soil properties and texture on page 276.
   a. Which soil type has the highest infiltration rate?
   b. Which soil type has the best water-holding capacity?
   c. Which soil type has the highest nutrient status?
   d. Which soil type has the best aeration?
   e. Which is the easiest soil to work?
4. Study Table 5.3 Structural quality of top soil on page 276.
   a. State how the quality of soil differs between a very good soil and a poor soil.
   b. Suggest how this may affect plant productivity.
5. Study Table 5.4 (page 276) which shows the force needed to pull a plough and clay content of soil.
   a. Describe the relationship between clay content and force needed to pull a plough.
   b. Suggest reasons to explain the relationship you have described.

6. Study Table 5.5 (page 277) which shows the influence of rainfall and texture on the average number of days in which soils can be worked.
   a. Describe the variations in the number of workdays for light soils:
      i. by month
      ii. by rainfall.
   b. Suggest reasons for your answers to a(i) and (ii).
   c. Comment on the variations in the number of days in which cultivation would be satisfactory for light, medium, and heavy soils.
   d. What is meant by a loam soil? Describe the differences between a sandy loam and a clay loam.

Big questions

Having read this section, you can now discuss the following big questions:

- What strengths and weaknesses of the systems approach and the use of models have been revealed through this topic?
- To what extent have the solutions emerging from this topic been directed at *preventing* environmental impacts, *limiting* the extent of the environmental impacts, or *restoring* systems in which environmental impacts have already occurred?
- What value systems can you identify at play in the causes and approaches to resolving the issues addressed in this topic?
- In what ways might the solutions explored in this topic alter your predictions for the state of human societies and the biosphere some decades from now?

Here are some questions you may want to consider in your discussions.

- How does the systems approach help our understanding of soils and soil processes?
- With respect to soils, how might the environmental value systems of a large-scale commercial farmer differ from that of a traditional subsistence farmer?
- How might the pressure on soils change over the next 20 years? Give reasons to support your answer.

5.2 Terrestrial food production systems

Significant ideas:

The sustainability of terrestrial food production systems is influenced by socio-political, economic, and ecological factors.

Consumers have a role to play through their support of different terrestrial food production systems.

The supply of food is inequitably available and land suitable for food production is unevenly distributed among societies; this can lead to conflict and concerns.

Big questions

As you read this section, consider the following big questions:

- Which strengths and weaknesses of the systems approach and of the use of models have been revealed through this topic?

- To what extent have the solutions emerging from this topic been directed at preventing environmental impacts, limiting the extent of the environmental impacts or restoring systems in which environmental impacts have already occurred?
- What value systems are at play in the causes and approaches to resolving the issues addressed in this topic?
- How are the issues addressed in this topic relevant to sustainability or sustainable development?

Knowledge and understanding:

- The sustainability of terrestrial food production systems is influenced by factors such as scale, industrialization, mechanization, fossil fuel use, seed/crop/ livestock choices, water use, fertilizers, pest control, pollinators, antibiotics, legislation, and levels of commercial versus subsistence food production.
- Inequalities exist in food production and distribution around the world.
- Food waste is prevalent in both less economical developed countries (LEDCs) and more economically developed countries (MEDCs), but for different reasons.
- Socio-economic, cultural, ecological, political, and economic factors can be seen to influence societies in their choices of food production systems.
- As the human population grows, along with urbanization and degradation of soil resources, the availability of land for food production *per capita* decreases.
- The yield of food per unit area from lower trophic levels is greater in quantity, lower in cost, and may require fewer resources.
- Cultural choices may influence societies to harvest food from higher trophic levels.
- Terrestrial food production systems can be compared and contrasted according to inputs, outputs, system characteristics, environmental impact, and socio-economic factors.
- Increased sustainability may be achieved through:
  - altering human activity to reduce meat consumption and increase consumption of organically grown and locally produced terrestrial food products
  - improving the accuracy of food labels to assist consumers in making informed food choices
  - monitoring and control of the standards and practices of multinational and national food corporations by governmental and intergovernmental bodies
  - planting of buffer zones around land suitable for food production to absorb nutrient run-off.

Sustainability of terrestrial food production systems

The sustainability of terrestrial food production systems is influenced by a number of factors including scale, industrialization, mechanization, fossil fuel use, seed/crop/livestock choices, water use, fertilizers, pest control, pollinators, antibiotics, legislation, and levels of commercial versus subsistence food production.

Despite the world's population rising from around 3 billion people in the 1950s to over 7 billion people in 2011, food production has managed to keep pace with population growth. But with increased standards of living in many places, there has been a change in diet away from grain and cereals towards increased consumption of meat and dairy products. Whether this growth can be sustained is another matter. Food crises in 2007–08 suggested that there were limits to the current system of food production.

Wasting food means losing life-supporting nourishment and precious resources, including land, water, and energy. These losses will be made worse by future population growth and the dietary trend away from grain-based foods towards consumption of animal products. As nations become more affluent in the coming decades through development, *per capita* calorific intake from meat consumption is set to rise 40 per cent by the middle of the century. Meat and animal products require significantly more resource to produce because only 10 per cent of energy is passed on at each trophic level (page 89).

CONCEPTS: Sustainability

Increases in global population to over 7 billion and changes in diet have put pressure on terrestrial food-production systems. Arable land is becoming limited due to increasing human settlement and urbanization, and soils are becoming degraded through intensive terrestrial farming: such factors can make food production systems unsustainable in the long run.

Over the last five decades, improved farming techniques and technologies have helped to significantly increase crop yields. New farmland (to support the projected rise in population and the increased standard of living) may have to come from marginal land (i.e. land that does not have great potential for farming) or from natural ecosystems.

Before the Second World War, farming was done within a network of small farms. After the war, there was concern in Europe about self-sufficiency: this led to significant changes in farming practice. Demands for increased production capacity through mechanization led to these small farms being combined into bigger farms, with fields combined to provide large uniform areas for agriculture. Greater intensification of products was achieved through increased fertilizer use. In most developed nations, much of the farming that is carried out today has been called **agribusiness** or agri-industrialization.

Throughout the second half of the 20th century, there was a shift from producing food for people's needs (subsistence farming) to producing food for commercial profit – a cultural factor within the capitalist societies where these intensive farming methods were introduced. Agribusiness occurs when food production is not to satisfy the community's needs but is to ensure profitable return for capital investment. The principal of agribusiness is to maximize productivity and profit in order to compete in a global market. It can be distinguished from traditional forms of food production, although the latter may assume either a subsistence or commercial form.

The main characteristics of agribusiness are:

- large-scale monoculture
- intensive use of fertilizers and pesticides
- mechanized ploughing and harvesting
- food production geared to mass markets including export.

Agribusiness supplies most of the products found in supermarkets. Many have travelled long distances from locations around the globe. Agribusiness in non-seasonal climates (e.g. parts of Africa) supply food throughout the year, so once seasonal crops are now available year-round in MEDCs.

Loss of crop rotation and natural ways of maintaining field fertility have led to substantial increase in fertilizer use. This form of modern agriculture has a great impact on the environment, with loss of biodiversity and increased run-off pollution (e.g. eutrophication, page 255). Genetically modified crops are currently used in some counties to increase yield. This may have a knock-on effect on wild populations if modified species cross-pollinate with wild ones. National political economies encourage agribusiness as a means to support gross national income and the high standards of living that their populations have come to expect.

CONCEPTS: Biodiversity

Increased agriculturalization has led to a loss of biodiversity as native habitats have been cleared.

Some farming in LEDCs is also agribusiness. Oil palm plantations in Malaysia, for example, are export-orientated and make use of high inputs of pesticides, machinery, and fossil fuels. Nevertheless, much agriculture in LEDCs suffers from low levels of technology, lack of capital, and uses high levels of labour.

Agricultural sustainability

The use of water for farming is causing concern. Over the past century, fresh water abstraction for human use has increased at more than double the rate of population growth. The demand for water in food production could reach 10–13 trillion cubic metres annually by the middle of the century. This is 2.5 to 3.5 times greater than the total human use of fresh water today. About 550 billion cubic metres of water is wasted globally in growing crops that never reach the consumer. Other problems include eutrophication and salinization.

In terms of energy use, on average 7–10 calories of input are required to produce one calorie of food. This varies dramatically, depending on the food, from 3 calories for plant crops to 35 calories in the production of beef. Since much of this energy comes from burning fossil fuels, food production systems contribute to global warming. Food processing also uses large amounts of electricity and/or fossil fuels. A beef burger typically requires between three and eight times more energy in its production and distribution than it delivers to the consumer as food.

Fertilizers and pesticides are a big component of modern food production. Fertilizers supply additional nutrients such as nitrogen or phosphorus, thus increasing crop yield. Pesticides kill insects that eat crops, again increasing the overall biomass of crop that is harvested. Of all the energy needed to produce wheat, approximately 50 per cent is needed to produce fertilizers and pesticides. Globally, fertilizer manufacturing consumes about 3–5 per cent of the world's annual natural gas supply. Agriculture currently consumes approximately 3.1 per cent of total global energy; this is divided into 2.5 per cent in MEDCs and 0.6 per cent in LEDCs. In many countries, grains including wheat, maize, and rice are too damp for direct transfer to storage, so they need to be dried. Drying large quantities of material requires substantial amounts of energy, particularly in the form of electricity or fossil fuels such as oil or gas.

A bee pollinating oilseed rape – the third-largest source of vegetable oil in the world

Currently 10–16 per cent of global crop production is lost to pests, although figures vary from country to country. In East Africa, for example, up to 60 per cent of maize and groundnut crops are lost to pests. Crop pests include insects, fungi, bacteria, viruses, and nematode worms.

To read more about the importance of pollinators, go to www.pearsonhotlinks.co.uk, enter the book title or ISBN, and click on weblink 5.4.

Around 250 000 species of flowering plants depend on bees for

The value of bee pollination in Western Europe is estimated to be 30 to 50 times the value of honey and wax harvests in this region. In Africa, bee pollination is sometimes estimated to be 100 times the value of the honey harvest, depending on the type of crop.

pollination. Many of these plants are crucial to world agriculture. Without pollinators, crops would not grow, and many fruits and vegetables would become scarce or prohibitively expensive. The steady decline of these insects over recent years raises significant concern about our ability to feed a growing population set to reach 9 billion by 2050.

Intensive animal farming (e.g. intensive cattle (page 293) and chicken farming) employs the widespread use of antibiotics. Because the animals are kept in tightly packed pens and are in close proximity to each other, disease can quickly spread through the farms. Antibiotics (which kill bacteria) ensure that the animals remain healthy. The use of antibiotics in farming has been criticized because it can lead to the emergence of antibiotic-resistant bacteria. Many scientists believe that antibiotics should be limited to treating human infection.

Levels of commercial and subsistence food production also affect the sustainability of terrestrial food production.

Inequalities exist in food production and distribution around the world.

Inequalities in global food supply

On average, there is enough food in the world for our population but there is an imbalance in the food supply. Many people in LEDCs are suffering from under-nourishment (their food intake does not contain enough energy) or malnutrition (their food intake lacks essential nutrients such as protein and minerals). Three-quarters of the world population is inadequately fed and around a sixth (1 billion) are going hungry. The majority of these people live in LEDCs. It is estimated that a child dies from hunger every 6 seconds. Food prices play a crucial role: a 10 per cent increase in food prices can lead to 40 million more people in food poverty. Yet in MEDCs there is a surplus of food, with markets producing too much food for the population to consume.

There has been an increased demand for food production in many societies around the world. Reduced death rates due to better medical care have led to increases in population growth. The increased wealth in MEDCs enables people to consume more; in many cases, more than they need. In Europe, the economics of food-production systems means that food production is a business, and the Common Agricultural Policy (CAP) subsidizes and guarantees prices no matter how much is produced.

There are concerns in MEDCs about food availability, stability of supply, and access to supplies. The result is that these countries often take protectionist measures to protect supplies. **Import tariffs** imposed by MEDCs make the import of food more expensive, which can have knock-on effects for exporting countries. In LEDCs, food production is used as a way to generate foreign currency, especially from cash crops such as sugar cane, so there is often an emphasis on export in these countries.

The current food crisis is partly a result of long-standing imbalances between rich and poor countries in international agricultural trade. Countries that rely on exports are more affected by global financial fluctuations (such as recessions) than those that are self-sufficient.

Huge domestic support and **export subsidies** provided by MEDCs to their farmers make farm products from LEDCs uncompetitive. For example, rice subsidies for farmers in the USA, have affected rice farmers in the Asia–Pacific region (Thailand, Vietnam, and India). Corn subsidies have also driven prices down, affecting farmers in the Philippines and China.

The rapid increase in food prices in the early 21st century were due, in part, to increased demands to use land for biofuel, leaving less land available for food crops. High meat consumption in MEDCs and increased meat and dairy consumption in LEDCs has meant a higher proportion of corn crops going to cattle feed than directly to feed human populations; this leads to higher corn prices. Increased oil prices also contribute to higher food prices due to increased transport costs. Despite these increases in prices, the overall cost of food in MEDCs is fairly inexpensive. Seasonal foods have generally disappeared as imports fill gaps. Modern technology and transport ensure that foodstuffs can be imported from all round the globe.

In LEDCs, however, many populations struggle to produce enough food, and generally food prices remain high (Figure 5.5). Political, economic, and environmental issues may all limit food production. The export-driven economies of many LEDCs may lead to crops being generated for cash (cash crops) rather than to feed the local population (e.g. in Kenya, vegetable crops often end up in MEDC supermarkets rather than feeding the local population, many of whom remain hungry). More recently, increased demands for biofuel by MEDCs means that LEDCs are increasingly allocating fertile land for the growth of biofuel crops, at the expense of using this land to grow food for their indigenous population. In India, for example, the *Jatropha* plant is grown as a biofuel because the plant produces seeds that are up to 40 per cent oil. The plant is grown on land once used for growing crops, pushing up the cost of food as land for edible crops becomes more limited.

Climate change has had more impact on LEDCs (e.g. increased incident of drought has reduced the amount of growing land) than MEDCs (Figure 5.6). Global warming could lead to tropical and subtropical countries like India facing short periods of super-high temperatures – with temperatures approaching 50 °C. These temperatures could completely destroy crops if they coincide with the flowering period.

Figure 5.5 Imbalances in global food supply combined with increases in food prices will affect LEDCs more than MEDCs.

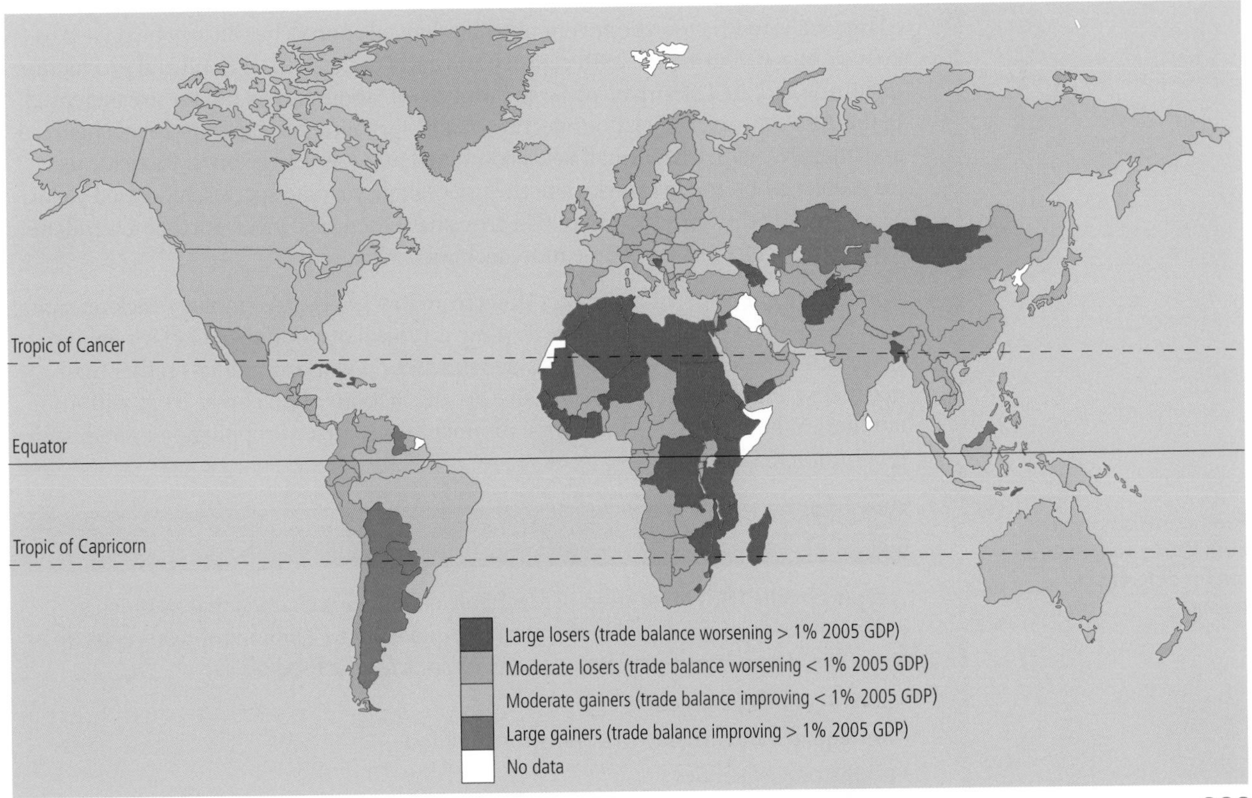

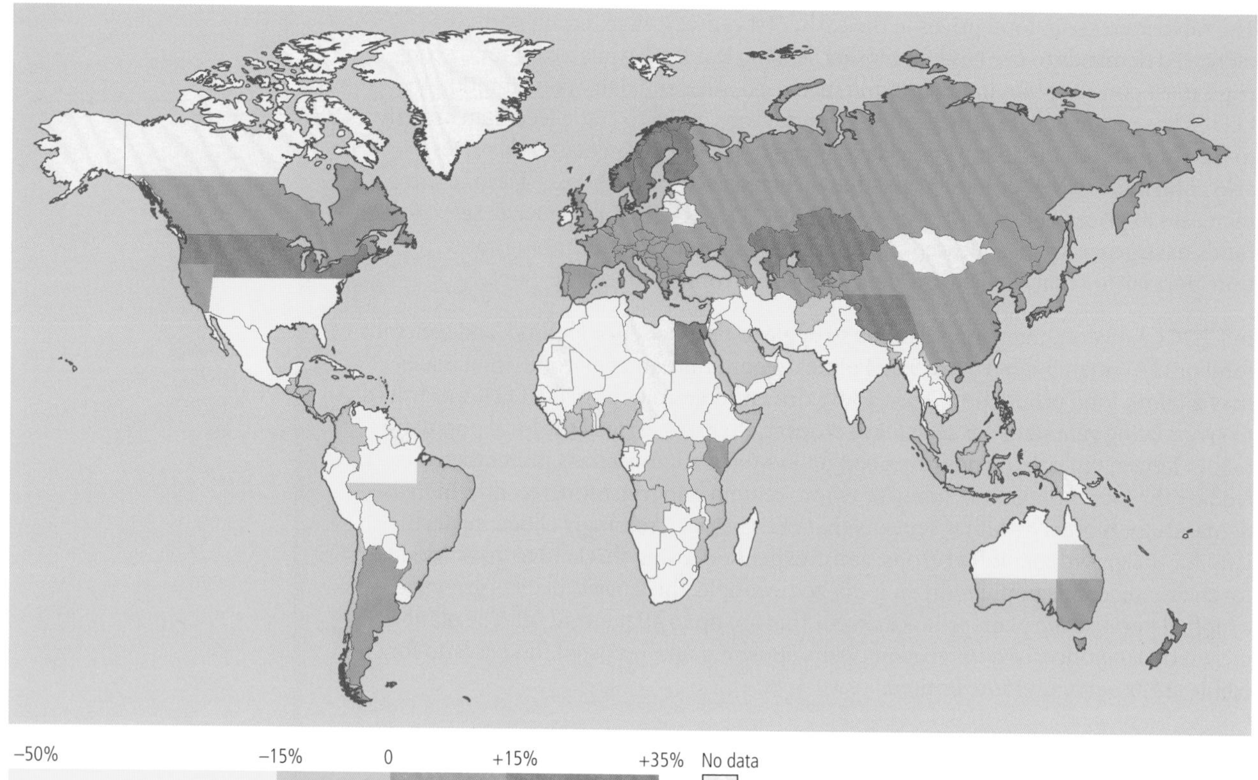

Figure 5.6 Projected changes in agricultural productivity by 2080 due to climate change

As more and more land is used for settlement and industry, there is an increasing need to intensify production on existing farmland. In MEDCs, food production is a complex process, involving high levels of technology, low labour, and high fuel costs; fertilizers and pesticides are factory produced and the product processing and packaging is on a large scale. In MEDCs, the advent of technological approaches has enabled yield to be maximized. During the 19th century and early in the 20th, agricultural production in Europe and the USA involved large numbers of labourers. As tractor use increased in the 20th century, farm labour decreased and agriculture became more mechanized and intensive, with many small fields combined into fewer large ones. Pesticide use (to protect crops and livestock), and the use of high-yielding species, increased yields. More recently the introduction of GM crops have increased yields further. Overall in MEDCs, agriculture has become more technocentric.

Agriculture in LEDCs, in contrast, suffers from low levels of technology, lack of capital, and uses high levels of labour. Rice farming is typical of tropical wet LEDCs, where rice is often the staple crop: there is a dependence on working animals rather than machinery, making it a labour-intensive process (labour often comes from within families). Whereas MEDCs have large monocultures, mixed cropping on a small scale is common in LEDCs.

Food waste occurs in both LEDCs and MEDCs, but for different reasons.

Food waste

A report by the UK's Institution of Mechanical Engineers claimed that as much as half of all the food produced in the world – equivalent to 2 billion tonnes – ends up as waste every year. The report lists various reasons for the food waste:

- poor agricultural practices
- inadequate infrastructure for transporting food
- poor storage facilities

- strict sell-by dates on supermarket food
- buy-one-get-one-free promotions on food in supermarkets
- western consumer demand for food that appears perfect (i.e. free from deformities).

The report, *Global Food: Waste Not, Want Not*, found that 30–50 per cent of the total amount of food produced around the world each year (about 4 billion tonnes) never makes it on to a plate: this amounts to 1.2–2 billion tonnes of food wasted.

Food waste fact-file

- Up to half of the food that is bought in Europe and the USA is thrown away by consumers.
- In the UK, as much as 30 per cent of vegetable crops are not harvested due to their failure to meet supermarkets' standards on appearance (i.e. nothing short of food that looks physically perfect appears on supermarket shelves).
- In the UK, about 7 million tonnes (worth about £10.2 billion) of food is thrown away from homes every year. It is estimated that this costs the average household £480 a year, which accumulates to £15 000–£24 000 over a lifetime. The average family in the UK spends 11 per cent of its expenditure on food.
- About 550 billion cubic metres of water is wasted globally in growing crops that never reach the consumer.
- It takes 20–50 times more water to produce 1 kg of meat than to produce 1 kg of vegetables.
- By 2050, we could need as much as 3.5 times the current total human use of fresh water to grow our food.
- The report claims that there is the potential to provide 60–100 per cent more food by cutting out waste. These reductions could be done while freeing up land, energy, and water resources which would previously have been used in wasted food production.
- In Ghana in 2008, there was a 50 per cent loss rate of stored maize from a total production of 1 million tonnes.
- In India, about 21 million tonnes of wheat annually perishes due to inadequate storage and distribution.
- In Pakistan, losses amount to 3.2 million tonnes (about 16 per cent of production) annually, because inadequate storage infrastructure leads to widespread rodent infestation problems.
- In Ukraine, there are 25–50 per cent losses annually. Typical grain production in Ukraine is around 24 million tonnes, so 6–12 million tonnes are lost annually in that country.

The nature of waste

LEDCs

In LEDCs, such as those of sub-Saharan Africa and South East Asia, wastage tends to occur primarily at the farmer–producer end of the supply chain. Inefficient harvesting, inadequate local transportation and poor infrastructure mean that produce is frequently handled inappropriately and stored under unsuitable farm site conditions.

As a result, mould and pests (e.g. rodents) destroy or at least degrade large quantities of food material. Substantial amounts of foodstuffs simply spill from badly maintained vehicles or are bruised as vehicles travel over poorly maintained roads.

As the development level of a country increases, the food-loss problem generally moves further up the supply chain with deficiencies in regional and national infrastructure having the largest impact. In South East Asian countries, for example, losses of rice range from 37 per cent to 80 per cent of the entire production, depending on development stage. The total is about 180 million tonnes annually. In China, a country experiencing rapid development, the loss is about 45 per cent. In less-developed Vietnam, rice losses amount to 80 per cent of production.

MEDCs

In MEDCs, consumerism, excess wealth, and mass marketing lead to wastage. More efficient farming practices and better transport, storage, and processing facilities ensure that a larger proportion of the food produced reaches markets and consumers. However, produce is often wasted through retail and customer behaviour. Major supermarkets, in meeting consumer expectations, often reject entire crops of perfectly edible fruit and vegetables at the farm because they do not meet exacting marketing standards for their physical characteristics, such as size and appearance. Globally, retailers generate 1.6 million tonnes of food waste annually in this way.

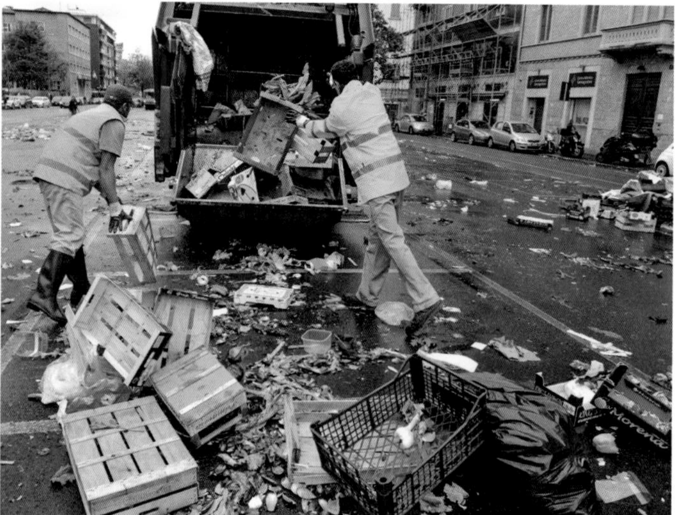

Food waste

Overall, wastage rates for vegetables and fruit are considerably higher than for grains. In the UK, a recently published study has shown that, of the potato crop, 46 per cent is not delivered to the retail market. The details revealed that 6 per cent is lost in the field, 12 per cent is discarded on initial sorting, 5 per cent is lost in store, 1 per cent is lost in post-storage inspection, and 22 per cent is lost due to rejection after washing. A similar survey in India showed that at least 40 per cent of all its fruit and vegetables is lost between grower and consumer due to lack of refrigerated transport, poor roads, and poor weather.

In MEDCs, approximately 25 per cent of all the food bought is discarded as waste, despite there often being nothing wrong with it. Most of the waste is generated at the consumer stage. Estimates show that the value of the food which goes to waste and is thrown away amounts to approximately US$250 billion globally.

Of the produce that does appear in the supermarket, commonly used sales promotions frequently encourage customers to purchase excessive quantities which, in the case of perishable foodstuffs such as vegetable and fruit, inevitably generates wastage in the home. Overall, 30–50 per cent of what has been bought in MEDCs is thrown away by the purchaser.

CHALLENGE YOURSELF

Think about ways in which you can reduce food waste. Discuss ideas within your class. Ideas for cutting food waste could include the following.

- Make meals from leftover food (e.g. refried vegetables; soup from a chicken carcass).
- Think before you shop: look in the fridge and have an idea of meals or recipe needs before you go shopping.
- Use teabags and leftovers to form compost for your garden, to recycle nutrients.
- If you have excess bread, use it to make panzanella, croutons, or bread-and-butter pudding.
- Use your freezer – if you know you are not going to use something, freeze it fresh for another day. A full freezer retains cold better than an empty one.
- Store food carefully: get into the habit of putting cereal, biscuits, and fresh nuts into tins or airtight containers.
- Remember that best-before dates may be over cautious. Often with non-meat or dairy products, you can use your common sense to check if they're still good to eat.

Controlling and reducing the level of wastage is frequently beyond the capability of the individual farmer, distributor, or consumer, since it depends on market philosophies, security of energy supply, quality of roads, and the presence of transport networks.

Links between social systems and food production systems

Socio-economic, cultural, ecological, political, and economic factors can be seen to influence societies in their choices of food production systems.

CONCEPTS: Environmental value systems

How people view the environment (their environmental value system or EVS) affects their choice of farming production system, and is affected by factors such as whether or not a society lives in close connection with the natural environment, and their dietary preferences.

You need to be able to discuss the links that exist between socio-cultural systems and food production systems.

Socio-economic factors in terrestrial food production include: farming for profit or subsistence; farming for export or local consumption; farming for quantity or quality; traditional or commercial farming.

There are many links between social systems and food production systems – we examine here **shifting cultivation**, wet rice agriculture (South East Asia).

Case study

Shifting cultivation

Shifting cultivation is also known as slash-and-burn agriculture because new land is cleared by cutting down small areas of forest trees and setting fire to them: the ash fertilizes the soil for a while and the clearing produced enables crops to be grown. Once the land in one area has been exhausted (e.g. minerals in the soil depleted), the farmer moves on to a new area (Figure 5.7). Old land can be returned to once the fertility has recovered.

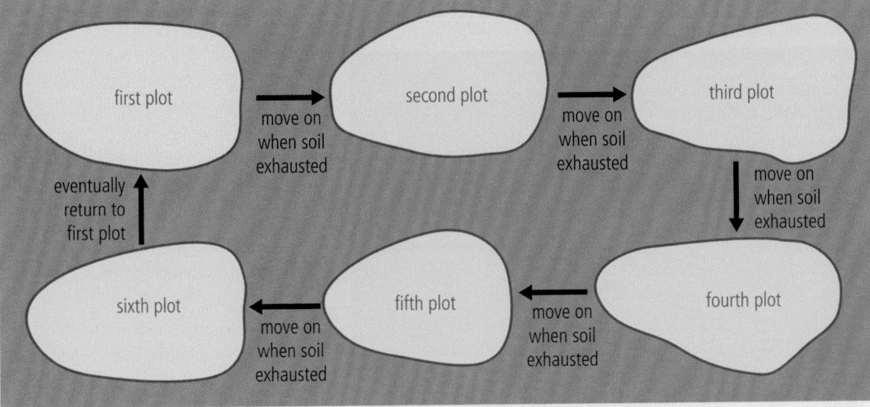

Shifting cultivation supports small communities and sometimes individual families.

Figure 5.7 Shifting cultivation follows a cycle where a sequence of clearings is used. Recognition that soil has become infertile leads farmers to shift and thus allow small pockets of forest to regenerate before returning to the plot as much as 50 years later. This allows soil fertility to be restored.

Slash-and-burn is practised in many tropical forest areas, such as the Amazon region, where yams, cassava, and sweet potatoes can be grown. This system is possible because low population densities (typical of these societies' biomes) can be supported by the food produced. If population densities increase too much, previously farmed land is returned to before soil fertility has been restored. The conditions of the forest encouraged shifting cultivation. It would be possible to clear only small areas of forest with the labour available (often from one family or small community), and hilly terrain meant that clearings often had to be on plains or along rivers (where fertility was better also). The socio-cultural features of people who use this method have developed in response to the farming system as well as shaping it. Shifting cultivation practice is bound to cultural practices and beliefs – traditions and rituals

continued

linked to choosing the site and carrying out the clearance. The plot cycle is used to recall past history by connecting events with the plots cleared at particular times. The people tend to be animists – believing that everything contains a spirit (or soul), including animals, plants, and trees. The spiritual role of forest is therefore a central feature of cultural life, leading to respect for trees and other species. Understanding how the forest works has led some shifting cultivators to adapt their practices to mimic the layering of the forest, where ground crops are protected from harsh sunlight and heavy downpours. As well as cultivation, land use includes forest materials for construction of homes and canoes, and for medicines.

People who live in close connection with nature, such as shifting cultivators in the Brazilian Amazon, show a closer connection between social systems and ecological systems (i.e. an ecocentric approach) than societies living away from natural systems, such as city dwellers. Urban capitalist elites in Brazil are more likely to view the interior of the country as a new frontier, and rainforest as a resource for development and cash (i.e. a technocentric approach). The lack of understanding of people disconnected from nature makes them more likely to underestimate the true value of natural resources (e.g. that a rainforest is worth more standing than cut down). They are also more likely to make decisions that produce wasteful and damaging actions (e.g. the construction of dams and whole-scale clearance of forest for timber or cattle ranching). Urban shanty dwellers who migrate to use deforested land are less likely to succeed than indigenous people as the areas they select are likely to have infertile soils. These issues are repeated globally, in different ecosystems and societies, but the underlying message remains the same.

Case study

Wet rice ecosystems of South East Asia

Rice can be grown in dry fields, but paddy field (wet rice) agriculture has become the dominant form of growing rice in South East Asia. It is an example of intensive subsistence farming, using high labour inputs but low technology. The high population densities in these countries lead to a high demand for food. Rice is in particular demand because it is a staple part of the diet and a central part of Asian culture. Soil fertility is good and supports the intensive nature of the agriculture. Paddy fields can be placed adjacent to rivers and areas that flood naturally, where annual inundation causes new deposits of silt in the fields and increased fertility. They can also be put on hills using terracing. The heavy clay soils created by river deposits are ideal for paddy fields – sandy and light-textured soils are not suitable as water drains away.

Terracing is used to grow rice in hills areas of South East Asia.

High rainfall in these regions facilitates this type of agriculture, allowing extensive field irrigation to be maintained throughout the year. Warm weather and intense sunlight allow high productivity all year round.

Recently, less land has been available for new expansion of rice farming in South East Asia. Declining soil fertility has also been a problem, and rice yields have been reaching their maximum. In the next 30 years, as populations grow, the security of smaller farms may depend on increasing diversification into higher value crops (e.g. vegetables and citrus), and into small livestock production and aquaculture (e.g. fish farming, pages 244–246).

General points

Socio-cultural factors influence tastes and the development of different food production systems.

- The desire for more organic food in Europe has led to the growth of organic farming.
- In MEDCs, there has been a growing trend for concern about animal welfare which has affected the farming methods adopted by some farms (e.g. free-range pigs and chickens rather than intensive battery farming).
- Educational levels determine the degree of exchange of ideas about new farming practices and the extent to which new technologies are applied. In Singapore, for example, the government has invested large amounts of time and money in promoting new technologies (e.g. hydroponics – growing plants in mineral solutions without soil). The shortage of available land on the island led to pressure to find alternative ways of growing food.
- Indirectly, socio-cultural factors such as land ownership, migration patterns, and attitudes to land in general have an impact on how land is used. Native American Indians did not believe that people could own land – they saw land as a communal commodity, so development was limited.
- Environmental constraints (e.g. rainfall, growing seasons, natural disasters, and soil fertility) influence choice of farming practices. Fertile soil and plenty of rainfall favour intensive crop production, and economic factors determine input costs (e.g. seeds, technology, and access to credit).
- Economic and technological factors are interconnected with socio-cultural features – they develop in response to farming systems as well as shape such systems.

Availability of land for food production

With the increases in human population, urbanization and degradation of soil resources, the availability of land for food production *per capita* decreases.

With a growing population demanding ever-more land to live on, less land is available to grow crops on. The more land is used for urban development, the less land is available for agriculture (Figure 5.8).

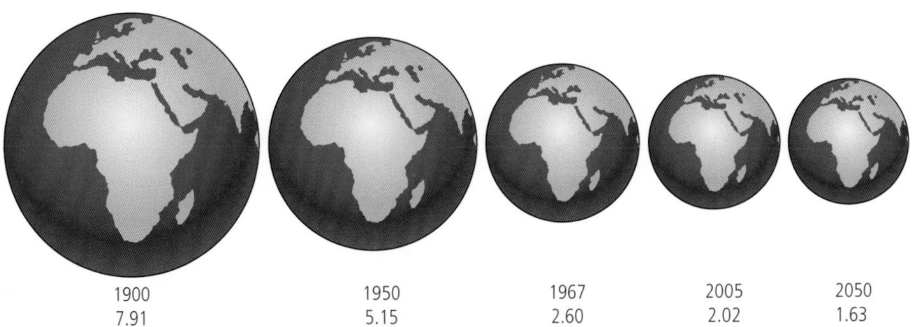

Figure 5.8 The number of hectares of land available per person to live on and to grow the food we require is decreasing.

Relatively little new land has been brought into agriculture over the past few decades: between 1967 and 2007, the area of land used for agriculture increased by only 8 per cent. The total area of agricultural land currently stands at approximately 4600 million hectares.

The intensive use of land has led to soil degradation, with several geographical areas particularly badly affected (Table 5.6).

Table 5.6 Human-induced land degradation

Region	Land area / 000s km²	Total affected by severe or very severe land degradation / 000s km²	Amount of severe or very severe land degradation due to agricultural activities / 000s km²
Sub-Saharan Africa	23772	5931	1996
North Africa and Near East	12379	4260	759
North Asia, east of Urals	21033	4421	1180
Asia and Pacific	28989	8407	3506
South and Central America	20498	5552	1795
North America	19237	3158	2427
Europe	6843	3274	727
World	134907	35003	12390

The problems associated with soil degradation are explored further in Section 5.3 (pages 297–307).

The yield of food per unit area from lower trophic levels is greater in quantity, lower in cost, and may require fewer resources than the higher trophic levels.

Efficiency of terrestrial production systems

As you have learned, the second law of thermodynamics (page 27) means that energy conversion through food chains is inefficient (pages 88–90 and Figure 5.9). Energy is lost by respiration and waste production at each level within a food web.

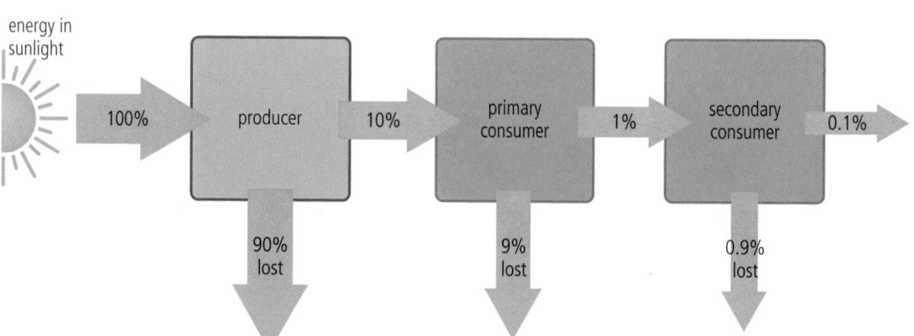

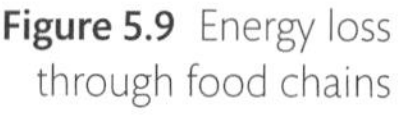
Figure 5.9 Energy loss through food chains

In terrestrial systems, most food is harvested from relatively low trophic levels (producers and herbivores). Systems that produce crops (arable) are more energy efficient that those that produce livestock. This is because in the former, crops are producers at the start of the food chain and contain a greater proportion of the Sun's energy than subsequent trophic levels (Figure 5.10).

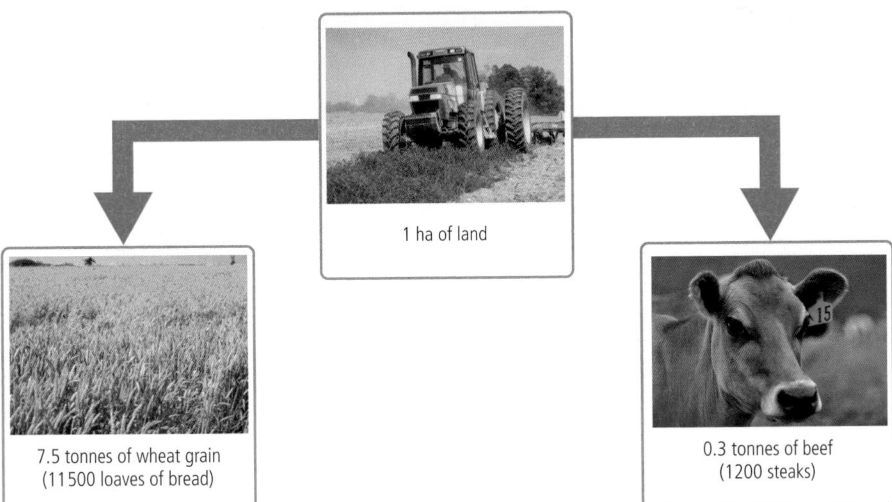

Figure 5.10 One hectare of land can produce a greater biomass of food in an arable system than in livestock farming.

Livestock farming demands extensive land use. One hectare of land can, for example, produce rice or potatoes for 19–22 people per annum, but the same area can produce only enough lamb or beef for one or two people. In terms of land use, agricultural food production based on livestock is far less efficient than that based on crops. This is largely because only about 3 per cent of the feed energy consumed by livestock remains in edible animal tissue.

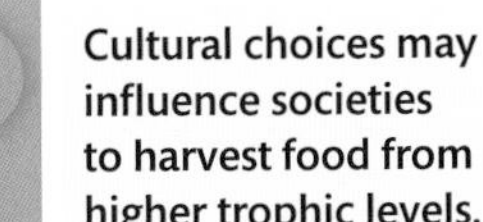

Cultural choices may influence societies to harvest food from higher trophic levels.

Cultural choices may influence societies to harvest food from higher trophic levels. In parts of the developing world, increased income has led to more meat and milk consumption. People earning around $2 a day do not consume much meat or milk, but those earning around $5 a day increase their intake of these products. In India, over the past 30–40 years there has been a six-fold increase, due to population growth and a two-fold *per capita* increase in meat consumption. More meat and dairy consumption in LEDCs will put further pressures on food production (i.e. corn to feed the animals) and lead to increases in grain prices.

TOK

It is estimated that for people in LEDCs to enjoy the same level of meat and dairy consumption as people in MEDCs, the latter will have to halve their meat and dairy intake. Is this something MEDC populations should be morally expected to do? Should legal agreements be put in place to force MEDCs to adopt this strategy? Can we all expect to continue to eat as much meat if it puts global food production at risk?

Despite the more efficient land use of arable systems, many cultures continue to use livestock as a part of their farming system. Taste and cultural demand play a role in this, and the animals provide a source of protein (essential for the human diet). Animals can convert into food vegetation that would not be available to humans directly. Additionally, the products from livestock are diverse (e.g. milk, meat, blood, wool, hide), and in many cultures the livestock are used as working animals.

In contrast to terrestrial systems, most food in aquatic systems (Chapter 4) is harvested from higher trophic levels where the total storages are much smaller. This is less energy efficient than crop production (i.e. crops capture energy directly from the primary source; fish are several steps away from primary production). Although energy conversions along an aquatic food chain may be more efficient than in a terrestrial chain (because fish store more biomass in flesh than bone), the initial fixing of solar energy by aquatic primary producers tends to be less efficient because the water absorbs and reflects light.

You need to be able to analyse tables and graphs that illustrate the differences in inputs and outputs associated with food production systems. You also need to be able to compare and contrast the inputs, outputs and system characteristics for two given food production systems.

Around 15 kg of grain and 100 000 dm^3 of water is required to produce 1 kg of beef. The energy necessary to produce one steak can feed around 40 people in grain-equivalents. About 10 per cent of global water consumed annually is used for cattle (in the USA, about half of total water consumption goes to cattle, including water used to grow the crops to feed them). A vegetarian diet makes better use of the Earth's resources (land, solar input, water) and can support more people than a meat-eating and dairy-based diet.

Flows should be drawn as arrows into and out of the food production system, and the food production system shown as a box (the storages). The arrows must go in the correct direction. Specific examples need to be given as inputs and outputs (e.g. 'pest-resistant crop seeds' *not* 'technology'; ' farming labour' *not* 'capital expenditure'; 'cattle feed or beef' *not* 'food').

Inputs, outputs, and environmental impacts of terrestrial food production systems

SYSTEMS APPROACH

In exams you may be asked to construct a simple diagram to show the inputs, outputs, and storages of terrestrial food production system. These would depend on the production method chosen. Figure 5.11 shows two examples.

Example 1 Beef production

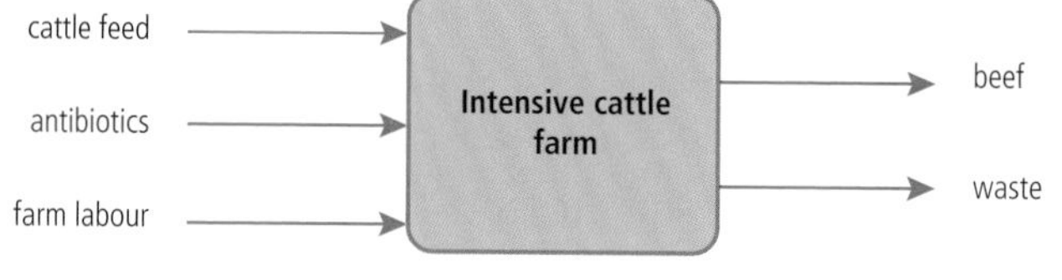

Example 2 Arable farming

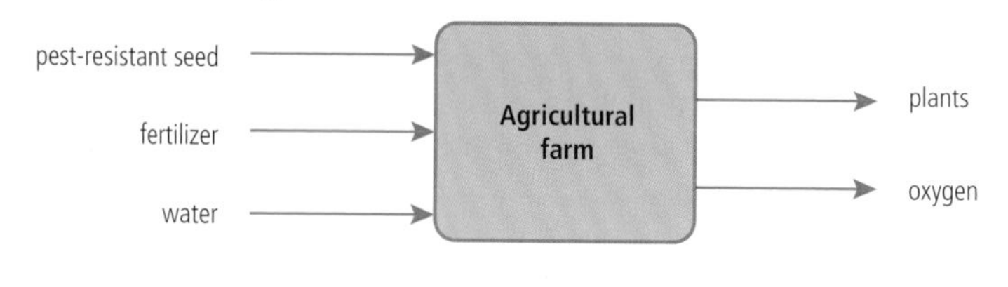

Figure 5.11 Inputs and outputs from two different food production systems

- *Inputs* to food production systems include: fertilizers (artificial or organic); water (irrigation or rainfall); pest-control (pesticides or natural predators); labour (mechanized and fossil-fuel dependent or physical labour), seed (GM or conventional); breeding stock (domestic or wild); livestock growth promoters (antibiotics or hormones versus organic or none).
- *Outputs* include: food quality, food quantity, pollutants (air, soil, water), consumer health, soil quality (erosion, degradation, fertility); common pollutants released from food production systems include fertilizers, pesticides, fungicides, antibiotics, hormones and gases from the use of fossil fuels; transportation, processing, and packaging of food may also lead to further pollution from fossil fuels.
- *System characteristics* include: diversity (monoculture versus polyculture).

Terrestrial farming systems can be divided into several types.

- **Commercial farming** is farming for profit – often of a single crop.
- **Subsistence farming** produces only enough to feed the farmer and his or her family, with none to sell for profit.

Both commercial and subsistence farming can be either intensive or extensive.

- **Intensive farms** generally take up a small area of land but aim to have very high output (through large inputs of capital and labour) per unit area of land.
- **Extensive farms** are large in comparison to the money and labour put into them (e.g. the cattle ranches of central Australia, where only a few workers are responsible for thousands of acres of land).

The efficiency of the system can be calculated by comparing outputs (e.g. marketable product) to inputs (fuel, labour, transport, fertilizer, dealing with waste products) per unit area of land.

Case study

North American cereal farming and subsistence farming in South East Asia

Cereal growing in North America and the Canadian prairies is an extensive commercial farming system. In terms of inputs, this system has high use of technology and fertilizers. Labour can be low per unit area as few workers are needed. Because large flat areas are farmed, mechanization is particularly effective. Outputs are low per hectare, due to the large areas of land in production and to yields being lower than when using intensive methods. On the other hand, outputs are high per farmer (because each owns a lot of land). Efficiency is medium, but environmental impact can be high as the farms require the clearance of natural ecosystems. This limits the habitat of wild native species and leads to loss of biodiversity; it may also result in soil erosion. In the future, GM technology may be used to improve crop yield.

Subsistence farms in South East Asia produce enough food to feed the family or small community working each farm, with surplus food being traded. Farmers typically use no machines: instead they use animals which can be fed and raised on the farm. This farming system uses a polyculture (many crops) approach rather than the monocultures grown on extensive farms. No artificial fertilizers are used: subsistence farmers rely on crop rotation to maintain the fertility of the ground, along with animal manure and compost to restore nutrients in the soil. Inputs are therefore low (hand tools and labour), as are outputs (food). In areas that are sparsely populated, subsistence agriculture is sustainable for long periods of time, and has a low impact on the environment. In more densely populated areas, subsistence agriculture may deplete the soil of nutrients, and damage the environment. Efficiency is high: subsistence farming can yield food energy up to 20 times the human energy invested. It is estimated that current world food production is still mainly produced from the subsistence multi-cropping system, and small farmers provide as much as 70 per cent of the food production in many tropical countries.

Nomadic herding is practised by 30 000–100 000 people (0.5–1.4 per cent of the world population); about 50 per cent of the world population is now urban.

The examples of contrasting food production systems given here are not meant to be prescriptive and you are encouraged to find out about appropriate local examples.

Case study

Intensive beef production in MEDCs and the Maasai tribal use of livestock

Maasai herding cattle in Tanzania

In intensive beef production, cattle are housed all year round and fed a diet of rolled barley mixed with a protein concentrate (often beans, soya, or rapeseed meal), fortified with vitamins and minerals. In the USA, cattle are put into pens containing up to 10 000 or 100 000 cows and fed corn for the last weeks of their lives, which can double their biomass before slaughter. Their movement within the pens is restricted. Intensive beef production is an energy inefficient form of farming, with yield as low as one tenth the level of energy as is invested in energy inputs. In terms of costs, however, it is very efficient, and significantly increases yield per acre, per person, per input, relative to extensive farming. There is not much space for the animals to move about, so they use less energy. This means less food is required, which leads to cheaper product. On the other hand, the animals are fed continuously for maximum growth and selective breeding has produced cows with high yield and good quality meat, which adds to overall costs. Inputs are therefore high (technology, heating, food) but so are the outputs (cost-effective production), although there may be hidden costs, such as transport. Environmental impact is high – energy usage releases greenhouse gases, and cows produce waste. Restraining animals in this way also has ethical implications.

The Maasai are an indigenous group living semi-nomadically in Kenya and parts of Tanzania. Their livestock are able to wander freely, herded by their owners (i.e. this is a nomadic form of farming). The Maasai diet is traditionally meat, milk, and blood supplied by their cattle. Once a month, blood is taken from living animals by inserting a small arrow into the jugular vein in the neck. The blood is mixed with milk for consumption. Virtually all social roles and status derive from the relationship of individuals to their cattle. This is an example of extensive subsistence farming – inputs are low (the animals are allowed to roam freely so fences and pens are not required, only human labour is used) and so are outputs (enough food to feed the community). As with other subsistence methods, efficiency is high and environmental impact is low (the Maasai use their natural environment to raise their animals). Socio-cultural factors can, however, lead to problems: for the Maasai: cattle equal wealth and quantity is more important than quality, and this has lead to overgrazing and desertification.

You need to be able to evaluate the relative environmental impacts of two given food production systems.

The environmental impacts of food production systems include:

- **soil degradation from erosion**
- **desertification**
- **eutrophication from agricultural run-off**
- **pollution from insecticides, pesticides, and fertilizers**
- **salinization from over-irrigation**
- **lowered water tables and over-abstraction of ground water**
- **loss of valuable habitats (e.g. wetlands drained for agriculture)**
- **disease epidemics from high-density livestock farming and monoculture.**

You need to be able to evaluate strategies to increase sustainability in terrestrial food production systems.

Increasing sustainability

CONCEPTS: Sustainability

Sustainability is the use and management of resources that allows full natural replacement of the resources exploited and full recovery of the ecosystems affected by their extraction and use. Sustainable farming methods should cause minimum impact to natural systems and involve the responsible use and management of global resources.

Increased sustainability may be achieved through:

- **altering human activity to reduce meat consumption**
- **increasing consumption of organically grown, seasonal, and locally produced food products**
- **planting of buffer zones**
- **improving the accuracy of food labels to assist consumers in making informed food choices**
- **monitoring and control of the standards and practices of multinational and national food corporations by governmental and intergovernmental bodies.**

Human attitude to eating meat

As a consumer, you have a role to play by selecting the food you eat from food-production systems that are sustainable. The question is – which food-production systems are sustainable, and how can we choose between them? Improving the accuracy of food labels in supermarkets would help consumers to make increasingly informed food choices. Buying locally produced food would minimize the food-miles used in transportation, limiting its ecological footprint (pages 437–441). Buying food that minimizes pesticide use also offers a more sustainable choice.

Per capita meat consumption has more than doubled in the past half-century.

More land is needed to grow meat than crops, therefore leading to greater habitat loss. Increased global meat consumption (pages 290–291) leads to a less sustainable future for farming. Altering human activity to reduce unsustainable meat consumption, and to increase consumption of locally produced terrestrial food products, should lead to a more sustainable use of land for food production. But should people in MEDCs, who have historically enjoyed high levels of meat consumption, be allowed to influence patterns of food consumption in developing countries where societies have only recently switched to higher levels of meat consumption?

Organic farming

Organic farming provides an ecocentric approach to farming by achieving an ecological balance that conserves soil fertility, prevents pest outbreaks, and takes a preventative rather than reactive approach to environmental issues. In other words, organic farming tends to avoid problems rather than having to solve them once they have emerged. In terms of environmental impact, organic farming is certainly sustainable at the local level, although it alone could not feed the world's growing population (i.e. it is not be sustainable at a global scale). Organic farming produces less food per unit of land and water than conventional agriculture (about 20–50 per cent less). While providing sufficient food locally, if scaled up to the levels needed to supply food globally, the lower yields would lead to greater pressure to convert land to agricultural use and produce more animals for manure to fertilize crops, both of which would be challenges to sustainability. Organic farming also prohibits the use of GM crops. GM crops can offer tolerance to drought, diseases and pests (reducing the need for farmers to use pesticides), and enhanced nutritional value (to fight malnutrition in LECDs). GM crops could provide one of a variety of different options for technological innovation in agriculture by increasing yields, reducing pest damage, and limiting environmental pollution. However, many non-governmental organizations (NGOs), such as Greenpeace, remain opposed to GM use for doctrinal and ideological reasons rather than scientific ones.

Buffer zones

Buffer zones of strips are areas of land containing native vegetation that are adjacent to or surround agricultural land. These areas support biodiversity that is absent from the arable land (which usually contain monocultures): these natural habitats support insect predators that limit crop pests, reducing pesticide use, and preserve species lost from the farmed area thereby supporting local food chains. Buffer zones also help to limit the run-off of fertilizers and pesticides into surrounding rivers and lakes, and control air and soil quality. Plants in the buffer zones trap sediment, and their roots hold soil particles together which reduce the effects of wind erosion. Many countries provide financial incentives for farmers to leave buffer zones on their land. Without such incentives, however, buffer zones may be unsustainable: local people, and the farmers themselves, may argue that this money would be better spent elsewhere.

Monitor and control

With uncertainty about future climate conditions (pages 379–381), sustainable farming requires governmental and intergovernmental bodies to monitor and control the standards and practices of multinational and national food corporations. The Commission on Sustainable Agriculture and Climate Change, for example, is working towards the integration of sustainable agriculture into national and international policies. The Commission has called for dramatically increased investments in sustainable agriculture over the next decade, including in national research and development budgets, land rehabilitation, economic incentives, and infrastructure improvement. It is only with national and international coordination that sustainable farming and food security for all will be achieved.

To learn more about the Commission on Sustainable Agriculture and Climate Change, go to www. pearsonhotlinks. co.uk, enter the book title or ISBN, and click on weblink 5.5.

CHALLENGE YOURSELF

Think about ways in which your own consumption of food could contribute to increased sustainability. Discuss these ideas within your class:

- plan meals carefully
- reduce waste – approximately one-third of food is thrown away each year
- eat local, in-season food
- cut back on dairy and red meat – fruit, vegetables, fish, and poultry have one-third the footprint of red meat and half that of dairy
- plan shopping trips and use the car less
- recycle packaging and compost organic scraps.

Exercises

1. Outline the issues involved in the imbalance in global food supply.
2. How does the nature of food waste vary between LEDCs and MEDCs?
3. Discuss the links that exist between socio-cultural systems and food-production systems.
4. Compare and contrast:
   a. the inputs and outputs of materials and energy (energy effciency); and
   b. the system characteristics.
   c. Evaluate the relative environmental impacts for two named food production systems.
5. Evaluate strategies to increase sustainability in terrestrial food-production systems.

Big questions

Having read this section, you can now discuss the following big questions:

- Which strengths and weaknesses of the systems approach and of the use of models have been revealed through this topic?
- To what extent have the solutions emerging from this topic been directed at *preventing* environmental impacts, *limiting* the extent of the environmental impacts, or *restoring* systems in which environmental impacts have already occurred?
- What value systems are at play in the causes and approaches to resolving the issues addressed in this topic?
- How are the issues addressed in this topic relevant to sustainability or sustainable development?

Here are some questions you may want to consider in your discussions.

- How can systems diagrams be used to show the impact of farming methods on natural systems? What are the limitations of such diagrams?
- How can the choice of farming system prevent environmental impacts, or limit the extent of environmental impacts?
- How do EVSs influence the choice of farming system?
- What are the issues relating to sustainable terrestrial food production? Is sustainable agriculture possible?

5.3 Soil degradation and conservation

Significant ideas

Fertile soils require significant time to develop.

Human activities may reduce soil fertility and increase soil erosion.

Soil conservation strategies exist and may be used to preserve soil fertility and reduce soil erosion.

Big questions

As you read this section, consider the following big questions:

- What strengths and weaknesses of the systems approach and the use of models have been revealed through this topic?
- To what extent have the solutions emerging from this topic been directed at *preventing* environmental impacts, *limiting* the extent of the environmental impacts, or *restoring* systems in which environmental impacts have already occurred?
- What value systems can you identify at play in the causes and approaches to resolving the issues addressed in this topic?
- In what ways might the solutions explored in this topic alter your predictions for the state of human societies and the biosphere some decades from now?

Knowledge and understanding:

- Soil ecosystems change through succession. Fertile soil contains a community of organisms that work to maintain functioning nutrient cycles and are resistant to soil erosion.
- Human activities which can reduce soil fertility include deforestation, intensive grazing, urbanization, and certain agricultural practices (irrigation, monoculture, etc.).
- Commercial industrialized food production systems generally tend to reduce soil fertility more than small-scale subsistence farming methods.
- Reduced soil fertility may result in soil erosion, toxification, salinization, and desertification.
- Soil conservation measures exist such as soil conditioners (e.g. organic materials and lime), wind reduction techniques (e.g. wind breaks, shelter belts), cultivation techniques (e.g. terracing, contour ploughing, strip cultivation), and avoiding the use of marginal lands.

Soil ecosystems

Soil ecosystems change through succession. Fertile soil contains a community of organisms that work to maintain functioning nutrient cycles and are resistant to soil erosion.

Some soils in areas of tropical rainforest have had the same climate for millions of years. This has allowed the soils to become very deep. Other soils, in temperate areas, have only developed over the last 10 000 years since the retreat of glaciers.

Fertile soils develop over a long time (Figure 5.12). However, time is not a causative factor. It does not cause soils to change but allows processes to operate to a greater extent, therefore allowing soils to evolve. The amount of time required for soil formation varies from soil to soil. Coarse sandstones develop soils more quickly than granites or basalts, and on glacial outwash a few hundred years may be enough for a soil to evolve. Thin soils are not necessarily young soils, nor are deep ones 'mature'. Phases of erosion and deposition keep some soils always in a state of change.

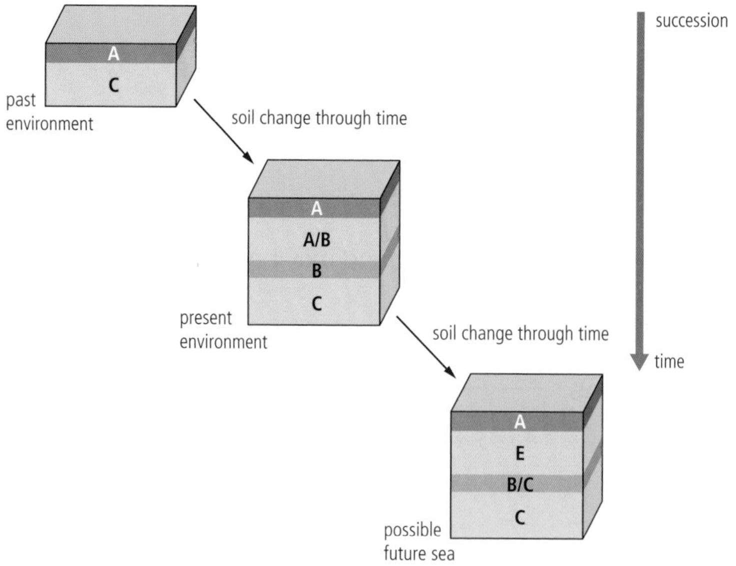

Figure 5.12 Soils as an open system. For an explanation of horizon A, E, B, and C, see pages 270–271.

In the top 30 cm of 1 ha of soil there are, on average, 25 tonnes of soil organisms:

- 10 tonnes of bacteria
- 10 tonnes of fungi
- 4 tonnes of earthworms
- 1 tonne of other soil organisms such as spring tails, mites, isopods, spiders, snails, lice, etc.

Organic matter is a basic component of soil although the influences of biotic factors range from microscopic organisms or bacteria to humans. Some influences are indirect, such as interception of precipitation by vegetation and the reduction of precipitation via evapotranspiration. Others are direct, such as the release of humic acids by decaying vegetation or the return of nutrients to the soil via litter decay (humification).

Animals too have an effect on soils. Earthworms alone represent from 50–70 per cent of the total weight of animals in arable soils. In 1 ha, 18 to 40 tonnes of soil is ingested each day by earthworms and passed on to the surface, this represents a layer up to 5 mm deep. Other animal activity is considerable:

- bacteria fix atmospheric nitrogen converting into a usable form for plant roots
- mycorrhizal fungi on tree roots take up soil nutrients and pass them directly to the tree
- decomposers break down litter releasing nutrients into the soil
- soil organisms help to mix the soil improving its structure
- animal burrows help to aerate the soil
- animal faeces return nutrient to the soil.

Human activity also has obvious effects, ranging from liming, fertilizer application, and mulching to mining, deforestation, agricultural practices, and gardening.

Fertile soils contain a community of organisms that work to maintain functioning nutrient cycles and are resistant to soil erosion. Figure 5.13 shows the natural nutrient cycle and one that is altered as a result of human activity, in this case farming.

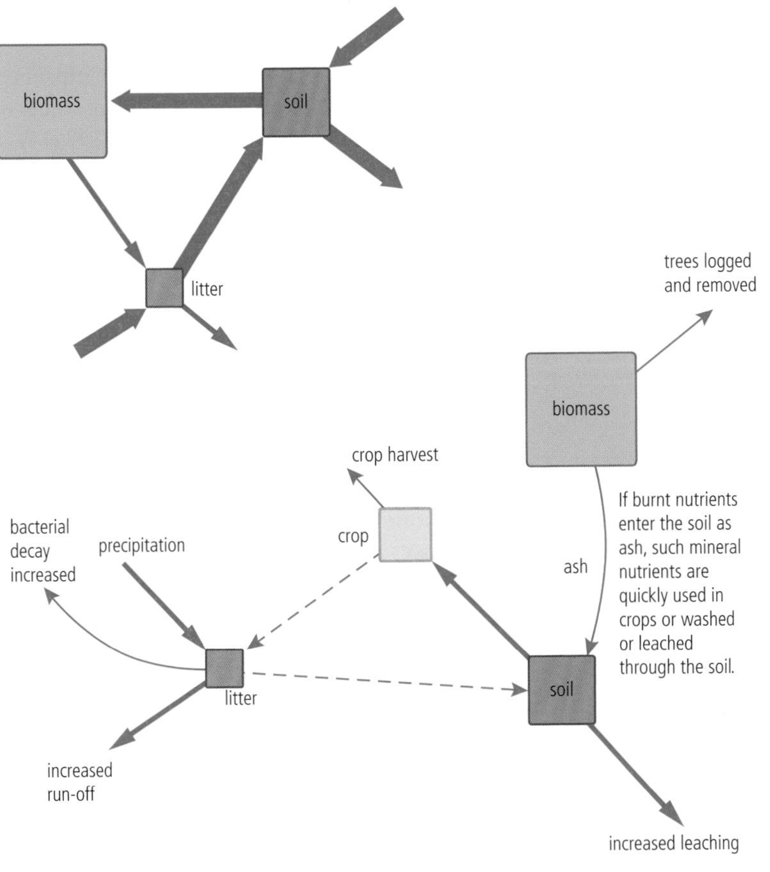

Figure 5.13 Nutrient cycling in a tropical rainforest – natural and following farming

Human activities which can reduce soil fertility include deforestation, intensive grazing, urbanization, and certain agricultural practices (irrigation, monoculture, etc.).

Rates of deforestation can be extremely high in places such as Nepal because of deforested steep slopes and high rainfall. The same rates of erosion were experienced on the South Downs in the UK following the conversion of land to arable, compaction by machinery and high rainfall.

Reduced soil fertility

Deforestation

Deforestation refers to the removal of some or all of a cover of trees. The greater the proportion of trees removed, the less interception that occurs, the more soil compaction by raindrop impact, and the more soil erosion that results (Figure 5.14). The potential for soil erosion increases with rainfall, and so the removal of tropical rainforests is particularly serious. (Chapter 4, pages 218–220, discusses the impact of deforestation.)

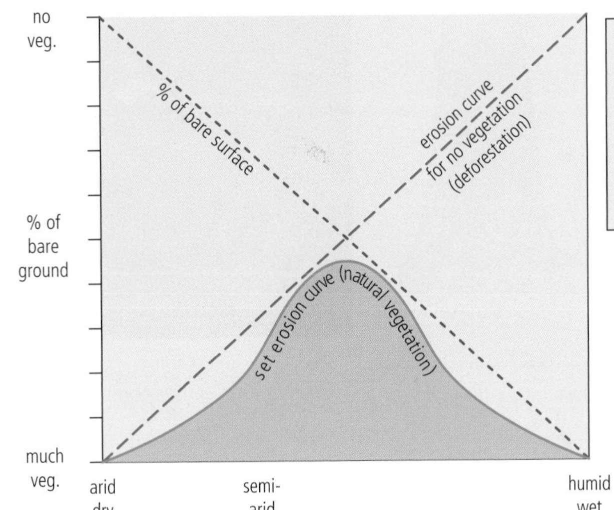

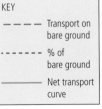

Rates of transport (and therefore denudation) tend to be highest in semi-arid areas and especially in more humid areas where vegetation is removed.

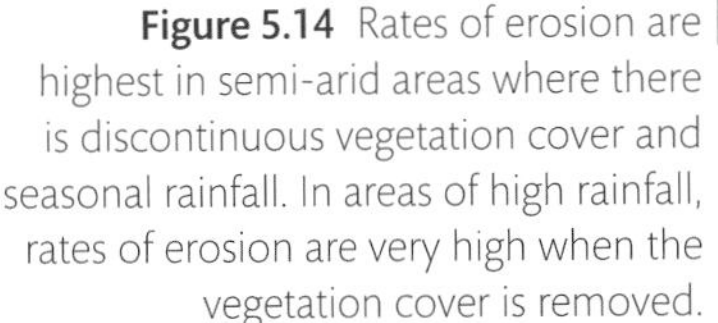

Figure 5.14 Rates of erosion are highest in semi-arid areas where there is discontinuous vegetation cover and seasonal rainfall. In areas of high rainfall, rates of erosion are very high when the vegetation cover is removed.

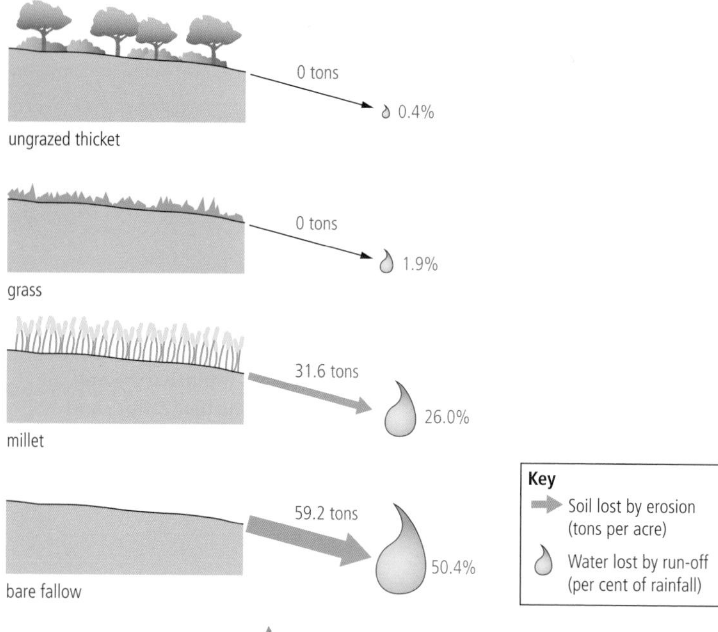

Figure 5.15 Impact of vegetation type and intensive grazing on soil erosion

Intensive grazing

Intensive grazing has two main impacts on vegetation cover. First, the greater the intensity of grazing, the more vegetation cover is removed. Second, large herds may destroy vegetation cover by trampling. This reduces interception, increases raindrop impact, and increases soil erosion. Third, grazers, especially large ones, may compact the soil, making it impermeable and thereby increasing the potential for soil erosion. In severe cases, such as around bore-holes in semi-arid regions, this may lead to desertification (i.e. the spread of desert-like conditions into previously productive regions). As Figure 5.15 shows, as long as there is a complete cover of vegetation, soil erosion is minimal and water loss is reduced. When the vegetation is removed, soil erosion increases, and water loss increases.

Urbanization

Studies from North America have shown that rates of soil erosion are greatest during phases of urbanization (Figure 5.16). This is because urbanization requires the removal of vegetation for construction, and there is often heavy machinery compacting the soil surface. This makes the soil impermeable and water is unable to infiltrate into the soil; instead it flows over the soil removing valuable, fertile topsoil.

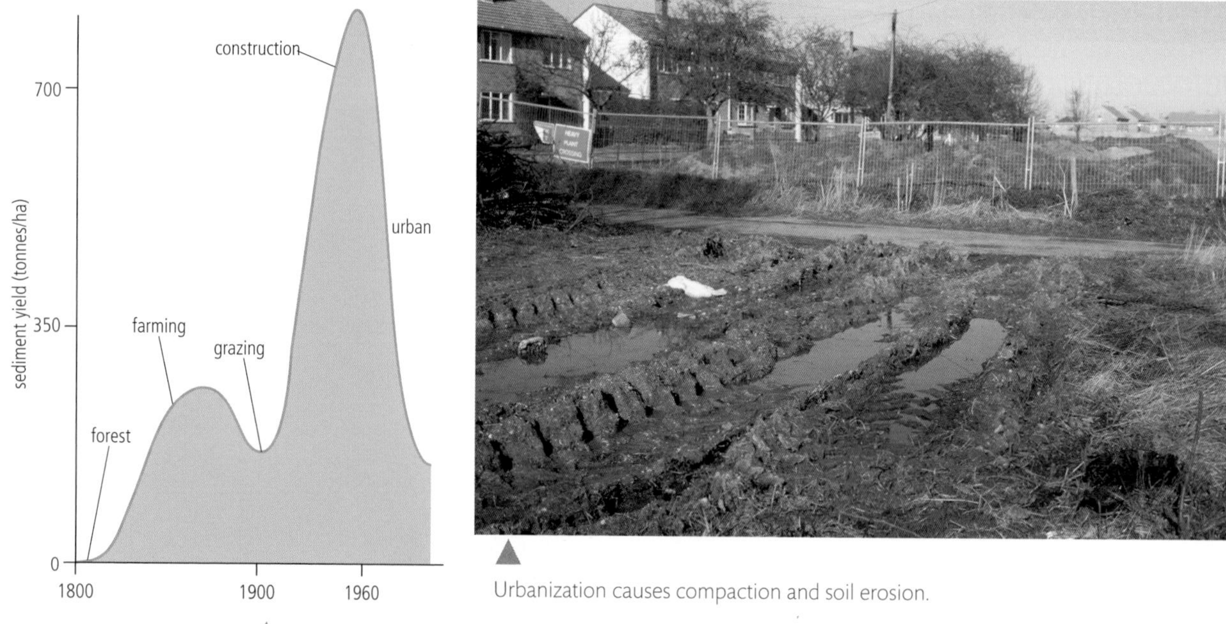

Figure 5.16 Soil erosion in North America: the effect of land use

Urbanization causes compaction and soil erosion.

Irrigation

Irrigation frequently leads to an increase in the amount of salt in the soil. This occurs when groundwater levels are close to the surface. In clay soils, this may be within 3 m of the surface but is less on sandy and silty soils. Capillary forces bring water to the surface where it may be evaporated leaving behind any soluble salts that it is carrying. This is known as *salinization.*

Some irrigation, especially of paddy fields for growing rice, requires huge amounts of water. As water evaporates in the hot sun, the salinity levels of the remaining water increase. This also occurs behind large dams.

In West Pakistan (Punjab), 15 per cent of irrigated land was uncultivatable by 1960 because of salinization.

In coastal and estuarine areas, saltwater intrusion is a major problem. Seawater is denser than fresh water. For example a column of 100 units of seawater could hold a column of 102.5 units of fresh water. Hence, groundwater does not lie flat over seawater but rests as a lens. The thickness of the lens is about 41 times the elevation of the water table above sea level. As the water pressure drops, due to over abstraction, the salt water will rise up by 40 units for every one unit that the fresh water falls.

Monoculture

Monoculture may lead to *soil exhaustion*. This is when a soil becomes depleted of a particular mineral. For example, following the **Green Revolution**, many high yielding varieties (HYVs) required large amounts of nutrients to grow. The practice of monoculture (growing a single type of crop) increases the demand for a certain set of nutrients. Growing the same crop year after year reduces the availability of these nutrients and may cause soil exhaustion. These lost nutrients can be replaced by using chemical and organic fertilizers but these are expensive. In the Caribbean, plantation monoculture of sugar and cotton has often resulted in soil exhaustion.

Commercialized food production systems

Commercial industrialized food production systems generally tend to reduce soil fertility more than small-scale subsistence farming methods.

In recent decades, commercial farming has become more geared towards the production of standardized products and attempts to achieve economies of scale. For example, fields have become larger to allow greater and easier use of machinery. There is greater use of pesticides and herbicides. Small-scale family farms have declined whereas large-scale agribusiness has expanded. During the recession, since 2007, consumers moved away from more expensive organic products to less expensive, mass-produced, highly processed foods. The result has been a deterioration of soil quality due to soil erosion, exhaustion, and toxification.

Case study

Commercial farming in East Anglia, UK

This area in the UK is an area of intensive arable farming, with large open fields and heavy use of agricultural chemicals. As a result, the once-fertile soil is much depleted and requires further use of fertilizers for commercial crops to be grown successfully. There are a number of options for soil conservation on commercial arable farms in East Anglia:

- avoid inappropriate weather conditions (e.g. heavy rain) for ploughing and harvesting
- add organic matter to the soil to increase water retention
- add clay to the soil to improve soil cohesion
- practise crop rotation so that soils do not become exhausted (this is less common now due to specialization in farming)
- use wind breaks to reduce the risk of wind erosion
- use cover crops to protect the soil in winter
- mulching – plough in the remains of the previous season's crop to improve nutrient retention in the soil
- leave some land fallow so that it can improve its fertility.

Large fields in the UK

Case study

Sierra de Santa Marta, Mexico

The Sierra de Santa Marta is a remote, mountainous region in the humid tropical state of Veracruz, Mexico. Soil erosion and soil fertility loss are major problems and result in:

- reduced agricultural productivity
- decreased availability of drinking water in nearby urban centres
- increased road maintenance
- falling hydroelectric potential
- a decline in the fishing industry in coastal lagoons.

Soil degradation can be severe when annual crops are grown on steep hillsides using practices that do not include cover crops or surface mulch. This is especially serious when fallow periods are reduced. In traditional shifting cultivation systems, the soil degradation occurring during the years of cultivation is offset by a fallow long enough to rebuild the soil's productive capacity. Such a system generally collapses with increasing land pressure, as fallow periods are reduced.

In Texizapan, the erosive ability of the natural environment, the high erodibility of the soil, and the limited soil cover provided by the annual crop leads to high rates of soil degradation. Perennial crops such as coffee, especially when grown under shade trees, generally provide better soil protection. However, in Veracruz, as in other areas in Central America, annual crops such as maize and beans provide most of the food and cash needs of the population. Resource-poor farmers are generally reluctant to stop growing these crops, even when others appear economically more attractive or environmentally less degrading.

Case study

A success story: Santa Rosa, Mexico

The Popoluca Indians of Santa Rosa, Mexico, practise a form of agriculture that resembles shifting cultivation, known as the milpa system. This is a labour-intensive form of agriculture, using fallow periods. It is a diverse form of polyculture with over 200 species cultivated, including maize, beans, cucubits, papaya, squash, water melon, tomatoes, oregano, coffee, and chilli. The variety of a natural rain forest is reflected by the variety of shifting cultivation. For example, lemon trees, peppervine, and spearmint are light seeking, and prefer open conditions not shade. Coffee, by contrast, prefers shade. The mango tree requires damp conditions.

The close associations that are found in natural conditions are also seen in the Popolucas' farming system. For example, maize and beans go well together, as maize extracts nutrients from the soil whereas beans return them. Tree trunks and small trees are left because they are useful for many purposes such as returning nutrients to the soil and preventing soil erosion. As in a rainforest the crops are multi-layered, with tree, shrub, and herb layers. This increases net primary productivity (NPP) per unit area, because photosynthesis is taking place on at least three levels (with the highest NPP in the forest canopy), and soil erosion is reduced because no soil or space is left bare. Animals include chickens, pigs, and turkeys. These are used as a source of food, and their waste is used as manure.

Thus, whereas there is widespread degradation in Veracruz, the Popolucas are able to maintain soil quality by working with nature.

Reduced soil fertility may result in soil erosion, toxification, salination, and desertification.

Results of reduced fertility

Soil erosion

Fertile soils are considered to be a non-renewable resource. This is because the rate at which they are being degraded and lost is faster than the rate at which they are being formed. The formation of a layer of 30 cm of soil takes between 1000 and 10 000 years. Rates of soil formation in humid tropical areas, such as Jamaica, are unlikely to exceed 1 mm per year. This is the equivalent of a catchment-wide figure of 10 t/ha/year. This is well below the rates of soil erosion from Jamaica and selected parts of the world (Table 5.7).

To learn more about soil degradation, go to www.pearsonhotlinks.co.uk, enter the book title or ISBN, and click on weblink 5.6.

Table 5.7 Rates of soil erosion (t/ha/year) from Jamaica and selected parts of the world

	Rates of soil erosion in tonnes per ha per year			
Jamaica	overall 97–280	forest 15–225	agriculture 24–294	yams 17–131
Other parts of the world	China 250	Nepal 70	Ethiopia 42	Burkina Faso 35
UK	South Downs 250	Norfolk 160	West Sussex 150	Shropshire 120

Clearly, in certain places soil is being destroyed at a far faster rate than it is being created.

CHALLENGE YOURSELF

Thinking skills 

Suggest reasons why rates of soil erosion may be higher in an MEDC, such as the UK, than in an LEDC such as Ethiopia, Burkina Faso, or Jamaica.

Case study

Australia is losing precious soil on a scale that is comparable to the desertification of Ethiopia, causing an economic disaster that now costs AU$2 billion (£1 billion) a year. But far more serious is the actual loss of the land, which will cost a further AU$2.5 billion to repair where possible. More than half of Australia's farming country is in need of treatment (Figure 5.17).

Years of neglect, coupled with a refusal to change inappropriate farming practices, have brought Australia to the edge of an environmental disaster.

The problem lies in the great age and fragility of the Australian landscape and the devastating toll that 200 years of European settlement has taken on the relatively fertile soil. The black soil plains of New South Wales and Queensland had a natural loss of only millimetres in thousands of years. In the 200 years since European settlement, clearing of vegetation, cultivation, grazing and construction has placed immense pressure on the land.

Although comprising less than a third of the total land area, Australia's non-arid zones are home to 98 per cent of the population and have to support all its grazing and crop production. Forty per cent of good farmland is badly eroded.

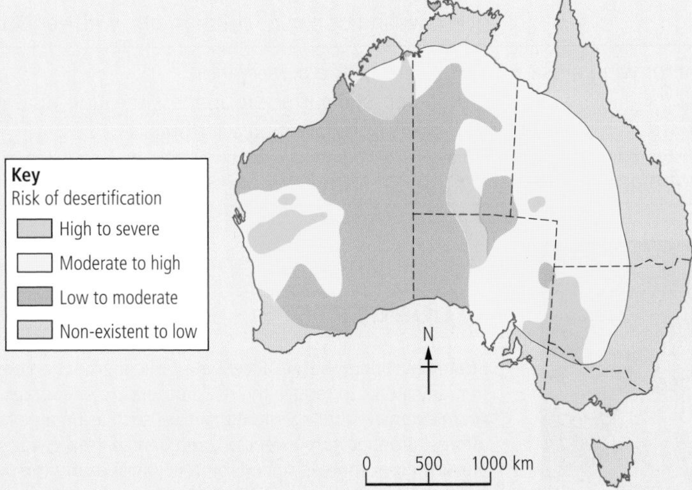

Figure 5.17 Desertification in Australia

Victoria is particularly badly hit by salinity, with 650 000 km² (406 000 square miles) affected in varying degrees. In Queensland 28 000 km² of land need repair after water erosion. Most of South Australia is prone to wind and gully erosion. In Western Australia, salinity and coastal erosion is the problem.

Up to 40 per cent of Australia's exports are produced directly from agriculture. Problems now include erosion, salinity, acidification, the effects of introduced animals, and chemical pollution from agrochemicals.

Australia has been described as 'an ecological disaster, characterized by a squalid history of greed, shortsightedness, and ignorance'.

More than 3 million hectares of China's farmland is too polluted with heavy metals and other chemicals to use for growing food. A key concern is cadmium, a carcinogenic metal that can cause kidney damage and other health problems and is absorbed by rice, the country's staple grain. Investigations by the Ministry of Environmental Protection found 'moderate to severe pollution' on 3.3 million hectares. The Chinese Government, in its 5-year development plan, 2010–15, promised to reduce heavy metal pollution and clean up contaminated areas. One possible approach is to plant trees or other vegetation that will absorb heavy metals from the soil but will not be consumed by humans.

Soil conservation measures exist. For example, soil conditioners (e.g. organic materials and lime), wind reduction techniques (e.g. wind breaks, shelter belts), cultivation techniques (e.g. terracing, contour ploughing, strip cultivation), and avoiding the use of marginal lands.

Soil conservation methods

Table 5.8 summarizes a number of soil conservation measures that can be used to reduce or prevent erosion of topsoil.

Table 5.8 Methods of soil conservation

Method of conservation	Action involves
revegetation	• deliberate planting • suppression of fire, grazing, etc. to allow regeneration
measures to stop bank erosion	• insert corrugated iron • concrete banks
measures to stop gulley enlargement	• planting of trailing plants, etc. • construction of weirs, dams, gabions, etc.
crop management	• maintaining a cover at critical times of the year • rotation • cover crops
slope run-off control	• terracing • deep tillage and application of humus • transverse hillside ditches to intercept run-off • contour ploughing • preservation of vegetation strips to limit field width
prevention of erosion from point sources such as roads, feedlots	• intelligent geomorphic location • channelling of drainage water to non-susceptible areas • covering of banks, cuttings, etc. with vegetation
suppression of wind erosion	• soil moisture preservation • increase in surface roughness through ploughing up clods or by planting windbreaks
avoid use of marginal land	• do not farm land that is too dry or infertile

Check dam, Eastern Cape, South Africa

CONCEPTS: Sustainability

Strategies for combating accelerated soil degradation are lacking in many areas. To reduce the risk, farmers are encouraged towards more extensive management practices such as organic farming, afforestation, pasture extension, and benign crop production. Nevertheless, there is a need for policymakers and the public to combat the pressures and risks to the soil resource.

Methods to reduce or prevent erosion can be mechanical (e.g. physical barriers such as embankments and wind breaks), or they may focus on vegetation cover and soil husbandry. Overland flow can be reduced by increasing infiltration.

Mechanical methods to reduce water flow

Mechanical methods include bunding, terracing, and contour ploughing. The key is to prevent or slow down the movement of rainwater down slopes. Contour ploughing takes advantage of the ridges formed at right angles to the slope to prevent or slow the downward accretion of soil and water.

On steep slopes and those with heavy rainfall (e.g. the monsoon in South East Asia), contour ploughing is insufficient and terracing is undertaken. The slope is broken up into a series of flat steps terraces) with bunds (raised levées) at the edge. The use of terracing allows areas to be cultivated that would not otherwise be suitable for cultivation.

Land around gullies and ravines can be fenced off, and planted with small trees and grass. Check dams can be built across gullies to reduce the flow of water and trap soil.

TOK

Despite a number of strategies, soil erosion and land degradation in the Caribbean have rarely been halted. Farming steep, unstable slopes is becoming progressively less productive.

Cropping and soil husbandry methods against water and wind damage

Preventing erosion by different cropping techniques largely focuses on:

- maintaining a crop cover for as long as possible
- keeping in place the stubble and root structure of the crop after harvesting
- planting a grass crop.

A grass crop maintains the action of the roots in binding the soil, and minimizes the action of wind and rain on the soil surface. Increased organic content allows the soil to hold more water, thus reducing mass movement and erosion, and stabilizing the soil structure. Soil organic matter is a vital component of productive and stable soils. It is an important source of plant nutrients, improves water retention and soil structure, and is important in terms of the soil's buffering capacity against many of the threats. In addition, to prevent damage to the soil structure, care should be taken to reduce use of heavy machinery on wet soils, and to minimize ploughing on soils sensitive to erosion. On acidic soil, ground lime can be added to make it more alkaline.

In areas where wind erosion is a problem, shelterbelts of trees or hedgerows are used. The trees act as a barrier to the wind and disturb its flow. Wind speeds are reduced which reduces the wind's ability to disturb the topsoil and erode particles.

Multicropping is also useful if it maintains a cover crop throughout the year. On the steepest slopes, cultivation is not recommended and the land should be forested or vegetated to maintain soil cover and reduce run-off.

Common measures to minimize wind erosion on light agricultural soils of Northern Europe

Table 5.9 summarizes some comments on soil conservation measures that can be used to reduce or prevent erosion by wind.

Table 5.9 Comments on some measures to curb wind erosion

Measure	Comment
Measures that minimize actual risk (short-term effect)	
autumn-sown varieties	need to be sown before the end of October to develop a sufficient cover
mixed cropping	after the main crop is harvested, second crop remains on the field
nursing or cover crop	more herbicides needed
straw planting	unsuitable on light sandy soils
organic protection layer (e.g. liquid manure, sewage sludge, sugar beet factory lime)	depends on availability, and regulations on the use of these products
synthetic stabilizers	unsuitable on peat soils
time of cultivation	depends on availability of labour and equipment
cultivation practice (e.g. minimum tillage, plough and press)	not suitable for all crop or soil types
Measures that lower the potential risk (long-term effect)	
smaller fields	increase in operational time and costs
change of arable land to permanent pasture or woodland	loss of agricultural production and farm income
marling (increasing the clay content to 8–10 per cent)	suitable material should be available close by
wind barriers	high investment cost, and loss of productive land; takes several years before providing full protection; level of protection reduces with distance from the shelter

Management of salt-affected soils

There are three main approaches in the management of salt-affected soils:

- flushing the soil with water and leaching the salt away
- application of chemicals (e.g. gypsum – calcium sulfate – to replace the sodium ions on the clay and colloids with calcium ions)
- reduction in evaporation losses to reduce the upward movement of water in the soil.

Summary of soil conservation methods

Socio-economic and ecological factors have been ignored for too long. An integrated approach to soil conservation is required in which non-technological factors such as population pressure, social structures, economy and ecological factors can determine the most appropriate technical solutions. There is a wide variety of possible solutions including: strip and ally cropping, rotation farming, contour planning, agroforestry, adjusted stocking levels, mulching, use of cover crops, and construction of mechanical barriers such as terraces, banks and ditches (Figure 5.18).

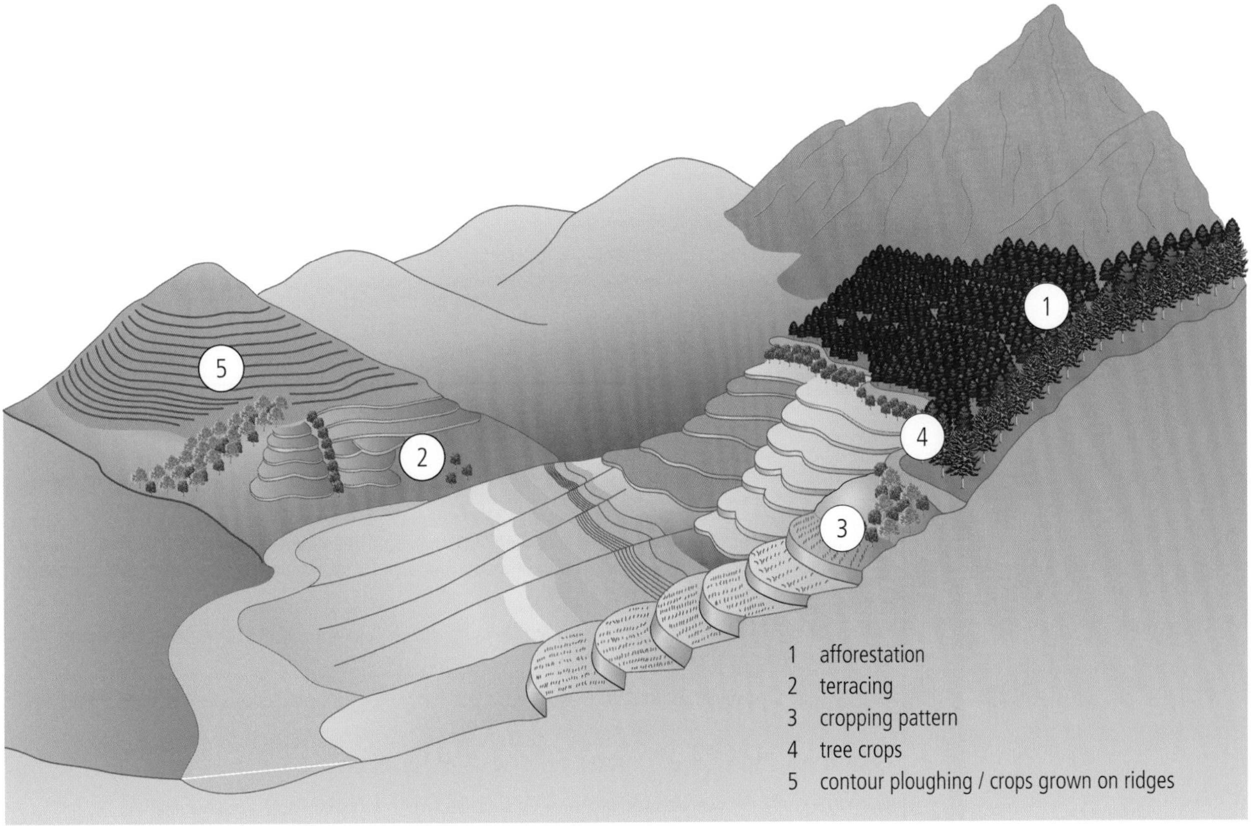

1 afforestation
2 terracing
3 cropping pattern
4 tree crops
5 contour ploughing / crops grown on ridges

Figure 5.18 Soil conservation methods

To learn more about soil conservation, go to www.pearsonhotlinks.co.uk, enter the book title or ISBN, and click on weblink 5.7.

Soil erosion is a major problem in the Caribbean on account of the steep topography, erodability of the soils, high rainfall, population pressure, and crop and soil management. Sandy acidic soils have low capacity for water storage, low organic material and low nutrient holding capacity. Liming of soils at 500–1000 kg per hectare can offset acidity. No tillage production (i.e. leaving the previous year's crops in the soil to protect it) increases soil organic matter. It can produce yields comparable with tillage agriculture. However, it can allow weeds to develop. The use of *Gliricidia* species (a small deciduous leguminous tree) can be an important source of nitrogen and phosphorus. Research has shown that *Gliricidia* trees can, within 2 years, produce 240 kg of nitrogen and 150 kg of phosphorus per hectare.

CHALLENGE YOURSELF

Apart from traditional methods of soil conservation, there are five other options.

1. Do not farm slopes steeper than 25°.
2. 'Do nothing' – but this will only lead to more poverty.
3. Agroforestry – but this can only supply a small number of people.
4. Move significant number of people off the land – but they may end up in slums in urban areas.
5. Develop off-farm incomes.

Evaluate each of the five options for resolving soil erosion.

To watch a video on the importance of soil, go to www.pearsonhotlinks.co.uk, enter the book title or ISBN, and click on weblink 5.8.

Exercises

1. a. Why have the last 200 years been so crucial in causing soil loss in Australia?
   b. What percentage of Australia is arid? How many people live there?
   c. List four problems that are facing Australian farmers. How much of Australia's exports come from the soil?
2. Outline soil conservation measures.
3. Evaluate soil management strategies in a named commercial farming system and in a named subsistence farming system.

Big questions

Having read this section, you can now discuss the following big questions:

- What strengths and weaknesses of the systems approach and the use of models have been revealed through this topic?
- To what extent have the solutions emerging from this topic been directed at *preventing* environmental impacts, *limiting* the extent of the environmental impacts, or *restoring* systems in which environmental impacts have already occurred?
- What value systems can you identify at play in the causes and approaches to resolving the issues addressed in this topic?
- In what ways might the solutions explored in this topic alter your predictions for the state of human societies and the biosphere some decades from now?

Here are some questions you may want to consider in your discussions.

- How might ecocentrists and technocentrists differ over methods of soil conservation?
- Could there be new methods of food production that may help feed the world's growing population?

Practice questions

1 The table below shows the impact of intensive farming on soil profiles.

Soil characteristic	Natural state	Intensive agriculture
organic content	A horizon – high (7%) B horizon – low (0%)	uniform (3–5%) in ploughed horizon
carbonates	A horizon – low/zero B/C horizon – maximum	uniform if limed and tilled
nitrogen	medium/low	high (nitrate fertilizers)
biological activity	high	medium
exchangeable cation balance	Ca 80% K 5% P 3% H 7%	Ca 70% K 10% P 12% H 4%

a Describe the main differences in organic content in natural soils and farmed soils. [2]

b Comment on the changes in nitrogen (nitrate) levels between natural soils and farmed soils. [3]

c Briefly explain why the level of carbonates in a natural soil varies between horizons whereas in an intensively farmed soil, the levels are uniform throughout the horizons. [3]

d Describe and suggest reasons for the differences in calcium (Ca), potassium (K), phosphorus (P), and hydrogen (H) in natural and farmed soils. [6]

2 a Suggest two reasons why food is in short supply in some societies. [2]

b Explain the relationships between population growth, social systems, and food production technologies. Refer to named contrasting countries in your answer. [9]

c Discuss the environmental problems caused by food production systems and suggest possible solutions. [10]

3 a Define the term *soil degradation*. [2]

b Choose an appropriate method to show the following data:

Causes of land degradation

Deforestation/fuelwood consumption	37%
Overgrazing	35%
Agricultural mismanagement	27%
Industry and urbanization	1%

[2]

c Comment on the types and causes of soil degradation. [4]

06 Atmospheric systems and societies

6.1 Introduction to the atmosphere

Opposite: To improve air quality, some urban areas have removed densely packed buildings and replaced them with trees and open spaces, as can be seen at La Rambla, Barcelona.

Significant ideas

The atmosphere is a dynamic system which is essential to life on Earth.

The behaviour, structure, and composition of the atmosphere influence variations in all ecosystems.

Big questions

As you read this section, consider the following big questions:

- To what extent have the solutions emerging from this topic been directed at *preventing* environmental impacts, *limiting* the extent of the environmental impacts, or *restoring* systems in which environmental impacts have already occurred?
- How are the issues addressed in this topic of relevance to sustainability or sustainable development?
- In what ways might the solutions explored in this topic alter your predictions for the state of human societies and the biosphere some decades from now?

Knowledge and understanding

- The atmosphere is a dynamic system (with inputs, outputs, flows, and storages) which has undergone changes throughout geological time.
- The atmosphere is a mixture of mainly nitrogen and oxygen, with smaller amounts of carbon dioxide, argon, water vapour, and other trace gases.
- Human activities impact atmospheric composition by altering inputs and outputs of the system. Changes in the concentrations of atmospheric gases (e.g. ozone, carbon dioxide, and water vapour) have significant effects on ecosystems.
- Most reactions connected to living systems occur in the inner layers of the atmosphere, which are the troposphere (0–10 km above sea level) and the stratosphere (10–50 km above sea level).
- Most clouds form in the troposphere and play an important role in the albedo effect for the planet.
- The greenhouse effect of the atmosphere is a natural and necessary phenomenon maintaining suitable temperatures for living systems.

Our atmosphere is a dynamic system

The atmosphere is a dynamic system (with inputs, outputs, flows and storages) which has undergone changes throughout geological time.

SYSTEMS APPROACH

Earth's atmosphere system can be seen as a closed system: the input is solar radiation (energy from the Sun), and the output is heat energy. Matter is recycled within the system. Although space ships and meteorites move tiny amounts of matter in to and out of the Earth system, they are generally discounted. Strictly speaking, closed systems do not occur naturally on Earth, but all the global cycles of matter (e.g. water and nitrogen cycles) approximate to closed systems.

The atmospheric system is an interactive system consisting of five major components:

- the atmosphere
- the hydrosphere (oceans, lakes, and rivers)
- the cryosphere (ice sheets, glaciers and snow)
- the land surface
- the biosphere (plants and animals).

The atmospheric system is driven by external forces, notably the Sun. The atmosphere is dynamic, present largely as a result of human activities. However, other forces are important too, such as sunspot activity, volcanic activity, continental drift, and large-scale changes in the Earth's axial tilt and its orbital path around the Sun.

The Earth's atmosphere is both influenced by the biosphere and influences the biosphere (Figure 6.1). The current atmosphere consists of nitrogen (78.1 per cent), oxygen (20.9 per cent), and carbon dioxide (0.4 per cent). If the effects of the biosphere were removed, it is estimated that the atmospheric composition would be 1.9 per cent nitrogen, 0 per cent oxygen and 98 per cent carbon dioxide. This illustrates the close relationship between the biosphere and the atmosphere. The current atmosphere has been strongly influenced, arguably created, by plant life. Without plants, oxygen levels would be 1000 times smaller than they currently are. Vegetation uses solar radiation to photosynthesize, thereby converting carbon dioxide into carbon and free oxygen gas. The carbon eventually combines with other elements to produce wood and leaves (i.e. biomass). Biomass absorbs atmospheric carbon dioxide throughout its lifetime and then releases carbon dioxide when it is burned. On Earth, most of the carbon dioxide is contained in rocks such as limestone, chalk, and coal (buried carbon). If as little as 0.3 per cent of this buried carbon were returned to the atmosphere, the atmospheric concentration of carbon dioxide would double. This would have major implications for climate change and the survival of human society.

Changes over geological time

There has been considerable fluctuation of the carbon dioxide level and of temperature throughout the history of the Earth (Figure 6.2). The dynamic system of the atmosphere has evolved over time. It continues to change as a result of natural and human-induced processes. Some 2 billion years ago, photosynthesizing algae in the oceans first released free oxygen into the atmosphere. Ocean microflora still supply

Figure 6.1 The climate system: components (in bold), processes and interactions (orange arrows), and some aspects that may change (white arrows)

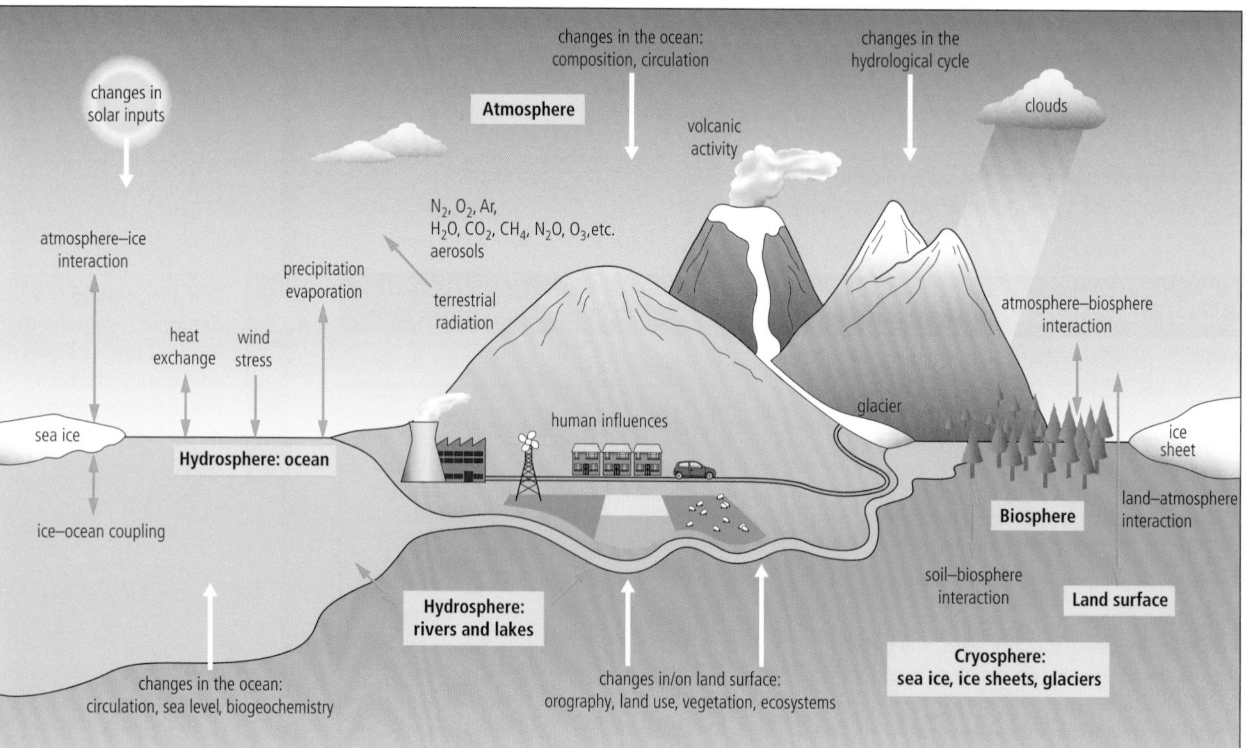

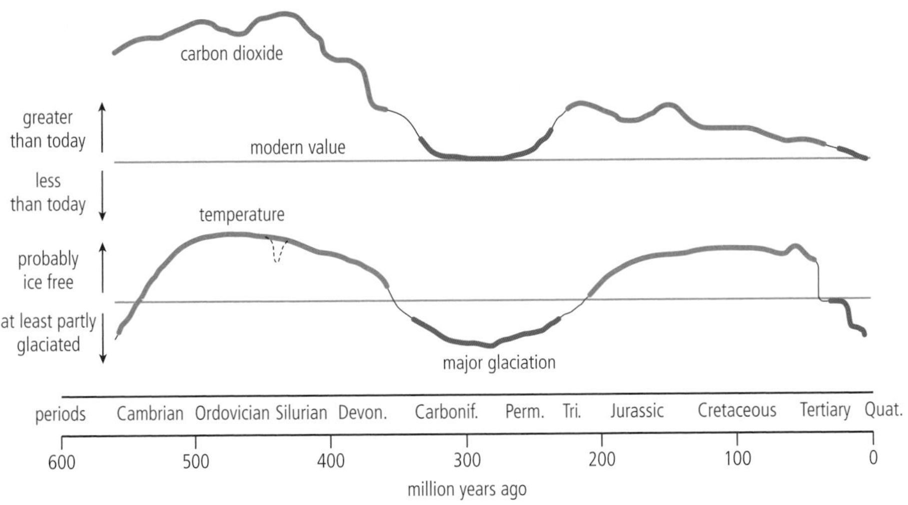

Figure 6.2 Carbon dioxide levels and average global temperature over the last 540 000 years

70 per cent of the world's oxygen and help maintain the protective ozone layer in the stratosphere.

The Earth's atmosphere today contains about 400 parts per million (ppm) carbon dioxide (CO_2) – this is 0.4 per cent of the atmosphere. In the last 600 million years of Earth's history, only the Carboniferous Period and our present age (the Quaternary Period) have had carbon dioxide levels of less than 400 ppm (Figure 6.2). However, levels of carbon dioxide are currently rising and the increase is thought to be entirely due to human activities. Humans have added up to 2.7 gigatonnes (Gt) of carbon to the atmosphere every year, increasing carbon dioxide levels from 280 ppm in pre-industrial times to 379 ppm today (although this is far less carbon dioxide than has been usual over geological time, it is still a significant rise in 160 years). Temperatures have been rising over this same period (Figure 6.3).

There are various explanations for the trend seen in the graph in Figure 6.3:

- the onset of global industrialization and the subsequent production of pollution derived from fossil fuels
- deforestation, particularly of rainforest
- volcanic activity
- sunspot activity.

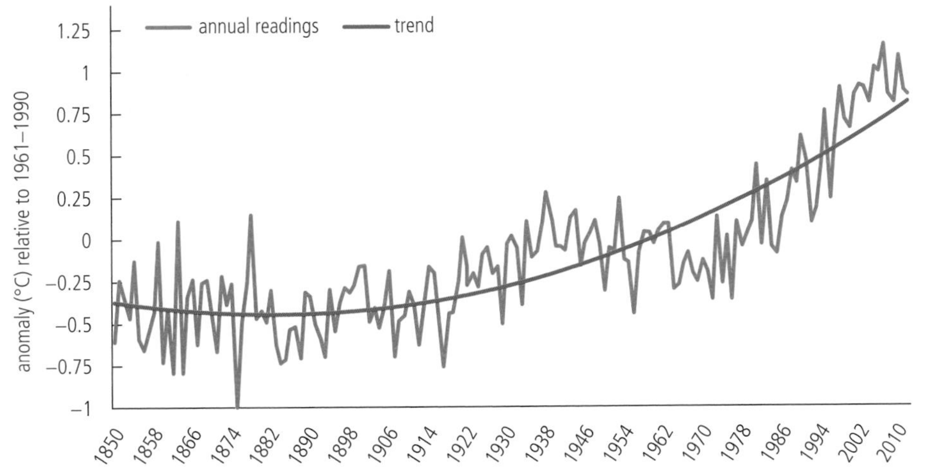

Figure 6.3 Mean global climate change from 1850 to 2010. The graph shows a smoothed curve of annual average temperature. There is an overall upward trend, accelerating in the last quarter of the graph. Before about 1935, temperatures were all below the average for 1961–90, and from about 1985 they have all been above the average.

The first two points assume a link between carbon dioxide emissions and temperature increase. The last two suggest possible natural phenomena that may have increased temperatures. Bearing in mind the large increase in carbon dioxide concentrations since the industrial revolution, most scientists make the assumption that the increase in temperature is caused by human activities, although as with comparisons between any two variables, correlation does not prove causation, especially when complex systems such as the atmosphere and biosphere are involved.

TOK

Global warming challenges views of certainty within the sciences. In the popular perception, global warming is having a negative impact on the world. There is, moreover, some confusion between the public perception of global warming and the greenhouse effect. The greenhouse effect is a natural process, without which there would be no life on Earth. The *enhanced* or *accelerated* greenhouse effect is synonymous with global warming. The enhanced greenhouse effect is largely due to human (anthropogenic) forces, although feedback mechanisms may trigger some natural forces, too. Lobby groups and politicians take views which suit their own economic and political ends. In the USA, the strength of the oil companies during the Bush Administration was seen by many as an example of economic groups, and the politicians they supported, choosing a stance which was not in the long-term environmental, social, or economic interest of the world. But it did benefit the oil companies and politicians.

Atmospheric gases

The atmosphere is predominantly a mixture of mainly nitrogen and oxygen, with smaller amounts of carbon dioxide, argon, water vapour and other trace gases.

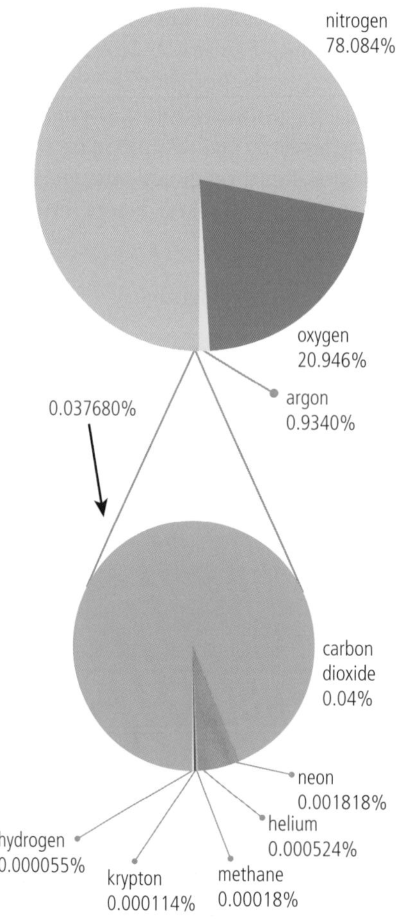

Figure 6.4 The composition of the Earth's atmosphere

The atmosphere contains a mix of gases, liquids, and solids (Table 6.1). Atmospheric gases in the lower atmosphere are held close to the Earth by gravity. These gases are relatively constant (Figure 6.4). Nevertheless, there are important spatial and temporal variations in atmospheric composition, and this causes variations in temperature, humidity, and pressure over time and between places. Besides nitrogen, oxygen, argon, and carbon dioxide, there are other important gases such as helium, ozone, hydrogen, and methane. These gases are crucial. For example, changes in the amount of carbon dioxide in the atmosphere is having an effect on global warming (pages 372–381) and the destruction of ozone is having an important effect on the amount of ultraviolet radiation reaching the Earth's surface (page 317).

The atmosphere also contains moisture. Most water vapour is held in lower 10–15 km of atmosphere. Above this, it is too cold and there is not enough turbulence or mixing to carry vapour upwards. In addition, there are solids such as dust, ash, soot, and salt. These allow condensation to occur which can cause cloud formation and precipitation. On a local-scale, large concentrations of solid particles causes an increase in fog, smog, haze, and/or precipitation.

Table 6.1 The composition of the atmosphere

Gas	Percentage by volume	Importance for weather and climate	Other functions/sources
Permanent gases			
nitrogen	78.09	• mainly passive	• needed for plant growth
oxygen	20.95	• mainly passive	• produced by photosynthesis • reduced by deforestation • needed for animal life
Variable gases			
water vapour	0.2–4.0	• source of cloud formation and precipitation • reflects/absorbs incoming radiation • keeps global temperature constant • provides majority of natural greenhouse effect	• essential for life • can be stored as ice/snow
carbon dioxide	0.04	• absorbs long-wave radiation from Earth and so contributes to greenhouse effect • increase due to human activity is a major cause of global warming	• used by plants for photosynthesis • increased by burning fossil fuels and by deforestation
ozone	<0.01	• absorbs incoming ultraviolet radiation	• reduced/destroyed by chlorofluorocarbons (CFCs)
Inert gases			
argon	0.93		
helium, neon, krypton	trace		
Gaseous pollutants			
Natural sulfur dioxide, nitrogen oxide, methane	trace	• affects radiation • causes acid precipitation	• industry, power stations, and car exhausts
Human-made CFCs	trace	• reduces the thickness of the ozone layer	• aerosols sprays and propellants • coolants
Non-gaseous component			
dust	trace	• absorbs/reflects incoming radiation • forms condensation nuclei necessary for cloud formation	• volcanoes • meteors • soil erosion by wind

NB Figures refer to dry air, so the variable amount of water vapour is not usually taken into consideration.

Variations in composition with altitude

Turbulence and mixing in the lower 15 km of the atmosphere produces fairly similar 'air'. At high altitudes, by contrast, marked concentrations of certain gases occur, for example, between 25 km and 35 km there is a concentration of ozone. Although this forms only a small percentage of the atmospheric gas, it is significant enough to lead to an increase in atmospheric temperature in this region and has an important role in stopping ultraviolet light reaching the Earth's surface.

The most significant concentrations of gases at altitude include:

- nitrogen at 100–200 km
- oxygen at 200–1100 km
- helium at 1100–3500 km
- hydrogen above 3500 km.

The concentrations of gases have an important effect on changes in temperature through the atmosphere (Figure 6.5). At the tropopause, there is a reversal in the temperature gradient. This acts as the upper limit of weather systems. In the stratosphere, the increase in temperature is related to the presence of ozone. Temperatures then fall in the mesosphere but increase again in the thermosphere.

Figure 6.5 Changes in temperature and composition with altitude

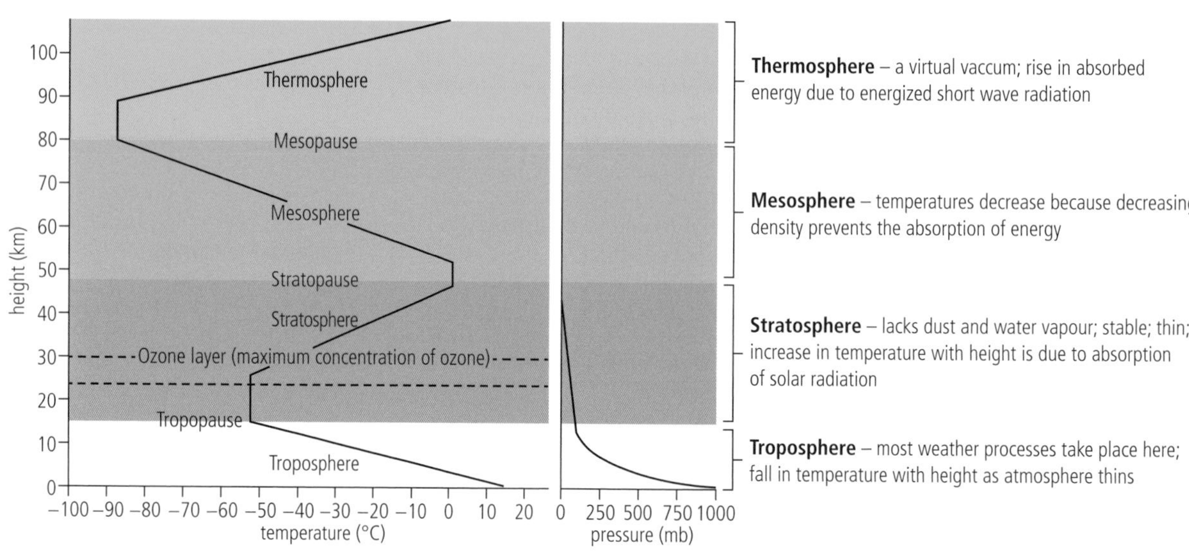

To learn more about the Earth's energy balance, go to www.pearsonhotlinks.co.uk, enter the book title or ISBN, and click on weblink 6.1.

The Earth's energy budget

The Earth's energy is mostly derived from the Sun. Insolation refers to incoming solar radiation. In some places, there may be important local sources of energy, such as in urban areas (heat islands) and tectonic regions (geothermal heat).

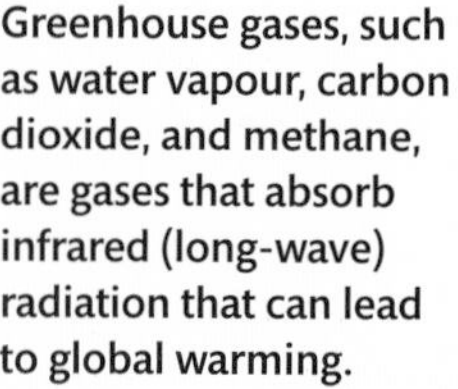

Greenhouse gases, such as water vapour, carbon dioxide, and methane, are gases that absorb infrared (long-wave) radiation that can lead to global warming.

CONCEPTS: Equilibrium

Solar energy (short-wave energy) is received by the Earth and its atmosphere and transformed in a number of processes (Figure 6.6). For every 100 units of solar energy reaching the Earth's atmosphere, 31 per cent is reflected back to space, and 69 per cent is absorbed by the Earth and the atmosphere. Some is reradiated from the Earth as long-wave radiation and is absorbed in the atmosphere by gases such as carbon dioxide, methane, and water vapour (i.e. greenhouse gases). These gases raise the Earth's temperature by about 30 °C - enough to support life on Earth.

Of the short-wave radiation which reaches the ground (46 units):

- 14 units are reradiated as long-wave radiation to the atmosphere and to space
- 10 units pass to the atmosphere by conduction (contact heating) or the lower atmosphere only because air is a poor conductor of heat
- 22 units are transferred by latent heat (heat energy used by a substance to change form but not temperature) - water is evaporated into the atmosphere, and releases heat on condensation.

Of the 46 units, 39 units are transferred to the atmosphere. Since the atmosphere only absorbs 23 units of short-wave radiation the atmosphere is largely heated from below.

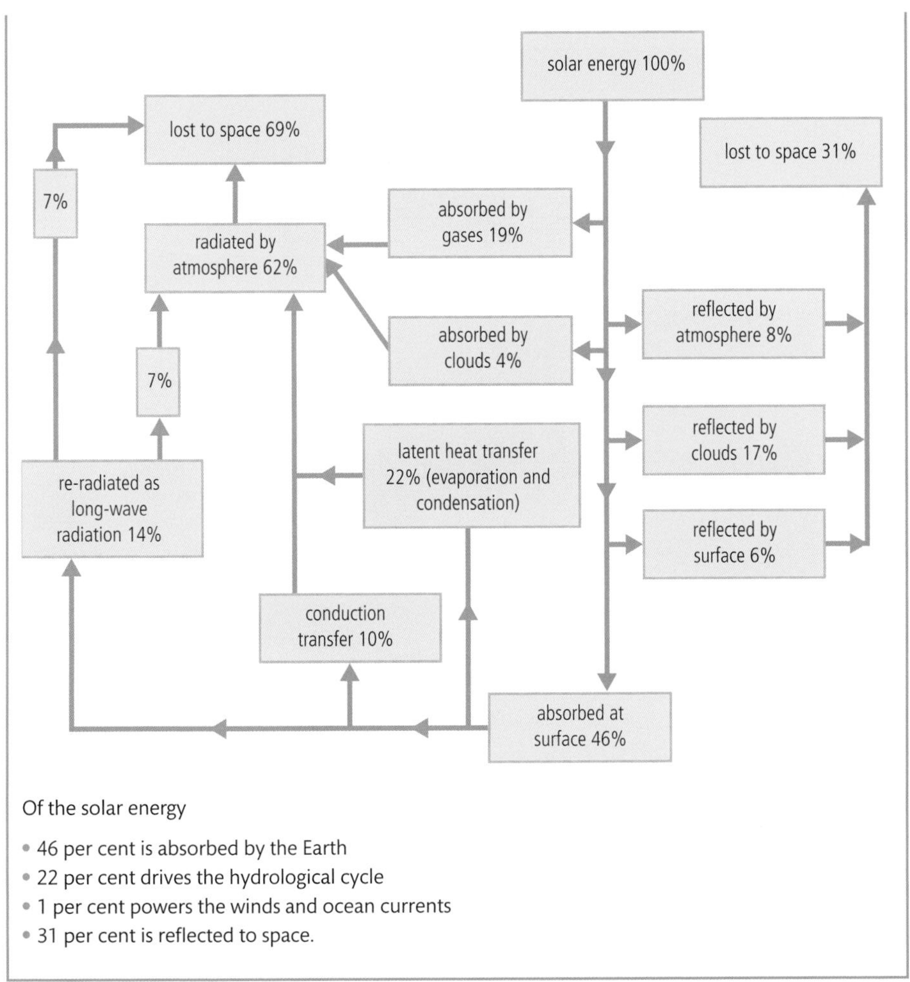

Figure 6.6 Earth's energy budget. The numbers refer to percentages of solar energy.

Human activities and atmospheric composition

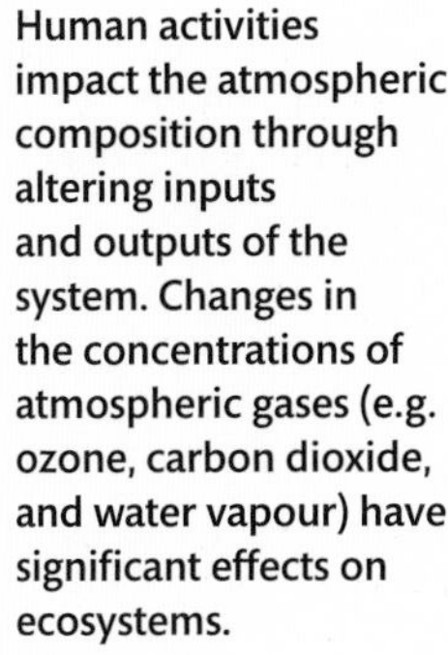

Human activities impact the atmospheric composition through altering inputs and outputs of the system. Changes in the concentrations of atmospheric gases (e.g. ozone, carbon dioxide, and water vapour) have significant effects on ecosystems.

Human activities have increased levels of greenhouse gases in the atmosphere. The best known example is the increase in carbon dioxide release by burning fossil fuels. But increases in levels of other gases (e.g. methane and nitrous oxide) are also linked to human activity. Changes in the amount of water vapour in the atmosphere – as a result of global warming – are important, as water vapour is the main greenhouse gas by volume. There is much concern that the increases in these gases in the atmosphere are causing an increase in average temperature of the Earth's atmosphere.

The increase in greenhouse gases in the atmosphere due to human activity (enhanced greenhouse effect) may be causing **global warming/climate change** (Figure 6.7). The main human activities releasing greenhouse gases are as follows.

Global warming is an increase in the average temperature of the Earth's atmosphere.

- Burning fossil fuels (coal, oil, gas) and releasing carbon dioxide.
- Deforestation affects carbon dioxide levels in a number of ways. Carbon dioxide is released through the breakdown of forest biomass and the increased rate of breakdown in organic content in soils (due to exposure to heat and water). Reduction in forest cover reduces the amount of carbon dioxide taken out of the atmosphere by photosynthesis. Deforestation is largely driven by pressures to free-up land for housing and agriculture, and to generate income from timber exports.

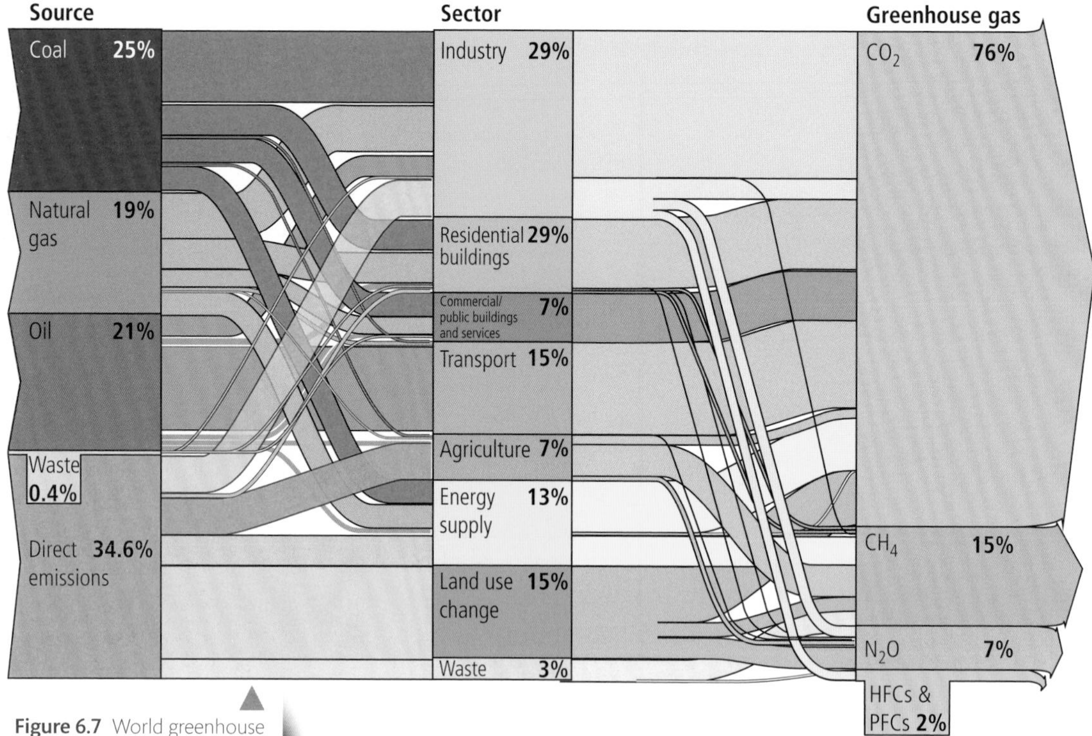

Figure 6.7 World greenhouse gas emissions, 2010

- Increased intensive cattle farming has led to increased methane levels: cows digest grass via fermentation and anaerobic bacteria in their gut. This releases methane as a waste product. Cattle ranching increases as the demands for beef increases.
- Rice farming in paddy fields creates waterlogged fields where anoxic conditions (low oxygen levels) enable anaerobic bacteria to thrive, also leading to increased methane release.
- Using fertilizers in agricultural systems has led to higher nitrous oxide (N_2O) concentrations when the fertilizers break down.

Most reactions connected to living systems occur in the inner layers of the atmosphere, which are the troposphere (0–10 km above sea level) and the stratosphere (10–50 km above sea level).

Most human activities that impact climate systems operate in the lower levels of the atmosphere. Most water vapour is concentrated in the troposphere. The use of CFCs affects ozone in the stratosphere. A reduction in stratospheric ozone due to use of CFCs can lead to a decline in photosynthesis in phytoplankton and plants, and has been linked with an increase in skin cancers in fish (pages 325–326).

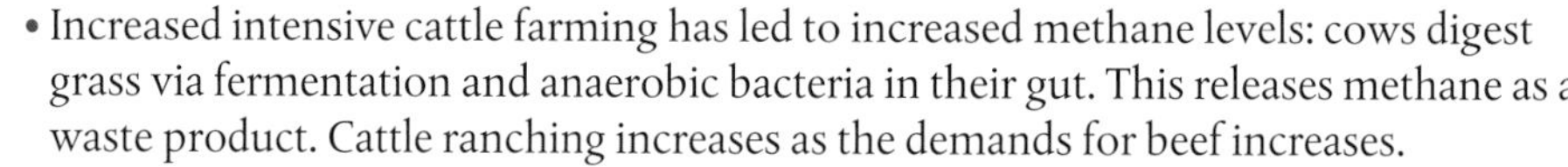

Many pollutants, such as ozone-depleting substances, are moved around by planetary winds and so may have impacts in regions in which they did not originate. CFCs, for example, concentrate over the Poles, especially the South Pole, an area in which there is very limited human activity.

Most clouds form in the troposphere and play an important role in the albedo effect for the planet.

Clouds

Clouds are formed of millions of tiny water droplets held in suspension. They are classified in a number of ways, the most important are:

- by form or shape, e.g. stratiform (layers) and cumuliform (rounded)
- by height, e.g. low (<2000 m), medium or alto (2000–700 m) and high (7000–1300 m).

There are a number of different types of cloud (Figure 6.8). High clouds consist mostly of ice crystals. Cirrus are wispy clouds, and include cirrocumulus (mackerel

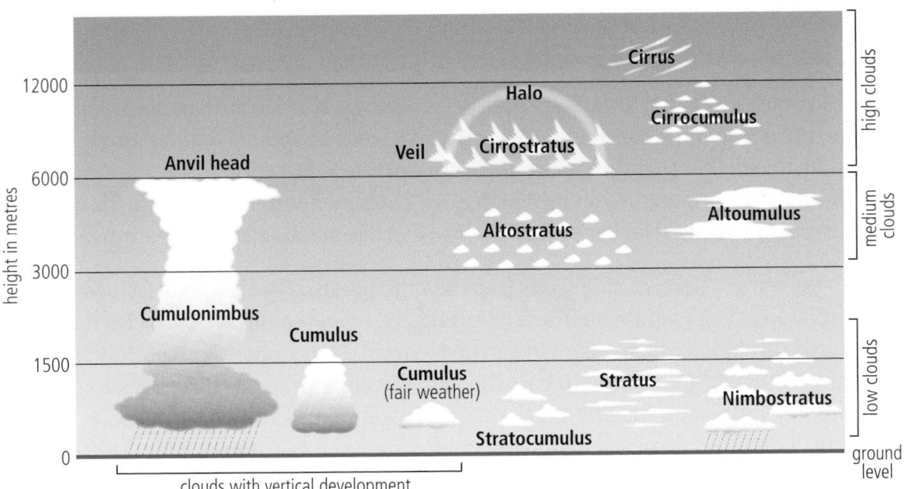

Figure 6.8 Cloud types

Cumulus clouds

Stratus clouds

sky) and cirrostratus (halo effect around Sun or moon). Alto or middle height clouds generally consist of water drops. They exist in temperatures of <0 °C. Low clouds indicate wet weather. Stratus clouds are dense, grey, low-lying ones. Nimbostratus are those which produce rain (*nimbus* means rain). Stratocumulus are long cloud rolls, and a mixture of stratus and cumulus. Clouds are an important part of the albedo of the Earth's atmosphere system (Table 6.2).

Table 6.2 Average albedo of various surfaces

Surface	Albedo
Earth systems (average)	0.31
Earth surface	0.14–0.16
global cloud	0.23
cumulonimbus	0.90
stratocumulus	0.60
cirrus	0.40–0.50
fresh snow	0.80–0.90
melting snow	0.40–0.60
sand	0.30–0.35
grass, cereal crops	0.18–0.25
deciduous forest	0.15–0.18
coniferous forest	0.09–0.15
tropical rainforest	0.07–0.15
water bodies*	0.06–0.10

*Increase when the Sun is at a low angle

You should be able to discuss the role of the albedo effect from clouds in regulating global average temperature.

CHALLENGE YOURSELF

Using a number of digital thermometers, investigate variations in temperature of different surfaces at approximately the same time (e.g. tarmac, grass, deciduous trees, bare ground, concrete, glass, and a bucket of water).

The greenhouse effect

The greenhouse effect is a natural and necessary process in which greenhouse gases allow short-wave radiation to pass through the atmosphere but trap a percentage of the outgoing long-wave radiation. It increases global temperatures by about 33 °C, thereby allowing life on Earth.

The enhanced greenhouse effect – or accelerated greenhouse effect – refers to the anthropogenic (human-induced) global warming as a result of increased use of greenhouse gases.

The preconditions for life to exist are a source of energy, water, nutrients, and warmth. The chemical reactions that support life processes work at optimal temperatures. Colder conditions mean that enzyme reactions do not operate at fast enough rates to support life. The average temperature on the Earth is 15 °C, and without this warmth, life could not exist on the planet. The temperature of the Earth is maintained by the atmosphere.

Within the atmosphere, certain gases trap the radiation that heats the surface. Short-wave ultraviolet (UV) light from the Sun is reflected from the surface of the Earth as infrared (IR) light (which has a longer wavelength). Atmospheric gases allow the incoming short-wave radiation to pass through but either trap or reflect back to Earth the outgoing long-wave radiation. The process is sometimes known as 'radiation trapping'. This effect is caused mainly by water vapour and carbon dioxide. Other gases involved are methane (CH_4), nitrous oxide (N_2O), and ozone (O_3). The gases create a 'thermal blanket' that maintains an average Earth temperature that can support life. Because these gases act in the same way that glass acts in a greenhouse, they are called **greenhouse gases** and the effect they have is called the greenhouse effect (Figure 6.9).

Although greenhouse gases make up only about 1 per cent of the atmosphere, they regulate our climate and make life possible. The term 'greenhouse effect' tends to be viewed negatively today, but greenhouse gases enable world temperatures to be warmer than they would otherwise be. Without them, the average temperature on Earth would be colder by approximately 30 °C – far too cold to sustain our current ecosystems.

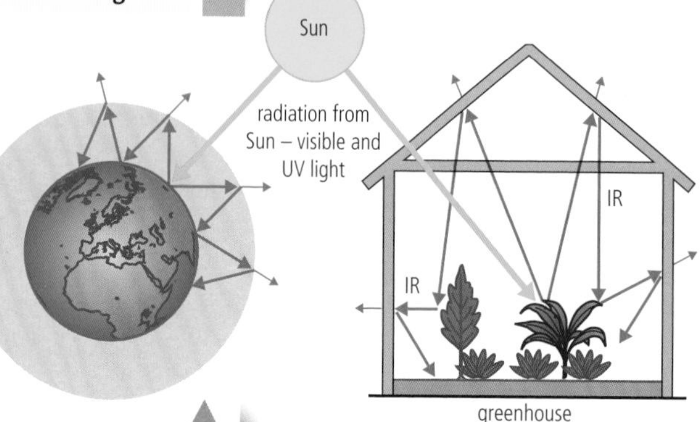

Figure 6.9 The greenhouse effect

You should distinguish between the greenhouse effect and the *enhanced* greenhouse effect. The greenhouse effect is natural and vital for life on Earth. The enhanced greenhouse effect refers to the additional greenhouse gases added by human activity that are leading to climate change (global warming) and having a serious impact on natural systems and societies.

You should be able to outline the role of the greenhouse effect in regulating temperature on Earth.

CHALLENGE YOURSELF

To what extent can weather and climate be considered a natural system if human activities are affecting it so much?

Exercises

1. Describe the composition of the atmosphere.
2. Outline the function of:
   a. nitrogen
   b. carbon dioxide.
3. Identify possible sources of dust in the atmosphere.
4. How and why does temperature vary with height in the atmosphere?
5. Study Figure 6.6, Earth's energy budget. Of incoming solar radiation:
   a. how much reaches the Earth's surface?
   b. how much is re-radiated as long-wave radiation?
   c. how much becomes latent heat transfer (evaporation and condensation)?
6. a. Compare the albedo of clouds with that of the Earth's surface.
   b. Suggest a reason to explain this.
7. Which clouds have the greatest albedo?
8. At what levels are cirrus clouds found?
9. What are the potential impacts of contrails (vapour trails from aircraft) on albedo? How may this affect global climate change?

10. Briefly outline the greenhouse effect.
11. Explain how human activity may affect the greenhouse effect.
12. Identify the main greenhouse gases in 2010 (Figure 6.7).

Big questions

Having read this section, you can now discuss the following big questions:

- To what extent have the solutions emerging from this topic been directed at *preventing* environmental impacts, *limiting* the extent of the environmental impacts, or *restoring* systems in which environmental impacts have already occurred?
- How are the issues addressed in this topic of relevance to sustainability or sustainable development?
- In what ways might the solutions explored in this topic alter your predictions for the state of human societies and the biosphere some decades from now?

Points you may want to consider in your discussions:

- Can there ever be a stable climate?
- To what extent are human influences on climate greater than natural influences?
- How and why is the greenhouse effect essential for human existence?

6.2 Stratospheric ozone

Significant ideas

Stratospheric ozone is a key component of the atmospheric system because it protects living systems from the negative effects of UV radiation from the Sun.

Human activities have disturbed the dynamic equilibrium of stratospheric ozone formation.

Pollution management strategies are being employed to conserve stratospheric ozone.

Big questions

As you read this section, consider the following big questions:

- To what extent have the solutions emerging from this topic been directed at *preventing* environmental impacts, *limiting* the extent of the environmental impacts, or *restoring* systems in which environmental impacts have already occurred?
- How are the issues addressed in this topic of relevance to sustainability or sustainable development?
- In what ways might the solutions explored in this topic alter your predictions for the state of human societies and the biosphere some decades from now?

Knowledge and understanding

- Some UV radiation from the Sun is absorbed by stratospheric ozone causing the ozone molecule to break apart. Under normal conditions, the ozone molecule will reform. This ozone destruction and reformation is an example of a dynamic equilibrium.

- Ozone-depleting substances (including halogenated organic gases such as chlorofluorocarbons, CFCs) are used in aerosols, gas-blown plastics, pesticides, flame retardants, and refrigerants. Halogen atoms (such as chlorine) from these pollutants increase destruction of ozone in a repetitive cycle so allowing more UV radiation to reach the Earth.
- UV radiation reaching the surface of the Earth damages human living tissues, increasing the incidence of cataracts, mutation during cell division, skin cancer, and has other effects on health.
- The effects of increased UV radiation on biological productivity include damage to photosynthetic organisms, especially phytoplankton which form the basis of aquatic food webs.
- Pollution management may be achieved by reducing the manufacture and release of ozone-depleting substances. Methods for this reduction include:
  - recycling refrigerants
  - developing alternatives to gas-blown plastics, halogenated pesticides, propellants, and aerosols
  - developing non-propellant alternatives.

  The United Nations Environment Programme (UNEP) has had a key role in providing information, and creating and evaluating international agreements, for the protection of stratospheric ozone.
- An illegal market for ozone-depleting substances continues and requires consistent monitoring.

  The *Montreal Protocol on Substances that Deplete the Ozone Layer* (1987) and subsequent updates is an international agreement for the reduction of use of ozone-depleting substances signed under the direction of UNEP. National governments complying with the agreement made national laws and regulations to decrease the consumption and production of halogenated organic gases such as chlorofluorocarbons (CFCs).

Some UV radiation from the Sun is absorbed by stratospheric ozone causing ozone molecules to break apart. Under normal conditions, the ozone molecules reform. This ozone destruction and reformation is an example of a dynamic equilibrium.

UV radiation and ozone

Ozone is essential for sustaining life. The highest concentration of ozone occurs in the upper part of the atmosphere, the stratosphere, where it is formed through the action of UV radiation on oxygen (Figure 6.10). The ozone layer shields the Earth from harmful radiation that would otherwise destroy most life on the planet (Figure 6.11).

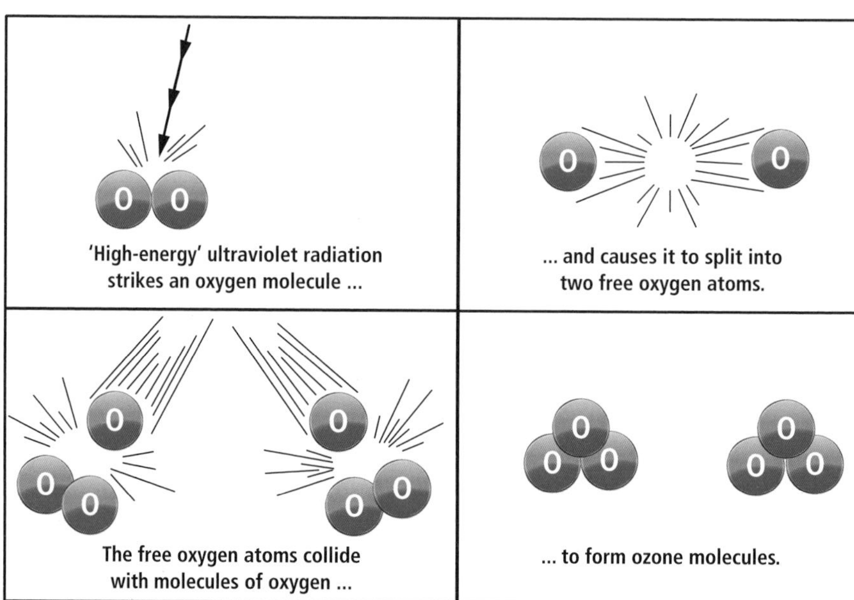

'High-energy' ultraviolet radiation strikes an oxygen molecule ...
... and causes it to split into two free oxygen atoms.
The free oxygen atoms collide with molecules of oxygen ...
... to form ozone molecules.

Figure 6.10 The formation of ozone in the stratosphere

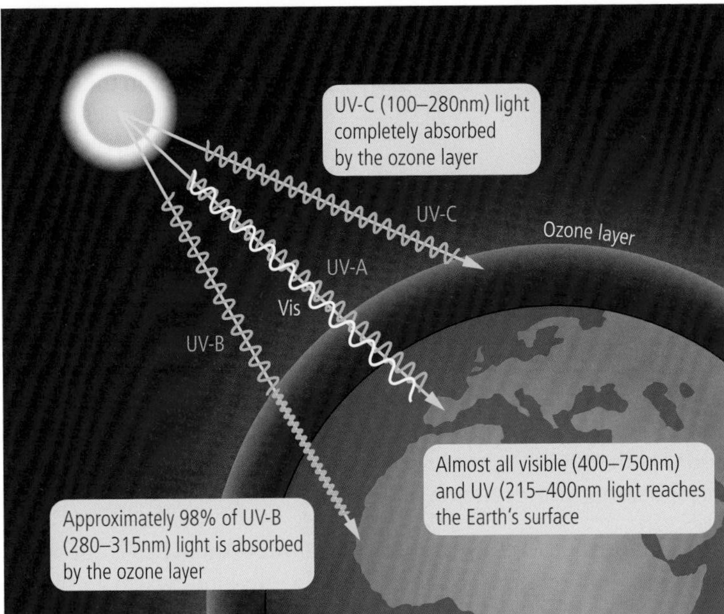

Figure 6.11 The role of ozone

The amount of ozone in the atmosphere is a small but it is a vital component. Ozone occurs because oxygen rising up from the top of the troposphere reacts under the influence of sunlight to form ozone. Most ozone is created over the equator and between the tropics because this is where solar radiation is strongest. However, winds within the stratosphere transport the ozone towards the polar regions where it tends to concentrate.

To learn more about stratospheric and ground-level ozone, go to www.pearsonhotlinks.co.uk, enter the book title or ISBN, and click on weblink 6.2.

Ozone has the vital role of absorbing UV radiation (wavelength 0.1–0.4 μm). It also absorbs some outgoing terrestrial radiation (wavelength 10–12 μm) – so it is a greenhouse gas. Ozone is constantly produced by sunlight and destroyed by nitrogen oxides in the stratosphere in a natural dynamic balance. The short-wave UV radiation breaks down oxygen molecules into two single oxygen atoms. The free oxygen atoms (O) combine with oxygen molecules (O_2) to form ozone (O_3). However, other mechanisms are at work to destroy the ozone. These include photochemical interactions with molecular oxygen, the oxides of nitrogen, chlorine, and bromine.

To take the ozone tour, go to www.pearsonhotlinks.co.uk, enter the book title or ISBN, and click on weblink 6.3.

Ozone-depleting substances

Under natural conditions, ozone exists in a steady-state or dynamic equilibrium, constantly forming and being destroyed. However, human activities may alter the balance of the equilibrium. There is now clear evidence that such activities have led to the creation of a 'hole' in the ozone layer over Antarctica. Levels of ozone have been falling since 1965. Scientists have identified the destructive role of oxides of nitrogen and chlorine, CFCs, and halogens. A halogen is any of a group of five non-metallic elements with a similar bonding: fluorine, chlorine, bromine, iodine, and astatine. They react with metals to produce a salt. Halogenated means a halogen atom has been added.

Ozone-depleting substances, including halogenated organic gases (e.g. CFCs), are used in aerosols, gas-blown plastics, pesticides, flame retardants, and refrigerants. Halogen atoms (e.g. chlorine) from these pollutants increase destruction of ozone in a repetitive cycle so allowing more ultraviolet radiation to reach the Earth.

In 1985, the British Antarctic Survey reported dramatic decreases in springtime atmospheric ozone compared with the previous 30 years – they had discovered the ozone hole. Nevertheless, the amount of ozone in the stratosphere fluctuates seasonally and over decades. Evidence from 2012 suggested the long-term decline in ozone was slowing down (Figure 6.12).

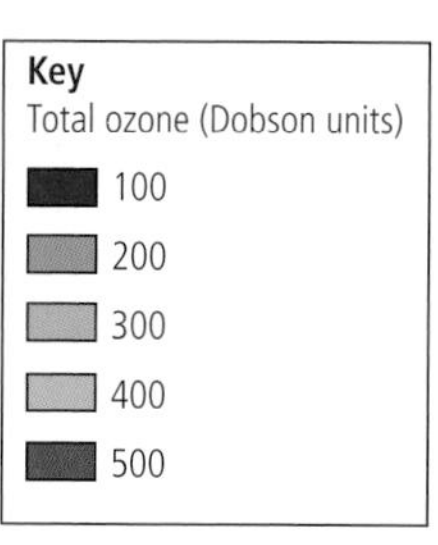

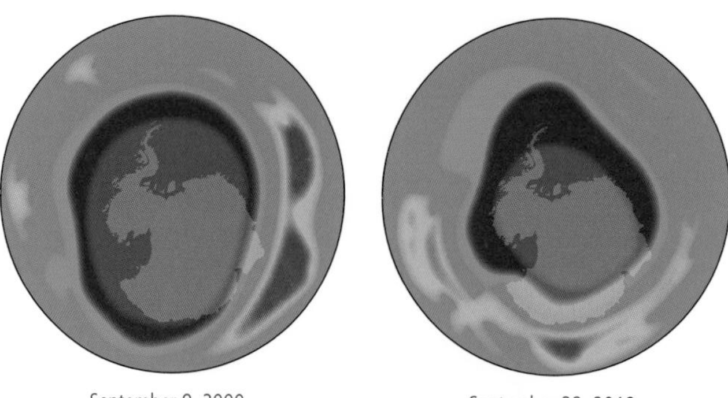

Figure 6.12 The ozone hole over Antarctica, 2000 and 2012

The ozone hole is an area of reduced concentration of ozone in the stratosphere, which varies from place to place and over the course of a year. There is a very clear seasonal pattern – each springtime in Antarctica (between September and October) there is a huge reduction in the amount of ozone in the stratosphere (Figure 6.13). The ozone layer above the whole of Antarctica now thins to between 40 per cent and 55 per cent of its pre-1980 levels, with up to a 70 per cent deficiency for short periods. As the summer progresses, the concentration of ozone recovers. So what causes the springtime depletion in ozone?

During winter in the southern hemisphere, the air over Antarctica is cut off from the rest of atmosphere by circumpolar winds – these winds block warm air from entering Antarctica. Therefore, the temperature over Antarctica becomes very cold, often down as far as −90 °C in the stratosphere. This allows the formation of clouds of ice particles. The ice particles offer surfaces on which chemical reactions can take place, involving chlorine compounds present in the stratosphere as a result of human activities. The reactions release chlorine atoms. In the presence of sunlight in the spring, the chlorine atoms destroy the ozone in a series of chemical reactions. Hence the hole in the ozone layer enlarges very rapidly in the spring. By summer, the ice clouds have evaporated and the chlorine is converted to other compounds such as chlorine nitrate. Thus the ozone hole diminishes, although it returns the following spring.

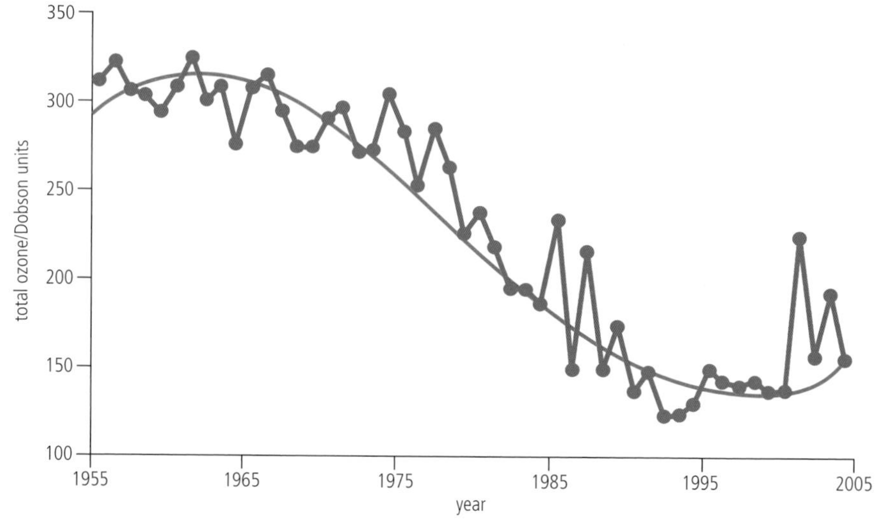

Figure 6.13 October ozone levels over Antarctica 1955–2005. The red line shows the trend in ozone levels whereas the blue line plots the actual concentrations measured.

Increasing measurements of CFCs correlate with declining ozone levels. Although there are important natural sources of chlorine (e.g. volcanoes and forest fires) which affect the ozone layer, the increases recorded are probably too large to be entirely natural. There are now ozone holes at both poles, the one over Antarctica stretching as far as Argentina. The balance of formation and destruction of stratospheric ozone is seriously affected by ozone-depleting substances (ODSs), notably those containing the halogens chlorine, fluorine, and bromine. Most ODSs do not occur naturally; they are industrial products or by-products and include CFCs, hydrochlorofluorocarbons

To learn more about the development of the ozone hole, go to www.pearsonhotlinks.co.uk, enter the book title or ISBN, and click on weblink 6.4.

(HCFCs), halons, and methyl bromide (bromomethane). These are usually very stable compounds – they persist in the atmosphere for decades as they travel upwards to the stratosphere. But once in the stratosphere and under the influence of UV radiation, they break down and release halogens. The halogen atoms act as catalysts for the reactions that destroy ozone molecules. Thus, ODSs greatly accelerate the rate of ozone destruction.

The size of the ozone hole is variable but always impressive. As early as 1987, it covered an area the size of continental USA and was as deep as Mount Everest. The maximum size of the ozone hole was 29.9 million km^{-2} in 2000. By 2014 it was down to 24.1 million km^{-2}. This is likely to be related to the success of the Montreal Protocol (pages 328–330).

Effects of ultraviolet radiation on human health

Even though the southern hemisphere ozone hole is no longer increasing, its recurrence each year exposes humans to the most harmful wavelengths of ultraviolet radiation (UV-B). In humans, exposure to UV-B is associated with eye damage (cataracts), sunburn, and skin cancers.

Ultraviolet radiation reaching the surface of the Earth damages human living tissues, increasing the incidence of cataracts, mutation during cell division, skin cancer, and other effects on health.

The effects of UV-B radiation on the eye may be acute (occurring after a short, intense exposure) or chronic (occurring after long-term exposure). The commonest acute effect is temporary blindness. In contrast, the effects of chronic exposure are irreversible leading to the development of cataracts and eventually blindness.

Acute exposure of the skin to UV-B radiation causes mutations during cell division, sunburn and, in the long term, skin cancers. Cancers occur when mutated or damaged cells in the body begin to divide and invade other areas, forcing out the healthy cells and tissues. The amount of UV-B radiation required to produce sunburn depends on latitude, time of day, and skin colour. Chronic exposure of the skin to UV-B radiation causes wrinkling, thinning, and loss of elasticity. Skin cancers occur most often and with high frequency in fair-skinned individuals living in sunny climates.

Monitoring of ozone concentrations suggests an annual loss of 1 per cent. As a result, there is a rise in the rate of skin cancers, estimated at 4 per cent.

The immune system can be damaged by UV-B radiation, leading to decreased immune responses to infectious agents and skin cancers. It is thought that UV-B radiation at a critical time during infection can increase the severity and duration of the disease. It is also believed that UV-B exposure during immunization can reduce the effectiveness of vaccinations. However, the full implications of the immunological effects are not well understood.

The effects of increased ultraviolet radiation on biological productivity include damage to photosynthetic organisms, especially phytoplankton, which forms the basis of aquatic food webs.

Stratospheric ozone depletion over mid-latitudes means that UV-B levels in southern Australia are likely to have risen by 10–15 per cent over the past 20 years. There is also a clear correlation between declining ozone levels and rising UV radiation on clear-sky days in the South Island of New Zealand. Melanoma incidence has risen 15 per cent in men and 12 per cent in women over the past decade. Melanoma incidence varies greatly from country to country – Australia has the highest rate with 39 cases per 100 000 people per year, followed by New Zealand with 34 cases per 100 000 people per year and the USA with 17 cases per 100 000 people per year. In contrast, north-west European countries have rates of between 4 and 10 cases per 100 000 people per year. There are also important variations within Australia – Queensland has a rate of up to 65 cases per 100 000 people per year, whereas New South Wales has 47 cases per 100 000 people per year.

In aquatic ecosystems, phytoplankton live in the upper layer of the water where there is sufficient light to support photosynthetic life. It is here that exposure to UV-B can occur. UV-B radiation may damage those organisms that live at the surface of the water during their early life stages. The effects of UV-B radiation are particularly significant on phytoplankton, fish eggs and larvae, zooplankton, and other primary and secondary consumers (Figure 6.14). Most adult fish are protected from UV-B radiation since they inhabit deep waters. Some shallow-water fish have been found to develop skin cancer.

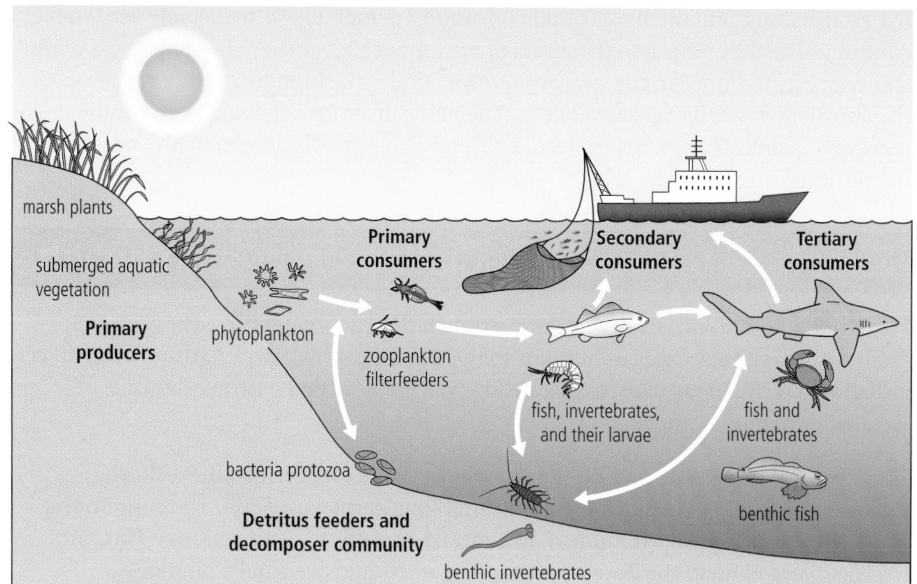

Figure 6.14 Because primary producers are sensitive to increased levels of UV-B, the depletion of ozone potentially threatens the whole marine food chain.

Animals suffer similar effects to humans from high UV-B levels. Aquatic fauna (e.g. frogs) and aquatic flora (e.g. phytoplankton) are particularly vulnerable to UV-B radiation. Recent studies of the effects of UV-B on phytoplankton have confirmed adverse effects on growth, photosynthesis, protein and pigment content, and reproduction.

Some crops and wild plants also suffer detrimental effects from increased UV-B radiation. Some crop varieties are UV-B-sensitive and produce reduced yield following an increase in UV-B. On the other hand, commercial forests, tree breeding, and genetic engineering may be used to improve UV-B-tolerant plants. The spread of the ozone hole over South America is believed to have had an impact on plant productivity there – herbaceous plants native to the southern tip of South America and the Antarctic Peninsula have been affected by the current high levels of UV-B radiation.

Reducing ozone-depleting substances (ODSs)

Pollution management may be achieved by reducing the manufacture and release of ODSs.

Sources of ODSs include refrigerants, gas-blown plastics, and the use of CFCs as propellants in aerosol cans. Methods for reducing the manufacture and release of ODSs include:

- recycling refrigerants
- developing alternatives to gas-blown plastics, halogenated pesticides, propellants, and aerosols
- developing non-propellant alternatives.

The United Nations Environment Programme (UNEP) has had a key role in providing information, and creating and evaluating international agreements for the protection of stratospheric ozone.

In the past, huge quantities of CFCs were used as propellants in aerosol sprays but most of this demand is now met by hydrocarbons and other technologies (e.g. pump action sprays). After aerosols, refrigeration was the next most important use for CFCs; today it seems likely that a combination of HFCs, ammonia, carbon dioxide, and hydrocarbon refrigerants will replace the CFCs. Nitrogen oxides also deplete the ozone layer and can be reduced through reduced use of fossil fuels, less-polluting combustion engines, and cleaner plane engines.

Besides replacement compounds, a range of alternative procedures have been implemented (e.g. trigger sprays have replaced aerosol propellants, leaking CFCs have been collected, some CFC waste has been incinerated, and old refrigerators have been locally or centrally collected). The recovery of ODSs from products still in use (e.g. old refrigerators and air conditioning units) is an important part of the response to stratospheric ozone depletion. Since 1993, Australia has collected more than 3000 tonnes of ODSs which have either been recycled, stored or destroyed. Nonetheless, significant amounts remain to be collected.

The phase-out of methyl bromide

Methyl bromide (MeBr) is an odourless, colourless gas that has been used as a soil pesticide to control pests across a wide range of agricultural sectors. Because MeBr depletes the stratospheric ozone layer, its production and import in the USA and Europe was phased out in 2005. Allowable exemptions to the **phase-out** include:

- the Quarantine and Preshipment (QPS) Exemption, to eliminate quarantine pests
- the Critical Use Exemption (CUE), designed for agricultural users with no technically or economically feasible alternatives.

Among the non-chemical alternatives, soilless cultivation, crop rotation, resistant varieties, and grafting are effective means of pest control.

The role of UNEP

In 1987, 24 countries were brought together by UNEP to sign the initial Montreal Protocol on Substances that Deplete the Ozone Layer. Now, 197 countries have signed the Protocol. In 1987, production of ODSs exceeded 1.8 million tonnes annually. By 2010, it had fallen to 45 000 tonnes. Nevertheless, the work of the Montreal Protocol is not finished. UNEP is working towards finally ending production of HCFCs by 2040.

The Montreal Protocol is considered to be a success as the production and consumption of ODSs has reduced by more than 95 per cent compared to 1986. However, illegal trade in ODSs has developed. It is believed that India and the Republic of Korea account for approximately 70 per cent of the total global production of CFCs. Countries in the region with a high consumption of CFCs include China, India, Malaysia, Pakistan, and the Philippines (Figure 6.15).

In many low-income countries, there is still a significant demand for CFCs as reliance on equipment using these chemicals remains high. The problem is made worse by the imports of used refrigeration and air-conditioning equipment.

Reasons for the illegal trade in ODSs are many:

- ODS substitutes are often costlier than CFCs
- updating equipment to enable use of alternative chemicals is generally expensive
- the lifetime of CFC-containing equipment is often long
- penalties in many countries for ODS smuggling are small.

The magnitude of the illegal trade in ODSs is between 7000 and 14 000 tonnes of CFCs annually. There are many inconsistencies in trade data between China, Indonesia, the Philippines, Malaysia, Vietnam, India, and Singapore. Meetings organized by UNEP Regional Office for Asia and the Pacific (ROAP) have identified the reasons for such trade data discrepancies.

In the USA, the Clean Air Act of 1990 aims to maximize recovery and recycling of ODSs (CFCs and HCFCs and their blends) during the servicing and disposal of air-conditioning and refrigeration equipment. US operations release some 111 000 tonnes of ozone-depleting refrigerants annually. In 1992, German scientists successfully manufactured hydrocarbons for use as alternatives to CFCs in fridges. However, the development of fridges that used hydrocarbons was initially slow. The environmental campaigner, Greenpeace, commissioned *greenfreeze* fridges that used hydrocarbons from a factory in east Germany. Now most of the east German fridge market uses *greenfreeze* fridges and the first UK *greenfreeze* factory opened in 1996.

To learn more about the Ozone Secretariat, go to www.pearsonhotlinks.co.uk, enter the book title or ISBN, and click on weblink 6.5.

An illegal market for ODSs continues and requires consistent monitoring.

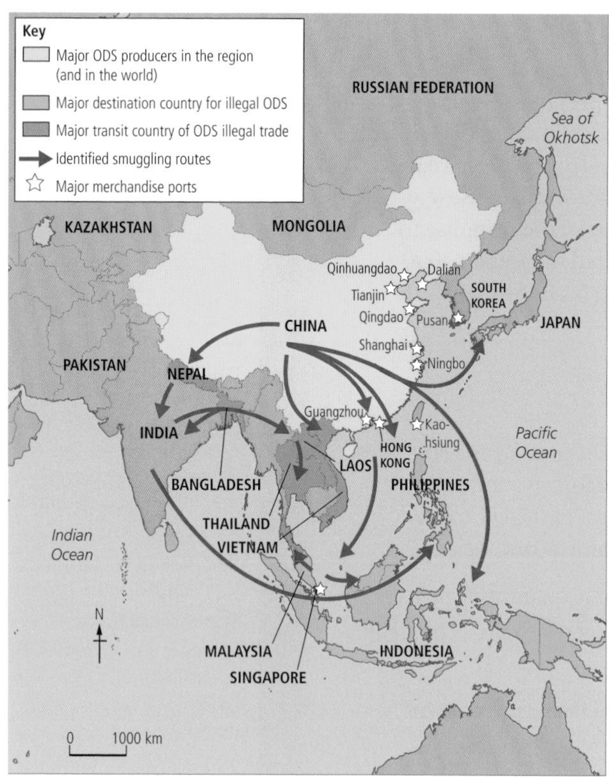

Figure 6.15 Illegal trade in ODSs

A number of activities have been encouraged at national, regional, and global scale aimed at improving the process of monitoring and controlling ODSs in order to combat the illegal trade in these chemicals. In 2001, the UNEP Division of Technology, Industry and Economy (DTIE) launched the Green Customs Initiative to encourage coordinated intelligence gathering, information exchange, guidance, and training among the partner organizations involved to counter illegal trade and environmental crime.

Since 2006, the South Asia–South East Asia and Pacific network countries have agreed on a mechanism of informal prior informed consent (iPIC) on export and import of CFCs to assist member countries to implement licensing systems effectively. The European Commission fully participates in the iPIC. Project Sky Hole Patching, a joint operation of customs administrations and international organizations in the Asia Pacific region, was launched in 2006. It monitors suspicious shipments of ODSs, which are imported, re-exported, or trans-shipped across international borders.

Information from the Philippines National Ozone Unit shows that in the Philippines, when the market price for CFC-12 was around US$6 per kilogram, the price for hydrofluorocarbon-134a (HFC-134a), an alternative for CFC-12, was around US$9 per kilogram. This price difference is the main catalyst behind many smuggling operations.

UNEP surveys in Asia found that retrofitting a mobile air-conditioning system to enable it to use HFC-134a in developing Asian countries could cost between US$100 and US$200. However, the cost of acquiring a 13.6 kg cylinder of CFCs, which contains enough refrigerant to service many such systems, is only about US$50.

National and international organizations and the reduction of ODSs

International cooperation between governments on the reduction of ODSs has been successful. Much of this has been organized by UNEP. In 1985, UNEP implemented the Vienna Convention for the Protection of the Ozone Layer. This aimed to protect human health and the environment against adverse effects resulting from human activities which modified or were likely to modify the ozone layer (Figure 6.16).

As a result of the 1987 Montreal Protocol on Substances that Deplete the Ozone Layer the hole in the ozone layer has been decreasing. By the end of 2002, industrialized countries had reduced their ODS consumption by more than 99 per cent and developing countries had reduced their consumption by slightly more than 50 per cent. A total phasing out worldwide is due by 2040. Total phase out in Europe occurred by 2000. Nevertheless, CFCs are persistent and long-lasting so their impact will continue for many decades. Production and consumption of CFCs, halons, and other ODSs have been almost completely phased out in industrialized countries and the timetable for banning the use of methyl bromide, a pesticide, has been agreed.

The *Montreal Protocol on Substances that Deplete the Ozone Layer* (1987) and subsequent updates is an international agreement for the reduction of use of ODSs signed under the direction of UNEP. National governments complying with the agreement made national laws and regulations to decrease the consumption and production of halogenated organic gases such as CFCs.

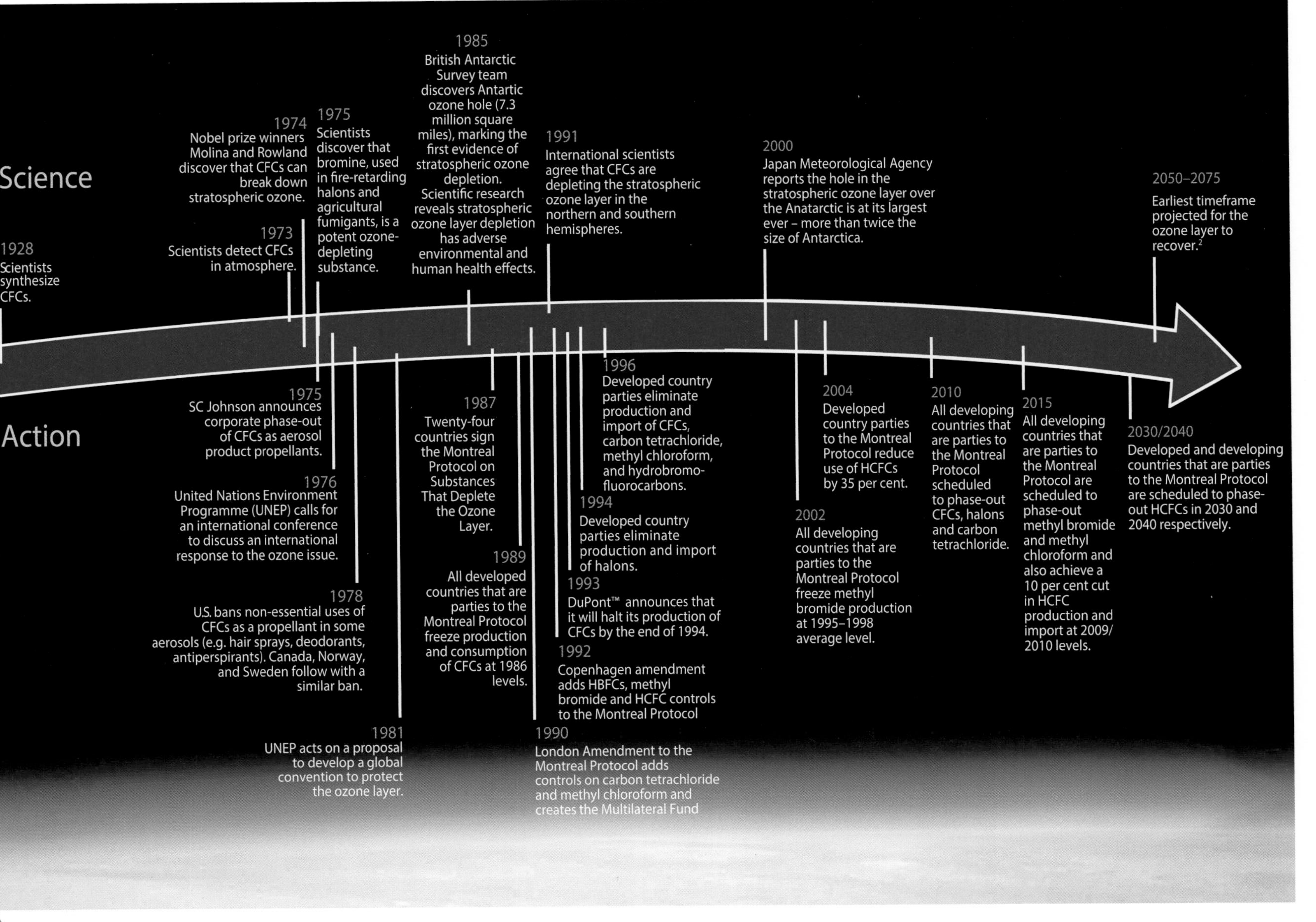

Figure 6.16 A time-line of the history and action related to stratospheric ozone

Developing countries have been given a longer time period over which to phase out their release of ODSs. Many developing countries have complained that there are many valuable CFCs which would help them develop. Although there has been a fund to help industry in developing countries switch from CFCs to ozone-friendly technologies, the funds are somewhat limited hence reduction of CFC-use in some developing countries has been limited.

The most widely used ODSs (CFC-11 and CFC-12) have often been replaced with HCFCs. Although these are greenhouse gases, their global warming potential (GWP) is less than that of CFCs. However, even greater GWP reductions can be achieved by replacing CFCs with substances such as hydrocarbons, carbon dioxide, water, and air. These substances contribute only minimally (or not at all) to GWP (Table 6.3).

Table 6.3 CFCs, HCFCs, and their replacements

Compound	Use	Atmospheric lifetime / years	ODP relative to CFC-11	GWP relative to CO_2 at 100 years	Current main substitute
Compounds already phased out in developed countries					
CFC-11	foam expander	50	1	4000	HCFC-141
CFC-12	refrigerant	102	1	8500	HCFC-134a
halon-1211	fire extinguishant	20	3.0	no data	dry powder
Compounds due to be phased out by the Montreal Protocol					
HCFC-	refrigerant	5.9	0.02	480	HFC-134a
Potential replacements, emissions controlled under Kyoto Protocol					
HFC-32	refrigerant	5.6	0	650	
HFC-125	refrigerant	32.6	0	2800	

Case study

Australia

Australia's obligations under the Montreal Protocol have been implemented at the national level. State legislation for ozone protection has been replaced by national legislation. For example, the New South Wales Ozone Protection Regulation 1997 was repealed in 2006.

The manufacture, import, and export of all major ODSs has been completely phased out in Australia, with the exception of HCFCs, which will be phased out in 2015. In addition, methyl bromide is used for quarantine and feedstock purposes. National regulations allow for limited categories of essential or critical use for halons, CFCs, and methyl bromide where no alternatives are available.

Transition substances, such as hydrofluorocarbons (HFCs), are strong greenhouse gases and contribute to climate change. Their concentrations are rising rapidly in the atmosphere, albeit from currently low concentrations. Australia is the first country to implement integrated control measures to manage both ODSs and their synthetic replacements that can also act as greenhouse gases. This is largely because they were one of the first countries to suffer the consequences of ozone depletion.

Future directions

Australia's response to phasing out the use of ODSs under the Montreal Protocol has been swift and effective. However, Australia accounts for less than 1 per cent of these global emissions. To assist this process, Australia participates in the Multilateral Fund for the Implementation of the Montreal Protocol (London, 1990) (which provides funds to help developing countries phase out their usage of ODSs).

The Montreal Protocol now covers all CFCs, carbon tetrachloride, and most halons. Consumption of these compounds is banned in developed countries. There is, however, some use of CFCs and halons in developing countries. A small volume of these chemicals is therefore manufactured for essential uses and developing countries. Because they are highly stable, ODSs will persist in the atmosphere for decades to come despite action to reduce their usage. There is, therefore, a lag between action addressing ODSs and recovery of the ozone layer. Based on projections of future levels of ODSs in the stratosphere, the recovery of the ozone layer over much of Australia is likely by 2049, and over Antarctica by 2065.

New chemicals depleting ozone

In 2014, it was reported that scientists had identified and measured four previously unknown compounds in the atmosphere – three CFCs and one HCFC – and warned of the existence of many more. The level of one of the compounds had doubled in just 2 years. Over 74 000 tonnes of the new gases have been released in the past 40 years. CFC-113a and the HCFC are accumulating rapidly. Until 2014, a total of 13 CFCs and HCFCs were known to destroy ozone but were controlled by the Montreal Protocol.

You should be able to evaluate the role of national and international organizations in reducing the emissions of ozone-depleting substances.

Despite the production of all CFCs having been banned since 2010, the concentration of one – CFC113a – is rising at an accelerating rate. The source of the chemicals is a mystery and it may be being used in the production of agricultural crop and soil pesticides.

Exercises

1. Describe the role of ozone in the absorption of ultraviolet radiation.
2. Explain the interaction between ozone and halogenated organic gases.
3. State the effects of ultraviolet radiation on living tissues and biological productivity.

Questions 4–6 refer to Table 6.3.

4. Identify the compound which has the highest ozone-depleting potential (ODP). State the level of its ODP relative to CFC-11.
5. Identify the two compounds that have the longest atmospheric lifetime in years. Give their lifetime in years.
6. Which compounds have the largest greenhouse warming potential (GWP)? State their GWP, at 100 years, relative to carbon dioxide.
7. Describe three methods of reducing the manufacture and release of ODSs.
8. Describe and evaluate the role of national and international organizations in reducing the emissions of ODSs.
9. What was the international treaty that led to a decline in the production of CFCs?
10. Outline the difficulties in implementing and enforcing international agreements.
11. Evaluate the effectiveness of international policies to reduce ODSs.

Big questions

Having read this section, you can now discuss the following big questions:

- To what extent have the solutions emerging from this topic been directed at *preventing* environmental impacts, *limiting* the extent of the environmental impacts, or *restoring* systems in which environmental impacts have already occurred?
- How are the issues addressed in this topic of relevance to sustainability or sustainable development?
- In what ways might the solutions explored in this topic alter your predictions for the state of human societies and the biosphere some decades from now?

Points you may want to consider in your discussions:

- Why has the Montreal Protocol been so successful? Will it continue to be successful?
- Outline the links between stratospheric ozone and sustainability/sustainable development.
- What are the greatest threats to stratospheric ozone likely to be, and from where, in the decades to come?

6.3 Photochemical smog

Significant ideas

The combustion of fossil fuels produces primary pollutants that may generate secondary pollutants and lead to photochemical smog, the levels of which can vary by topography, population density, and climate.

Photochemical smog has significant impacts on societies and living systems.

Photochemical smog can be reduced by decreasing human reliance on fossil fuels.

Big questions

As you read this section, consider the following big questions:

- To what extent have the solutions emerging from this topic been directed at *preventing* environmental impacts, *limiting* the extent of the environmental impacts, or *restoring* systems in which environmental impacts have already occurred?
- How are the issues addressed in this topic of relevance to sustainability or sustainable development?
- In what ways might the solutions explored in this topic alter your predictions for the state of human societies and the biosphere some decades from now?

Knowledge and understanding

- Primary pollutants from the combustion of fossil fuels include carbon monoxide, carbon dioxide, black carbon or soot, unburned hydrocarbons, oxides of nitrogen, and oxides of sulfur.
- In the presence of sunlight, secondary pollutants are formed when primary pollutants undergo a variety of reactions with other chemicals already present in the atmosphere.
- Tropospheric ozone is an example of a secondary pollutant, formed when oxygen molecules react with oxygen atoms that are released from nitrogen dioxide in the presence of sunlight.
- Tropospheric ozone is highly reactive and damages plants (crops and forests), irritates eyes, creates respiratory illnesses, and damages fabrics and rubber materials. Smog is a complex mixture of primary and secondary pollutants, of which tropospheric ozone is the main pollutant.
- The frequency and severity of smog in an area depends on local topography, climate, population density, and fossil fuel use.
- Thermal inversions occur due to lack of air movement when a layer of dense, cool air is trapped beneath a layer of less dense, warm air. This causes concentrations of air pollutants to build up near the ground instead of being dissipated by 'normal' air movements.
- Deforestation and burning may also contribute to smog.
- Pollution management strategies include:
  - altering human activity to consume less fossil fuel – example activities include the purchase of energy-efficient technologies, the use of public or shared transit, and walking or cycling
  - regulating and reducing pollutants at point of emission by government regulation or taxation

- using catalytic converters to clean primary pollutants from car exhaust
- regulation of fuel quality by governments
- adopting clean-up measures such as reforestation, re-greening, and conservation of areas to sequester carbon dioxide.

Source and impact of tropospheric ozone

In the troposphere (lower atmosphere), ozone is considered a pollutant. It is formed here as the result of pollution by volatile organic compounds (VOCs), carbon monoxide, carbon dioxide, black carbon or soot, unburned hydrocarbon, oxides of nitrogen (NO_x), and oxides of sulfur. VOCs are organic chemical compounds able to evaporate into gases and take part in photochemical reactions. There are many of them, including methane, ethane, and alcohol.

Primary pollutants from the combustion of fossil fuels include carbon monoxide, carbon dioxide, black carbon or soot, unburned hydrocarbons, oxides of nitrogen, and oxides of sulfur.

Hydrocarbons (from unburned fuel) and nitrogen monoxide (NO) are given off when fossil fuels are burned. Nitrogen monoxide (aka nitric oxide) reacts with oxygen to form nitrogen dioxide (NO_2), a brown gas that contributes to urban haze and smogs.

Unlike other pollutants, tropospheric ozone is not directly emitted from human-made sources in large quantities. Ozone occurs naturally in the upper atmosphere, but the chemical reactions between volatile organic compounds (VOCs), nitrogen oxides, and sunlight can produce tropospheric ozone **photochemical smog**.

To learn more about trends in ground level ozone, go to www.pearsonhotlinks.co.uk, enter the book title or ISBN, and click on weblink 6.6.

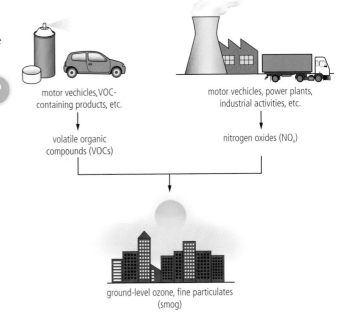

The main sources of VOCs and nitrogen oxides are road transport, solvent release from drying paints, glues, or inks, and petrol handling and distribution. Nitrogen is present in fuels and the air. At the high temperature of the internal combustion engine, nitrogen is oxidized to nitric oxide and some nitrogen dioxide. Once the exhaust gases leave the engine, nitric oxide is oxidized to nitrogen dioxide.

Sunlight and primary pollutants

Nitrogen dioxide can absorb sunlight and break up to release oxygen atoms that combine with oxygen in the air to form ozone (Figure 6.17).

In the presence of sunlight, secondary pollutants are formed when primary pollutants undergo a variety of reactions with other chemicals already present in the atmosphere. Tropospheric ozone is an example of a secondary pollutant.

The photochemical reactions between the nitrogen oxides and VOCs in sunlight may take hours or days to produce ozone. Because the reactions are photochemical, ozone concentrations are greatest during the day, especially during warm, sunny, stable conditions. Above 20 °C, reactions are accelerated.

Ozone formation may take a number of hours, by which time the polluted air has drifted into surrounding suburban and rural areas. Hence, ozone pollution may be greater outside the city centre.

Figure 6.17 The causes of tropospheric ozone

Smog is a mix of primary and secondary pollutants, of which the main pollutant is tropospheric ozone.

Tropospheric ozone is highly reactive and damaging to plants and animals.

The effects of tropospheric ozone

Tropospheric ozone damages plants (crops and forests), irritates eyes, creates respiratory illnesses, and damages fabrics and rubber materials. It is also the main pollutant in smog.

Effects on forests and crops

Tropospheric ozone harms sensitive vegetation, such as black cherry and ponderosa pine during the growing season. When sufficient ozone enters the leaves of a plant, it:

- interferes with the ability of sensitive plants to produce and store food
- damages the leaves of trees and other plants, harming the appearance of vegetation in urban areas, national parks, and recreation areas.

Continued ozone exposure over time can lead to increased susceptibility of sensitive plant species to disease, damage from insects, effects of other pollutants, competition, and harm from severe weather.

Tropospheric ozone pollution has also been suggested as a cause of the dieback of German forests (which had previously been linked with acid rain).

To learn more about the health effects of ground-level ozone, go to www.pearsonhotlinks.co.uk, enter the book title or ISBN, and click on weblink 6.7.

Effects on humans

Ozone can harm lung tissues, impair the body's defence mechanism, increase respiratory tract infections, and aggravate asthma, bronchitis, and pneumonia. Even at relatively low levels, coughing, choking, and sickness increase. The long-term effects include premature ageing of the lungs. Children born and raised in areas where there are high levels of ozone can experience up to a 15 per cent reduction in their lung capacity.

Effects on materials

High levels of ozone can damage fabrics and rubber materials.

Tropospheric ozone is a cause of poor air quality in the UK. The relatively high temperature and sunshine level in southern England, combined with additional VOCs and nitrogen oxides blown over from Europe, mean that ozone concentrations are higher there than elsewhere in the UK. Ground-level ozone affects plant photosynthesis and growth and so may significantly reduce crop yields. Crop exposure to ozone levels above 40 parts per billion (ppb) tends to be more common in southern England and so affects extensive areas of arable farming. An ozone level of 40 ppb is the threshold used by the United Nations Economic Commission for Europe to measure crop damage.

The frequency and severity of smog in an area depends on local topography, climate, population density, and fossil fuel use.

Ozone and smog

High levels of ozone are more common in low-lying areas and depressions, where polluted air tends to collect. Places such as Mexico City and Athens, which are surrounded by high ground, tend to have poor air quality because cold air sinks from the higher ground, preventing the dispersion of pollutants.

Photochemical smog is associated with certain climates – in particular, high air pressure systems. This is because winds in a high pressure system are usually weak. Hence pollutants remain in the area and are not dispersed. Poor air quality often persists for many days because stable high pressure conditions generally prevail for a few days. In some climates, notably Mediterranean ones, stable high pressure conditions persist all season, hence poor air quality can remain for months. In monsoonal areas, smog only occurs in the dry season.

Tropospheric ozone is also associated with large populations. In general, the larger the population, the greater the number of vehicles. As vehicles are one of the main sources of nitrogen oxides, it follows that in larger cities there is greater potential for the production of tropospheric ozone. In societies that continue to use fossil fuels, the chances of tropospheric ozone being produced remain high. In contrast, in cities where there is greater reliance on sustainable forms of transport (e.g. Curitiba in Brazil), the risk of tropospheric ozone is reduced. Places which have high pressure, a large population, continue to use fossil fuels, and surrounded by high ground (e.g. Mexico City, pages 338–339) are associated with very poor air quality.

Background levels of ground-level ozone have risen substantially over the last century. There is evidence that the pre-industrial near ground-level concentrations of ozone were typically 10–15 ppb. The current annual mean concentrations are approximately 30 ppb over the UK. The number of hours of high ozone concentrations tends to increase from north to the south across the country. Concentrations can rise substantially above background levels in summer heat waves when there are periods of bright sunlight with temperatures above 20 °C, and light winds. Once formed, ozone can persist for several days and can be transported long distances.

Thermal inversions

Smogs are associated with temperature inversions (Figure 6.18). These may be more common in winter because of high levels of sulfur dioxide and other pollutants resulting from increased heating of homes, offices and industries. Under cold conditions, vehicles operate less efficiently (until they have warmed up). This inefficient operation releases larger amounts of carbon monoxide and hydrocarbons. Urban areas surrounded by high ground are especially at risk from winter smog. This is because cold air sinks in from the surrounding hills, reinforcing the inversion.

Figure 6.18 Temperature inversion and winter smog

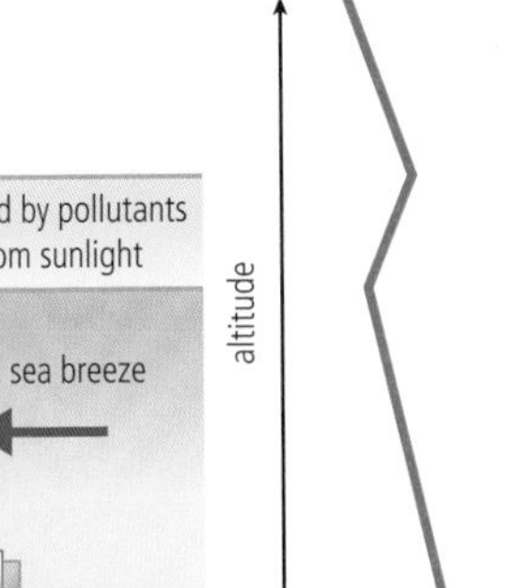

Deforestation and burning

Deforestation and burning, may also contribute to smog. People have used fire for a variety of reasons:

- to clear land for agriculture, attract game, and deprive game of cover
- to improve grazing by removing dead vegetation and encourage new grass and herbaceous species
- to drive away wild animals
- to manipulate plant and animal communities.

Thermal inversions occur due to lack of air movement when a layer of dense, cool air is trapped beneath a layer of less dense, warm air. This causes concentrations of air pollutants to build up near the ground instead of being dissipated by 'normal' air movements.

Fire is important in areas where there is a long dry season or where lightning is common. It can be from natural causes such as lightening or be deliberately set by humans. Use of fire by humans created the heathland and moorland ecosystems of the UK, for example, as well as the savannahs and temperate grasslands of the mid-latitudes. Much of Europe's heathlands were formerly forests, but burning and grazing have prevented regeneration.

Natural causes include lightning, fermentation of litter, sparks caused by boulders. In the USA >50 per cent of the fires are natural, whereas in Australia and the south of France, almost all of the fires are man-made. They also occur more frequently in semi-arid and savannah areas than temperate and cool temperate areas.

Case study

South East Asian fires, 1997

During 1997, forest fires raged across parts of Indonesia, Kalimantan, and Sumatra. The smog these created covered parts of Indonesia, Malaysia, Brunei, Singapore, and parts of the southern Philippines and southern Thailand (Figure 6.19). In some places, visibility was less than 1 m. Up to 70 million people in six countries were affected by smog created by the fires, which could burn underground for decades.

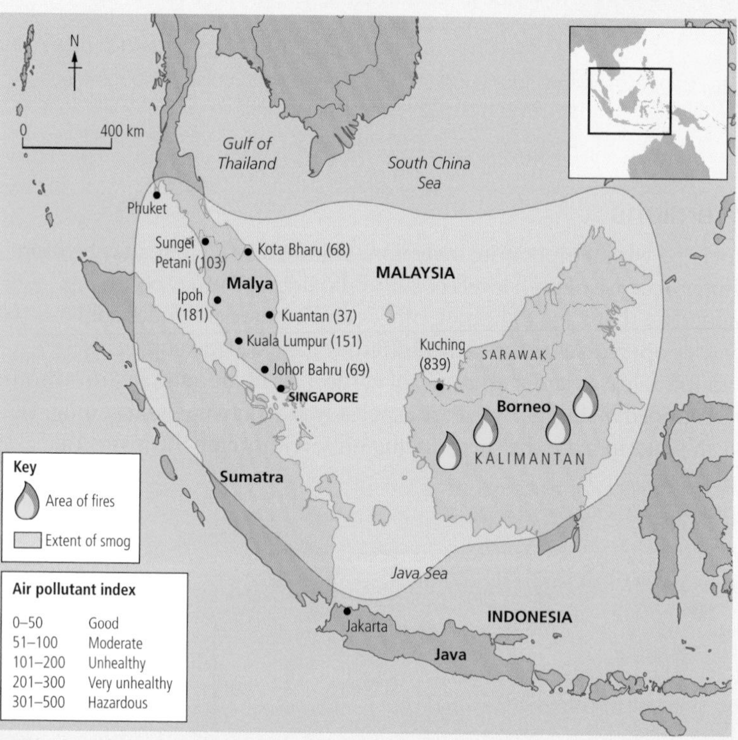

Figure 6.19 The Asian 'brown haze'

The fires had many effects. Over 275 people died from starvation in the Indonesian province of Irian Java. Others died from cholera caused by a lack of clean water. The most badly affected people in Sarawak were the indigenous communities. Living in remote forests, cut off from medical care and access to bottled water, they received very little help from the government or from aid organizations. Over 60 000 Malaysians and Indonesians, mostly children and the elderly, were treated for smog-related illnesses. Schools in Sarawak were forced to close down; hospitals struggled to cope with the increase in the number of throat infections, diarrhoea, conjunctivitis, and other eye problems.

The effect of the smog has been likened to smoking three packets of cigarettes each day. Some doctors claim that the effects are even worse – the sulfide and carbon gases obstruct the airways and destroy the lungs. The young and the elderly are especially vulnerable.

Many of the Indonesian fires, which affected up to 4.5 million hectares of rain forest and plantations were started deliberately by plantation owners as a cheap way of clearing the land. The Indonesian government blamed 176 plantation companies for causing the fires, but took limited action against any of them. Many of these companies have western consumers.

The fires had a negative effect on economic activities. Tourism dropped almost immediately. Several domestic flights were cancelled because of the thick smog. Ships without radar navigation were advised not to sail in the Strait of Malacca, which separates Malaysia from Indonesia. Crop yields were down and many food-exporting countries, such as Thailand and the Philippines, had to import coffee and rice.

Economic losses

Economic losses caused by urban air pollution can be significant.

Economic losses include losses due to tourism, the cost of clean-up strategies, decreased worker productivity, decreased crop productivity, the cost of healthcare, and the cost of replacing materials.

A case study in 2013 showed that there were approximately 200 000 early deaths every year in the USA due to air pollution. In California, between 2005 and 2007, if clean air standards had been applied, 30 000 admissions to hospital would have been avoided, saving an estimated US$200 million. Urban air quality is predicted to become the top environmental cause of early death by 2050. In Beijing, China, air pollution is believed to have caused US$3.7 billion in economic losses. The World Bank suggests that the cost of air pollution to China is worth 3.8 per cent of the country's GDP. As ozone levels decrease, worker productivity increases.

A new study in the USA has found that exposure to ground-level ozone is associated with an increased risk of death. Increases in the ozone level contribute to thousands of deaths every year. The risk of death is similar for adults of all ages but slightly higher for people with respiratory or cardiovascular problems. The increase in deaths occur at ozone levels below the clean air standards of the Environmental Protection Agency (EPA).

National air quality and mortality data from 95 large urban areas in the USA for years 1987–2000 were used to investigate whether daily and weekly exposure to ground-level ozone was associated with mortality. The researchers adjusted for particulate matter, weather, seasonality, and long-term trends. An increase of 10 ppb in the daily ozone levels for the previous week was associated with a 0.52 per cent increase in daily mortality. This corresponds to 3767 additional deaths annually in the 95 urban areas studied. According to the study, if ozone levels in the USA were reduced by about a third, about 4000 lives each year would be saved.

Smog in Beijing, China

Environmental philosophies (and policies) may change when certain administrations are under the spotlight. In order to clean up Beijing for the 2008 summer Olympics and during International Olympic Committee pre-visits, a series of policies was enforced (e.g. banning cars, closing factories, re-locating a steel works) so that air quality was above average for the duration of the Games when the world's gaze was firmly fixed on Beijing.

Pollution management strategies

There are a number of ways in which photochemical smog can be managed:

- reduction in the burning of fossil fuels
- greater use of energy-efficient technologies such as hybrid/electric cars
- increased use of public transport rather than use of private cars
- car pooling schemes (e.g. liftshare.com)
- increased use of bicycles or walking (e.g. Living Streets' Walk to School campaign)
- use of catalytic convertors to reduce emissions of NO_x
- greater enforcement of emissions standards (e.g. the Zero Emissions network in east London)
- clean-up measures such as re-forestation, re-greening, conservation areas (e.g. La Rambla, Barcelona)
- public information (e.g. by TV, radio, smart boards) regarding air quality.

La Rambla, Barcelona – a former slum area that is now a conservation area

In general, reducing the emissions of fossil fuels will have much greater success, and will be cheaper, than trying to clean up the pollution after it has happened.

In Mexico City, the average visibility has decreased from 100 km in the 1940s to about 1.5 km in the 2000s. Levels of nitrogen dioxide regularly exceed international standards by two to three times, and levels of ozone are twice as high as the maximum allowed limit for one hour a year. This occurs several times every day.

The average altitude of Mexico City is 2240 m above sea level. Consequently, average atmospheric pressure is roughly 25 per cent lower than at sea level. The lowered partial pressure of oxygen (pO_2) has significant effect on transport – fuel combustion in vehicle engines is incomplete and results in higher emissions of carbon monoxide and other compounds such as hydrocarbons and VOCs.

The most important air pollutants in Mexico City include particulates (PM_{10}), ozone (O_3), sulfur dioxide (SO_2), nitrogen oxides (NO_x), hydrocarbons, and carbon monoxide (CO). Intense sunlight turns these into photochemical smog. In turn, the smog prevents the Sun from heating the atmosphere enough to penetrate the inversion layer blanketing the city (Figure 6.20).

Figure 6.20 Air quality in Mexico City

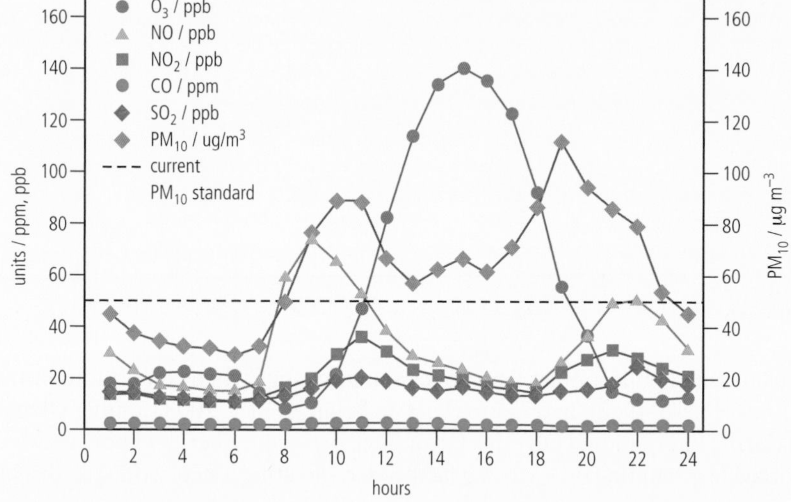

The most serious pollutants are PM_{10} and ozone. In the late 1990s, ozone levels exceeded standards on almost 90 per cent of days and PM_{10} exceeded standards on 30–50 per cent of days. Research suggests that reducing PM_{10} would yield the greatest health and financial benefits. Reducing both ozone and PM_{10} by 10 per cent would save US$760 million a year. In human terms, this would result in over 33 000 fewer emergency visits and over 4000 fewer hospital admissions for respiratory distress in 2010. In addition, it would lead to more than 260 fewer infant deaths each year.

Photochemical smog over Mexico City

The main programmes to combat air pollution in the Mexico City Metropolitan Area are as follows:

- Reduce the use of private vehicles: the government has implemented a one-day-stop programme called *Hoy no circula* (Today my car doesn't move). Anyone found driving when they shouldn't be has their plates taken away and must pay a fine of 20 days' pay based on the Mexico City minimum wage.
- Stopping days are randomly distributed to encourage car owners to use public transport and/or adopt car-pooling.
- Control of vehicle conditions: the enforcement of engine maintenance standards. Change of fuels: only small changes in gasoline quality have been accepted so far.
- Two major programmes already working within the Mexico City Metropolitan Area:
  - reduction of lead and sulfur in fuels
  - compulsory implementation of catalytic converters.

Other initiatives to improve the city's air quality over the past two decades – such as moving refineries beyond its boundaries and introducing cleaner buses – are having some effect. Between 1990 and 2012, levels of ozone fell from 43 ppb to 27 ppb; sulfur dioxide from 55 ppb to 5 ppb, and carbon monoxide from 84 ppb to 10 ppb.

Most people travel around Mexico City using public transport (74 per cent). Although private transport only accounts for about a quarter of the population, it uses about three-quarters of the total energy. More than 3 million vehicles – 30 per cent of them more than 20 years old – use Mexico City's roads. By expanding the city's Metro system and investing in the Ecobici bike hire scheme – which is used for about 26 000 journeys a day – the city hopes to reduce people's dependence on cars.

CONCEPTS: Sustainability

A growing number of *azoteas verdes* (green roofs) are springing up around Mexico City as part of the city's attempts to clean its air pollution. The *azotea verde* of the botanical garden's office, for example, is planted with stonecrop. This can withstand the Mexico City summer while it produces oxygen and filters some of the carbon dioxide and heavy metal particles from the air. The roof also helps regulate the temperature of the offices below and soaks up rainwater to keep the building dry.

The *azoteas verdes* project had green roofs on hospitals, schools and government buildings with a total area of over 20 000 m^2 in 2014. The green roofs do far more than simply purify the air: they reduce the 'heat island effect', help educate children about nature, and speed up the recovery of hospital patients by creating a more pleasant environment.

However, despite the measures taken so far, the combination of Mexico City's size (20 million), the number of cars, industries, its topography, and its altitude mean that photochemical smog is likely to remain a problem. The national debt and the poverty experienced by so many of its population mean that a number of more costly policies to improve the environment cannot be implemented.

You should be able to evaluate pollution management strategies for reducing photochemical smog.

Exercises

1. State the source and outline the effect of tropospheric ozone.
2. Outline the formation of photochemical smog.
3. Comment on the potential impacts of photochemical smog.
4. Describe and evaluate pollution management strategies for urban air pollution.

CHALLENGE YOURSELF

Many government organizations provide data on air quality.

The US Embassy in Beijing provides a range of data for air quality at weblink 6.8 (see hotlinks box opposite).

Air quality for Oxford, UK can be found at weblink 6.9.

Over a period of 2 weeks, monitor variations in air quality. How, for example, does air quality vary hourly? When is air quality best? When is it worst? How does this vary with the type of weather system the area is experiencing (high pressure, low pressure). How does air quality vary between central areas and suburban areas or nearby rural areas?

To view air-quality data for Beijing, go to www.pearsonhotlinks.co.uk, enter the book title or ISBN, and click on weblink 6.8

To view air quality data for Oxford, click on weblink 6.9.

Big questions

Having read this section, you can now discuss the following big questions:

- To what extent have the solutions emerging from this topic been directed at *preventing* environmental impacts, *limiting* the extent of the environmental impacts, or *restoring* systems in which environmental impacts have already occurred?
- How are the issues addressed in this topic of relevance to sustainability or sustainable development?
- In what ways might the solutions explored in this topic alter your predictions for the state of human societies and the biosphere some decades from now?

Points you may want to consider in your discussions:

- Outline the solutions to ground-level ozone. Why is there still ground-level ozone?
- Comment on the links between sustainability and photochemical smog.
- Suggest how photochemical smog is likely to change in the decades to come.

6.4 Acid deposition

Significant ideas

Acid deposition can impact living systems and the built environment.

The pollution management of acid deposition often involves cross-border issues.

Big questions

As you read this section, consider the following big questions:

- To what extent have the solutions emerging from this topic been directed at *preventing* environmental impacts, *limiting* the extent of the environmental impacts, or *restoring* systems in which environmental impacts have already occurred?
- How are the issues addressed in this topic of relevance to sustainability or sustainable development?
- In what ways might the solutions explored in this topic alter your predictions for the state of human societies and the biosphere some decades from now?

Knowledge and understanding

- The combustion of fossil fuels produces sulfur dioxide and oxides of nitrogen as primary pollutants. These gases may be converted into secondary pollutants of dry deposition (such as ash and dry particles) or wet deposition (such as rain and snow)
- The possible effects of acid deposition on soil, water and living organisms include:
  - direct effects (e.g. acid on aquatic organisms and coniferous forests)
  - indirect toxic effects (e.g. increased solubility of metal ions – such as aluminium ions – on fish)
  - indirect nutrient effects (e.g. leaching of plant nutrients).
- The impacts of acid deposition may be limited to areas downwind of major industrial regions but these areas may not be in the same country as the source of emissions.
- Pollution management strategies for acid deposition could include:
  - altering human activity (e.g. through reducing use of, or using alternatives to fossil fuels; international agreements and national governments may reduce pollutant production)

  - regulating and monitoring release of pollutants (e.g. through the use of scrubbers or catalytic converters that remove sulfur dioxide and oxides of nitrogen from coal-burning power plants and cars).
- Clean-up and restoration measures may include spreading ground limestone in acidified lakes or recolonization of damaged systems but the scope of these measures is limited.

The formation of acid deposition

Sulfur dioxide (SO_2) and nitrogen oxides (NO_x) are emitted from industrial areas, vehicles and urban areas. Some of these oxides fall directly to the ground as **dry deposition** (dry particles, aerosols, and gases) close to the source (Figure 6.21).

The longer the sulfur dioxide and oxides of nitrogen remain in the air, the greater the chance they will be oxidized to sulfuric acid (H_2SO_4) and nitric acid (HNO_3). These acids dissolve in cloud droplets (rain, snow, mist, hail) and reach the ground as **wet deposition.** Wet deposition can be carried thousands of kilometres downwind from the source.

The dissolved acids consist of sulfate ions (SO_4^{2-}), nitrate ions (NO_3^-), and hydrogen ions (H^+). These ions form **acid rain.** Transport of pollutants over 500 km is most likely to produce strong acid rain. Until about the 1960s, acid rain was a local phenomenon. However, the construction of tall smokestacks (e.g. the 381 m superstack at Sudbury, Ontario) forced the pollutants higher in the atmosphere and resulted in acid rain falling at greater distances from the original source.

Rainwater is normally a weak carbonic acid with a pH of about 5.5. Acid rain is a more acidic because of the addition of sulfur dioxide and the oxides of nitrogen. Any rain with a pH below 5.5 is termed acid rain.

The combustion of fossil fuels produces sulfur dioxide and oxides of nitrogen as primary pollutants. These gases may be converted into secondary pollutants of dry deposition (such as ash and dry particles) or wet deposition (such as rain and snow).

Figure 6.21 The formation of acid deposition

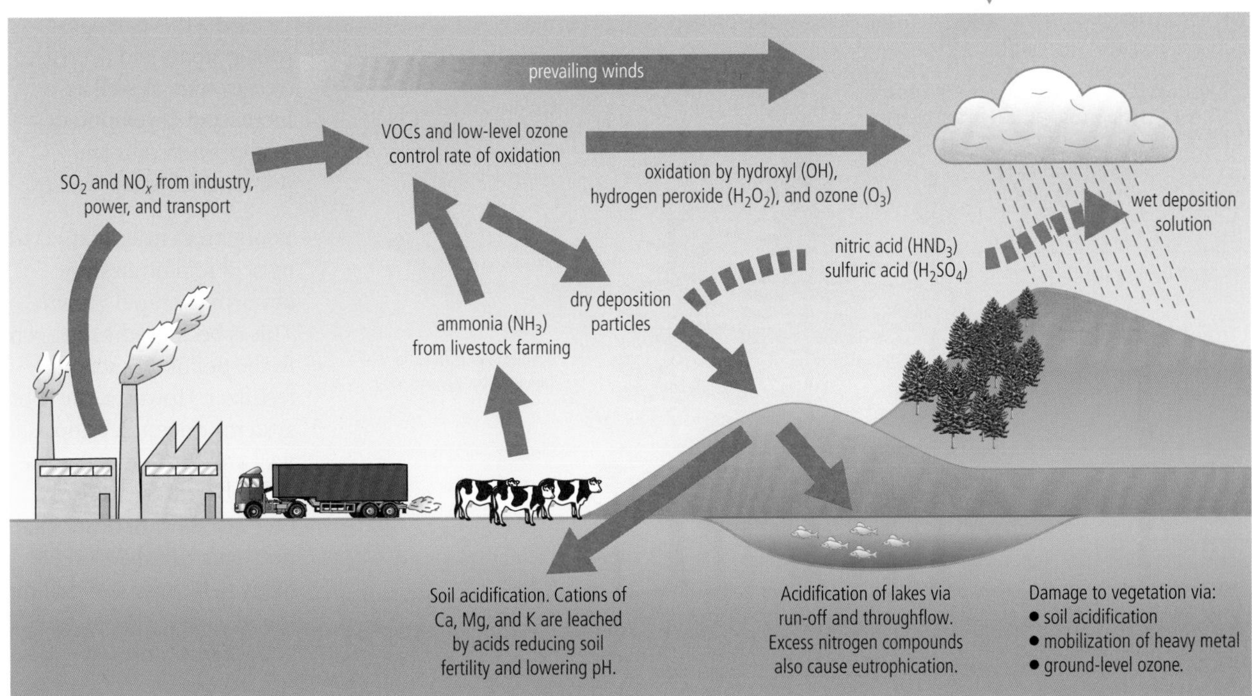

The possible effects of acid deposition on soil, water and living organisms include:
- **direct effects (e.g. acid on aquatic organisms and coniferous forests)**
- **indirect toxic effects (e.g. increased solubility of metal ions – such as aluminium ions – on fish)**
- **indirect nutrient effects (e.g. leaching of plant nutrients).**

The pH scale is logarithmic. This means that pH 6.0 is 10 times more acidic than pH 7.0; natural rainwater at pH 5.5 is about 25 times more acidic than distilled water at pH 7.0. Acid rain is frequently more than 20 times more acidic than natural rainwater. Rain over Scandinavia commonly has a pH of 4.2–4.3.

The most acidified water on Earth were the pools of water located on Chances Peak, Soufriere Hills, Montserrat. Due to an ongoing volcanic eruption from 1995, and the associated sulfur emissions, the pH of the water was recorded at 1.5.

The first effects of acid rain were noted in Scandinavian lakes in the 1960s. Over 18 000 lakes in Sweden are acidified, 4000 of them are seriously affected. Fish stocks in about 9000 Swedish lakes, mostly in the south and the centre of the country, are badly affected. In the Eastern USA and Canada, over 48 000 lakes are too acidic to support fish.

Direct effects of acid rain

Acidified lakes

Acidified lakes are characterized by:

- an impoverished species structure
- visibility several times greater than normal
- white moss spreading across the bottom of the lake
- increased levels of dissolved metals such as cadmium, copper, aluminium, zinc, and lead (so these metals become more easily available to plants and animals).

The effect on trees

Acid rain severely affects trees and forests (Figure 6.22). It breaks down fats in the foliage and damages membranes which can lead to plant death. Sulfur dioxide interferes with the process of photosynthesis. Coniferous trees seem to be most at risk from acid rain. These trees do not shed their needles at the end of every year: on a healthy conifer, needles can be up to 7 years old, but trees affected by acid rain often have needles from only the last 2 or 3 years. This means the tree has far fewer needles than normal. If a conifer loses over 65 per cent of its needles, it will probably die. Besides this, where soil pH is below 4.2, aluminium is released – this damages root systems and decreases tree growth, as well as increasing development of abnormal cells and premature loss of needles.

Young trees in soils affected by acid rain often show abnormally rapid growth. This is because the nitrogen in the pollutants acts as a fertilizer. However, the root systems do not develop as well and the trees are more easily blown over. Also, they are short of other vital nutrients and the wood is likely to be very soft making the trees more prone to attacks from insects.

Damaged trees are more likely to be affected by pests, diseases, frost and/or drought.
Needles and leaves turn yellow and drop off.
Acid rain percolates into the soil.
Top soil becomes more acidic.
Nutrients such as magnesium and calcium are lost through leaching.
Branches and leaves at the top of the tree have reduced growth and die.
Needles and leaves are damaged by SO_2 and NO_x.
Seedlings fail to grow.
Decomposition of dead organic matter slows down.
When the pH is less than 4.5, iron and aluminium are released into groundwater. Root hairs are damaged and so the tree receives fewer nutrients.

Figure 6.22 The effect of acidification on trees

Damaged conifers are easily recognizable. The extremities of the trees die, especially the crown which is most exposed. Needles drop, so the tree looks very thin. Branches on some trees droop. In most cases, acid rain does not kill the tree, but it is an added pressure on the tree which is then more likely to suffer damage from insects, fungi, frost, wind, and drought. Although deciduous trees generally do not suffer as much, research is showing that their growth is also affected.

The low pH of soil and the presence of metals may cause damage to root hairs (used by the tree to absorb nutrients). The tree loses vitality, growth is retarded, there is an inability to cope with stress (such as frost, drought, and pests); the tree becomes susceptible to injury. Needles turn brown, fall off, and finally whole branches snap away. In parts of Germany, more than 50 per cent of the spruce trees are dead or damaged.

CONCEPTS: Biodiversity

The most obvious environmental effect of acid rain has been the loss of fish in acidified lakes. Many species of fish are not able to survive in acidic water (Figure 6.23). It is the extreme pH values which cause most damage to plants and animals. Very often organisms are exposed to extremely low pH during the most sensitive part of their life cycle (for fish this may be the fry stage). These short periods coincide with snow melt and the accompanying acid surge. At these times, the water also has a high metal content.

	pH 6.5	pH 6.0	pH 5.5	pH 5.0	pH 4.5	pH 4.0
Trout						
Bass						
Perch						
Frogs						
Salamanders						
Clams						
Crayfish						
Snails						
Mayfly						

Figure 6.23 pH tolerance levels for selected aquatic organisms

Indirect toxic effect

As acidification increases, there is increased solubility of certain metals. Iron and aluminium have increased mobilization when pH is 4.5 or less. High levels of aluminium ions in fresh water have a toxic effect on fish. In particular, aluminium affects fish gills making respiration more difficult. It also affects the circulatory system and may affect kidneys and bones.

Indirect nutrient effects

One of the most important health effects of acid water is the result of its ability to flush trace metals from soil and pipes. Some wells in Sweden have aluminium levels up to 1.7 mg dm^{-3}, the World Health Organization safe limit is 0.2 mg dm^{-3}. This has been identified as a possible cause of Alzheimer's disease in people. High levels of mercury accumulated in fish can cause serious health problems when the fish are eaten by people.

Figure 6.24 shows the effect of soil acidification on nutrient availability. With increasing acidity, many nutrients such as

Figure 6.24 The effect of soil acidification on nutrient availability (the thickness of the bar represents relative availability)

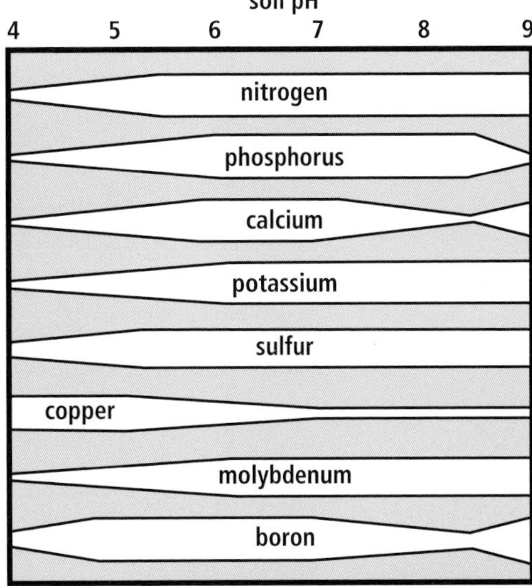

nitrogen and phosphorus become unavailable to plants, but copper becomes more available in acidic soils. Calcium, magnesium, iron, and aluminium can all be leached from acidic soil.

Increasing acidity leads to falling numbers of fungi, bacteria and earthworms. Earthworms generally do not tolerate soils with a pH less than 4.5, and so acidification may lead to less mixing (by earthworms) in a soil.

The impacts of acid deposition may be limited to areas downwind of major industrial regions but these areas may not be in the same country as the source of emissions.

Distribution of acid deposition

Acidification is largely related to human activity. Many countries produce the pollutants and they may be deposited many hundreds of kilometres from their point of origin (Figure 6.25). However, there are variations within areas receiving acid rain. Lime-rich soils and rocks are better able to absorb and neutralize the acidity.

In the 1980s and 1990s, the areas most affected by acid rain included Sweden, Norway, eastern North America, Germany, Belgium, the Netherlands, Scotland, countries of the former Yugoslavia, Austria, and Denmark.

These areas have a number of features in common:

- they are industrialized belts
- they are downwind of dense concentrations of fossil-fuel power stations, smelters, and large cities
- they are upland areas with high rainfall
- they contain lots of forest, streams, and lakes
- they have thin soils.

The main trends in the distribution of acid deposition in the last 20 years have been due to increased sulfur emissions in newly industrializing countries (NICs) and LEDCs. In China, the worst affected areas have been in the south – the Pearl River delta and central and eastern areas of Guangxi. In South Africa, coal-burning power stations and large metalworking industries are concentrated in the Eastern Transvaal Highveld; rainfall in the area has, on average, a pH of 4.2 and has been recorded at pH 3.7.

In contrast, in MEDCs there has been a dramatic decrease in sulfur dioxide emissions in 1990–2000, especially in the European Union (decreased 52 per cent) and North America (decreased 17 per cent).

To learn more about acid rain in Canada, go to www.pearsonhotlinks.co.uk, enter the book title or ISBN, and click on weblink 6.10.

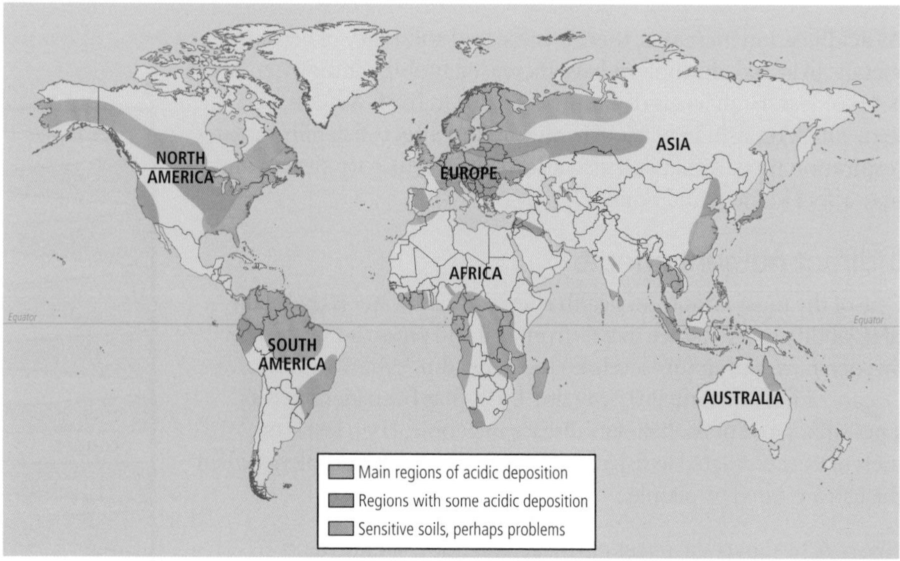

Figure 6.25 Distribution of acid deposition

There are also natural causes of acidification: bog moss secretes acid, heathers increase soil acidity, and conifer plantations acidify soils. Litter from conifers is acidic and not easily broken down, specialized decomposing bacteria are required. The sitka spruce is not native to the UK but was planted in great numbers. The absence of the decomposing bacteria led to the accumulation of an acid humus layer in the soil.

Volcanoes are also important sources of sulfur dioxide and nitrogen dioxide. For example, before the eruption of the Soufrière volcano in 1995, Montserrat had some of the finest cloud forest in the Caribbean. By 1996, vegetation loss from acid rain, gases, heat, and dust was severe and the lake at the top of Chances Peak was recorded at pH 1.5.

▲ Acidified vegetation due to volcanic eruption – Plymouth, Montserrat

Some environments are able to neutralize the effects of acid rain. This is referred to as buffering capacity. Chalk and limestone areas are very alkaline and can neutralize acids effectively. The underlying rocks over much of Scandinavia, Scotland, and northern Canada are granite. They are naturally acid, and have a very low buffering capacity. It is in these areas that there is the worst damage from acid rain.

To learn more about acid rain, go to www.pearsonhotlinks.co.uk, enter the book title or ISBN, and click on weblink 6.11.

Pollution management strategies

Given the causes of acid deposition – industrial and transport emissions – it is not easy to target all potential polluters. But it is expensive and time-consuming to treat extensive areas already affected by acid deposition.

Pollution management strategies for acid deposition could include:

- **altering human activity (e.g. through reducing use of, or using alternatives to fossil fuels; international agreements and national governments may reduce pollutant production)**
- **regulating and monitoring release of pollutants (e.g. through the use of scrubbers or catalytic converters that remove sulfur dioxide and oxides of nitrogen from coal-burning power plants and cars).**

Prevention

Prevention is better than cure. Prevention has a number of options:

- reducing use of fossil fuel (this requires a government initiative in order to switch to nuclear or hydro-power)
- reduce the number of private cars on the road and increase the number of people using public transport or park-and-ride schemes
- switch to low sulfur fuel (oil/gas plus high-grade coal).

Methods for regulating and monitoring the release of pollutants include:

- removing sulfur before combustion (expensive for coal, cheaper for oil)
- reducing sulfur oxides released on combustion (fluidized bed technology, FBT)
- burning coal in the presence of crushed limestone in order to reduce the acidification process
- removing sulfur from waste gases after combustion (flue-gas desulfurization, FGD)
- allowing decomposition of plants to return nutrients to the soil and offset the acidification process.

Both FBT and FGD are well developed and effective but they are very expensive. FBT brings the flue gases into contact with a sulfur-absorbing chemical, such as limestone, which can capture up to 95 per cent of the sulfur pollutants in power stations. FGD removes sulfur dioxide from exhaust flue gases in power stations. There are three methods of FGD:

- wet scrubbing – uses alkaline scrubbers such as limestone to absorb the sulfur dioxide
- spray dry scrubbing – uses hydrated lime to form a mix of calcium sulfate/sulfite
- scrubbing with a sodium sulfite solution.

To learn more about acid rain and its management, go to www.pearsonhotlinks.co.uk, enter the book title or ISBN, and click on weblink 6.12.

In the UK, 46 per cent of nitrogen oxides come from power stations, and 28 per cent come from vehicle exhausts. Emission from power stations can be reduced by FGD and special boilers which reduce the amount of air present at combustion. Car exhaust emissions can be reduced by different types of engine or exhaust, lower speed limits, and more public transport.

To learn more about the Convention on Long-Range Transboundary Air Pollution, go to www.pearsonhotlinks.co.uk, enter the book title or ISBN, and click on weblink 6.13.

As acid deposition is often a transboundary issue, legislation has to involve many countries. The 1979 *Convention on Long-Range Transboundary Air Pollution* was crucial in the clean-up of acidification in Europe.

- It brought together polluter and polluted.
- It set clear targets for pollution reduction.
- It made polluters recognize their international environmental responsibilities.

The 1999 Gothenburg Protocol to abate acidification, eutrophication and ground-level ozone commits countries to reduce their emissions of sulfur dioxide and oxides of nitrogen.

You should be able to evaluate pollution management strategies for acid deposition.

Clean up and restoration

Clean-up and restoration measures may include spreading ground limestone in acidified lakes or recolonization of damaged systems but the scope of these measures is limited.

Adding crushed limestone can limit the impact of acidification. This process, called liming, has been used extensively in lakes in Norway and Sweden but has not been used much in the USA. It is an expensive process, and has to be done repeatedly. It allows fish to remain in a lake, and enables aquatic ecosystems to function. However, it is a short-term solution. As long as energy power plants, vehicles and industries continue to emit oxides of nitrogen and sulfur dioxide, the problem will remain.

Exercises

1. Outline the processes leading to the formation of acidified precipitation.
2. Describe three possible effects of acid deposition on soil, water and living organisms.
3. Explain why the effect of acid deposition is regional rather than global.
4. Describe and evaluate pollution management strategies for acid deposition.
5. This photo shows acidification caused by a natural process.

a. Suggest one natural cause of acidification.
b. Distinguish between wet and dry deposition.
c. Outline the impacts of acidification of vegetation.

Big questions

Having read this section, you can now discuss the following big questions:

- To what extent have the solutions emerging from this topic been directed at *preventing* environmental impacts, *limiting* the extent of the environmental impacts, or *restoring* systems in which environmental impacts have already occurred?
- How are the issues addressed in this topic of relevance to sustainability or sustainable development?
- In what ways might the solutions explored in this topic alter your predictions for the state of human societies and the biosphere some decades from now?

Points you may want to consider in your discussions:

- To what extent is acidification 'yesterday's problem'? Why has acidification declined in certain regions?
- Examine the relationship between acidification and sustainability.
- In what ways is acidification likely to change over the next decades?

Practice questions

1 **a** State how incoming solar radiation (insolation) differs from the Earth's outgoing radiation. [1]

b List three greenhouse gases. [1]

c Describe the role of greenhouse gases in maintaining mean global temperatures. [4]

d Define the term *albedo.* [1]

e Explain how human activity has had an impacts on greenhouse gases in the atmosphere. [2]

2 The figure below shows how human activity can destroy ozone in the atmosphere.

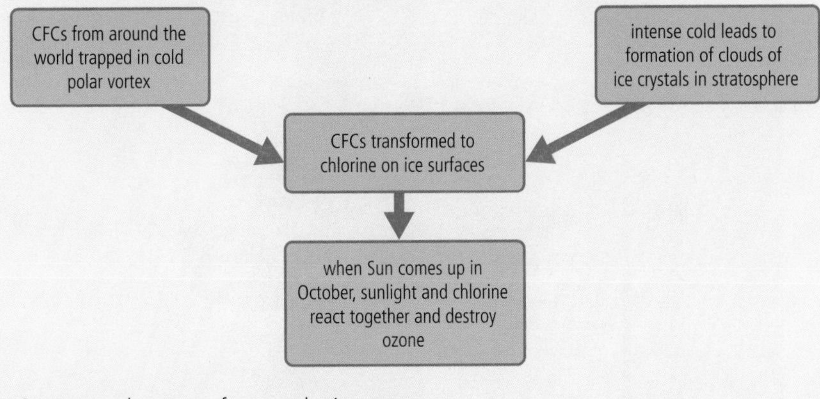

How ozone is destroyed by CFCs

a State one advantage of stratospheric ozone. [1]

b Identify two sources of CFCs. [2]

c Explain how CFCs can lead to a decline in ozone. [3]

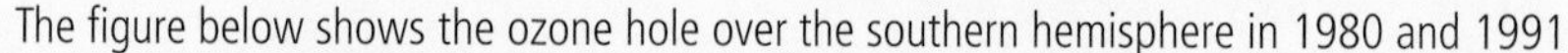

The figure below shows the ozone hole over the southern hemisphere in 1980 and 1991.

Ozone hole over the Antarctic (1980 and 1991)

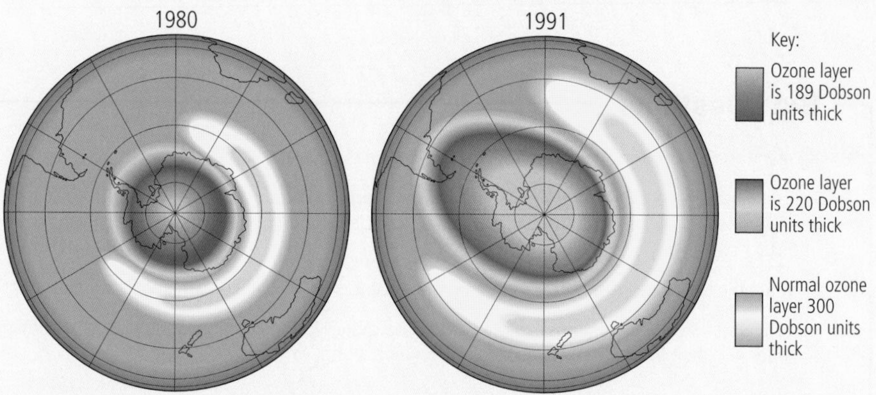

d What is meant by the term *ozone hole*? [1]

e Describe the changes in the ozone hole over the southern hemisphere between 1980 and 1991. [3]

f Outline the potential impacts of UV radiation on ecosystems and human health. [4]

3 The maps below show air quality in China and the USA at approximately the same time.

Key to the maps

AQI	Air Pollution Level	Health Implications
0–50	Good	Air quality is considered satisfactory, and air pollution poses little or no risk.
51–100	Moderate	Air quality is acceptable; however, for some pollutants there may be a moderate health concern for a very small number of people who are unusually sensitive to air pollution.
101–150	Unhealthy for Sensitive Groups	Members of sensitive groups may experience health effects. The general public is not likely to be affected.
151–200	Unhealthy	Everyone may begin to experience health effects; members of sensitive groups may experience more serious health effects.
201–300	Very Unhealthy	Health warnings of emergency conditions. The entire population is more likely to be affected.
300+	Hazardous	Health alert: everyone may experience more serious health effects.

Air quality in China: 13.21 GMT on 21 October 2014

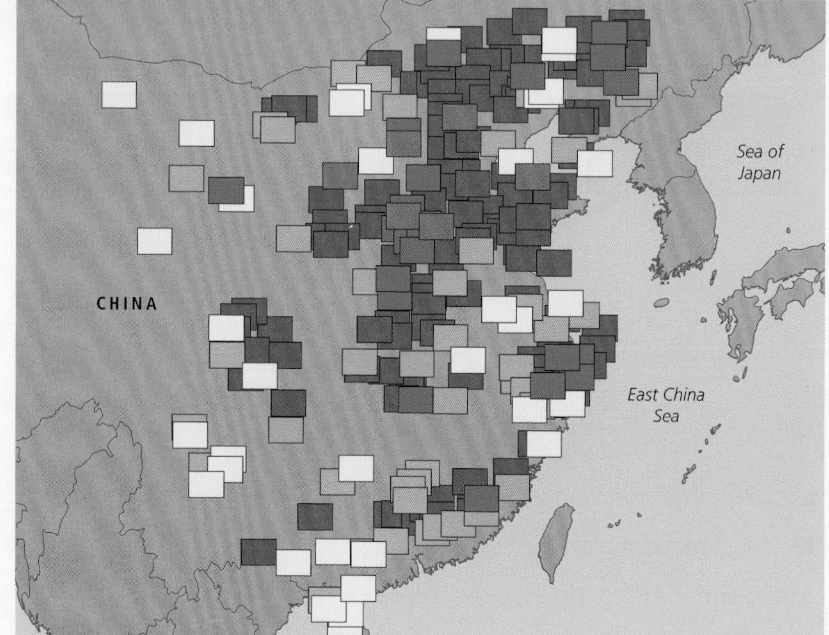

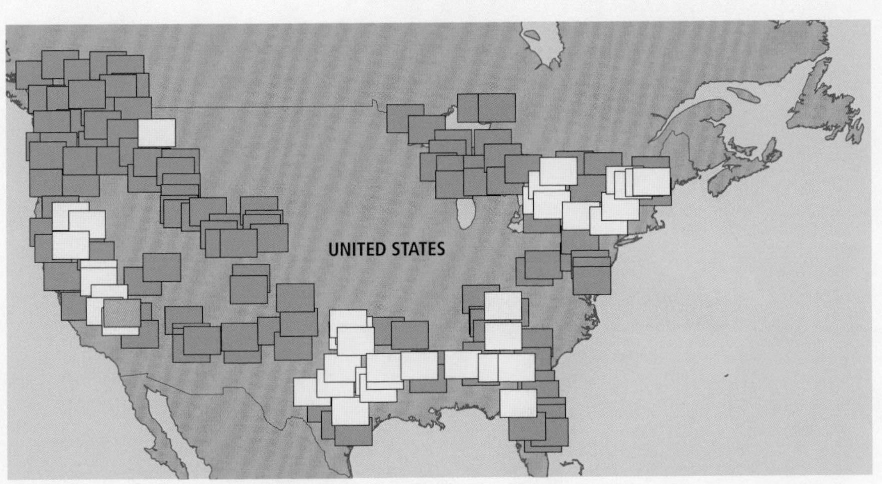

Air quality in the USA: 13.23 GMT on 21 October 2014

a **i** State the range of values for mainland China and the USA. [2]

ii Describe the main pattern of air quality in both countries. [2]

b Suggest reasons for the patterns you have identified in part (a). [3]

c Outline how the burning of fossil fuels can cause photochemical smog. [4]

d Examine the human factors which affect the successful implementation of pollution management strategies. [5]

4 Study the figure below.

	pH 6.5	pH 6.0	pH 5.5	pH 5.0	pH 4.5	pH 4.0
Trout						
Bass						
Perch						
Frogs						
Salamanders						
Clams						
Crayfish						
Snails						
Mayfly						

pH tolerance levels for selected aquatic organisms

a Identify the two organisms (i) most able and (ii) least able to tolerate acidic water. [2]

b Describe the impact of acidification on vegetation. [3]

c Explain why acidification is a regional rather than a global problem. [4]

d Evaluate the success of international agreements to reduce acidification. [4]

07 Climate change and energy production

7.1 Energy choices and security

Opposite: Climate change due to energy production is one of the greatest threats to humanity. Here is one solution – the Svartsengi geothermal plant near Keflavik, Iceland

Significant ideas

There is a range of energy sources available to societies that vary in their sustainability, availability, cost, and socio-political implications.

The choice of energy source is controversial and complex. Energy security is an important factor in making energy choices.

Big questions

As you read this section, consider the following big questions:

- What strengths and weaknesses of the systems approach and the use of models have been revealed through this topic?
- To what extent have the solutions emerging from this topic been directed at *preventing* environmental impacts, *limiting* the extent of the environmental impacts, or *restoring* systems in which environmental impacts have already occurred?
- What value systems can you identify at play in the causes and approaches to resolving the issues addressed in this topic?
- How does your own value system compare with others you have encountered in the context of issues raised in this topic?
- How are the issues addressed in this topic of relevance to sustainability or sustainable development?
- In what ways might the solutions explored in this topic alter your predictions for the state of human societies and the biosphere some decades from now?

Knowledge and understanding

- Fossil fuels contribute to the majority of humankind's energy supply and they vary widely in the impacts of their production and their emissions; their use is expected to increase to meet global energy demand.
- Sources of energy with lower carbon dioxide emissions than fossil fuels include renewable energy (solar, biomass, hydropower, wind, wave, tidal, and geothermal) and their use is expected to increase. Nuclear power is a low-carbon, low-emission, non-renewable resource but is controversial due to radioactive waste and the potential scale of any accident.
- Energy security depends on an adequate, reliable and affordable supply of energy that provides a degree of independence. An inequitable availability and uneven distributions of energy sources may lead to conflict.
- The energy choices adopted by a society may be influenced by availability, sustainability, scientific and technological developments, cultural attitudes, and political, economic, and environmental factors. These in turn affect energy security and independence.
- Improvements in energy efficiencies and energy conservation can limit growth in energy demand and contribute to energy security.

Fossil fuels contribute to the majority of humankind's energy supply and they vary widely in the impacts of their production and their emissions; their use is expected to increase to meet global energy demand.

Range of energy resources

Energy can be generated from both non-renewable and renewable resources. Non-renewable energy supplies include fossil fuels (e.g. coal, gas, and oil). These cannot be renewed at the same rate as they are used; this results in depletion of the stock. Nuclear

You should be able to evaluate the advantages and disadvantages of different energy sources.

power can be considered non-renewable because the source of the fission process is uranium, which in a non-renewable form of natural capital.

Renewable energy sources include solar, hydroelectric, geothermal, biomass, and tidal schemes. They can be large scale (e.g. country-wide schemes of energy generation) or small scale (microgeneration) within houses or communities. Renewable energy resources are sustainable because there is no depletion of natural capital.

The majority of the world's fuel comes from non-renewable sources and this is unlikely to change much by 2030 (Figure 7.1).

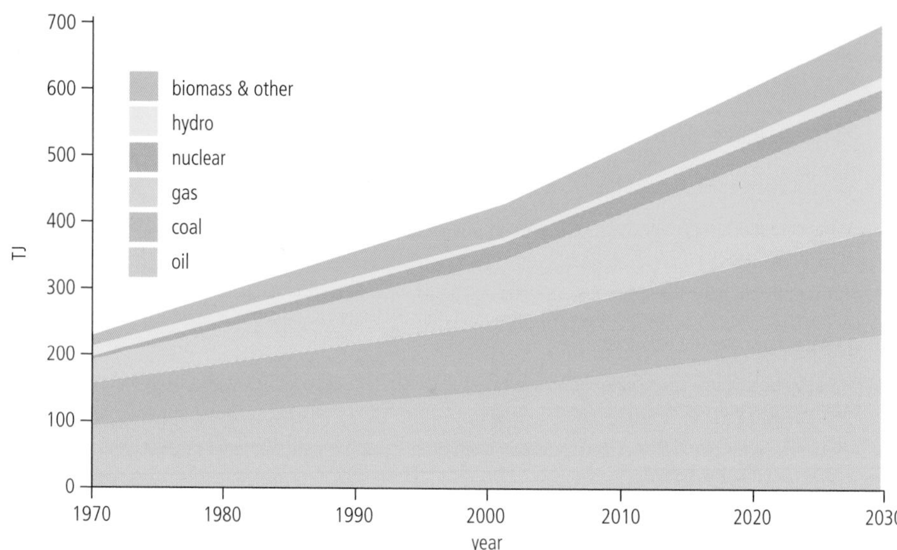

Figure 7.1 The world's fuel sources, 1970–2030

Fossil fuels

To learn about the BP Statistical Review of Energy (updated annually), go to www.pearsonhotlinks.co.uk, enter the book title or ISBN, and click on weblink 7.1.

From the industrial revolution onward, transport and energy generation have been founded on fossil fuel technology. Easily mined sources of coal led to early forms of transport based on this fuel. Processing of fossil fuels to produce petroleum led to the invention of the combustion engine, and this technology has continued to dominate until the present day. The growth of fossil fuel technology has been accompanied by a general unawareness of the effects on the environment. Pollution and global warming were not factors that were considered when fossil fuels were adopted as the primary source of energy generation. More recently, our growing awareness of environmental problems linked to fossil fuel use has put more emphasis on renewable forms of energy.

Energy consumption is much higher in MEDCs than LEDCs. The economies of MEDCs have been based on energy generation built on fossil fuel use, whereas energy demands in LEDCs have traditionally been much lower due to less available technology and reliance on natural resources (wood burning or other biomass sources).

Renewable fuels

Renewable sources of energy have been slow to grow globally. There are several reasons for this. Non-renewable sources of energy (e.g. gas) are generally cheaper than renewables. Gas is cheap because it is relatively plentiful, can be burned directly without the need for refining, and the technology is already in place to access the gas and burn it in existing gas-fired power stations. Renewables such as wind power often require high set-up costs (e.g. the installation of new wind turbines) and may still be unreliable.

In future, the cost of non-renewable energy is likely to be much higher (a possible exception is shale gas, pages 354–356). This is because stocks will become depleted and the easiest and most accessible resources will have already been mined. Only resources which are difficult to access (and therefore more costly to reach) will remain. The increasing scarcity of non-renewable resources will push costs up, and environmental taxes to compensate for global warming will also make fossil fuels more expensive. Therefore, in the future, renewable sources of energy will become more attractive and increased use is likely. Adoption of sustainable energy will have a significant beneficial effect on the planet.

Advantages and disadvantages of fossil fuels

Fossil fuels are formed when dead animals and plants decompose in anoxic conditions (where oxygen is absent), are covered by silt and mud, and are subjected to heat and pressure over tens of thousands of years. The term 'fossil' refers to the fact that the fuels are made from preserved dead organisms. Gas and oil are largely made from oceanic organisms (e.g. plankton) and coal from land plants (mainly from trees growing during the Carboniferous Period – carboniferous means coal-bearing).

All fossil fuels release energy in the same way (Figure 7.2). The original source of energy contained within fossil fuels is the Sun. Photosynthesis traps the energy in plant matter, converting it to chemical energy. This store of energy can remain below ground for millions of years, until humans mine it. Because the fuels are made from dead organisms, they contain much carbon – they lock-up extensive amounts of carbon beneath the ground. Thus, burning fossil fuels releases a great deal of carbon in the form of carbon dioxide.

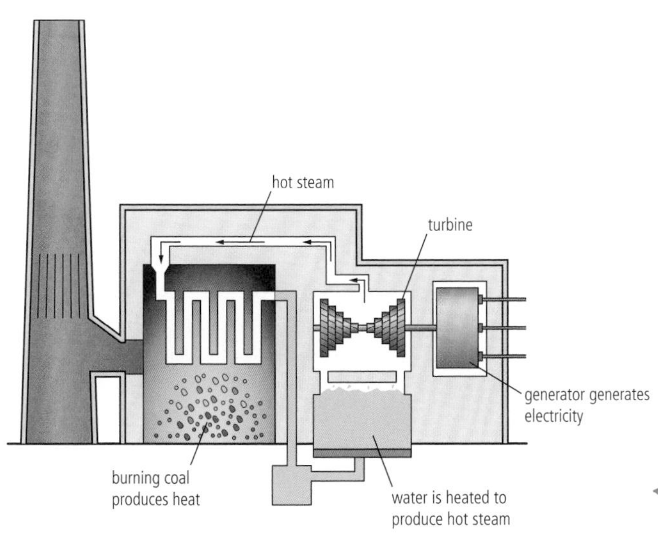

Figure 7.2 Coal and the production of electricity

Coal-fired power stations turn chemical energy in the coal into (about) 40 per cent electricity and 60 per cent waste heat. The clouds emitted from their cooling chimneys are formed by condensed water vapour created by this method of energy generation.

Advantages

The advantages of fossil fuels are that they are relatively cheap and plentiful. At the same time, advanced technologies have been developed to allow safe extraction and the technology already exists for their use (e.g. the combustion engine). The technology for controlling pollution from these fuels also exists. At present, no other energy source is close to replacing the amount of energy generated by fossil fuels. Oil and gas have a further advantage in that they can be delivered over long distances by pipeline.

In the US town of Centralia, Pennsylvania, an exposed coal seam caught fire in 1962. It burned for 17 years before a petrol station dealer found the ground temperature was 77.8 °C and other residents found their cellar floors too hot to touch. Smoke began to seep from the ground. The town was abandoned, residents relocated, and buildings bulldozed in 2002. The coal is still burning and may do so for 250 years.

Disadvantages

The two main disadvantages of fossil fuels are their contribution to climate change, and their unsustainability. They are the most important contributor to the build-up of carbon dioxide in the atmosphere and consequently the most important contributor to global warming. Use of fossil fuels is unsustainable because it implies liquidation of a finite stock of the resource: we can extend the lifetime of this resource, through the use of shale gas and tar sands, but it is ultimately unsustainable.

Other disadvantages are that these fuels will become increasingly difficult to extract, and extraction may become more and more potentially dangerous as mines get deeper and oil-rigs are placed further out to sea (e.g. the Deepwater Horizon oil platform in the Gulf of Mexico). Oil spillages from tankers and burst pipelines can severely damage natural ecosystems, and it is very expensive to clear up this sort of pollution. Coal extracted from underground mines causes minimal disturbance at the surface, but open-cast mining clears habitat from the surface and can cause extensive environmental damage.

Fossil fuel consumption and shale gas

Fossil fuel consumption is largest in MEDCs. However, fossil fuel consumption in LEDCs is expected to increase in future because of increasing population, income, and technological development. Coal is not easily transported over long distances, it is mainly consumed where it is locally available. Oil and gas can be consumed far from their source of extraction because they can be piped.

The availability of energy still relies extensively on fossil fuels, which account for around 80 per cent of global energy consumption. Consumption varies country by country, and by region (Figure 7.3). The biggest consumers are the USA, China, and Europe (together accounting for more than half of all fossil fuel consumption).

Figure 7.3 (a) Production and (b) consumption of oil by region

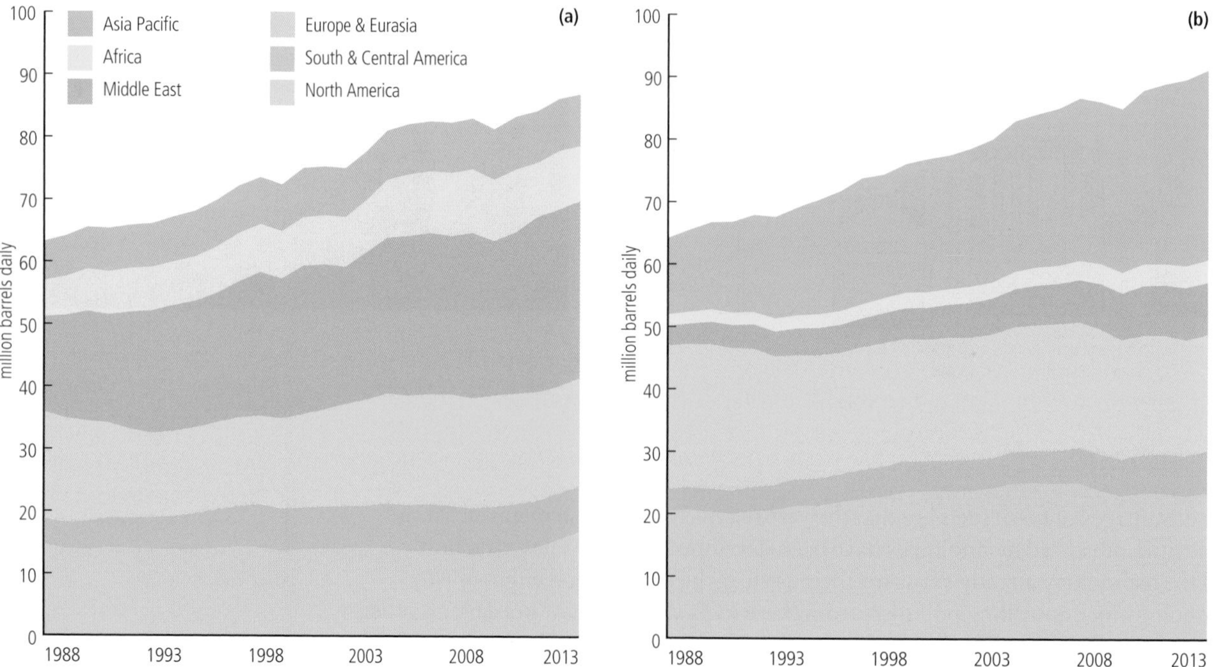

Just as nuclear scientists in the 1950s and 1960s believed that nuclear energy was going to be the answer to the world's energy needs, oil and gas producers believe that gas derived from shale could bring a plentiful supply of low-cost energy. Shale gas could transform the pattern of energy trade in the world. Nevertheless, it has its critics and there may be problems related to the extraction of shale gas.

To watch an excerpt from *The High Cost of Cheap Gas* – a film about the impacts of fracking, go to www.pearsonhotlinks.co.uk, enter the book title or ISBN, and click on weblink 7.2.

Shales are one of the most common forms of sedimentary rocks on Earth. Significant reserves have been found in China, Argentina, USA, and South Africa (Figure 7.4). A new wave of gas producers may emerge. Shale has the potential to change domestic economies. In the USA, President Obama has suggested it could support 600 000 jobs.

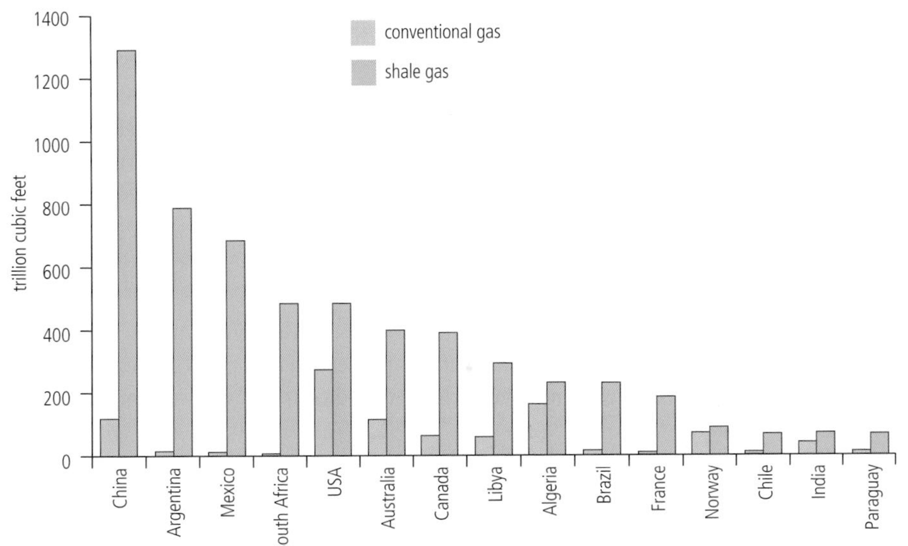

Figure 7.4 Estimated technically recoverable shale gas in relation to conventional supplies, top 15 countries

Shale is also having a geopolitical influence. The world's gas trade has long been dominated by Russia, Qatar, and Algeria. The gas pipelines that link Russia's Siberian gas fields with Europe, and the liquid natural gas (LNG) tankers that carry LNG from the Persian Gulf and South East Asia have created a network of dependency, that may be about to change. For example, Poland and Ukraine have begun investigating their resources, partly to reduce their dependence on Russian sources of natural gas. Russia's domination of the European gas market is no longer assured. China claims to have 25 trillion cubic metres of potentially recoverable shale (about one-fifth of global shale resources), enough to meet consumption for 200 years. If China were to end its reliance on coal and switch to gas, which is cleaner burning, it would have a significant impact on carbon emissions.

However, as with the nuclear dawn, there are potential drawbacks with shale. Hydraulic fracturing (fracking) may pollute groundwater and soil, release methane and trigger earthquakes. For critics such as Josh Fox, maker of the film *Gasland*, the risks are great and the techniques used are contentious. In contrast, a review by the Massachusetts Institute of Technology concluded that with over 20 000 shale wells drilled in the last 10 years, the environmental record of shale gas development has for the most part been 'a good one'.

In 2010, the USA replaced Russia as the world's largest gas producer. This happened because of a combination of large supplies of raw material, a well-developed service industry to drill the wells and provide the equipment, and a network of pipelines to transport the gas. In China, another potential contender, conditions are not so beneficial. Many exploratory projects are in the earthquake-prone Sichuan province. China also lacks the extensive pipeline network needed to get the gas to market.

In addition, large volumes of water are needed in the fracking process – and China faces increasing water shortages. Dozens of wells have been drilled but so far the results have been mixed. Some of China's shale resources are deeply buried in heavily faulted areas. In addition, China's deposits have a higher clay content than the brittle marine shales of the USA, making them more difficult to frack and less productive. Nevertheless, China is the world's biggest energy user, and shale could be the answer to reducing China's growing dependence on imported energy. China has set itself a target of 6.5 billion cubic metres of annual output by 2015, equivalent to 2–3 per cent of its projected gas production, and 60 billion cubic metres by 2020.

USA shale gas exports may well be cheap, as well as plentiful. Canada also has ambitious plans for the export of shale gas. Japan has seen an increased demand for energy imports following the Fukushima nuclear disaster in 2011. As one energy specialist said 'if shale gas proves as plentiful around the world as it is in the USA, it could not only displace coal, wind, and uranium to dominate global electricity generation but also replace oil as the main fuel for transport vehicles.' However, the world is yet to be convinced that hydraulic fracking is completely safe environmentally.

Renewable and alternative energy sources

Sources of energy with lower carbon dioxide emissions than fossil fuels include renewable sources and nuclear power.

The use of renewable sources of energy (solar, biomass, hydropower, wind, wave, tidal, and geothermal) is expected to increase. Nuclear power is a low-carbon, low-emission, non-renewable resource but is controversial due to radioactive waste and potential scale of any accident.

Advantages and disadvantages of renewables

General advantages of renewable sources of energy are that they do not release pollutants such as greenhouse gases or chemicals that contribute to acid rain. Because they are renewable, they will not run out.

There are several restrictions that currently limit large-scale use of renewable energy sources. Fossil fuel resources are still economically cheaper to exploit, and the technologies to harness renewable sources are not available on a large scale. Inertia within cultures (e.g. the USA's car culture) and traditions of both MEDCs and LEDCs means that non-renewable resources are favoured (although certain renewable energy supplies have always been widely used in LEDCs). The locations available for renewable energy sources are often limited by politics – for example, for wind turbines are often not exploited because people living nearby do want their environment 'spoilt' by the presence of wind turbines. All these factors mean that renewable resources are not able to meet current demand.

Hydro-electric power

Hydro-electric power (HEP) uses turbines which can be switched on whenever energy is needed, so it is a reliable form of energy generation (Figure 7.5). Dams are used to block the flow of water so forming large artificial lakes which can be used for leisure purposes, food sources and irrigation as well as electricity generation. Once the construction is completed, HEP schemes are relatively cheap to run. However, there are several disadvantages to HEP. Vast areas may be flooded involving loss of habitats, farmland, and displacement of people, and dams may restrict the flow of sediment thereby affecting ecosystems or farming downstream. They may also lead to increased erosion rates downstream when the flow of natural river systems are disrupted. The cost of building dams is high, and dams may eventually silt-up rendering them unusable.

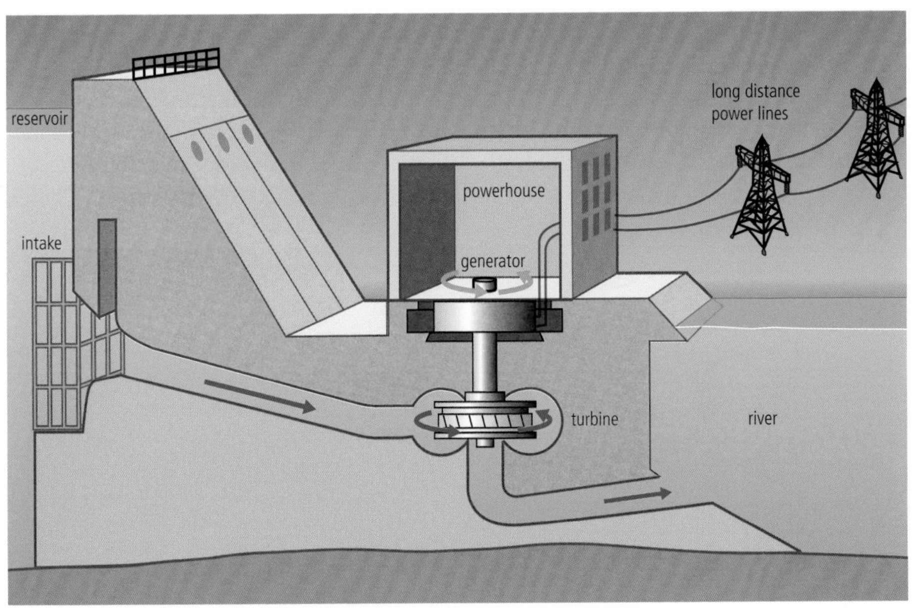

Figure 7.5 Hydroelectric energy is generated when water contained in an artificially made reservoir (created by damming a river) is allowed to flow though a turbine under immense pressure. The water turns the propellers which cause rotation in the turbine shaft, which generates electricity in the turbine's motor.

Case study

The Narmada River Dam Project, India

In India, biomass is a traditional source of energy. A huge proportion of the population relies on local sources of firewood for energy because it is the most readily available source and is inexpensive. Technology such as solar-powered stoves is neither available nor affordable. The Indian government, in a drive to develop economically, has sought to harness other sources of cheap energy to stimulate industrial development. In particular, the government is promoting hydroelectric power, which historically has sometimes been extremely controversial for social and environmental reasons.

The most controversial dam development in India is the Narmada River Dam Project. Plans were initiated in the 1940s by the country's first prime minister Jawaharlal Nehru. Legal and logistical problems delayed the start of the project until 1979. The plan involves the construction of some 3200 dams of varying sizes on the Narmada River (Figure 7.6).

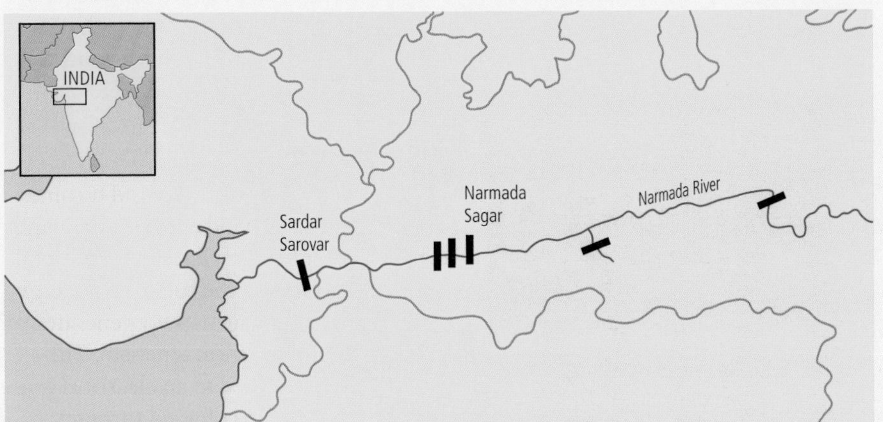

Figure 7.6 The location of dams in the Narmada River, India

The Sardar Sarovar Dam is the biggest dam on the river and its construction has been fiercely opposed. 200 000 people could be displaced by the project, and major damage caused to the ecosystems of the region. Those in favour of the project say that it will supply water to 30 million people and irrigate crops to feed another 20 million people. In October 2000, the Indian Supreme Court gave a go-ahead for the construction of the Sardar Sarovar Dam, saying that the benefits of the project outweigh negative environmental and social impacts. In 2014, the Narmada Control Authority approved a series of changes in the final height from 80 m to 163 m in depth. The project is expected to be completed by 2025.

Tidal power

Tidal power produces energy by using the ebbing or flooding tide to turn turbines which produce energy (Figure 7.7). The major limitations of this method are that a good tidal range is required to generate sufficient energy, together with the right shape of coastline to channel water through the turbines. Such installations may interfere with navigation and can have impact on wildlife. They are expensive to set up.

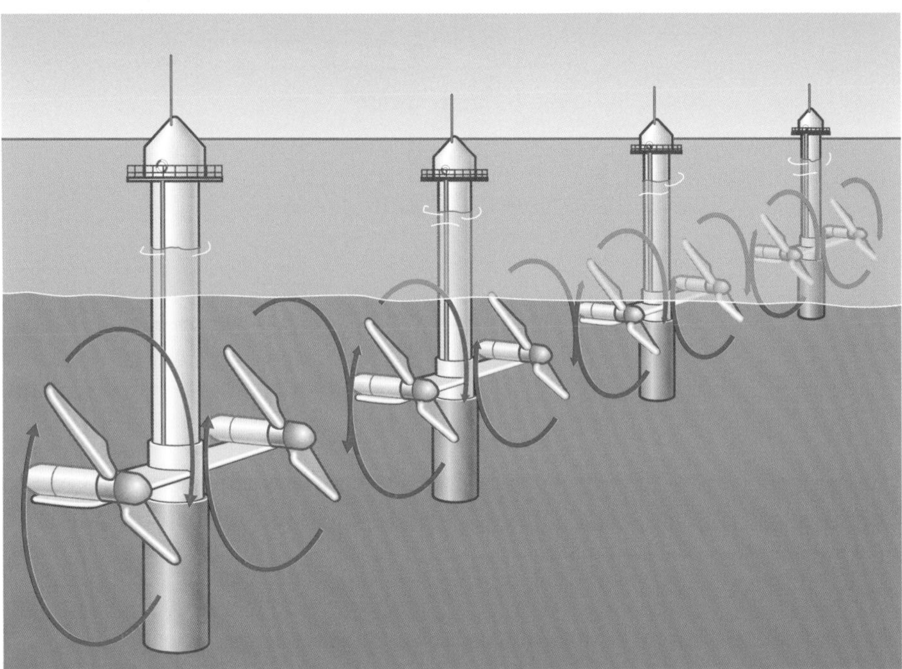

Figure 7.7 The turbines generate power as the tide comes in and again as it goes out

Solar energy

At present, it is very expensive to turn solar energy into high-quality energy needed for manufacturing (compared to using fossil fuels). However, passive solar energy (combined with insulation) is much cheaper for heating homes than fossil fuels. Solar energy has the disadvantage that its usefulness is limited in northern countries during winter months.

Solar panels are large flat panels made up of many individual solar cells

Wind power

Wind power is produced by wind turbines driven by available wind energy – the wind turns the rotor blades which rotate a metal shaft which transfers the rotational energy into a generator. The generator generates electricity using electromagnetism. The energy is then supplied to an electrical grid. The major limitations of wind turbines are that if there is no wind, no energy is generated. Thus, placement of the turbines is critical: they need to be in areas of consistent high wind.

Biofuel

Biofuel energy is produced by burning plant material to produce heat (Figure 7.8). Other forms of biofuel energy transform plant matter into ethanol which is then used as a fuel, or use methane digestion methods to convert biomass to methane which is then burned to generate electricity. The disadvantages of these techniques are that they produce emissions and require large amounts of land to grow the biofuel crop. Biofuel crops may take up land once used for growing food crops, thus pushing up the price of food, and disadvantaging local people who cannot get enough food to live. Biofuel crops are often planted at the expense of natural ecosystems, where new land clearance to create space for the biofuel crop has destroyed the natural ecosystem.

A wind turbine, Aero Island, Denmark

To learn about the wind farm in the Thames estuary, go to www.pearsonhotlinks.co.uk, enter the book title or ISBN, and click on weblink 7.3.

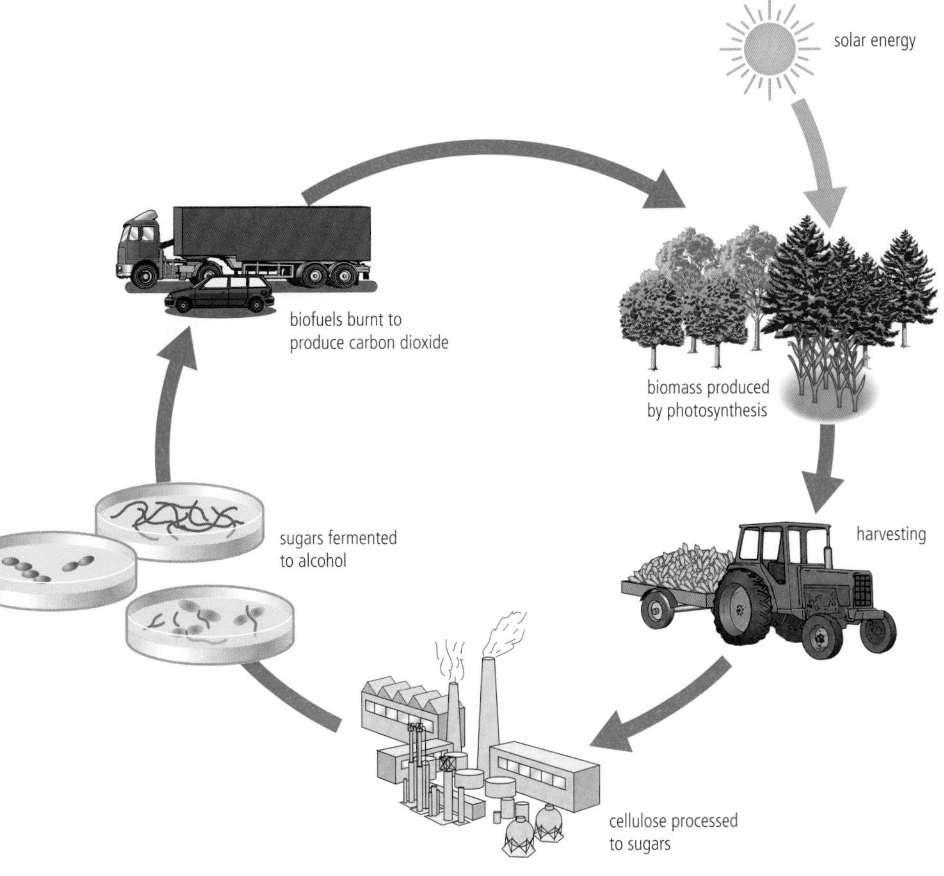

Figure 7.8 Biofuels are seen as a green form of power generation because, although they produce carbon dioxide, the gas is recycled (biofuel crops absorb carbon dioxide when they photosynthesize).

Wastes

Energy can be obtained from wastes. Organic waste decomposes and gives off methane gas which can be burned. Waste can also be burned directly to generate energy, for example burning straw. Advantages are that the resource used is readily available and its use does not deplete natural capital. At the same time, a useful purpose is being served by waste that would otherwise have to be disposed of in some other way. Disadvantages are that the burning adds to global warming gases in the atmosphere (although it could be argued that decomposition of the waste would do this in any case).

Geothermal energy

Energy can be obtained from residual heat in the ground. Water is pumped into pipes beneath the ground and the geothermal heat from the ground heats the water which can then be used to heat buildings. The pipes do not have be buried at great depth to be effective, although deeper burial allows greater heat capture. This method of heat transfer is low impact and does not release any form of pollution. The pipes can be arranged in various formations (Figure 7.9).

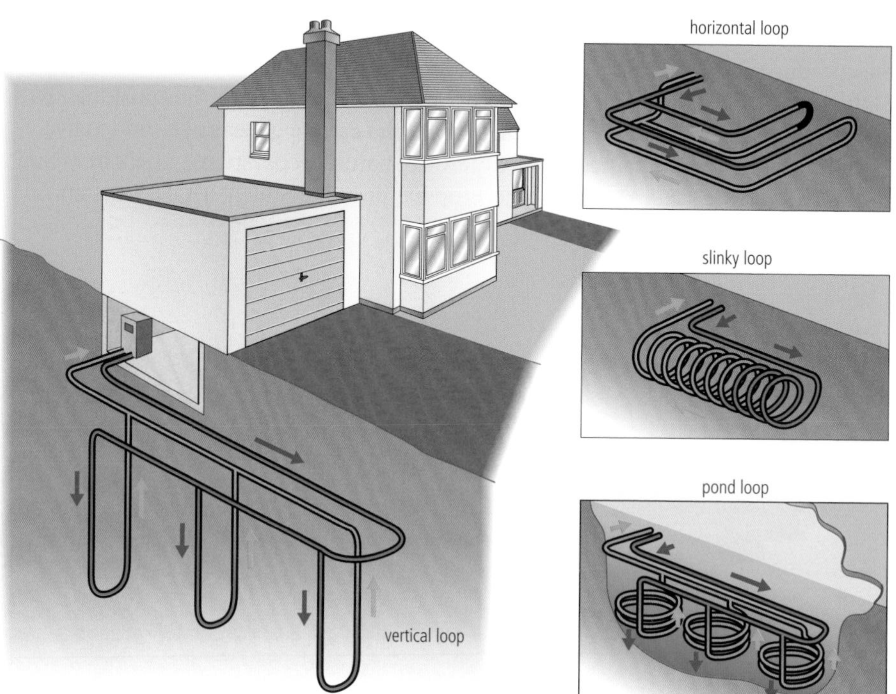

Figure 7.9 Designs for pipe layout to supply geothermal energy to the home

CONCEPTS: Sustainability

In *Renewable Energy – A Global Perspective*, Mohamed El-Ashry (Senior Fellow, UN Foundation) argues that 2008, seemingly the peak of the global financial crisis, was the best year for renewables. In just 1 year, all forms of grid-connected solar power grew by 70 per cent. Wind power grew by 29 per cent, and solar hot water increased by 15 per cent. According to the *Financial Times*, more than 50 per cent of total added power capacity in 2008 in both the USA and Europe was renewable – more than new capacity for oil, gas, coal, and nuclear combined.

National investments in renewable energy also changed. In 2006, Germany and China were the global leaders in new capacity investment, with the USA far behind. But a massive increase in wind power investment in the USA allowed it to become the global leader in 2008. Spain, China, and Germany were not far behind. Spain moved up to second place thanks to its large investments in solar power. Brazil was fifth, due to large investments in biofuels. The global recession might turn out to be a blessing in disguise for renewable energy, because governments of the world's largest economies have, for the first time, provided direct financial support. Governments did not invest just for energy security and climate change. They recognized the economic benefits of clean energy. At the end of 2008 and in early 2009, a number of national governments announced plans to greatly increase public finance of renewables and other low-carbon technologies. Many of these announcements were directed at economic stimulus and job creation, with millions of new 'green jobs' targeted.

Einstein's equation $E = mc^2$ related energy and matter, where the amount of energy in matter (E) was calculated by multiplying mass (m) by the speed of light squared (c^2), generating enormous numbers for E from even small quantities of matter.

Advantages and disadvantages of nuclear power

The equations of Albert Einstein first alerted scientists to the possibility of generating huge amounts of energy from splitting atoms. Fission technology was first developed in 1945 and used in atomic bombs at the end of the Second World War. It was then used in generating atomic energy (Figure 7.10). When enough fissionable material (e.g. uranium or plutonium) is brought together, and the process initiated, a chain reaction occurs that splits atoms releasing a tremendous amount of energy.

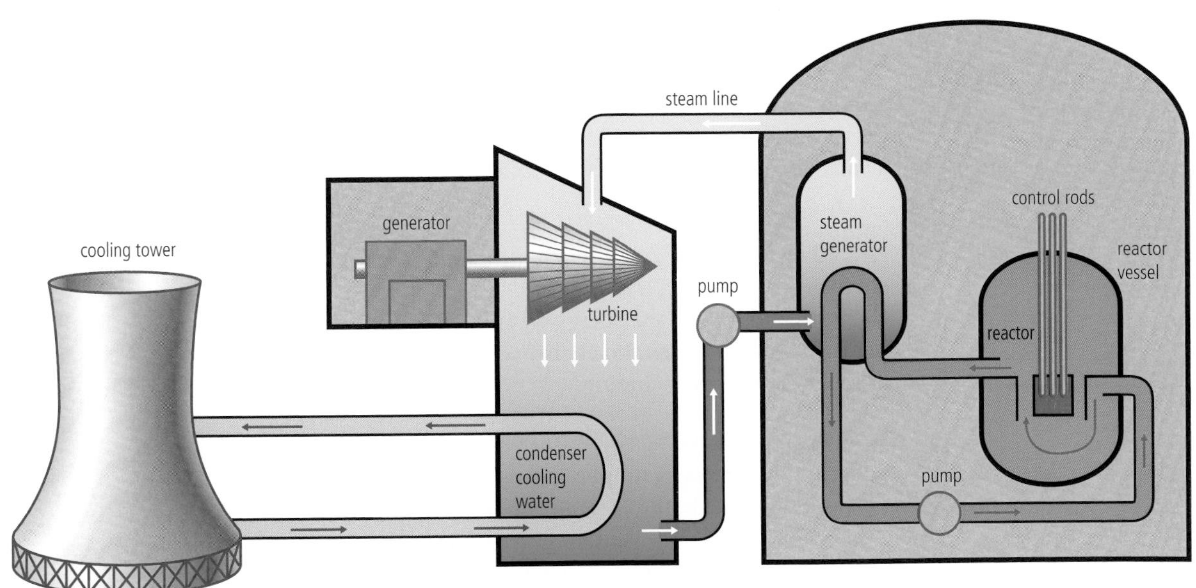

Figure 7.10 When a fission reaction takes place, a large amount of heat is given off. This heats water around the nuclear core and turns it to steam. The steam passes over the turbine causing it to spin, which turns a large generator, creating electricity. The steam is then cooled by cold water from the cooling tower travelling through the condenser below the turbine. The drop in temperature condenses the steam back into water, which is pumped back to the reactor to be reheated and continue the process.

Nuclear power plants produce radioactive wastes, including some that can remain dangerous for tens of thousands of years. Radioactivity is the result of nuclear changes in which unstable (radioactive) isotopes emit particles and energy continuously until the original isotope is changed into a stable one. When people are exposed to such radiation, the DNA in their cells can be damaged by mutation. If mutation occurs in body (somatic) cells, cancers, miscarriages, and burns can be caused. If the mutation occurs in reproductive cells (eggs and sperm), genetic defects can appear in subsequent generations.

The advantages of nuclear power generation are as follows.

- It does not emit carbon dioxide and so does not contribute to global warming.
- The technology is readily available.
- A large amount of electrical energy is generated in a single plant.
- It is very efficient, especially in comparison to fossil fuels: 1 kg uranium contains 20 000 times more energy than 1 kg coal.

Nuclear power generation has the following disadvantages.

- The waste from nuclear power stations is extremely dangerous and remains so for thousands of years. How best to dispose of this is still an unresolved problem.
- The associated risks are high. It is impossible to build a plant with 100 per cent reliability, and there will always be a small probability of failure (e.g. the Chernobyl and Fukushima-Daiichi disasters). The more nuclear power plants (and nuclear waste storage shelters) are built, the higher is the probability of a disastrous failure somewhere in the world. The potential of nuclear power plants to become targets for terrorist attack has been pointed out by opponents of this type of energy generation.
- The energy source for nuclear power is uranium, which is a scarce and non-renewable resource. Its supply is estimated to last for only the next 30–60 years depending on actual demand.
- The time frame needed to plan and build a new nuclear power plant is 20–30 years: uptake of nuclear power will therefore take time.

Energy security depends on adequate, reliable, and affordable supply of energy that provides a degree of independence.

Energy security

Energy security refers to a country's ability to secure all its energy needs, whereas energy insecurity refers to a lack of security over energy sources. Inequitable availability and uneven distribution of energy sources may lead to conflict (Figure 7.11).

According to the analyst Chris Ruppel (2006), the period from 1985 to 2003 was an era of energy security, and since 2004 there has been an era of energy insecurity. He claims that following the energy crisis of 1973 and the Iraq War (1990–91), there was a period of low oil prices and energy security. However, insecurity has since arisen for a number of reasons, including:

- increased demand, especially by NICs
- decreased reserves as supplies are used up
- geopolitical development: countries such as Venezuela, Iran, and Russia have 'flexed their economic muscle' in response to their oil resources and the decreasing resources in the Middle East and North Sea
- global warming and natural disasters such as Hurricane Katrina, which have increased awareness about the misuse of energy resources
- terrorist activity (e.g. in Nigeria and Iraq)
- the conflict between Russia and the Ukraine.

Energy insecurity can cause and be the result of geopolitical tension. For most consumers, a diversified energy mix is the best policy, rather than depending on a single supplier.

Figure 7.11 (a) Nuclear energy consumption and (b) hydro-energy consumption by region

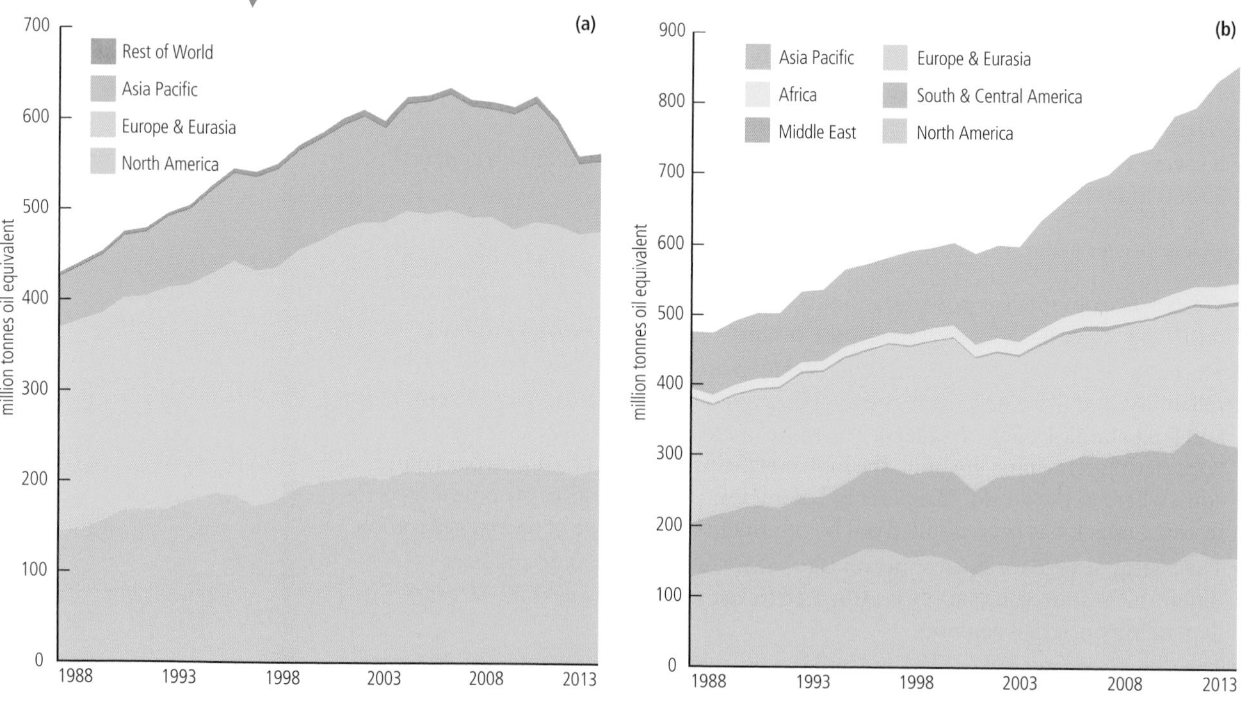

The potential for conflict

The Middle East controls about 60 per cent of the world's remaining oil reserves. Saudi Arabia alone controls over 20 per cent (Figure 7.12). On the other hand, the USA possesses less than 2 per cent of the world's oil reserves yet consumes over 200 million barrels of oil per day. This means that the USA has to source much of its oil from overseas, notably the Middle East. This gives the Middle East an economic and political advantage – countries that want oil have to stay on friendly terms with those that supply it. (There are obvious exceptions such as the US–British invasion of Iraq, and the Iraqi invasion of Kuwait.)

Countries that depend on the Middle East for their oil need to:

- help ensure political stability in the Middle East
- maintain good political links with the Middle East
- involve the Middle East in economic cooperation.

On the other hand, the situation is also an incentive for rich countries to increase energy conservation or develop alternative forms of energy. There is a need to reassess other energy sources such as coal, nuclear power and renewables, and use energy less wastefully.

Figure 7.12 Oil reserves (the area of each country is shown in proportion to its percentage of world oil reserves)

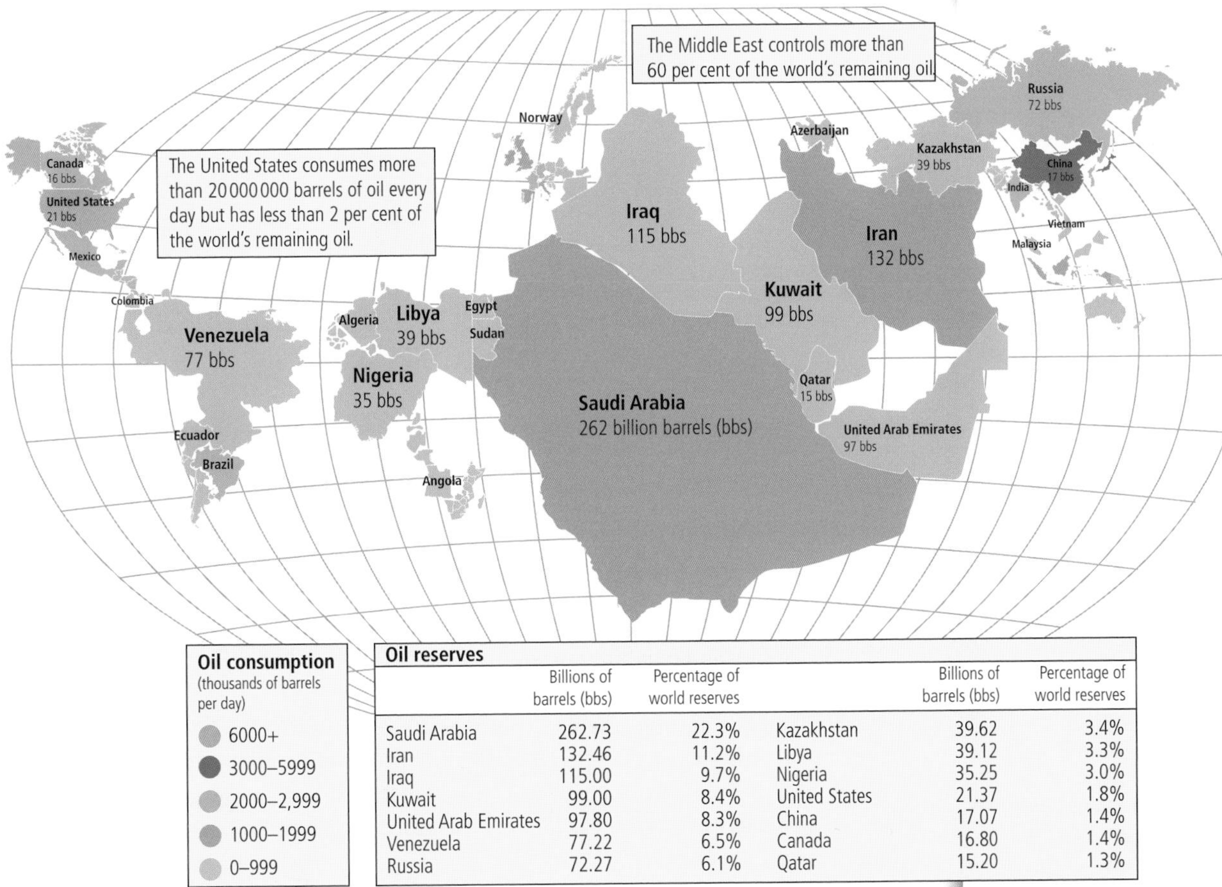

Oil reserves

	Billions of barrels (bbs)	Percentage of world reserves		Billions of barrels (bbs)	Percentage of world reserves
Saudi Arabia	262.73	22.3%	Kazakhstan	39.62	3.4%
Iran	132.46	11.2%	Libya	39.12	3.3%
Iraq	115.00	9.7%	Nigeria	35.25	3.0%
Kuwait	99.00	8.4%	United States	21.37	1.8%
United Arab Emirates	97.80	8.3%	China	17.07	1.4%
Venezuela	77.22	6.5%	Canada	16.80	1.4%
Russia	72.27	6.1%	Qatar	15.20	1.3%

The scramble for the Arctic

Case study

Another potential conflict is over the oil and gas reserves in the Arctic Ocean (Figure 7.13). Scientists believe rising temperatures could leave most of the Arctic ice free in summer months in a few decades' time. This would improve drilling access.

Denmark is trying to prove that a detached part of the underwater Lomonosov Ridge is an extension of Greenland, which is Danish territory. Russia has staked a claim by sending a submarine to plant a flag some 4 km below the North Pole. In 2008 Canada, Denmark, Norway, Russia, and the United States met in Greenland to discuss how to divide up the resources of the Arctic Ocean.

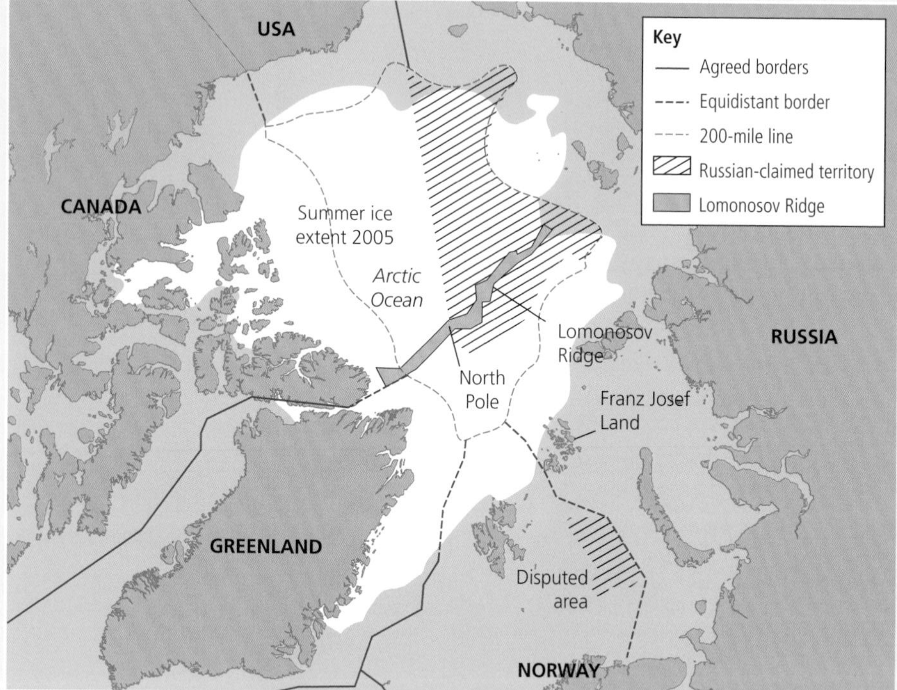

Figure 7.13 Territorial claims over the Arctic

According to the US Geographical Survey, the Arctic could hold a quarter of the world's undiscovered gas and oil reserves. This amounts to 90 billion barrels of oil and vast amounts of natural gas. Nearly 85 per cent of these deposits, they believe, are offshore. The five countries are racing to establish the limits of their territory, stretching far beyond their land borders.

Environmental groups have criticized the scramble for the Arctic, saying it will damage unique animal habitats, and have called for a treaty similar to that regulating the Antarctic, which bans military activity and mineral mining.

Under the 1982 UN Law of the Sea Convention, coastal states own the seabed beyond existing 370 km zones if it is part of a continental shelf of shallower waters. While the rules aim to fix shelves' outer limits on a clear geological basis, they have created a tangle of overlapping Arctic claims.

Factors which affect the choice of energy generation

The energy choices adopted by a society may be influenced by factors such as the availability and sustainability of resources, scientific and technological developments, cultural attitudes, and political, economic, and environmental factors. These in turn affect energy security and independence.

There are many important factors to consider in the choice of energy resources by societies. These include the following:

- The availability and reliability of supply – the UK used to have coal, then it had oil, but it has limited potential for solar or geothermal energy.
- Sustainability of supply – there are perhaps 40 years' worth of oil, 140 years' worth of coal, but an infinite supply of solar and geothermal energy in the world.
- Scientific and technological development – LEDCs use less energy and more basic energy (e.g. fuelwood) whereas MEDCs use more energy and more expensive forms (e.g. nuclear and oil).
- Political factors – in 1973 the Organization of Petroleum Exporting Countries (OPEC) raised the price of oil, causing other countries to develop their own cheaper resources.
- Economic factors such as cost of production, distribution, and use mean that nuclear power or tidal energy may be too expensive for many countries.

You should be able to discuss the factors that affect the choice of energy resources of two different societies. These could be two types of country (MEDC/LEDC) or could be two different societies within the same country (e.g. indigenous population and advanced-urban population in Brazil or Australia).

- Cultural attitudes – increased awareness of the problems of global warming or the risks associated with nuclear power may cause nations to change their energy choices.
- Environmental factors – certain climates allow for the use of certain types of energy such as solar or wind power; colder climates require more heating, warmer climates more air-conditioning.

The choice of energy sources adopted by different countries often has an historical basis. Large oil, coal, and gas reserves in certain countries (e.g. the UK) made fossil fuels an obvious choice for exploitation in those countries. Energy generation may also depend on economic, cultural, environmental, and technological factors.

Oil use in MEDCs is almost 50 per cent greater than in LEDCs, and fossil fuels in MEDCs account for 85 per cent of energy use as opposed to 58 per cent in LEDCs. The use of nuclear power is five times more important in MEDCs than in LEDCs. Biomass use in LEDCs is more than 10 times that in MEDCs. In some instances, both LEDCs and MEDCs have similar levels of use of a resource; examples are coal (25 per cent), HEP, geothermal, and solar power (6 per cent in LEDCs and 7 per cent in MEDCs) (Figure 7.14).

There are various explanations for these observed patterns. Oil is used extensively to produce petroleum products, and difference in oil use between LEDCs and MEDCs can be explained by the more prevalent use of cars in MEDCs. Biomass is very important in LEDCs as fuel for cooking, whereas MEDCs use gas or electricity (i.e. fossil fuels). The relatively small contribution of nuclear power may be due to the problems of disposing of nuclear fuel and the cost of nuclear technology. The relatively small proportion of nuclear power generation in MEDCs may also be affected by the general distrust of the industry in certain countries. Cultural fears, based on perception of nuclear accidents and waste, have made this a politically unpopular choice in many countries.

There are various factors are currently restricting the use of renewable energy sources. Fossil fuel resources are still economically cheaper to exploit, and the technology to harness renewable sources is currently not available on a large scale. Culture and tradition means that non-renewable resources are favoured, and the locations for renewable energy sources are often limited by available sites and local political issues. The low uptake of renewable energy globally means that renewables are not able to meet current demand. However, recession can change things.

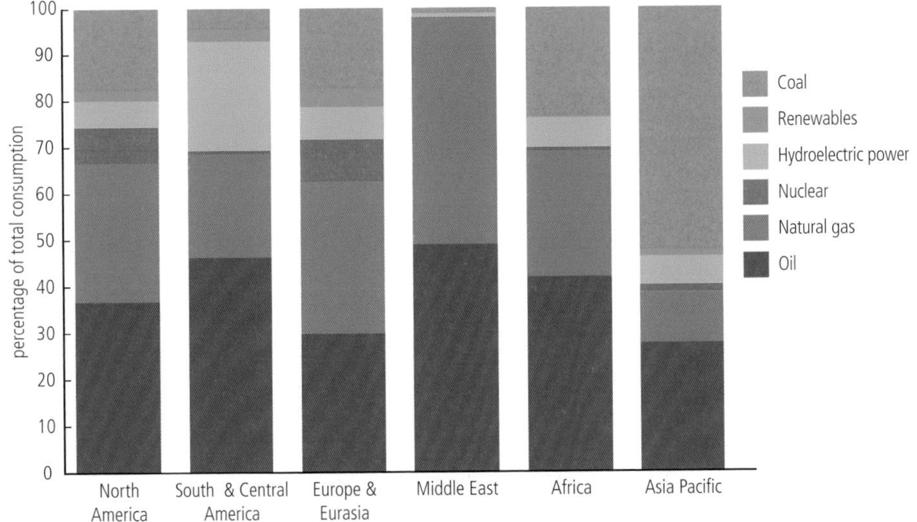

Figure 7.14 Relative contributions of different sources of commercial energy by world region

TOK

Nodding donkey, Brunei oilfield

Peak oil production refers to the year in which the world or an individual oil-producing country reaches its highest level of production, with production declining thereafter (Figure 7.15).

We depend on oil for many things: we use it for fuel, transport and heating, as a raw material in the plastics industry, and for fertilizer in food production. As oil production decreases after peak oil, so will all of these, unless we can find new materials and alternatives.

Peak oil varies country by country. The peak of oil discovery occurred in the 1960s, and by the 1980s the world was using more oil than was being discovered. Since then the gap between use and discovery has been increasing, and many countries have passed their peak oil production. However, reliable data is hard to come by, and some data is jealously guarded.

The International Energy Agency suggests that global peak oil will occur between 2013 and 2037. In contrast, the US Geological Survey suggests it will not occur until 2059. M King Hubbert, who popularized the theory of peak oil, predicted that it would occur in 1995 'if there were no changes in contemporary trends'. The Association for the Study of Peak Oil (ASPO) suggests it will be 2011. They claim that in 1950 the world consumed 4 billion barrels of oil per annum and the average discovery was 30 billion barrels per annum. Now, they say, the figures are reversed: new discoveries are around 4 billion barrels per year compared with consumption of 30 billion barrels.

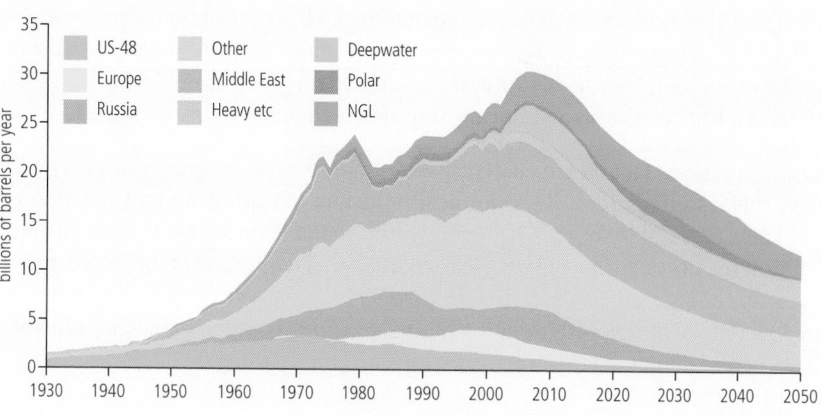

Figure 7.15 Peak oil

CONCEPTS: Environmental value systems

Environmental value systems are important in determining energy usage. A technocentric worldview may lead to continued use of fossil fuels in the belief that pollution can be minimized through technological solutions. A more ecocentric philosophy would see humanity living more within its means, and making better use of renewable sources of energy.

Changes in methods of energy generation employed can result from changing costs of production and from changes in social perspectives on established fuel supplies, which in turn may lead to shifts in environmental philosophy. In countries that rely on fossil fuels, the costs of exploitation have increased as the most easily accessible reserves have been used up, thus alternative sources been sought. The increasing cost of fossil fuels will change peoples' views of them.

At the same time, changing awareness of the environmental implications of fossil fuel exploitation (e.g. global warming) has led to a shift in attitude towards renewable energy sources (e.g. wind power), despite aesthetic and environmental implications, and increased demand for renewable, non-polluting sources. This has led to greater investment and research into alternatives (e.g. wind and tidal power).

Energy conservation

Improvements in energy efficiencies and energy conservation can limit growth in energy demand and contribute to energy security.

There are a number of possibilities for greater energy conservation. These include greater efficiency and the use of alternatives. Greater efficiency could be achieved through:

- smart meters
- enhanced environmental standards.

Improved building design could include:

- reduced energy use and emissions of carbon dioxide
- reduction of waste
- improved thermal efficiency of walls and windows
- reduction of heat loss between inner and outer walls
- energy-efficient domestic appliances
- improved daylighting by larger windows.

Energy saving is the quickest, most effective and cost-effective way of reducing greenhouse gas emissions. It also reduces the use of scarce resources.

Case study

Beddington Zero Energy Development (BedZED)

Beddington Zero Energy Development (BedZED) is an environmentally friendly housing development near Wallington, in the London Borough of Sutton. The 99 homes, and 1405 m^2 of work space were built in 2000–02.

Because of BedZED's low energy-emission concept, cars are discouraged; the project encourages public transport, cycling, and walking, and has limited parking space. It Is close to the tramline that runs between Croydon and Wimbledon.

Monitoring conducted in 2003 found that BedZED had achieved these reductions in comparison to UK averages:

- heating requirements were 88 per cent less
- hot-water consumption was by 57 per cent less
- electric power usage was 25 per cent less (and 11 per cent of the power used was produced by solar panels)
- car mileage of residents was 65 per cent less.

BedZED

BedZED has achieved this through a combination of:

- a zero energy import policy – renewable energy is generated on site by 777 m^2 of solar panels; tree waste is also used
- energy efficiency – houses face south, are triple glazed, and have high thermal insulation
- water efficiency – most rain water falling on the site is collected and reused
- low-impact materials – building materials were selected from renewable or recycled sources within 35 miles of the site, to minimize the energy required for transportation
- waste recycling
- encouraging eco-friendly transport – public transport, car-sharing, cycling.

Exercises

1. Outline the range of energy resources available to society.
2. Evaluate the advantages and disadvantages of two contrasting energy sources.
3. Discuss the factors that affect the choice of energy sources adopted by different societies.

Big questions

Having read this section, you can now discuss the following big questions:

- What strengths and weaknesses of the systems approach and the use of models have been revealed through this topic?
- To what extent have the solutions emerging from this topic been directed at *preventing* environmental impacts, *limiting* the extent of the environmental impacts, or *restoring* systems in which environmental impacts have already occurred?
- What value systems can you identify at play in the causes and approaches to resolving the issues addressed in this topic?
- How does your own value system compare with others you have encountered in the context of issues raised in this topic?
- How are the issues addressed in this topic of relevance to sustainability or sustainable development?
- In what ways might the solutions explored in this topic alter your predictions for the state of human societies and the biosphere some decades from now?

Points you may want to consider in your discussions:

- How does the systems approach help our understanding of energy choices and security?
- Why do countries lack energy security?
- Why do some societies continue to use fossil fuels despite the damage they do to the environment?
- How do environmental value systems affect the choice of energy supply?
- Compare your environmental value systems with at least two other value systems in relation to energy consumption.
- To what extent are global energy choices sustainable?
- How might energy choices evolve in the next decade?

7.2 Climate change: causes and impacts

Significant ideas

Climate change has been a normal feature of the Earth's history, but human activity has contributed to recent changes.

There has been a significant debate about the causes of climate change.

Climate change causes widespread and significant impacts.

Big questions

As you read this section, consider the following big questions:

- What strengths and weaknesses of the systems approach and the use of models have been revealed through this topic?
- To what extent have the solutions emerging from this topic been directed at *preventing* environmental impacts, *limiting* the extent of the environmental impacts, or *restoring* systems in which environmental impacts have already occurred?
- What value systems can you identify at play in the causes and approaches to resolving the issues addressed in this topic?

- How does your own value system compare with others you have encountered in the context of issues raised in this topic?
- How are the issues addressed in this topic of relevance to sustainability or sustainable development?
- In what ways might the solutions explored in this topic alter your predictions for the state of human societies and the biosphere some decades from now?

Knowledge and understanding

- Climate describes how the atmosphere behaves over relatively long periods of time whereas weather describes the conditions in the atmosphere over a short period of time.
- Weather and climate are affected by ocean and atmospheric circulatory systems.
- Human activities are increasing levels of greenhouse gases (e.g. carbon dioxide, methane, and water vapour) in the atmosphere, which lead to:
  - an increase in the mean global temperature
  - increased frequency and intensity of extreme weather events
  - the potential for long term change in climate and weather patterns
  - rise in sea level.
- The potential impacts of climate change may vary from one location to another and may be perceived as either adverse or beneficial. These impacts may include changes in water availability, distribution of biomes, and crop growing areas, loss of biodiversity and ecosystem services, coastal inundation, ocean acidification, and damage to human health.
- Both negative and positive feedback mechanisms are associated with climate change and may involve very long time lags.
- There has been significant debate due to conflicting environmental value systems surrounding the issue of climate change.
- Global climate models are complex and there is a degree of uncertainty regarding the accuracy of their predictions.

Climate and weather

The term *climate* refers to the average and extreme states of the atmosphere over a period of not less than 30 years. It includes variables such as temperature, rainfall, winds, humidity, cloud cover, and pressure. In contrast, *weather* refers to the state of the atmosphere at any particular moment in time. However, we usually look at the weather over a period of a few days to a week. The same variables are considered as for climate.

Climate describes how the atmosphere behaves over relatively long periods of time whereas weather describes the conditions in the atmosphere over a short period of time.

Climate and weather are affected by a number of factors such as atmospheric circulation, ocean circulation, latitude, altitude, distance from the sea, prevailing winds, aspect, and human activities.

To explore which companies are most responsible for emissions of greenhouse gases, go to www.pearsonhotlinks.co.uk, enter the book title or ISBN, and click on weblink 7.4.

TOK

The use of records covering a period of 30 years is considered adequate and so many climate statistics are based on the period 1980–2010. However, there are a number of arguments against using a 30-year period:

- the database is too short
- 1980–2010 has been a period of climate change and so is not a representative period
- it is impossible to create a 50-year maximum (the maximum size of event that would be expected once every 50 years), or 100-year return event (the size of an event that would occur, on average, only once every 100 years) from a record set of 30 years.

Weather and climate are affected by ocean and atmospheric circulatory systems.

Ocean circulatory systems

Warm ocean currents move water away from the equator, whereas cold ocean currents move water away from cold regions towards the equator (Figure 7.16). The major currents move huge masses of water over long distances. The warm Gulf Stream, for instance, transports 55 million cubic metres of water per second. Without the Gulf Stream, the temperate lands of north-west Europe would be more like the sub-Arctic. In addition, there is the Great Ocean Conveyor Belt. This deep, global-scale circulation of the ocean's waters effectively transfers heat from the tropics to colder regions, such as northern Europe (for more on this see page 224).

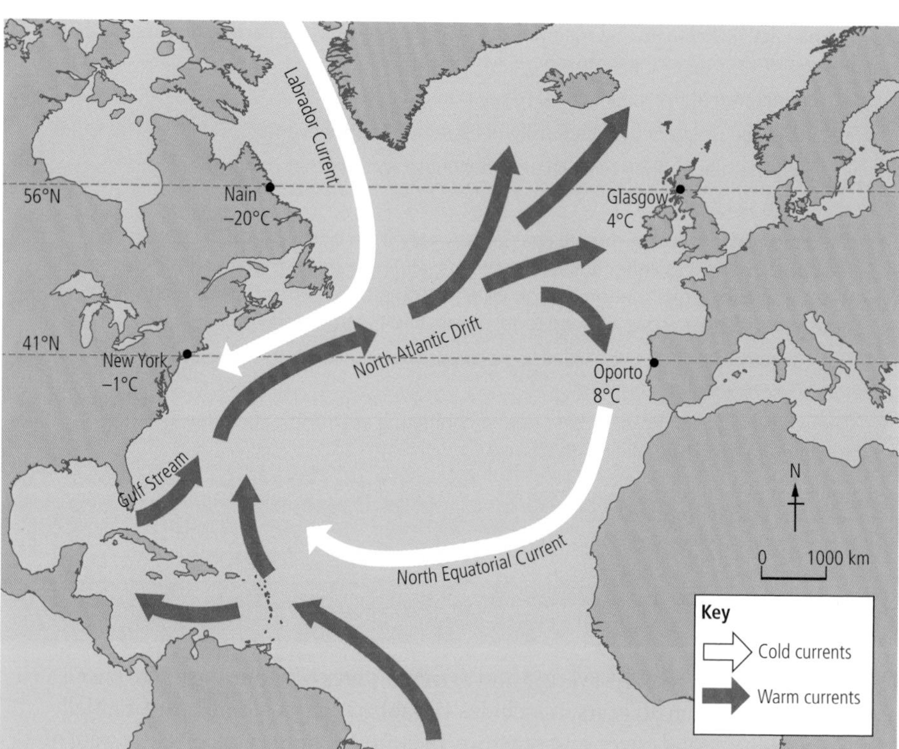

Figure 7.16 Warm and cold ocean currents in the north Atlantic Ocean

Specific heat capacity

Specific heat capacity is the amount of energy it takes to raise the temperature of 1 g of substance by 1 °C. It takes more energy to heat up water than it does to heat land. However, it takes longer for water to lose heat. Hence, land is hotter than the sea by day, but colder than the sea by night (Figure 7.17). Places that are close to the sea are cool by day, but mild by night. With increasing distance from the sea, this effect is reduced.

Surface ocean currents

Surface ocean currents are caused by the influence of prevailing winds blowing across the sea. The dominant pattern of surface ocean currents (known as gyres) is roughly circular. The pattern of these currents is clockwise in the northern hemisphere and anti-clockwise in the southern hemisphere. The main exception is the circumpolar current that flows around Antarctica from west to east. There is no equivalent current in the northern hemisphere because of the distribution of land and sea. Within the circulation of the gyres, water piles up into a dome. The effect of the rotation of the Earth is to cause water in the oceans to push westward; this piles up water on the western edge of ocean basins rather like water slopping in a bucket. The return flow is often narrow, fast-flowing currents like the Gulf Stream.

The effect of surface ocean currents on temperatures depends on whether the current is cold or warm. Warm currents from equatorial regions raise the temperature of polar areas (with the aid of prevailing westerly winds). However, the effect is only noticeable in winter. For example, the North Atlantic Drift (the northern extension of the Gulf Stream) raises the winter temperatures of north-west Europe. Some areas are more than 24 °C warmer than the average for their line of latitude. By contrast, there are other areas which are made colder by ocean currents. Cold currents such as the Labrador Current off the north-east coast of North America may reduce summer temperature, but only if the wind blows from the sea to the land.

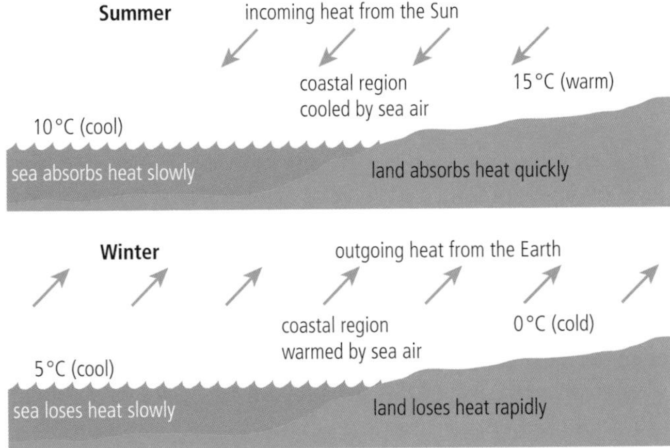

Figure 7.17 The effect of the sea on temperatures in coastal margins

Atmospheric circulatory systems

Air motion

The basic cause of air motion is the unequal heating of Earth's surface. Variable heating of the Earth causes variations in pressure and this in turn sets the air in motion. Most of the energy received by the Earth is in the tropical areas whereas there is a loss of energy from more polar areas. The major equalizing factor is the transfer of heat by air movement.

Pressure variations

Pressure is measured in millibars (mb) and is represented on maps by isobars, lines of equal pressure. On maps, pressure is adjusted to mean sea level (MSL). MSL pressure is 1013 mb, although the mean range is from 1060 mb in the Siberian winter high pressure system to 940 mb (although some intense low pressure storms may be much lower). The trend of pressure change is of more importance than the actual reading itself. Decline in pressure indicates wetter weather and rising pressure indicates drier weather.

Figure 7.18 General circulation models: (a) Hadley's three-cell model (Chapter 2, page 103) and (b) Palmen's model

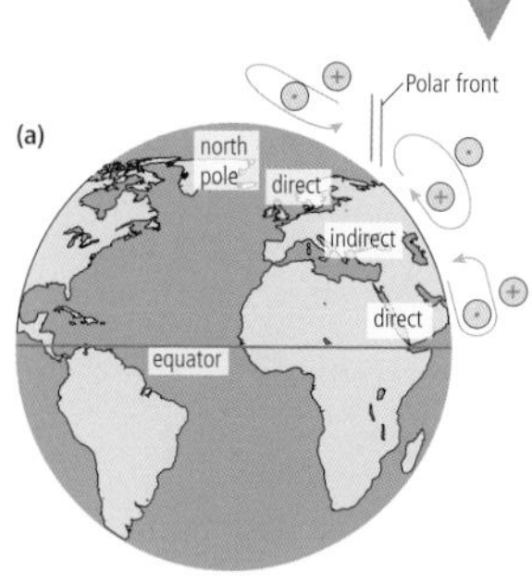

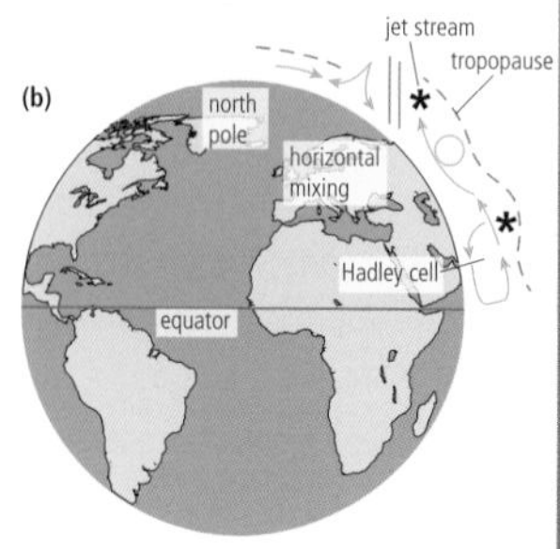

General circulation models

In 1735, George Hadley described the operation of the Hadley cell to explain atmospheric circulation. He suggested that direct heating of low latitudes forces air to rise by convection, the air then travels towards the poles but sinks at the subtropical anticyclone (high pressure belt). Hadley suggested that similar cells might exist in mid-latitudes and high latitudes. William Ferrel refined Hadley's ideas by suggesting that air in a Hadley cell rotates and interlinks with a mid-latitude cell, also rotating. These cells in turn rotate the polar cell (Figure 7.18a). The most recent models have refined the basic principles and include air motion in the upper atmosphere, in particular jet streams (very fast thermal winds) (Figure 7.18b).

Greenhouse gases and human activities

Human activities are increasing levels of greenhouse gases (e.g. carbon dioxide, methane, and water vapour) in the atmosphere, which leads to:

- **an increase in the mean global temperature**
- **increased frequency and intensity of extreme weather events**
- **the potential for long-term change in climate and weather patterns**
- **rise in sea level.**

The quantity of greenhouse gases emitted by any individual country depends on its economy, level of development, and societal expectations. Figure 7.19 show that greenhouse gas emissions in the USA are dominated by carbon dioxide. This reflects the high-energy demands of the USA. Transport, a lifestyle with expectations of air-conditioning, and other high-energy demands at home and work all lead to the high fuel economy seen there. On the other hand, emissions such as methane are lower, due to an absence of rice growing and large-scale cattle ranching, again reflecting the culture and environment of the USA. All greenhouse gases have global warming potential (GWP) (Table 7.1).

Table 7.1 Characteristics of greenhouse gases

Gas	Lifetime in the atmosphere (years)	100-year global warming potential (GWP)	Proportion of greenhouse gas emissions in CO_2 equivalent / %
carbon dioxide	5–200	1	77
methane	10	23	14
nitrous oxide	115	296	8
hydrofluorocarbons (HFCs)	1–250	10–12 000	0.5
perfluorocarbons (PFCs)	2 500	5 500	0.2
sulfur hexafluoride (SF)	3 200	22 200	1

To learn more about carbon dioxide and global warming, go to www.pearsonhotlinks.co.uk, enter the book title or ISBN, and click on weblink 7.5.

The GWP of carbon dioxide is defined as 1. Thus the GWP of all other gases is a measure of how much more a greenhouse gas contributes to global warming than carbon dioxide. Gases with very large GWPs are known as high GWP gases. They are hydrofluorocarbons (HFCs), perfluorocarbons (PFCs), and sulfurhexafluoride (SF), all of which are released by human activities. Over the past 100 years, atmospheric concentrations of these gases have risen dramatically. They are emitted from a variety of sources, including air conditioning and refrigeration equipment, where they were introduced as a replacement for chlorofluorocarbons (CFCs). In the 1990s, CFCs were

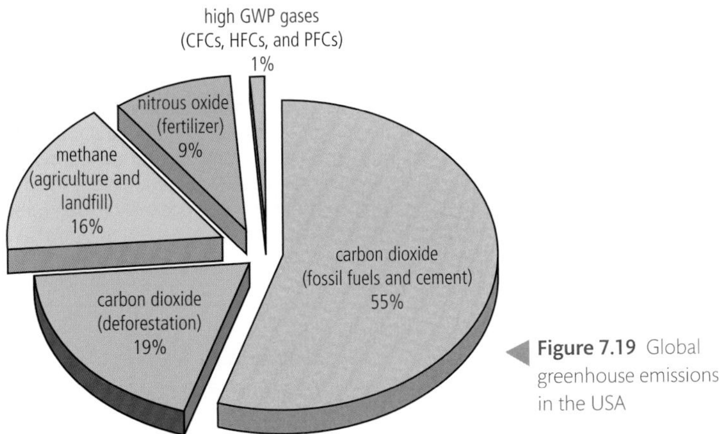

Figure 7.19 Global greenhouse emissions in the USA

phased out of use when their ozone-depleting character was recognized. Unfortunately, the replacement of CFCs with these other gases has led to other problems.

Annual production of greenhouse gases (in carbon equivalent) is approximately 55 billion tonnes, of which 35 billion tonnes is carbon dioxide from the burning of fossil fuels. Some 3.5 billion tonnes comes from land use changes (deforestation, clearing land for pasture).

Studies of cores taken from ice packs in Antarctica and Greenland show that the level of carbon dioxide remained stable at about 270 ppm from around 10 000 years ago until the mid-19th century. By 1957, the concentration of carbon dioxide in the atmosphere was 315 ppm and it has since risen to about 400 ppm in 2014. Most of the extra carbon dioxide has come from burning fossil fuels, especially coal, although some of the increase may be due to the deforestation of the rainforests. Much of the evidence for the greenhouse effect comes from ice cores dating back 160 000 years. These show that the Earth's temperature closely parallels the amount of carbon dioxide and methane in the atmosphere. For every 1 tonne of carbon burned, 4 tonnes of carbon dioxide are released. By the early 1980s, 5 gigatonnes (Gt) of fuel were burned every year. Roughly half the carbon dioxide produced is absorbed by natural sinks, such as vegetation and plankton. (1 Gt is 1000 million tonnes.)

The effects of global warming

We have already seen how biome distribution is influenced by temperature patterns, so any change in those patterns is likely to be followed by new global distribution of biomes poleward. Temperature increases are likely to have serious knock-on effects on the Earth's climate and ecosystems, and thus on human society (Table 7.2).

Table 7.2 Impact of temperature increase on aspects of the environment and society

Feature	Effect or impact
Environmental features	
ice and snow	melting of polar ice caps and glaciers
coastlines	increase in sea level causing coastal flooding
water cycle	increased flooding; more rapid circulation
ecosystems	change in biome distribution and species composition (e.g. poleward and altitudinal migration)
Societal features	
water resources	severe water shortages and possibly wars over supply
agriculture	may shift towards poles (away from drought areas)
coastal residential locations	relocation due to flooding and storms
human health	increased disease (e.g. risk of malaria)

Climate change in the geological past can show how we might expect biomes to move with changes in global temperature in the future (Figure 7.20). Models suggest a north/south shift in biomes relative to the equator (a latitudinal shift). Biomes will also move up slopes (altitudinal shift) as on mountains (Figure 7.21). Low-lying biomes such as mangroves may be lost due to changes in sea level.

The potential impacts of climate change may vary from one location to another and may be perceived as either adverse or beneficial. These impacts may include changes in water availability, distribution of biomes, and crop-growing areas, loss of biodiversity and ecosystem services, coastal inundation, ocean acidification, and damage to human health.

CONCEPTS: Biodiversity

Species composition in ecosystems is also likely to change. Climate change in the past has happened over long periods of time, allowing adaptation of animals to new conditions. Current increases in temperature are happening very rapidly so there is little time for organisms to adapt. Some organisms will be able to migrate to new areas where the conditions they need are found, but many face insurmountable obstacles to migration (e.g. rivers and oceans) or even no suitable habitat, and will become extinct (Figure 7.21). Tropical diseases can be expected to spread as warmer conditions are found in higher latitudes.

50–60 million years ago

cool temperate
temperate forest
subtropical forest
tropical forest
approximate coastline

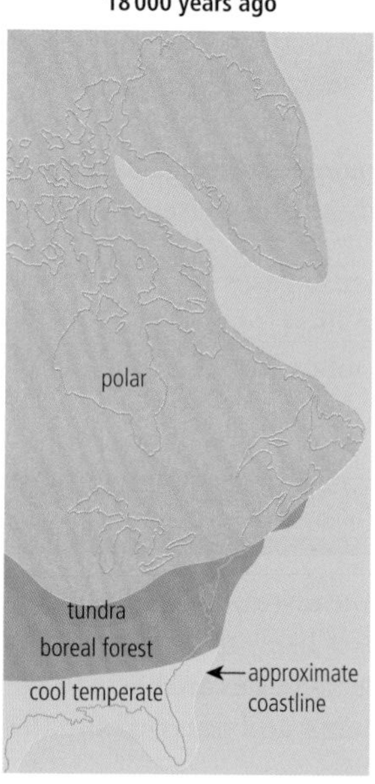

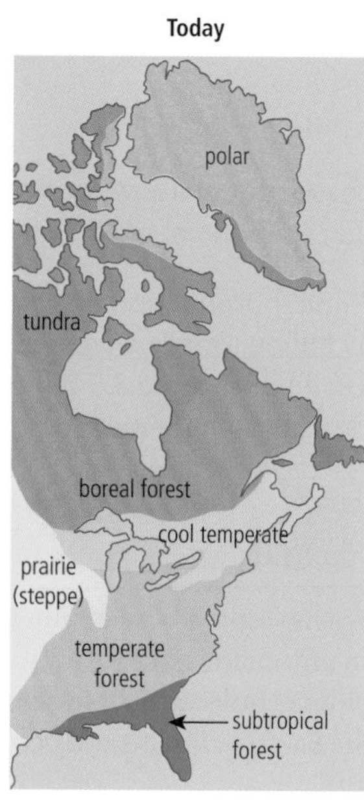

Figure 7.20 In the most recent globally warm period 50–60 million years ago, the Arctic was free of ice and subtropical forests extended northwards to Greenland. During the Pleistocene glaciations (18 000 years ago) these areas were covered by ice sheets. In the last 18 000 years, temperatures have increased and tundra and temperate forest biomes have shifted north. With further increases, all biomes are likely to move further poleward, with the probable disappearance of tundra and boreal forests.

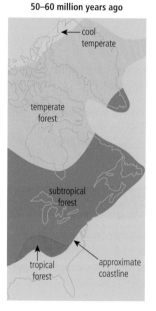

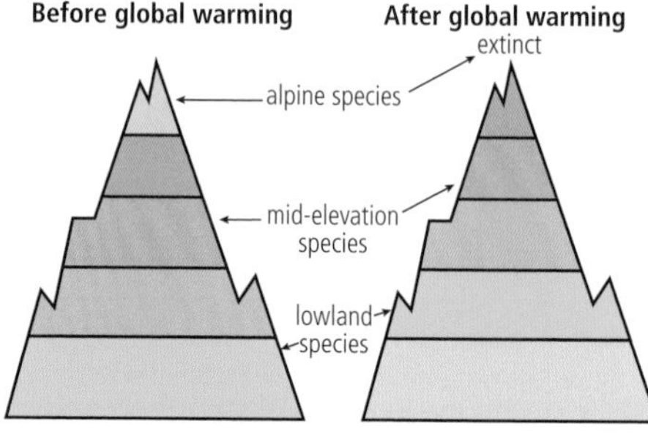

Figure 7.21 Alpine (mountain) species are particularly at risk, because zonation will move up the mountain.

To learn more about the impacts of climate change on different cities, go to www.pearsonhotlinks.co.uk, enter the book title or ISBN, and click on weblink 7.6.

Change in climate can lead to changes in weather patterns and rainfall (in both quantity and distribution). Climates may become more extreme and more unpredictable. An increase in more extreme weather conditions (e.g. hurricanes) can be expected as atmospheric patterns are disturbed.

Agriculture will be affected. Drought reduces crop yield, and the reduction in water resources will make it increasingly difficult for farmers in many areas to irrigate fields. Changes in the location of crop-growing areas can be expected, with movements north and south from the equator: recent models predict dramatic changes to the wheat-growing regions of the USA, with many becoming unviable by 2050 (Figure 7.22). This would have serious knock-on effects on the economy. Crop types may need to change and changing water resources will either limit or expand crop production depending on the region and local weather patterns.

Tourism is also likely to change as global warming changes weather patterns. Summer seasons may be extended and coastal resorts selling sun, sea, and sand may develop further north. Winter sports holidays, however, may be stopped by lack of snow and ice. Reduced precipitation in some areas may make some currently popular resorts uneconomic due to lack of water resources.

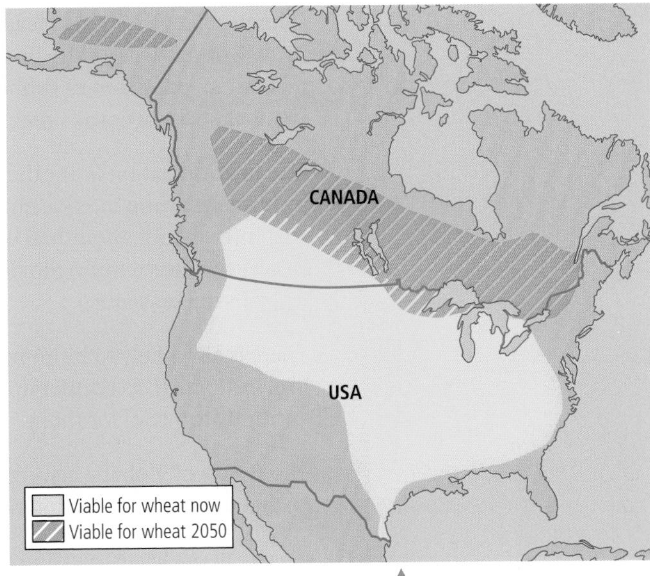

Figure 7.22 Scientists project a northward shift of wheat-growing in North America.

The impacts outlined above will indirectly lead to social problems such as hunger and conflict hunger, which will have implications for levels of economic development (Figure 7.23). National resource bases will change, which will drive economic, social, and cultural change. These issues are more likely to affect LEDCs than MEDCs because LEDCs are technologically and economically less able to cope. Moreover, a greater percentage of the population in LEDCs is already vulnerable to the effects of climate change (e.g. in Bangladesh, 20 per cent of gross domestic product (GDP) and 65 per cent of the labour force is involved in agriculture which would be threatened by floods in low-lying areas). Coastal flooding, caused by the melting of the polar ice caps and the thermal expansion of the oceans, will particularly affect countries that have land below sea level (e.g. the Netherlands) and may lead to economic and social stress due to loss of land and resources. LEDCs are also more likely to have weak infrastructure, communications and emergency services, which will also make them less able to respond to the effects of climate change.

Climate change may also lead to a reduction in biodiversity as species change their distribution in response to changes in climate. Some species – especially high altitude and high latitude species, have fewer options for migration, and so are more endangered.

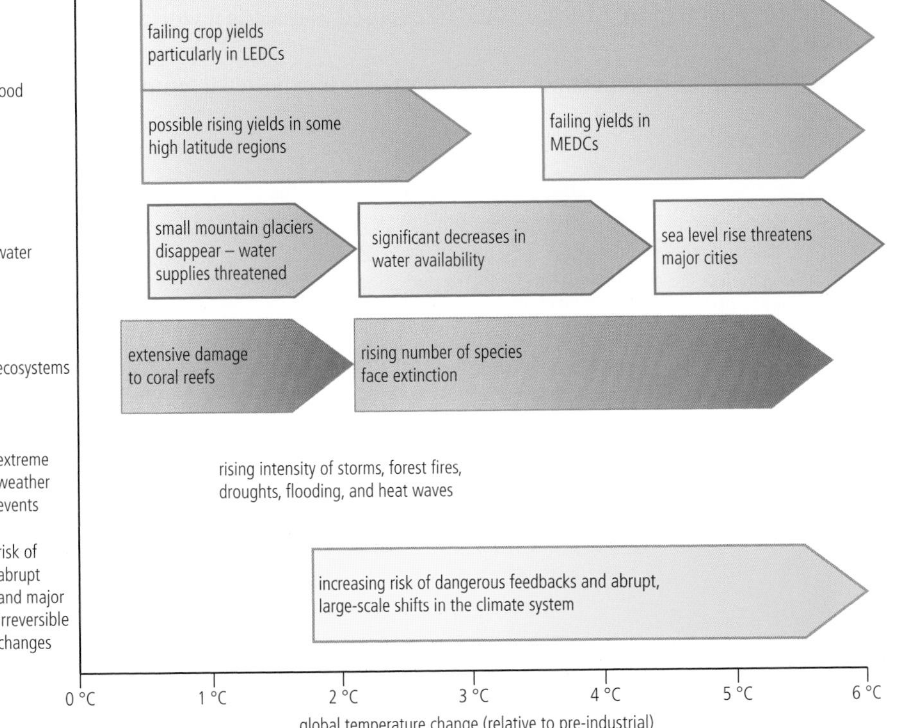

Figure 7.23 The projected impacts of climate change

Climate change may also lead to a loss in ecosystem services. Ecosystems provide a range of services (e.g. primary productivity, pollination, flood control, climate regulation, provision of timber products) and these may be placed at risk if there are significant changes to climate.

Ocean acidification is another problem. The ongoing decrease in ocean pH is caused by human carbon dioxide emissions caused by burning fossil fuels. The oceans currently absorb about half of the carbon dioxide produced by burning fossil fuels. However, when carbon dioxide dissolves in seawater, it turns to carbonic acid, and acidifies the seawater.

There is also likely to be an impact of human health. Diseases such as malaria are likely to spread, as temperatures rise and allow mosquitoes to breed in areas that are currently too cool for them.

The effects of global warming are very varied. Much depends on the scale of the changes. Impacts could include the following:

- A rise in sea levels, causing flooding in low-lying areas such as the Netherlands, Egypt, and Bangladesh; up to 200 million people could be displaced.
- Four million square kilometres of land, home to a twentieth of the world's population, is threatened by floods from melting glaciers.
- An increase in storm activity such as more frequent and intense hurricanes is likely (because of more atmospheric energy).
- Reduced rainfall over the USA, southern Europe, and the Commonwealth of Independent States (CIS) is likely.
- Changes in agricultural patterns (e.g. a decline in the USA's grain belt, but an increase in Canada's growing season) will occur.
- A 35 per cent drop in crop yields across Africa and the Middle East is expected if temperatures rise by 3 °C.
- An estimated 200 million more people could be exposed to hunger if world temperatures rise by 2 °C, 550 million if temperatures rise by 3 °C.
- Around 60 million more Africans could be exposed to malaria if world temperatures rise by 2 °C.
- Extinction of up to 40 per cent of species of wildlife is expected if temperatures rise by 2 °C.

Up to 4 billion people could be affected by water shortages if temperatures increase by 2 °C.

These are the likely impacts of climate change on the UK for the 2020s to 2050s.

- Temperatures are expected to increase at a rate of about 0.2 °C per decade; higher rates of increase will occur in the south-east, especially in summer. It will be about 0.9 °C warmer than the average of 1961–90 by the 2020s and about 1.6 °C warmer by the 2050s. This temperature change is equivalent to about a 200 km southward shift of the UK climate.
- Annual precipitation over the UK as a whole is expected to increase by about 5 per cent by the 2020s and by nearly 10 per cent by the 2050s; winter precipitation increases everywhere but mostly over the southern UK.
- The contrast in the UK's climate is likely to become exaggerated: the currently dry south-east will tend to become drier and the moist north-west will get wetter. Drought in the south-east and flooding in the north-west will both become more common.
- Sea level is expected to rise at a rate of about 5 cm per decade. This is likely to be increased in southern and eastern England by the sinking land, whereas in the north it will be offset by rising land (as a result of glacial melt).
- Extreme tidal levels will be experienced more frequently. For some east-coast locations, extreme tides could occur 10 to 20 times more frequently by the 2050s than they do now.
- By 2050, the UK will be subject to more intense rainfall events and extreme windspeeds, especially in the north. Gale frequencies will increase by about 30 per cent.

TOK

The poor are more vulnerable to global warming than the rich. However, on average, people in rich countries produce a larger amount of greenhouse gases per person than people in poor countries. Is this morally just?

Feedback and global warming

Both negative and positive feedback mechanisms are associated with climate change and may involve very long time lags.

You have already looked at feedback mechanisms and their pivotal role in creating equilibrium in ecosystems (Chapter 1, pages 29–36). Feedback mechanisms also play a key role in controlling the Earth's atmosphere, and any changes to these mechanisms are likely to have implications for the climate.

You should be able to discuss the feedback mechanisms that would be associated with a change in mean global temperature.

Positive feedback

- Some people believe that the impacts of global warming may be greatest in tundra environments. These are the regions of seasonal ice cover at the edges of permanent glaciers and ice sheets. Moreover, it is believed that the effects will be most noticeable in terms of winter warming. Melting of the polar ice caps results in less ice and lowers planetary albedo. Albedo is the amount of incoming solar energy reflected back into the atmosphere by the Earth's surface. Since ice is more reflective than water, less ice means less reflection. Lowering albedo increases the amount solar energy absorbed at the Earth's surface, and leads to increase in temperature (Figure 7.24).
- Rotting vegetation trapped under permafrost in the tundra releases methane that is unable to escape because of the ice covering. Increased thawing of permafrost will lead to an increase in methane levels as the gas escapes, adding to global warming gases in the atmosphere and thereby increasing mean global temperature (Figure 7.25).

Other mechanisms of positive feedback include:

- increased carbon dioxide released from increased biomass decomposition due to rising temperatures, especially in forest regions leads to further increase in temperature, as greenhouse gases are added to the atmosphere
- tropical deforestation increases warming and drying, causing a decline in the amount of rainforest
- increased forest cover in high latitudes decreases albedo and increases warming.

Feedback mechanisms associated with global warming tend to involve very long time lags. By the time effects appear, the mechanisms responsible may have already gone past the tipping point (the point of no return), and attempts to alleviate the problem may be doomed to fail.

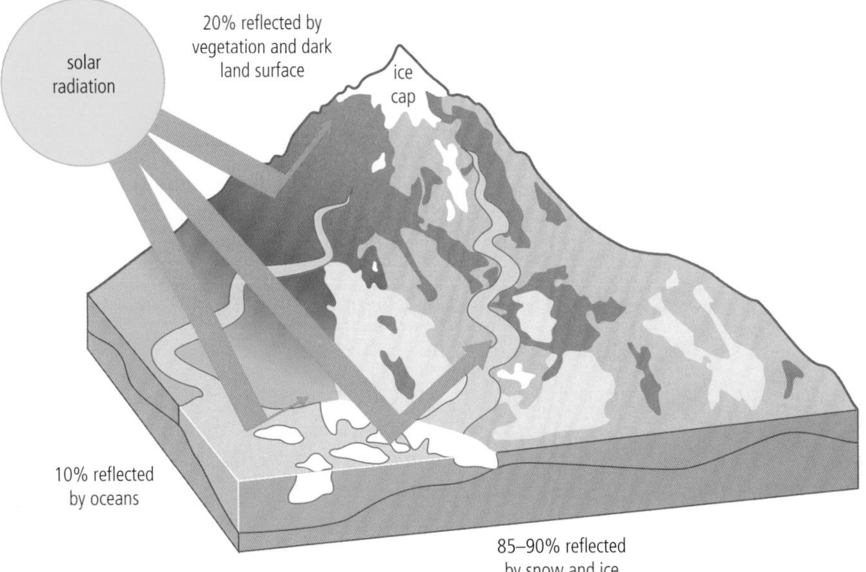

Figure 7.24 Melting ice reduces the planet's albedo

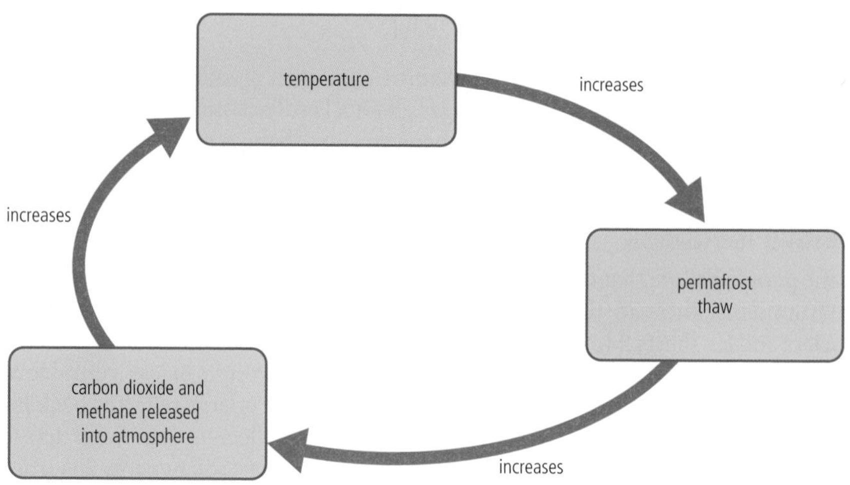

Figure 7.25 A positive feedback mechanism enhancing climate change. Such mechanisms are often linked to tipping points (i.e. the point at which the system actually becomes unstable and forms a new equilibrium).

Negative feedback

- Increased evaporation in low latitudes, due to higher levels of precipitation, will lead to increased snowfall on the polar ice caps, reducing the mean global temperature.
- Increase in carbon dioxide in the atmosphere leads to increased plant growth by allowing increased levels of photosynthesis. Increased plant biomass and productivity would reduce atmospheric concentrations of carbon dioxide.

CONCEPTS: Equilibrium

Increased evaporation in tropical and temperate latitudes leading to increased snowfalls in polar areas would be an example of negative feedback. There is some evidence to suggest this has happened in parts of Norway. Growth of glaciers and ice caps, albeit localized, can reduce mean temperatures. In addition, the amount of ice melting from the surface of the Greenland ice sheet in 2002 threatened a rise in sea levels and a return of very cold winters to the UK. Increased melting of the Greenland ice could shut off the currents of the Gulf Stream, allowing depressions to dump snow rather than rain on the UK and thus leading to a much colder continental climate. This would be comparable to the situation on the eastern seaboard of Canada, which is at the same latitude as the UK but without the mitigating effects of a warm ocean current like the Gulf Stream. Were this to happen, the sea could freeze and snow lie for weeks or months instead of a day of two. Thus, there is uncertainty as to whether global warming will lead to an increase or decrease in temperatures over the UK.

Since 2002, large areas of the Greenland ice shelf, previously too high and too cold to melt, have begun pouring billions of gallons of fresh water into the northern Atlantic. The Greenland ice sheet's maximum melt area increased on average by 16 per cent from 1979 to 2002. In particular, the north and north-east part of the ice sheet experienced melting up to an elevation of 2000 m.

Warm water from the tropics currently travels north past UK shores and warms the western coastlines of Europe as far north as Norway before sinking to the bottom of the ocean and returning south. It has been noted that this deepwater convection in the North Atlantic is slowing down. In past studies, changes in the North Atlantic circulation have been implicated in starting and stopping northern hemisphere glacial phases.

Both sea ice and glacier ice cool the Earth, reflecting back into space about 80 per cent of springtime sunshine and 40–50 per cent during the summer melt. But winter sea-ice cover slows heat loss from the relatively warm ocean to the cold atmosphere. Without large sea-ice masses at the poles to moderate the energy balance, warming increases.

Other mechanisms of negative feedback include:

- burning leads to increased aerosols and thus reduced solar radiation at the surface thereby causing cooling
- increased evaporation increases cooling.

Arguments about global warming

There has been significant debate due to conflicting environmental value systems surrounding the issue of climate change.

The arguments for the human influence on global climate change seem very persuasive, but there are also other non-human-related factors which can affect global climate. These include:

- greenhouse gases produced by a range of natural phenomena such as
  - volcanic activity producing greenhouse gases
  - methane released by animals and peat bogs
  - sunspot activity (variations in the Sun's radiation)
- volcanic ash and dust blocking out solar radiation
- Earth's tilt and variation in orbit around the Sun leading to seasonal and regional changes in temperature
- changes in albedo due to position and extent of ice sheets
- changes in albedo due to variations in cloud cover
- ocean currents leading to warming or cooling
- natural fluctuations in atmospheric circulation (e.g. El Niño and La Niña)
- bush fires releasing carbon into the atmosphere.

Contrasting perceptions of global warming

Al Gore

The former US Vice president, Al Gore, won the 2007 Nobel Peace Prize 'for efforts to build up and disseminate greater knowledge about man-made climate change, and to lay the foundations for the measures that are needed to counteract such change'. In his book, *An Inconvenient Truth*, Gore states:

> *Our climate crisis may at times appear to be happening slowly, but in fact it is happening very quickly – and has become a true planetary emergency. The Chinese expression for crisis consists of two characters. The first is a symbol for danger; the second is a symbol for opportunity. In order to face down the danger that is stalking us and move through it, we first have to recognize that we are facing a crisis. So why is it that our leaders seem not to hear such clarion warnings? Are they resisting the truth because they know that the moment they acknowledge it, they face a moral imperative to act? Is it simply more convenient to ignore the warnings? Perhaps, but inconvenient truths do not go away just because they are not seen. Indeed, when they are not responded to, their significance doesn't diminish; it grows.*

David Guggenheim's film *An Inconvenient Truth* was made by following Al Gore on the lecture circuit as he campaigned to raise public awareness of global warming.

CONCEPTS: Environmental value systems

Deep ecologists and **self-reliant soft ecologists** stress the importance of nature and the operation of natural processes. In contrast, technocentrics believe that technology can provide solutions to the problems created by human actions.

Campaigns such as Al Gore's book and film actively promote the environmental movement and reach a global audience. The science and evidence about climate change would have reached people not previously exposed to this information due to their background or education. Such publicity can influence personal viewpoints in a way not possible prior to the advent of mass communication.

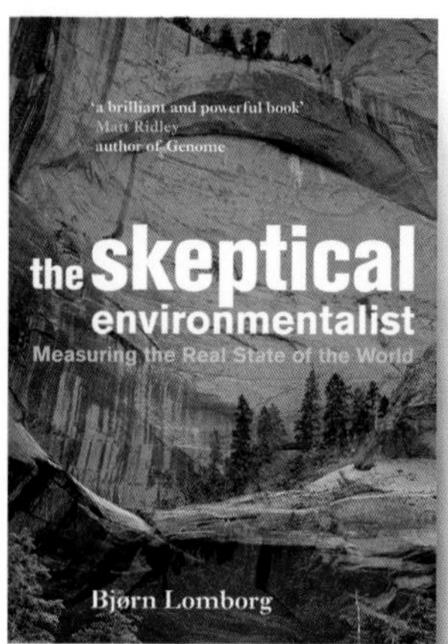

The Sceptical Environmentalist: Measuring the Real State of the World was written by the Danish environmentalist Bjorn Lomborg.

You should be able to evaluate contrasting viewpoints on the issue of climate change.

Bjorn Lomborg

In his book, Bjorn Lomborg argues that many global problems such as aspects of global warming, over-population, biodiversity loss, and water shortages are unsupported by statistical analysis. He argues that many of the problems are localized and are generally related to poverty, rather than being of global proportions.

Regarding global warming, he accepts that human activity has added to global temperature increases. However, he outlines a number of uncertainties (e.g. the simulation of future climate trends) and some weaknesses in the collection of data worldwide. Nevertheless, he finds issues relating to the politics and policy responses to global warming. For example, he concludes that the cost of combating global warming will be disproportionately borne by poor countries. He also believes that the Kyoto Protocol and various carbon taxes are among the least efficient ways of dealing with global warming. Instead he argues that a global cost–benefit analysis should be carried out before deciding on how to deal with global warming.

The Stern Report

The report by Sir Nicholas Stern, commissioned by the UK government, analysed the financial implications of climate change, and has a simple message:

- climate change is fundamentally altering the planet
- the risks incurred by inaction are high
- time is running out.

The report states that climate change poses a threat to the world economy and it will be cheaper to address the problem than to deal with the consequences. The global warming argument had seemed to be a dispute between the scientific case to act and the economic case not to. Now, economists are urging action.

The Stern report says doing nothing about climate change – the business-as-usual approach – would lead to a reduction in global *per capita* consumption of at least 5 per cent, now and forever. In other words, global warming could deliver an economic blow of between 5 per cent and 20 per cent of GDP to world economies because of natural disasters and the creation of hundreds of millions of climate refugees displaced by sea level rise. Dealing with the problem, by comparison, will cost just 1 per cent of GDP, equivalent to £184 billion.

The main points of the Stern Report are as follows.

- Carbon emissions have already increased global temperatures by more than 0.5 °C.
- With no action to cut greenhouse gases, we will warm the planet another 2–3 °C within 50 years.
- Temperature rise will transform the physical geography of the planet and the way humans live.
- Floods, disease, storms, and water shortages will become more frequent.
- The poorest countries will suffer earliest and most.
- The effects of climate change could cost the world between 5 per cent and 20 per cent of GDP.
- Action to reduce greenhouse gas emissions and the worst of global warming would cost 1 per cent of GDP.
- With no action, each tonne of carbon dioxide emitted will cause damage costing at least US$85.
- Levels of carbon dioxide in the atmosphere should be limited to the equivalent of 450–550 ppm.

- Action should include carbon taxes, new technology, and robust international agreements.

Complexity of the problem

There are a number of reasons for the complexity of the problem of climate change.

- It is an issue on a huge scale, which includes the atmosphere, the oceans, and the land mass of the whole planet.
- The interactions between these three factors (atmosphere, oceans, and land mass) are many and varied.
- It includes natural as well as anthropogenic forces.
- Not all the feedback mechanisms are fully understood.
- Many of the processes are long-term and the impact of many changes may not yet have occurred. For example, is global dimming reducing the effects of global warming? Are there specific tipping points that may be crossed?

Uncertainty of climate models

There are still people who question whether global warming really exists, and whether or not current changes are normal variation in the Earth's climate. There is also considerable uncertainty with climate models having both high and low estimates.

Global climate models are complex and there is a degree of uncertainty regarding the accuracy of their predictions.

For those who accept that global warming is occurring, there are two main questions.

- By how much is the planet warming?
- Where will the impacts of global warming be felt most?

SYSTEMS APPROACH

After the 9/11 attacks on the World Trade Center, the US air fleet was grounded for 3 days in the interests of national security. In the 3-day absence of vapour trails, the temperature rose by an average of 1.1 °C. Air pollution also has a cooling effect. Scientists who discovered the phenomenon called it **global dimming**. It is possible that global dimming has been masking what would be even faster global warming than is currently occurring.

Scientists have shown that from the 1950s to the early 1990s, the level of solar energy reaching the Earth's surface had dropped 9 per cent in Antarctica, 10 per cent in the USA, 16 per cent in parts of the UK, and almost 30 per cent in Russia. This was all due to high levels of pollution at that time. Natural particles in clean air provide condensation nuclei for water. Polluted air contains far more particles than clean air (e.g. ash, soot, sulfur dioxide) and therefore provides many more sites for water to bind to. The droplets formed tend to be smaller than natural droplets. Therefore, polluted clouds contain many more smaller water droplets than naturally occurring clouds. Many small water droplets reflect more sunlight than a fewer larger ones (Figure 7.26), so polluted clouds reflect far more light back into space, so preventing the Sun's heat from getting through to the Earth's surface.

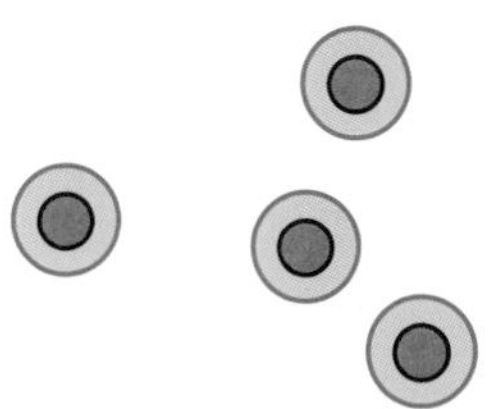

Water distributed around large natural particles forms a few large droplets with moderate reflectivity which eventually fall as rain.

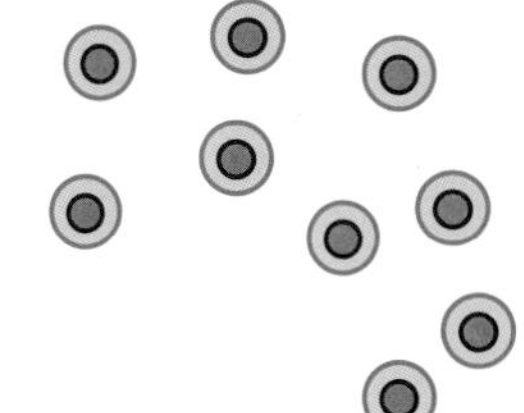

The same amount of water distributed around small polluting particles forms many small droplets with increased reflectivity. It does not fall as rain.

Figure 7.26 Global dimming: a few large droplets in the atmosphere hold the same amount of water as a greater number of smaller ones. However, the total surface area is less for the larger particles, so they reflect less sunlight than the smaller ones.

Exercises

1. Distinguish between weather and climate.
2. Explain how human activities have affected weather and climate.
3. Outline the impact of ocean currents on climate.
4. Briefly explain the impact of rising levels of greenhouse gases.
5. Explain how climate change may lead to (a) positive feedback and (b) negative feedback.

Big questions

Having read this section, you can now discuss the following big questions:

- What strengths and weaknesses of the systems approach and the use of models have been revealed through this topic? How does a systems approach help our understanding of climate change.
- To what extent have the solutions emerging from this topic been directed at *preventing* environmental impacts, *limiting* the extent of the environmental impacts, or *restoring* systems in which environmental impacts have already occurred? Evaluate the success of the Kyoto Protocol in stabilizing global climate change.
- What value systems can you identify at play in the causes and approaches to resolving the issues addressed in this topic? Explain why there are still uncertainties regarding global climate change.
- How does your own value system compare with others you have encountered in the context of issues raised in this topic? Evaluate measures of mitigation and adaptation.
- How are the issues addressed in this topic of relevance to sustainability or sustainable development? Can sustainable development be achieved without a solution to global climate change?
- In what ways might the solutions explored in this topic alter your predictions for the state of human societies and the biosphere some decades from now? Outline the obstacles to tackling global climate change.

Points you may want to consider in your discussions:

- How does a systems approach help our understanding of climate change?
- To what extent do we already know the solutions to climate change?
- How will we find them/why have they not been implemented?
- Why are some sectors of society in denial of climate change? Do you agree with them? Give reasons to support your answer.
- Examine the links between climate change and sustainability.
- Is climate change inevitable? Why?

7.3 Climate change – mitigation and adaptation

Significant ideas

Adaptation attempts to manage the impacts of climate change.

Mitigation attempts to reduce the causes of climate change.

Big questions

As you read this section, consider the following big questions:

- What strengths and weaknesses of the systems approach and the use of models have been revealed through this topic? How does a systems approach help our understanding of climate change?
- To what extent have the solutions emerging from this topic been directed at *preventing* environmental impacts, *limiting* the extent of the environmental impacts, or *restoring* systems in which environmental impacts have already occurred? Evaluate the success of the Kyoto Protocol in stabilizing global climate change.
- What value systems can you identify at play in the causes and approaches to resolving the issues addressed in this topic? Explain why there are still uncertainties regarding global climate change.
- How does your own value system compare with others you have encountered in the context of issues raised in this topic? Evaluate measures of mitigation and adaptation.
- How are the issues addressed in this topic of relevance to sustainability or sustainable development? Can sustainable development be achieved without a solution to global climate change?
- In what ways might the solutions explored in this topic alter your predictions for the state of human societies and the biosphere some decades from now? Outline the obstacles to tackling global climate change.

Knowledge and understanding

- Mitigation involves reduction and/or stabilization of greenhouse gas (GHG) emissions and their removal from the atmosphere.
- Mitigation strategies to reduce GHGs in general may include:
  - reduction of energy consumption
  - reduction of emissions of nitrogen oxides and methane from agriculture
  - use of alternatives to fossil fuel
  - geo-engineering.
- Mitigation strategies for carbon dioxide removal (CDR techniques) include:
  - protecting and enhancing carbon sinks through land management (e.g. United Nations - Reduction of Emissions from Deforestation and Forest Degradation in Developing Countries (UN-REDD) programme)
  - using biomass as fuel source
  - using carbon capture and storage (CCS)
  - enhancing carbon dioxide absorption by the oceans through either fertilizing oceans with nitrogen, phosphorus, iron (N/P/Fe) to encourage the biological pump, or increasing upwellings to release nutrients to the surface.
- Even if mitigation strategies drastically reduce future emissions of GHGs, past emissions will continue to have an effect for some time.
- Adaptation strategies can be used to reduce adverse affects and maximize any positive effects. Examples of adaptations include flood defences, vaccination programmes, desalinization plants, and planting of crops in previously unsuitable climates.
- Adaptive capacity varies from place to place and can be dependent on financial and technological resources. MEDCs can provide economic and technological support to LEDCs.
- There are international efforts and conferences to address mitigation and adaptation strategies for climate change (e.g. Intergovernmental Panel on Climate Change (IPCC), National Adaptation Programmes of Action (NAPAs), United Nations Framework Convention on Climate Change (UNFCCC)).

Pollution management strategies for global warming involve mitigation and adaptation.

Mitigation and adaptation

Mitigation involves reduction and/or stabilization of greenhouse gas (GHG) emissions and their removal from the atmosphere (Table 7.3). Politicians agreed to try to limit the increase in average global temperature to no more than 2 °C above the pre-industrial mean temperature.

Adaptation refers to efforts to live with the consequences of climate change. Adaption includes measures such as protecting cities from storm surges and protecting crops from high temperatures and droughts.

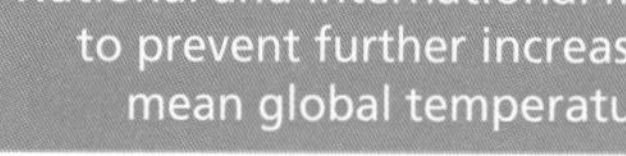

Table 7.3 Pollution management strategies for global warming

National and international methods to prevent further increases in mean global temperature	Ways in which individuals can contribute to the reduction of greenhouse gas emissions
• controlling the amount of atmospheric pollution • reducing atmospheric pollution • stopping forest clearance • increasing forest cover • developing alternative renewable energy sources • improving public transport • setting national limits on carbon emissions • developing carbon dioxide capture methods • recycling programmes	• grow your own food • eat locally produced foods • use energy-efficient products rather than traditional ones • reduce your heating – insulate your home • unplug standby appliances when not in use • turn off lights • reduce the use of air conditioning and refrigerants • use a manual lawnmower rather than an electric or diesel one • turn off taps • take a shower rather than a bath • walk more • ride a bike • use public transport • use biofuels • eat lower down the food chain (vegetables rather than meat) • buy organic food • get involved in local political action.

To learn about plans to use giant 'hoovers' to clear up China's smog, go to www.pearsonhotlinks.co.uk, enter the book title or ISBN, and click on weblink 7.7.

Mitigation

There is a limit to how much we can adapt. Hence, mitigation is essential. Nevertheless, it is important to adapt because climate change is happening and will continue to happen even if mitigation is highly successful. There is a lag time in the warming and it will take us some considerable time at the global scale to bring greenhouse gas emissions under control.

Mitigation strategies to reduce GHGs in general may include:

- **reduction of energy consumption**
- **reduction of emissions of nitrogen oxides and methane from agriculture**
- **use of alternatives to fossil fuels**
- **geo-engineering.**

Reduction of energy consumption

Climate change mitigation is an enormous priority. One of the main ways to reduce emissions of greenhouse gases is to consume less energy. This can be done in a variety of ways (e.g. increase the use of public transport, car pooling, and energy conservation). Since about three-quarters of greenhouse gas effect is due to carbon dioxide, the main mitigation priority should be to reduce emissions of carbon dioxide. Most carbon dioxide emissions come from burning fossil fuels, so the reduction of

energy-related carbon dioxide emissions is the main priority on the mitigation agenda. The second is land-use change (deforestation), and third is emissions of methane and nitrogen oxides.

There are many possible trajectories of future carbon dioxide emissions. The business-as-usual trajectory will result in emissions of about 60 billion tons of carbon dioxide in 2040. This is because the world economy is growing rapidly, and as it grows it uses more and more fossil fuels. With this trajectory, temperatures could increase by as much as 4–7 °C above pre-industrial levels.

If carbon dioxide levels could be held to below 450 ppm, it would be likely (but not certain) to contain the rise in temperature to less than 2 °C limit. Such a trajectory would be very tricky to accomplish.

The term *decarbonization* refers to a large reduction of carbon dioxide per value of gross world product. Since most of the carbon dioxide comes from burning fossil fuels, a sharp reduction in the use of fossil fuels or a large-scale system to capture and sequester the carbon dioxide is needed.

California, USA, has committed by law to reducing its carbon emissions by 80 per cent by 2050.

There are three key steps of to decarbonization.

1. *Energy efficiency* to achieve much greater output per unit of energy input. Much can be saved in heating, cooling, and ventilation of buildings, and electricity use by appliances.
2. *Reduce the emissions of carbon dioxide per mega-watt hour of electricity generated.* This involves increasing dramatically the amount of electricity generated by zero-emission energy, such as wind and solar power, while reducing the production of energy based on fossil fuels. It may also involve carbon capture and sequestration.
3. *Fuel shift*, from direct use of fossil fuels to electricity based on clean primary energy sources. This kind of substitution of fossil fuels by clean energy can happen in many sectors. Internal combustion engines in automobiles could be replaced by electric motors. Battery powered vehicles could be recharged on a renewable power grid.

Reductions of emissions of oxides of nitrogen and methane from agriculture

Agriculture is a major source of greenhouse gases, in particular oxides of nitrogen from fertilizers and methane from livestock. Agriculture would have less of an impact on global climate change if less chemical fertilizer were used and if there was less intensive livestock farming.

Alternatives to fossil fuels

There are many approaches to low carbon energy (e.g. photovoltaic (PV) cells to convert light energy into electrical energy). There is great potential to expand wind power in the US Midwest and the north-east desert regions of North Africa, northern Europe, and parts of central and western China.

Some of the areas that have the highest potential for renewable energy are very distant from centres of population, and their source of energy is intermittent. Hence, large-scale renewable energy will require the construction of new transmission lines. For example, the DESERTEC project (Figure 7.27) is designed to link North Africa, the Middle East, and Europe into a single grid. This system would tap the strong solar and wind potential of North Africa and the Arabian Peninsula both to supply energy for these economies and to export the surplus to Europe. The idea is potentially a key solution to Europe's unsolved challenge of deep decarbonization and an enormous boost to the economies of North Africa and the Middle East.

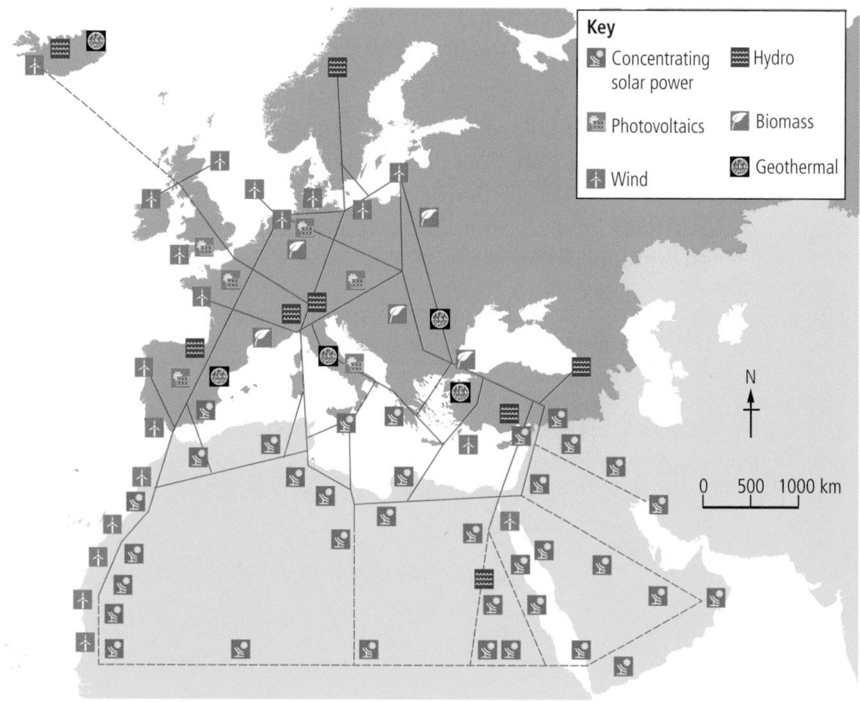

Figure 7.27 DESERTEC

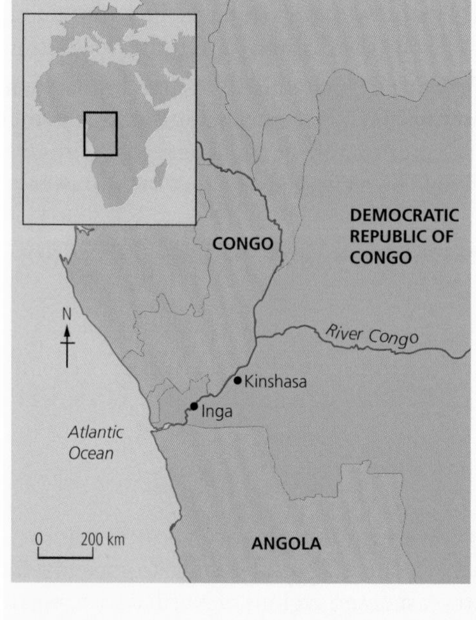

Figure 7.28 The location of the proposed Great Inga Dam, DR Congo

The Grand Inga hydroelectric scheme on the river Congo could produce 40 gigawatts of hydroelectric power. This is more than twice the amount of the world's largest HEP scheme (The Three Gorges Dam in China), and would double the amount of energy in Africa (Figure 7.28). The scheme is a joint one involving DR Congo, Burundi, Rwanda, and South Africa. Finance is to be provided by the African Development Bank, the Chinese Development Bank and the World Bank. Nevertheless, there is opposition from environmental groups, poor management, and uncertainty in one of the world's least developed and corrupt countries. Environmentalists warn of loss of species and agricultural land, coastal erosion, and the release of methane gas.

Geoengineering

Some scientists have suggested the use of sulfate aerosol particles into the air in order to dim the incoming sunlight and thereby cool the planet in order to offset the warming effects of carbon dioxide (global dimming, page 381). Another idea is to place giant mirrors in space in order to deflect some incoming solar radiation. These are fairly radical, and perhaps unworkable, ideas.

Carbon dioxide removal (CDR) techniques

Mitigation strategies for carbon dioxide removal (CDR techniques) include:

- protecting and enhancing carbon sinks through land management (e.g. United Nations – Reduction of Emissions from Deforestation and Forest Degradation in Developing Countries (UN-REDD) programme)
- using biomass as fuel source
- using carbon capture and storage (CCS)
- enhancing carbon dioxide absorption by the oceans through either fertilizing oceans with nitrogen, phosphorus, iron (N/P/Fe) to encourage the biological pump, or increasing upwellings to release nutrients to the surface.

UN-REDD Programme

The UN-REDD Programme is the United Nations Initiative on Reducing Emissions from Deforestation and Forest Degradation (REDD) in low-income countries. The programme was launched in 2008 and involves the Food and Agriculture Organization (FAO), the United Nations Development Programme (UNDP) and the United Nations Environment Programme (UNEP). In 2007, it was estimated that the loss of forests, through deforestation, forest degradation, and land-use changes, contributed approximately 17 per cent of global greenhouse gases.

The UN-REDD Programme supports nationally led REDD+ processes in over 50 countries, in Africa, Asia-Pacific, and Latin America. It provides support in two ways

- direct support to the design and implementation of UN-REDD National Programmes
- complementary support to national REDD+ action.

REDD stresses the role of conservation, the sustainable management of forests and the increase of forest carbon stocks. By June 2014, total funding had reached almost US$200 million. Norway is the leading donor country.

Biomass

Biomass can be used as a fuel source. There are many ways in which biomass can be grown and converted into fuels. The problem with biofuels, however, is that in many cases the production of the biomass for fuel competes with food production. Biomass production has driven up food prices while doing little to reduce net carbon dioxide emissions. Biomass can be a sustainable source of fuel as long as the annual use of biomass does not exceed the annual production of biomass for fuel.

Carbon capture and sequestration (CSS)

Currently, when fossil fuels are burned, the carbon dioxide enters the atmosphere, where it may reside for decades or even centuries. A potential solution is to capture the carbon dioxide instead of allowing it to accumulate in the atmosphere. Two main ways to do this have been proposed. The first is to capture the gas at the site where it is produced (e.g. the power plant), and then to store it underground in a geologic deposit (e.g. an abandoned oil reservoir). The second is to allow the gas to enter the atmosphere but then to remove it directly from the atmosphere using specially designed removal processes (e.g. collecting the carbon dioxide with special chemical sorbents that attract it). This latter approach is called direct air capture of carbon dioxide.

There are many technical and policy issues about the feasibility and cost-effectiveness of large-scale CCS technologies. First, how costly will it be to capture carbon dioxide on a large scale? How costly will it be to ship the carbon dioxide by a new pipeline network and then store the carbon dioxide in some safe, underground geologic deposit? If the carbon dioxide is put underground, how certain are we that the carbon dioxide will stay where it is put, rather than returning to the surface and then into the atmosphere? Tens of billions of tons of carbon dioxide would have to be captured and stored each year for CCS to play the leading role in addressing carbon dioxide emissions. Is there enough room for all this carbon? There is relatively little research and development underway to test the economic and geologic potential for large-scale CCS.

Enhancing carbon dioxide absorption

Carbon dioxide absorption can be increased by fertilizing the ocean with compounds of iron, nitrogen, and phosphorus. This introduces nutrients to the upper layer of the oceans, increases marine food production, and removes carbon dioxide from the atmosphere. In some cases it may trigger an algal bloom. The algae trap carbon dioxide and sink to the ocean floor.

Sperm whales transport iron from the deep ocean to the surface during prey consumption and defecation. Increasing the number of sperm whales in the Southern Ocean could help remove carbon from the atmosphere.

There are certain issues regarding ocean fertilization. The 2007 London Dumping Convention stated that there were concerns over 'the potential for large scale iron fertilization to have negative impacts on the marine environment and human health'.

Even if mitigation strategies drastically reduce future emissions of GHGs, past emissions will continue to have an effect for some time.

In some locations, upwelling currents bring nutrients to the surface (e.g. off the coast of Peru). These support large-scale fisheries and also help to lock up carbon. Artificial upwelling can be produced by devices that help pump water to the surface. Ocean wind turbines can also cause upwellings. These can then support plankton blooms which help lock up carbon. However, these are costly to build and run.

A new global deal on climate change is unlikely to be enforced much before 2020, and with the low oil prices of 2014–5, the potential for climate change is clear.

Adaptation strategies can be used to reduce adverse affects and maximize any positive effects. Examples of adaptations include flood defences, vaccination programmes, desalinization plants, and planting of crops in previously unsuitable climates.

Adaptation

It is possible to reduce human emissions of greenhouse gases substantially. The technologies are within reach, and measures such as energy efficiency, low-carbon electricity, and fuel switching (e.g. electrification of buildings and vehicles) are all possible. Nevertheless, even with these, carbon dioxide will continue to rise for a number of decades. By the time that the oceans warm, they are likely to add a further 0.6 °C to global temperatures. Thus, as well as trying to mitigate climate change, humanity needs to adapt to climate change as well. For example, in agriculture, crop varieties must be made more resilient to higher temperatures and more frequent floods and droughts. Cities need to be protected against rising ocean levels and greater likelihood of storm surges and flooding. The geographic range of some diseases, such as malaria, will spread as temperatures rise. More widespread vaccination programmes will be needed to deal with the spread of such diseases. To cope with changes in the supply of (and demand for) water, more desalinization plants will be required. These are expensive and some LEDCs may struggle to meet the demand for fresh water.

Case study

The Thames Barrier, London

The Thames Barrier protects London from the most severe form of tidal flooding. Before the construction of the Thames Barrier, an area of 116 km^2 was at risk. Much of London is built on the natural floodplain of the River Thames. Without flood defences, 420 000 homes on the Thames tidal floodplain would have a 0.1 per cent annual risk of flooding. This amounts to a flood risk property value of £80 billion. The risk from tidal flooding is expected to increase with rising tide levels. A gradual sea level rise of 4 mm per year is expected as a result of global warming. In addition, the south-east of England is sinking. London is 30 cm lower than it was at the end of the Second World War.

While the risk of tidal flooding from the River Thames is significant, the probability is low because the Thames Barrier and a number of other defences including the Barking and Dartford Creek Barriers provide London with a higher level of protection than any other part of the UK. The Thames Barrier became operational in October 1982. On average, it is closed three times per year. But during the winter of 2000/01 there were 24 closures.

Although the risk remains small, it is estimated that it will double between now and 2030. From 2030, the protection offered by the barrier will continue to decline unless improvements are made. By 2030, it is forecast that the barrier will have to close about 30 times per year to maintain the standards of tidal defence in the Thames Estuary. With closures this frequent, shipping would be severely disrupted. This has serious implications for London's ambitions to revitalize use of the River Thames for freight and passenger transport.

Similarly, the damage caused by Superstorm Sandy in New York in 2012 (estimated at US$19 billion) has called for improvements in storm barriers, which could cost as much as US$22 billion. The potential from more intense storms is driving up the size and the cost of protective barriers.

The Thames Barrier

Policy changes

Carbon dioxide imposes high costs on society (including future generations) but those who emit the carbon dioxide do not pay for the social costs that they cause. The result is the lack of a market incentive to shift from fossil fuels to the alternatives. Options here include carbon taxes, carbon trading, and carbon offset schemes.

Adaptive capacity varies from place to place and can be dependent on financial and technological resources. MEDCs can provide economic and technological support to LEDCs.

Carbon taxes

One option is for users of fossil fuels to have an extra 'carbon tax' equal to the social cost of the carbon dioxide emitted by the fuels. This would raise the costs of coal, oil, and gas compared with wind and solar, for example, thereby shifting the energy use towards the low-carbon options. Economists have suggested a carbon tax on the order of US$25–100 per tonne.

Some countries are introducing such taxes to encourage producers to reduce emissions of carbon dioxide. These environmental taxes can be implemented by taxing the burning of fossil fuels (coal, petroleum products such as gasoline and aviation fuel, and natural gas) in proportion to their carbon content. These taxes are most effective if they are applied internationally, but are also valuable nationally.

Carbon trading

Carbon trading is an attempt to create a market in which permits issued by governments to emit carbon dioxide can be traded. In Europe, carbon permits are traded through the Emissions Trading System (ETS). Governments set targets for the amount of carbon dioxide that can be emitted by industries; they are divided between individual plants or companies. Plants that exceed that limit are forced to buy permits from others that do not. The system works by putting a limit on total emissions. Critics argue that the targets are too generous.

Carbon offset schemes

Carbon offset schemes are designed to neutralize the effects of the carbon dioxide human activities produce by investing in projects that cut emissions elsewhere. Offset companies typically buy carbon credits from projects that plant trees or encourage a switch from fossil fuels to renewable energy. They sell credits to individuals and companies who want to go 'carbon neutral'. Some climate experts say offsets are dangerous because they dissuade people from changing their behaviour.

To learn more about carbon offset schemes, go to www.pearsonhotlinks.co.uk, enter the book title or ISBN, and click on weblink 7.8.

The effectiveness of reducing carbon dioxide emissions and the implications for economic growth and national development, vary depending on the level of development of the country in question. MEDCs have greater economic resources to help solve the problem (e.g. in developing alternative sources of energy).

There are international efforts and conferences to address mitigation and adaptation strategies for climate change (e.g. International Panel on Climate Change (IPCC), National Adaptation Programmes of Action (NAPAs), United Nations Framework Convention on Climate Change (UNFCCC)).

To learn more about the UN Framework Convention on Climate Change, go to www.pearsonhotlinks.co.uk, enter the book title or ISBN, and click on weblink 7.9.

The politics of carbon dioxide mitigation and adaptation

There are many obstacles to a low-carbon world: technological, economic, and political. Political obstacles are found nationally and internationally. The fossil fuel industry is the most powerful lobby group in the USA. Coal, oil, and gas interests have managed a veto on climate control regulations in the USA. The main obstacle to a global agreement on climate change remains the bargaining power of the major fossil fuel countries such as the USA, Canada, China, Russia, and the Middle East.

In 1992, at the Rio de Janeiro Earth Summit, the world's governments adopted the UN Framework Convention on Climate Change (UNFCC). Its main objective states:

> *The ultimate objective of this Convention and any related legal instruments that the Conference of the Parties may adopt is to achieve, in accordance with the relevant provisions of the Convention, stabilization of greenhouse gas concentrations in the atmosphere at a level that would prevent dangerous anthropogenic interference with the climate system. (Article 2).*

UNFCC went into effect in 1994 but has failed in its attempt to slow down greenhouse gas emissions. The Kyoto Protocol, signed in 1997, was the first major attempt to implement the treaty. MEDCs were required to cut their carbon emissions by 20 per cent by 2012, compared with 1990 emissions. LEDCs were not obliged to meet specific targets. The USA did not sign, and although Canada and Australia signed, they did not implement the treaty. At the same time, emissions soared in China and other rapidly industrializing nations. The USA argued that signing would give China a competitive edge in world trade.

The UNFCC encouraged MEDCs to lead the way in climate change mitigation. This was because:

- they have the technology
- they are better able to bear the costs of low-carbon energy developments
- they have caused a disproportionate amount of historic carbon dioxide
- LEDCs need time to develop their economies.

The Kyoto Protocol

The Toronto Conference of 1988 called for the reduction of carbon dioxide emissions by 20 per cent of 1988 levels by 2005. Also in 1988, the UNEP and the World Meteorological Organization (WMO) established the IPCC. In 1997, at an international and intergovernmental meeting in Kyoto, Japan, 183 countries around the world signed up to an agreement that called for the stabilization of greenhouse gas emissions at safe levels that would avoid serious climate change. The agreement is known as the Kyoto Protocol and aimed to cut greenhouse gas emissions by 5 per cent of their 1990 levels by 2012. It is currently the only legally binding international agreement that seeks to tackle the challenges of global warming.

The Kyoto Protocol came into force in 2005 and was due to expire in 2012, but was extended. Within the agreement, countries were allocated amounts of carbon dioxide they were allowed to emit. These permitted levels were divided into units – and countries with emission units to spare are allowed to sell them to countries that would otherwise go over their permitted allowance – this carbon trading now works like any other commodities market and is known as the carbon market.

The coordinating body of the Kyoto Protocol is the Conference of Parties (COP). It meets every year to discuss progress in dealing with climate change. Several of these climate conferences were focused on working out a framework for climate change negotiations after 2012, when the protocol expired. The COP meeting in 2008 agreed on the principles of financing a fund to help the poorest nations cope with the effects of climate change, and also approved a mechanism to incorporate forest protection into efforts. In 2009, a meeting in Copenhagen (the 15th COP meeting) failed to reach global climate agreement for the period from 2012, as a follow-up and successor to the Kyoto Protocol.

The use of alternative energy sources is also encouraged by the Kyoto Protocol. Avoidance of fossil fuels and the greater use of hydroelectric, solar, and wind power are actively encouraged, as these do no emit greenhouse gases. Nuclear power has been adopted by some countries (e.g. France) as a method of 'clean energy' generation, although the problems of disposal of the radioactive waste means this method of energy generation remains controversial. Incidents such as the Fukushima Daiichi disaster in Japan have made nuclear power less attractive for many.

The success of international solutions to climate change depends on:

- the extent to which governments wish to sign up to international agreements
- whether governments are preventive (i.e. act before climate change gets out of hand) or reactive (i.e. respond once the problem becomes obvious).

The majority of carbon emission targets in the Kyoto Protocol relate to MEDCs. In the future, it will be essential that LEDCs are brought into the agreement as they will be responsible for an increasing proportion of carbon dioxide emissions (Figure 7.29). Economic and social considerations make this a difficult area to negotiate – LEDCs rightly say that development in the MEDCs occurred on the back of an unenvironmental use of resources (largely, energy generation through burning fossil fuels). Allowing for development while capping greenhouse gas emissions calls for a new approach to energy generation (i.e. renewable rather than fossil fuel) and it is only through fresh thinking that solutions will be found.

The complexity of these issues was one of the reasons why the Copenhagen Summit failed to reach an agreement between all the parties.

Total 1990 emissions: 6 billion tons

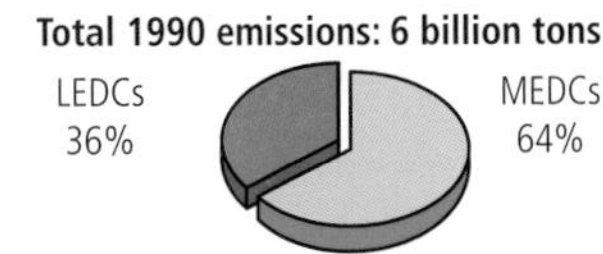

Total 2015 emissions: 8.45 billion tons

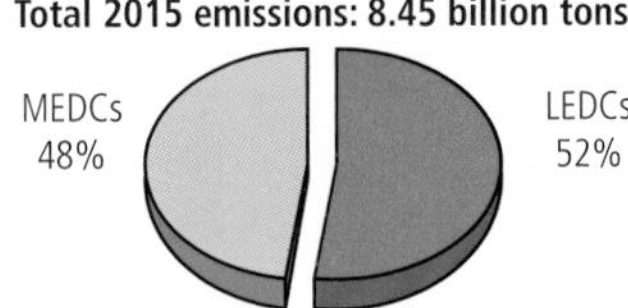

Total 2100 emissions: 19.8 billion tons

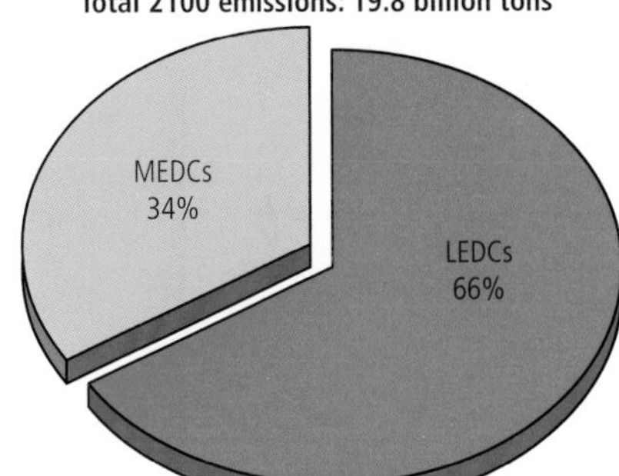

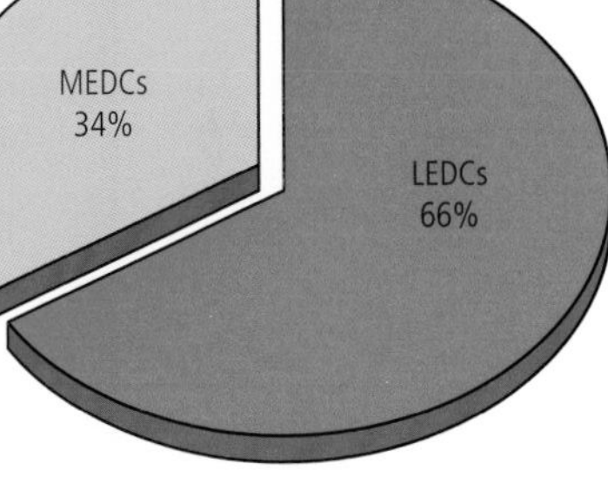

Figure 7.29 Changing patterns in carbon emissions

CONCEPTS: Environmental value systems

It may be difficult to change the expectations of those who live in MEDCs: a comfortable home, one's own transport, access to cheap flights around the globe, and general high-energy culture.

People in LEDCs think it is unreasonable to expect their countries to curb emissions until they have caught up with standards in MEDCs. But ultimately, solutions must be found if irreversible climate change is to be avoided.

The Paris Agreement

In 2014, politicians met in Lima, Peru to finalize a draft for the 2015 Paris Agreement, in which it is hoped that the world's leaders will be able to sign a binding agreement on climate change.

The 2015 Paris Summit holds hope for a new deal on climate change. The reasons for such hope were varied:

- President Obama of the USA is keen to cut carbon emissions by between 26 per cent and 28 per cent of 2005 levels by 2025
- China's President Xi Jinping offered a time-scale for peak emissions by 2030
- the EU agreed a 40 per cent cut in greenhouse emissions by 2030 compared with 1990.

Although there is still no ongoing obligation for NICs such as Brazil, India, Russia, and China to cut emissions, they have accepted the need for a cap. At Christmas 2014, Pope Francis, spiritual leader of the world's 1 billion Catholics promised that he would emphasize the responsibility of all Catholics to take action on scientific as well as moral grounds.

To learn about the 'artificial volcano' that scientists are experimenting with in the hope of reducing climate change, go to www.pearsonhotlinks.co.uk, enter the book title or ISBN, and click on weblink 7.10.

CONCEPTS: Environmental value systems

Technocentrists would argue that humankind can continue to emit greenhouse gases because technological solutions will be found to solve the problem. People with an ecocentric point of view would put the counter-argument that reducing emissions is the first essential step to combating climate change.

The Intergovernmental Panel on Climate Change

The IPCC is the international body for assessing the science related to climate change. It was set up by the WMO and UNEP to provide policymakers with regular assessments of climate change, its impacts and future risks. IPCC assessments provide a scientific basis for governments at all levels to develop climate-related policies.

Responding to the 2014 IPCC report on climate change, the UN secretary general, Ban Ki-moon, said 'Leaders must act. Time is not on our side.' He said that quick, decisive action would build a better and sustainable future, while inaction would be costly.

The report states that:

- it is economically affordable
- that carbon emissions will ultimately have to fall to zero
- global poverty can only be reduced by halting global warming
- carbon emissions, mainly from burning coal, oil, and gas, are currently rising to record levels, not falling.

The IPCC considers carbon capture and storage (CCS) – the unproven technology which aims to bury carbon dioxide underground – extremely important. Abandoning nuclear power or deploying only limited wind or solar power increases the cost of emission cuts by just 6–7 per cent. The report also states that behavioural changes, such as dietary changes that could involve eating less meat, can have a role in cutting emissions.

National adaptation programmes of action

The main content of national adaptation programmes of action (NAPAs) is a list of ranked priority adaptation activities and projects. NAPAs focus on urgent and immediate needs – those for which further delay could increase vulnerability or lead to increased costs at a later stage. NAPAs use existing information, are action oriented and country driven. The steps for the preparation of the NAPAs include:

- synthesis of available information
- assessment of vulnerability to current climate and extreme events
- identification of key adaptation measures as well as criteria for prioritizing activities
- selection of a prioritized short list of activities.

By 2008, the UNFCCC had received NAPAs from 39 developing countries.

Case study

Bangladesh's updated NAPA, 2009

Bangladesh is already vulnerable to climate change. It is expected that climate change will not only bring changes to the climate but also more natural hazards that will lead to changes in physical, social, and economic systems (e.g. sea level rise, higher and more erratic rainfall, and impacts on biodiversity, agriculture, health, water, and sanitation).

Climate change could affect over 70 million people in Bangladesh on account of its geographic position, low elevation, high population density, poor infrastructure, high levels of poverty, and high dependence on natural resources. Coastal areas are more vulnerable than other areas, especially to hurricane activity and storm surges. It is predicted that for a 45 cm rise of sea level, 10–15 per cent of the land could be flooded by 2050, resulting in over 35 million climate refugees. Such impacts would make it hard for Bangladesh to achieve its MDGs. Moreover, the OECD and World Bank estimate that 40 per cent of the aid given to Bangladesh may be climate sensitive or at risk.

Different areas will require different action plans. Coastal areas may be affected by cyclone activity and saline intrusions; floodplains in the central areas are prone to river floods; the north-west regions are prone to drought; and the hilly north-east is subject to erosion and landslides (Tables 7.4 and 7.5).

Table 7.4 Causes of impacts, vulnerable areas, and impacted sectors

Climate and related elements	Critical vulnerable areas	Most impacted sectors
temperature rise and drought	• north-west	• agriculture (crop, livestock, fisheries) • water • energy • health
sea level rise and salinity intrusion	• coastal area • islands	• agriculture (crop, livestock, fisheries) • water (water logging, drinking water, urban) • human settlement • energy • health
floods	• central region • north-east region • char land	• agriculture (crop, livestock, fisheries) • water (urban, industry) • infrastructure • human settlement • health • disaster • energy
cyclone and storm surge	• coastal and marine zone	• marine fishing • infrastructure • human settlement • life and property
drainage congestion	• coastal area • urban • south-west	• water (navigation) • agriculture (crop)

continued

Table 7.5 Intensity of impacts on different sectors due to climate change

	Physical vulnerability context							
	Extreme temperature	Sea level rise		Drought	Flood		Cyclone and storm surges	Erosion and deposition
		Coastal inundation	Salinity intrusion		River flood	Flash flood		
crop agriculture	+++	++	+++	+++	+	++	+++	-
fisheries	++	+	+	++	++	+	+	-
livestock	++	++	+++	-	-	+	+++	-
infrastructure	+	++	N/A	_	++	+	+	+++
industries	++	+++	++	-	++	+	+	-
biodiversity	++	+++	++	-	++	-	+	-
health	+++	+	+++	-	++	-	++	-
human settlement	-	-	-	-	-	-	+++	+++
energy	++	+	-	-	+	-	+	-

Adaptation measures

Research and knowledge management

1. Model climate change scenario.
2. Model the likely hydrological impacts of climate change on the Ganges–Brahmaputra–Meghna river system.
3. Monitor and research the impacts of climate change on ecosystems and biodiversity.
4. Research the likely impacts of climate on the macro-economy and key sectors of Bangladesh (e.g. livelihoods and food security).
5. Research the linkages between (a) climate change and poverty; and (b) climate change, poverty and health to increase the resilience of poor and vulnerable households to climate change.

Agriculture, fisheries and livestock

1. Develop climate change resilient cropping systems (e.g. research to develop crop varieties, which are tolerant of flooding, drought, and salinity).
2. Development and protection of dry season fish refuge.
3. Diversification of aquaculture techniques in the flood-prone north-central region of Bangladesh.
4. Reduce stresses in livestock and poultry due to temperature extremes.

Health

1. Implement surveillance systems for existing and new disease risks.
2. Development of Strategy for Alternative Sources of Safe Drinking Water and Sanitation Programme in areas at risk from climate change (e.g. coastal areas, flood- and drought-prone areas).

Building climate resilient infrastructure

1. Repair and rehabilitate existing infrastructure (e.g. coastal embankments, river embankments and drainage systems, urban drainage systems).
2. Plan, design, and construct urgently needed new infrastructure (e.g. cyclone shelters, coastal and river embankments, and water management systems; urban drainage systems, river erosion control works, flood shelters) to meet the challenges of climate change.

Disaster management

1. Strengthen the capacity to manage natural disasters.
2. Strengthen the cyclone, storm surge and flood early warning systems to enable better forecasting.

Livelihood

1. Increase the resilience of vulnerable groups, including women and children, through development of community-level adaptation, livelihood diversification, better access to basic services, and social protection (e.g. safety nets, insurance).

Biodiversity

1. Undertake community participation afforestation programme.

Exercises

1. Describe how human activities add to greenhouse gases.
2. Discuss the potential effects of increased mean global temperature.
3. Discuss the feedback mechanisms that would be associated with an increase in mean global temperature.
4. Describe and evaluate pollution management strategies to address the issue of global warming.
5. Outline the arguments surrounding global warming.
6. Evaluate contrasting human perceptions of the issue of global warming.

CHALLENGE YOURSELF

Communication skills 

Organize a class debate: Should developing nations be allowed to have larger carbon emissions per person so that they can catch up levels of development in richer countries?

Big questions

Having read this section, you can now discuss the following big questions:

- What strengths and weaknesses of the systems approach and the use of models have been revealed through this topic? How does a systems approach help our understanding of climate change?
- To what extent have the solutions emerging from this topic been directed at *preventing* environmental impacts, *limiting* the extent of the environmental impacts, or *restoring* systems in which environmental impacts have already occurred? Evaluate the success of the Kyoto Protocol in stabilizing global climate change.
- What value systems can you identify at play in the causes and approaches to resolving the issues addressed in this topic? Explain why there are still uncertainties regarding global climate change.
- How does your own value system compare with others you have encountered in the context of issues raised in this topic? Evaluate measures of mitigation and adaptation.
- How are the issues addressed in this topic of relevance to sustainability or sustainable development? Can sustainable development be achieved without a solution to global climate change?
- In what ways might the solutions explored in this topic alter your predictions for the state of human societies and the biosphere some decades from now? Outline the obstacles to tackling global climate change.

Points you may want to consider in your discussions:

- How do models and a systems approach help our understanding of climate mitigation and adaptation?
- How far do we already know the answers to climate mitigation and adaptation?
- In what ways do different people/societies consider climate mitigation and adaptation?
- What do you think is the best way forward? Justify your answer.
- Which is more sustainable – mitigation or adaptation?
- How might the solutions to climate change evolve in the future?

Practice questions

1 a Using the figure below, describe the changes in oil reserves between 1993 and 2013. [4]

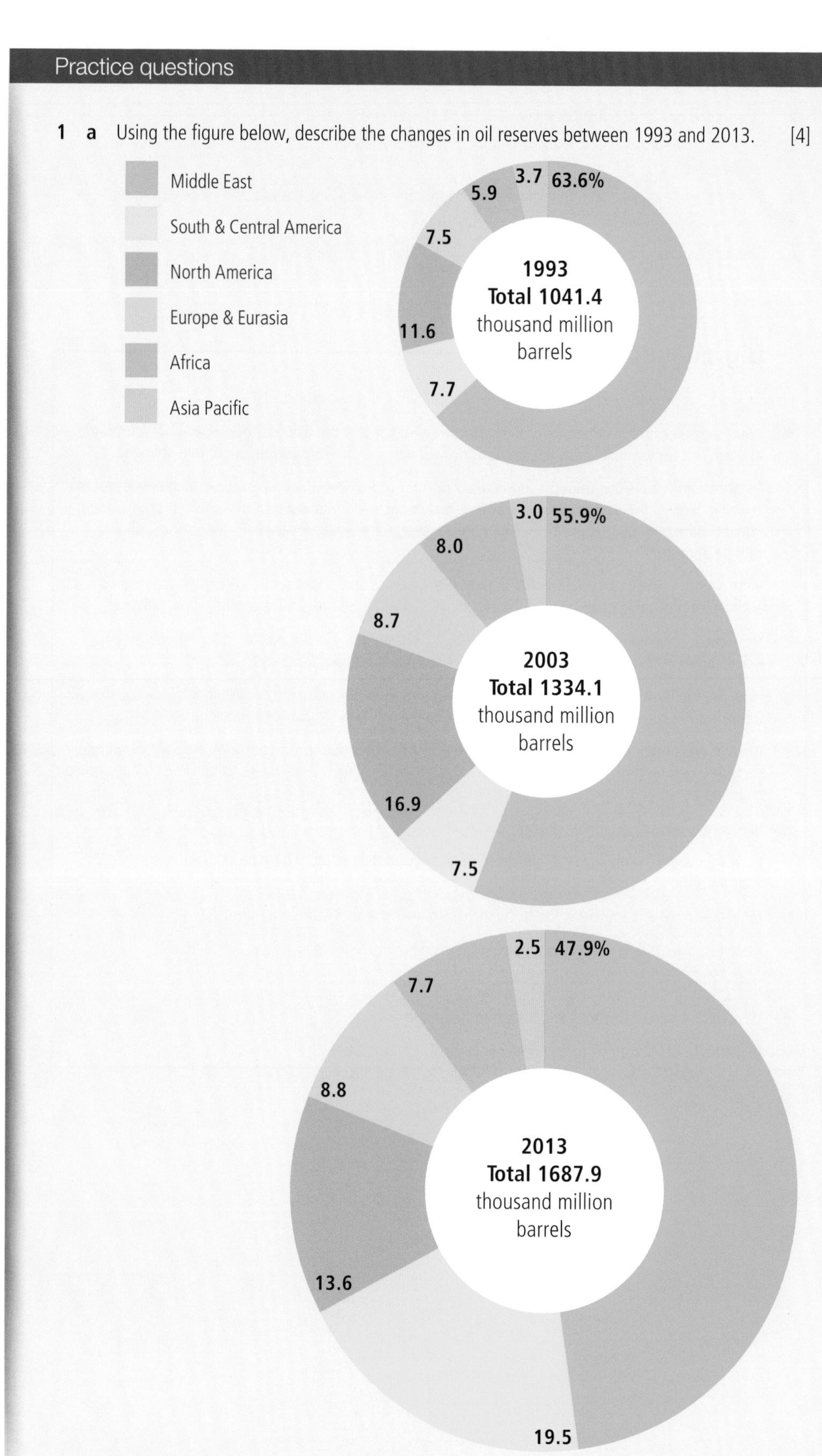

b The map below shows variations in consumption of oil *per capita* (tonnes) in 2013.

Comment on the variations shown on the map. [4]

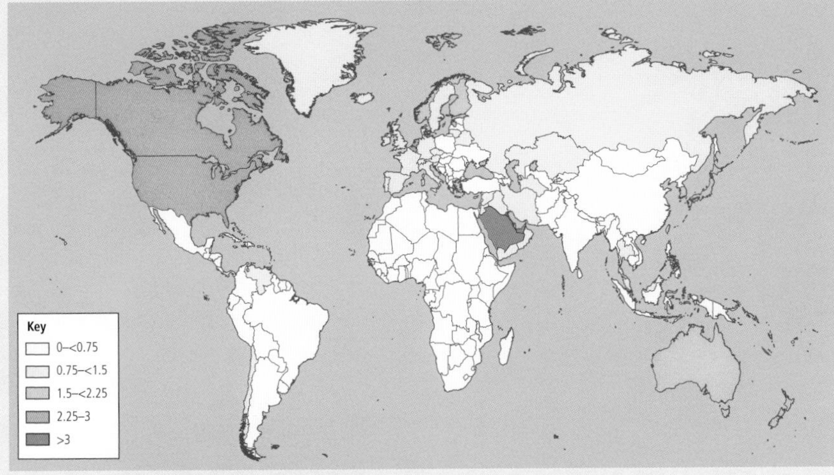

c Evaluate the advantages and disadvantages of two contrasting energy sources and discuss the economic factors that affect the choice of these energy sources by different named societies. [9]

2 a i Explain how global warming may lead to changes in biotic components of ecosystems. [3]

ii Suggest why these changes are significant for humans. [3]

b Discuss why there is uncertainty about predictions concerning global warming. [6]

c In 2007, the Intergovernmental Panel on Climate Change (IPCC) released its findings on global warming. The IPCC stated that the climate was definitely warming and was most likely caused by human activities. Evaluate this point of view. [6]

3 Justify the argument that changes need to be made in individual lifestyles to effectively address the issues of global warming. [6]

4 Describe ecocentric and technocentric responses to global warming, and justify which may be more effective in reducing the impacts of global warming. [7]

08 Human systems and resource use

8.1 Human population dynamics

Opposite: As populations continue to grow, more people live in urban areas (such as in Pushkar in India, as seen in this photo). This creates environmental stress and requires the use of a wide range of resources.

Significant ideas

A variety of models and indicators are used to quantify human population dynamics.

Human population growth rates are impacted by a complex range of changing factors.

Big questions

As you read this section, consider the following big questions:

- What strengths and weaknesses of the systems approach and the use of models have been revealed through this topic?
- To what extent have the solutions emerging from this topic been directed at *preventing* environmental impacts, *limiting* the extent of the environmental impacts, or *restoring* systems in which environmental impacts have already occurred?
- What value systems can you identify at play in the causes and approaches to resolving the issues addressed in this topic?
- How does your own value system compare with others you have encountered in the context of issues raised in this topic?
- How are the issues addressed in this topic of relevance to sustainability or sustainable development?
- In what ways might the solutions explored in this topic alter your predictions for the state of human societies and the biosphere some decades from now?

Knowledge and understanding

- Demographic tools for quantifying human population include crude birth rate (CBR), crude death rate (CDR), total fertility rate (TFR), doubling time (DT), and natural increase rate (NIR).
- Global human population has followed a rapid growth curve but there is uncertainty as to how this may be changing.
- As the human population grows, increased stress is placed on all of Earth's systems.
- Age/sex pyramids and the demographic transition model (DTM) can be useful in the prediction of human population growth. The DTM is a model which shows how a population transitions from a pre-industrial stage with high CBR and CDR to an economically advanced stage with low or declining CBR and low CDR.
- Influences on human population dynamics include cultural, historical, religious, social, political, and economic factors.
- National and international development policies may also have an impact on human population dynamics.

Demographic variables

Birth rates

Demographic tools for quantifying human population include crude birth rate (CBR), crude death rate (CDR), total fertility rate (TFR), doubling time (DT) and natural increase rate (NIR).

In the USA in 2012, there were 3 952 841 births out of a total population of 313 914 040; this gives a **crude birth rate (CBR)** of 12.6 per thousand per year. In Mauritius in 2013, there were 17 917 births out of a population of 1 331 155; this gives a CBR of 13.46 per thousand per year. Globally, there are major variations in the CBR, with the highest rates in poorer countries and lower rates in rich countries (Figure 8.1).

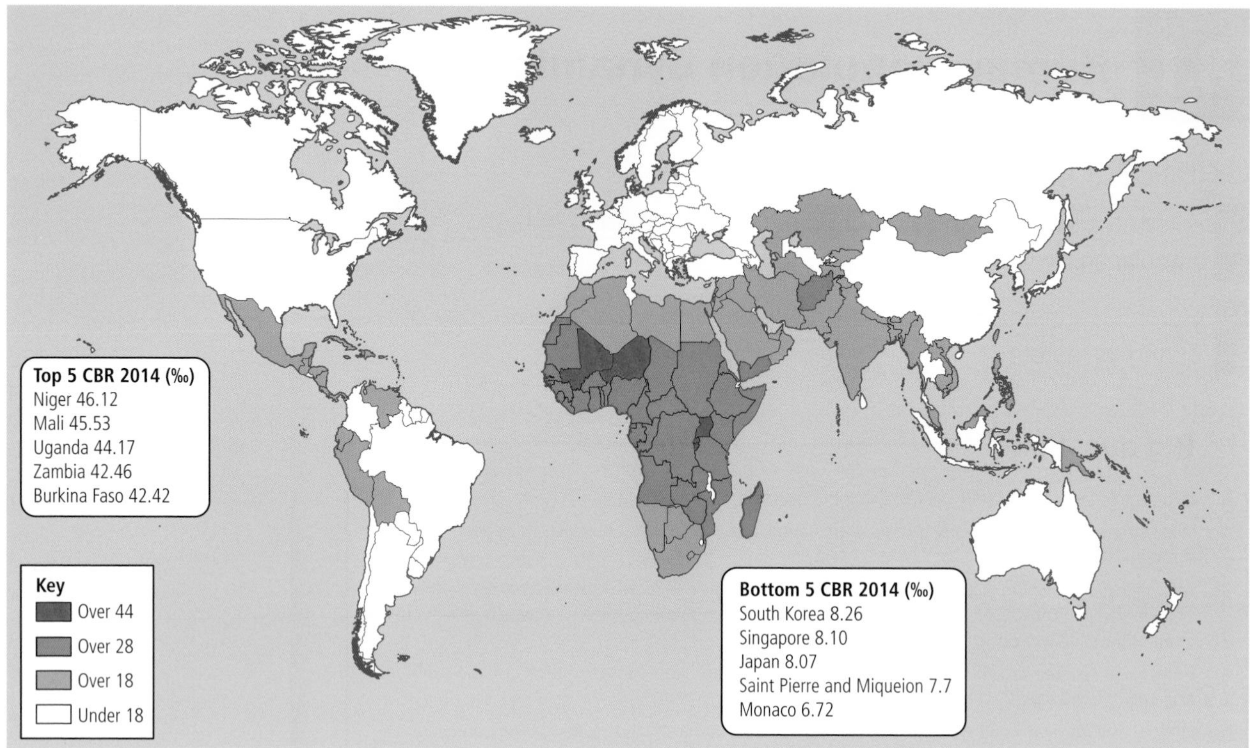

Figure 8.1 Global variations in CBR, 2014

CBR is easy to calculate and the data are readily available.

$$\text{CBR} = \frac{\text{total number of births}}{\text{total population}} \times 1000$$

Fertility

The **total fertility rate (TFR)** is the average number of births per woman of child-bearing age (Table 8.1).

Table 8.1 World's highest and lowest fertility rates

Highest in 2013	1971	2013	Lowest in 2013	1970	2013
Niger	7.4	7.6	Taiwan	3.9	1.1
South Sudan	6.9	7.0	Portugal	3.0	1.2
Somalia	7.2	6.6	Singapore	3.2	1.2
Chad	6.5	6.6	South Korea	4.5	1.2
Congo, Dem. Rep.	6.2	6.6	Moldova	2.6	1.2
Central African Republic	6.0	6.2	Poland	2.3	1.2
Angola	7.3	6.2	Bosnia- Herzegovina	2.7	1.3
Mali	6.9	6.1	Spain	2.9	1.3
Burundi	7.3	6.1	Greece	2.4	1.3
Zambia	7.4	6.0	Hungary	2.0	1.3
			Slovakia	2.4	1.3
			Romania	2.9	1.3

(Population Reference Bureau, 2014 World Population data sheet)

In general, highest fertility rates are found among the poorest countries, especially in sub-Saharan Africa, and very few LEDCs have made the transition from high birth rates to low birth rates. Most MEDCs have brought the birth rate down. In MEDCs, fertility rates have fallen as well.

Changes in fertility are a combination of both socio-cultural and economic factors. While there may be strong correlations between these sets of factors and changes in fertility, it is impossible to prove the linkages or to prove that one set of factors is more important than the other.

To learn more about global demographic trends, go to www.pearsonhotlinks.co.uk, enter the book title or ISBN, and click on weblink 8.1.

Level of education and material ambition

In general, the higher the level of parental education, the fewer the children. Poor people with limited resources often have large families. Affluent people can afford large families. Middle-income families with high aspirations but limited means tend to have the smallest families. They wish to improve their standard of living, and limit their family size to achieve this.

Political factors and family planning

Most governments in LEDCs have introduced some programmes aimed at reducing birth rates. Their effectiveness is dependent on:

- a focus on general family planning not specifically birth control
- investing sufficient finance in the schemes
- working in consultation with the local population.

Where birth controls have been imposed by government, they are less successful (except in China). In the MEDCs, financial and social support for children is often available to encourage a pro-natalist approach. However, in countries where there are fears of negative population growth (as in Singapore), more active and direct measures are taken by the government to increase birth rates.

Economic prosperity

The correlation between economic prosperity and the birth rate is not total, but there are links. As gross national product (GNP) *per capita* increases, so the birth rate decreases (Figure 8.2).

Economic prosperity favours an increase in the birth rate, while increasing costs lead to a decline in the birth rate. Recession and unemployment are also linked with a decline in the birth rate. This is related to the cost of bringing up children. Surveys

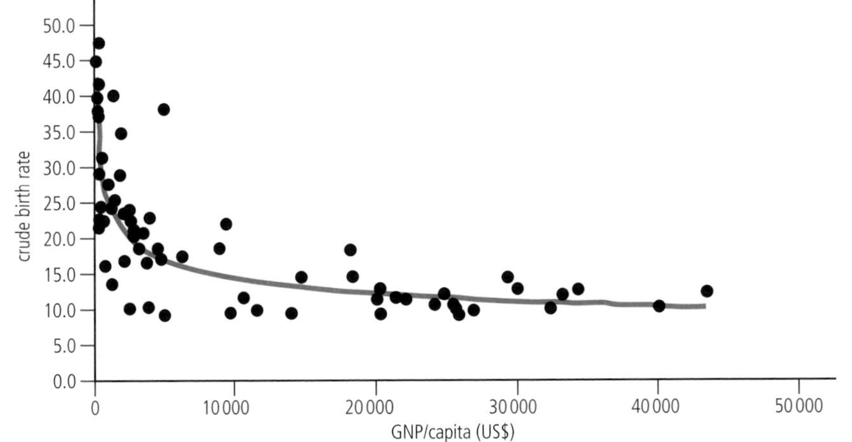

Figure 8.2 GNP *per capita* and CBR

have shown that the cost of bringing up children in the UK can be over £200 000, partly through lost earnings on the mother's part. Whether the cost is real or perceived (imagined) does not matter. If parents believe they cannot afford to bring up children, or that having more children will reduce their standard of living, they are less likely to have children. At the global scale, a strong link exists between fertility and the level of economic development; the UN and many NGOs believe that a reduction in the high birth rates in LEDCs can only be achieved by improving the standard of living in those countries.

To learn more about the world population data sheet, go to www.pearsonhotlinks.co.uk, enter the book title or ISBN, and click on weblink 8.2.

The need for children

High infant mortality rates increase the pressure on women to have more children. Such births offset the high mortality losses and are termed replacement births or compensatory births. In some agricultural societies, parents have larger families to provide labour for the farm and as security for the parents in old age. This is much less important now as fewer families are engaged in farming, and many farmers are labourers not farm owners.

Death rates

The **crude death rate (CDR)** is the number of deaths per thousand people in a population. In the USA in 2011, there were 2 513 171 deaths among a population of 311 575 900; this gives a CDR of 8.06 per thousand. However, the CDR is a poor indicator of mortality trends – populations with a large number of aged (as in most MEDCs) have a higher CDR than countries with more youthful populations (e.g. Denmark CDR = 10 per thousand and Mexico CDR = 5 per thousand). To compare mortality rates, we use **age-specific mortality rates (ASMR)**, such as the **infant mortality rate (IMR)**.

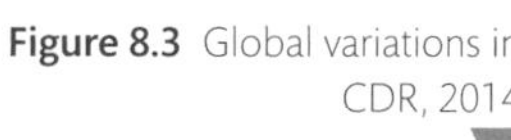

Figure 8.3 Global variations in CDR, 2014

The pattern of mortality in MEDCs differs from that in LEDCs (Figure 8.3). In the former, as a consequence of better nutrition, healthcare, and environmental conditions,

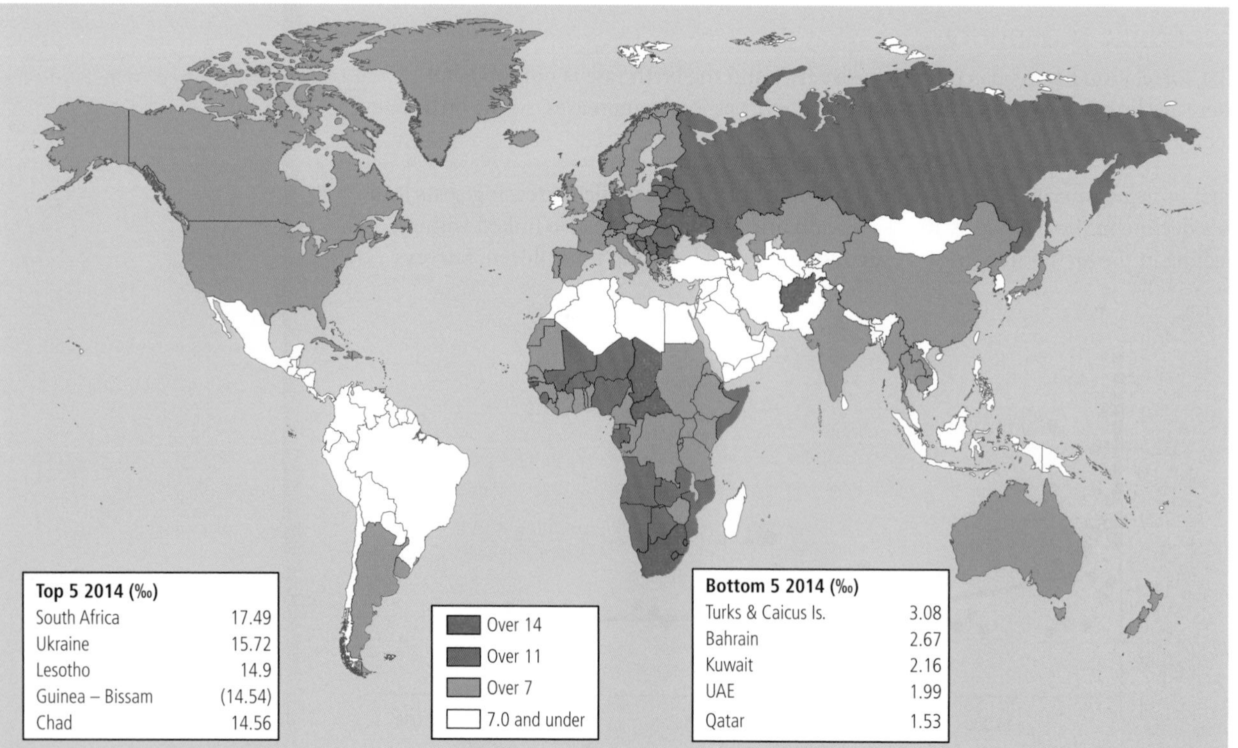

the death rate falls steadily to a level of about 9 per thousand. In many very poor countries, high death rates are still common although the crude death rate has shown a decrease over the past few decades because of improvements in food supply, water, sanitation and housing. This trend, unfortunately, has been reversed as a consequence of AIDS, particularly in sub-Saharan Africa.

At both the global and local scale, variations in mortality rates occur for the following reasons.

- Age structure – Some populations (e.g. those in retirement towns and especially in the older industrialized countries) have very high life expectancies and this results in a rise in the CDR. Countries with a large proportion of young people, have much lower death rates (e.g. Mexico has about 30 per cent of its population under the age of 15 years and has a CBR of 5 per thousand).
- Social class – Poorer people within any population have higher mortality rates than the more affluent. In some countries (e.g. South Africa) this is also reflected in racial groups.
- Occupation – Certain occupations are hazardous (e.g. the military, farming, oil extraction, and mining). Some diseases are linked to specific occupations (e.g. mining and respiratory disease).
- Place of residence – In urban areas, mortality rates are higher in areas of relative poverty and deprivation, such as inner cities and shanty towns. This is due to overcrowding, pollution, high population densities, and stress. In rural areas where there is widespread poverty and limited farm productivity, mortality rates are high (e.g. in rural north-east Brazil, life expectancy is 27 years shorter than in the richer south-east region).
- Child mortality and IMR – While the CBR shows small fluctuations over time, the IMR can show greater fluctuations and is one of the most sensitive indicators of the level of development. This is because:
  - high IMRs are only found in the poorest countries
  - the causes of infant deaths are often preventable
  - IMRs are low where there is safe water supply and adequate sanitation, housing, healthcare, and nutrition.

As a country develops, the major forms of illness and death change. LEDCs are characterized by a high proportion of infectious and contagious diseases such as cholera, tuberculosis, gastroenteritis, diarrhoea, and vomiting. These are often fatal. In MEDCs, fatal diseases are more likely to be degenerative diseases such as cancer, strokes, or heart disease. The change in disease pattern from infectious to degenerative is known as the *epidemiological transition model.* (epidemiology is the study of diseases). Such a change generally took about a century in MEDCs but is taking place faster in LEDCs.

In South Africa, the IMR varies with race. Whites have a higher income and a better standard of living; they have a lower IMR (10–15 per thousand) than the other racial groups (blacks 50–100 per thousand, coloureds 45–55 per thousand).

The cause and age of infant death also vary with race:

- for whites, **neonatal deaths** (0–7 days) and **perinatal deaths** (8–28 days) are more likely to be due to birth deformities
- for blacks, such deaths are more likely to be due to low birth weight, gastroenteritis, pneumonia, and jaundice, occurring between 8 and 365 days, the **post-neonatal period**.

South Africa has a high IMR, especially in poor, rural communities.

You need to be able to calculate values of CBR, CDR, TFR, DT, and NIR.

CHALLENGE YOURSELF

Use the hotlink below to the *CIA World Factbook* to find out the current birth rate, death rate, population size and other demographic indicators for China.

To learn more about the *CIA World Factbook*, go to www.pearsonhotlinks.co.uk, enter the book title or ISBN, and click on weblink 8.3.

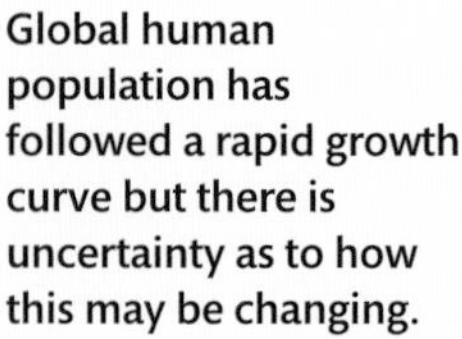

Global human population has followed a rapid growth curve but there is uncertainty as to how this may be changing.

Natural increase

Natural increase rate (NIR) is calculated by subtracting CDR from the CBR (and dividing by 10 if the CBR and CDR are expressed per thousand rather than per cent). It is usually expressed as a percentage.

Doubling time (DT) is the number of years needed for a population to double in size, assuming the natural growth rate remains constant. It is found by dividing 70 (years) by the rate of natural increase (per cent). It is expressed in years.

$$\text{NIR (\%)} = \text{CBR} - \text{CDR}$$

$$\text{DT (years)} = \frac{70}{\text{NIR}}$$

Human population growth

The world's population is growing very rapidly. Most of this growth is quite recent. The world's population doubled between 1804 and 1922, between 1922 and 1959, and between 1959 and 1974. It is thus taking less and less time for the population to double, although growth has slowed down since 1999 (Figure 8.4). Up to 95 per cent of population growth is taking place in less economically developed countries (LEDCs). However, the world's population is expected to stabilize at about 8.5 billion following a peak at 11 billion.

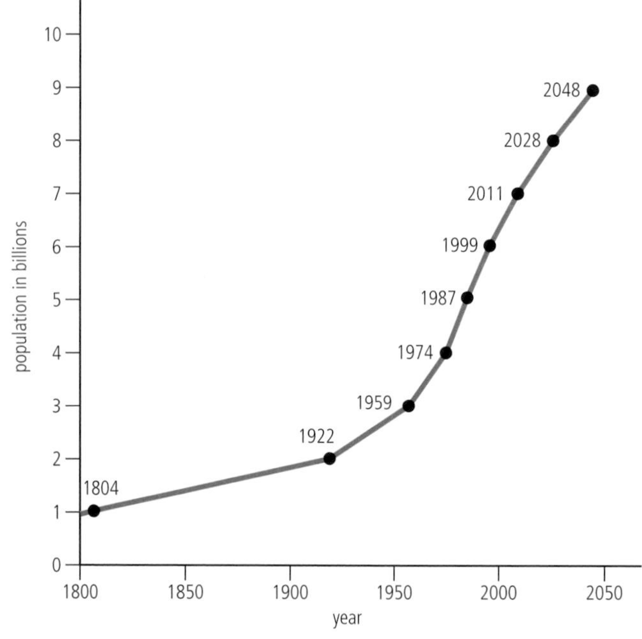

Figure 8.4 Exponential growth in world population

Exponential growth is a growth rate that is increasingly rapidly (an accelerating rate of growth).

In 1990, the world's women were, on average, giving birth to 3.3 children in their lifetime. By 2002, the figure had dropped to 2.6 children – slightly above the level needed to ensure replacement of the population. If these trends continue, the level of fertility for the world as a whole will drop below replacement level before 2050. The projections also suggest that AIDS, which has killed more than 20 million people in the past 20 years, will lower the average life expectancy at birth in some countries to around 50 years. AIDS continues to have its greatest impact in the developing countries of Asia, Latin America, and especially sub-Saharan Africa. Botswana and South Africa are among the countries that may see population decline because of AIDS deaths.

- The world's population is likely to peak at 11 billion in 2100.
- North America (USA and Canada) will be one of only two regions in the world with a population still growing in 2100. It will have increased from 314 million today to 454 million, partly because first-generation immigrant families tend to have more children. The other expanding region is Latin America where the population is forecast to increase from 515 million to 934 million.
- Despite disease, war, and hunger, the population of Africa will grow from 784 million today to 1.6 billion in 2050. By 2100, it will be 1.8 billion, although it will have begun to decline. By the end of the century, more than a fifth of Africans will be over 60, a higher proportion than in western Europe today.
- The China region will see its population shrink significantly by 2100, from 1.4 billion to 1.25 billion. When China reaches its population peak of about 1.6 billion (by 2020), it will have more literate people than Europe and North America combined. This is because of its education programme.
- India will overtake China as the world's most populous nation by 2020.
- A tenth of the world's population is over 60. By 2100, that proportion will have risen to a third.

To learn more about population simulations, go to www.pearsonhotlinks.co.uk, enter the book title or ISBN, and click on weblink 8.4.

Should we predict changes in population growth when many people are free to choose how many children they have, while in other countries, the government tries to control how many children people have?

As the human population grows, increased stress is placed on all of Earth's systems.

Human population growth stresses water systems, agricultural systems, and energy systems. The impact of exponential growth is that a huge amount of extra resources are needed to feed, house, clothe, and look after the increasing number of people. However, it can be argued that the resource consumption of much of the world's poor population (i.e. those in LEDCs) is much less than the resource consumption of populations in MEDCs where population growth rates are much lower.

There is a need to develop more sustainable agricultural systems, more sustainable energy systems and more sustainable water systems. However, without accurate population projections, it is difficult to know exactly how large the demand for these products will be.

The price to be paid for a shrinking world population is an increase in the number of elderly people in the world. **Life expectancy** is increasing but social security systems are not.

Age/sex pyramids

Age/sex pyramids and the demographic transition model (DTM) can be useful in the prediction of human population growth. The DTM is a model which shows how a population transitions from a pre-industrial stage with high CBR and CDR to an economically advanced stage with low or declining CBR and low CDR.

Population structure or composition refers to any *measurable* characteristic of the population. This includes the age, sex, ethnicity, language, religion, and occupation of the population. These are usually shown by population pyramids.

Population pyramids can tell us a great deal of information about the age and sex structure of a population (Figure 8.5). For example:

- a wide base indicates a high birth rate
- narrowing base suggests falling birth rate
- straight or near vertical sides indicate a low death rate
- concave slopes characterize a high death rate
- bulges in the slope suggest baby booms or high rates of immigration or in-migration (e.g. excess males aged 20–35 years will be economic migrants looking for work; excess elderly, usually female, will inundate retirement resorts)
- 'slices' in the slope indicate emigration or out-migration or age-specific or sex-specific deaths (epidemics, war).

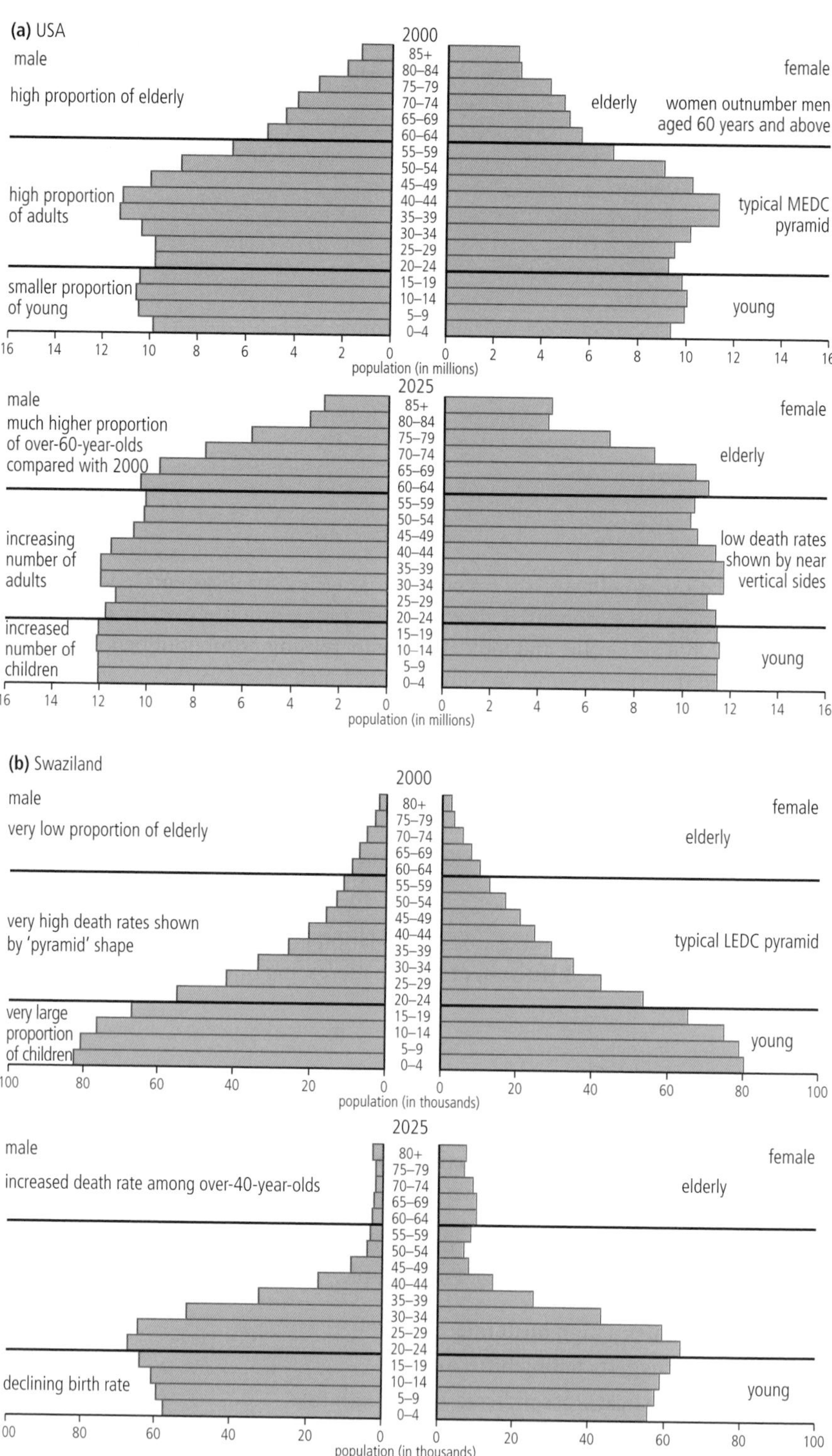

Figure 8.5 Population pyramids for (a) the USA and (b) Swaziland, 2000 and 2025

The demographic transition model

The general d**emographic transition model (DTM)** shows the change in population structure from LEDCs to MEDCs (Figure 8.6).

The DTM suggests that death rates fall before birth rates, and that the total population expands. However, the DTM is based on the data from just three countries – England,

Stage 2
Early expanding:
- birth rate remains high but the death rate comes down rapidly
- population growth is rapid
- Afghanistan and Sudan are at this stage
- UK passed through this stage by 1850

Stage 3
Late expanding:
- birth rate drops and the death rate remains low
- population growth continues but at a smaller rate
- Brazil and Argentina are at this stage
- UK passed through this stage in about 1950

Stage 4
Low and variable:
- birth rates and death rates are low and variable
- population growth fluctuates
- UK and most developed countries are at this stage

Stage 1
High and variable:
- birth rates and death rates are high and variable
- population growth fluctuates
- no countries, only some indigenous (primitive) tribes still at this stage
- UK at this stage until about 1750

High birth and death rates
Parents want children:
- for labour
- to look after them in old age
- to continue the family name
- prestige
- to replace other children who have died

People die from:
- lack of clean water
- lack of food
- poor hygiene and sanitation
- overcrowding
- contagious diseases
- poverty

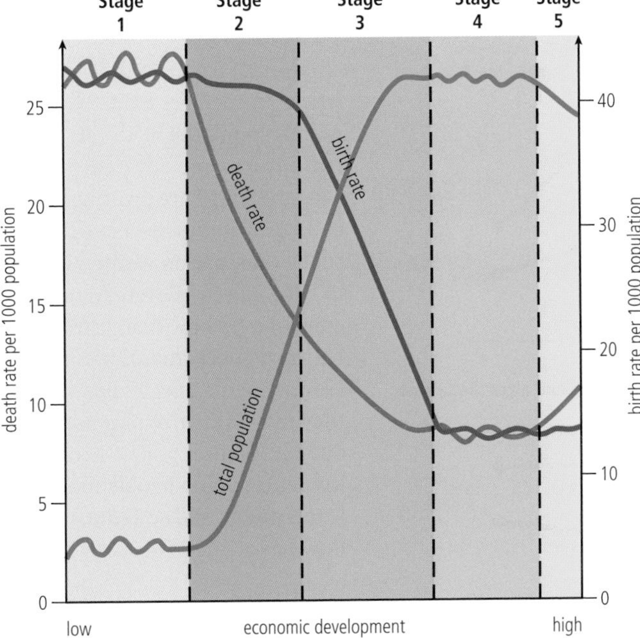

Stage 5
Low declining:
- the birth rate is lower than the death rate
- the population declines
- Japan and Sweden are in this stage

Low birth and death rates
Birth rates decline because:
- children are very costly
- the government looks after people through pensions and health services
- more women want their own career
- there is more widespread use of family planning
- as the infant mortality rate comes down there is less need for replacement children

Death rates decline because:
- clean water
- reliable food supply
- good hygiene and sanitation
- lower population densities
- better vacations and healthcare
- rising standards of living

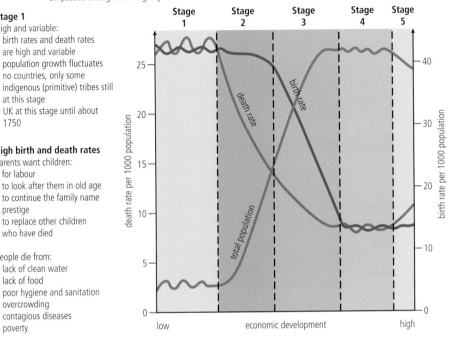

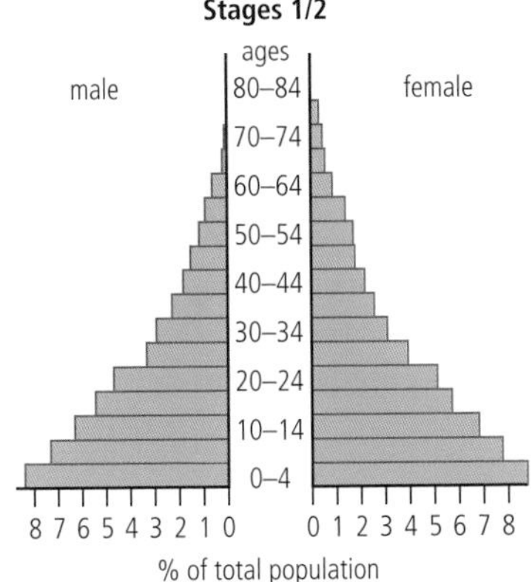

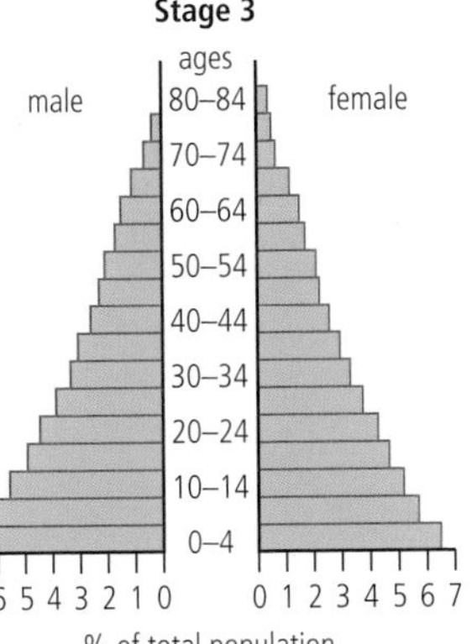

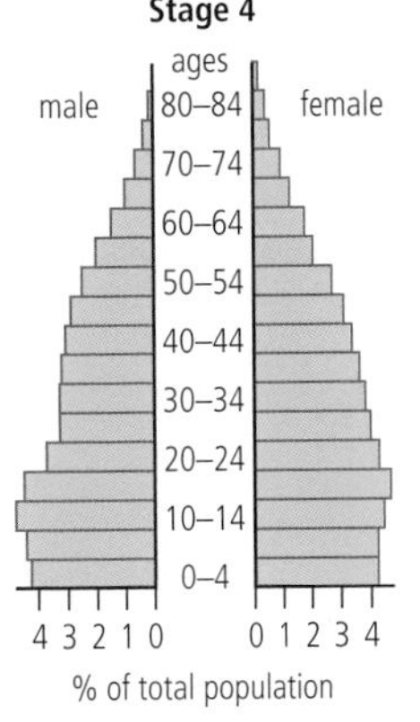

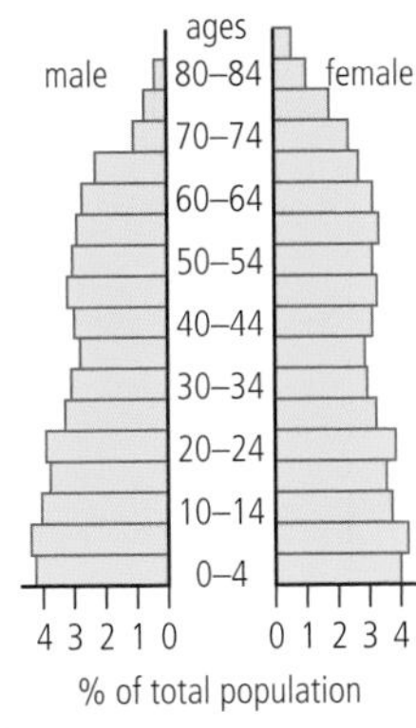

Figure 8.6 The demographic transition model

Wales, and Sweden. Not only is the time-scale for the DTM in these countries longer than in many LEDCs, there are other types of DTM (Figure 8.7). For example, Ireland's DTM was based on falling birth rates and rising death rates as a result of emigration following the 1845–49 famine. The DTM in Japan shows a period of population expansion before the Second World War, followed by population contraction once the country's expansionist plans could not be fulfilled. Other nations have experienced a similar drop in birth rates and death rates (e.g. former Yugoslavia).

The general DTM suggests that countries progress through recognized stages in the transition from LEDC to MEDC.

Influences on human population dynamics include cultural, historical, religious, social, political, and economic factors.

Population dynamics

The range of factors which affect population growth is varied and differs with different scales. For example, national or regional change in population takes into account

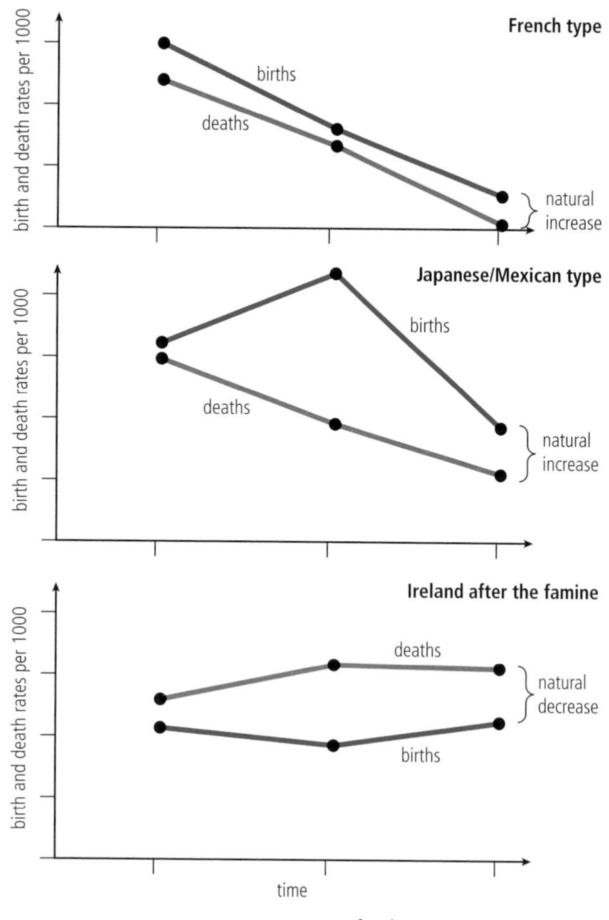

Figure 8.7 Alternative demographic transitions

in-migration and out-migration, whereas global population change does not take migration into account at all.

Factors influencing the birth rate include cultural, historical, religious, social, political, and economic factors. For example, in some cultures, especially agricultural ones, there is an advantage in having more children to work on the land. In contrast, in cultures where women are employed in the workforce, such as in Singapore, the birth rate may be very low. Religious reasons may include beliefs about family planning. Most religions are pro-natalist, although many 'Catholic countries' in Europe have low birth rates. There may be social pressure on women in more traditional societies to bear more children. Many women may not have control over how many children they have. Increasingly, national governments are influencing population size. Governments may be pro-natalist or anti-natalist, and they may influence population dynamics by having an open migration policy. Economic factors also have an influence. Some people may feel that the cost of childcare is too much, and so reduce the number of children that they have.

Similarly, the death rate is affected by many factors. These include the age-structure of the population, availability of clean water, sanitation, adequate housing, reliable food supply, prevalence of disease, provision of healthcare facilities, type of occupation, natural hazards, civil conflict and war, and chance factors. Social and economic factors have a major influence on death rates – poor people are far more vulnerable to the risk of early death, due to a combination of poor living conditions, poor diet, lack of access to clean water, and sanitation.

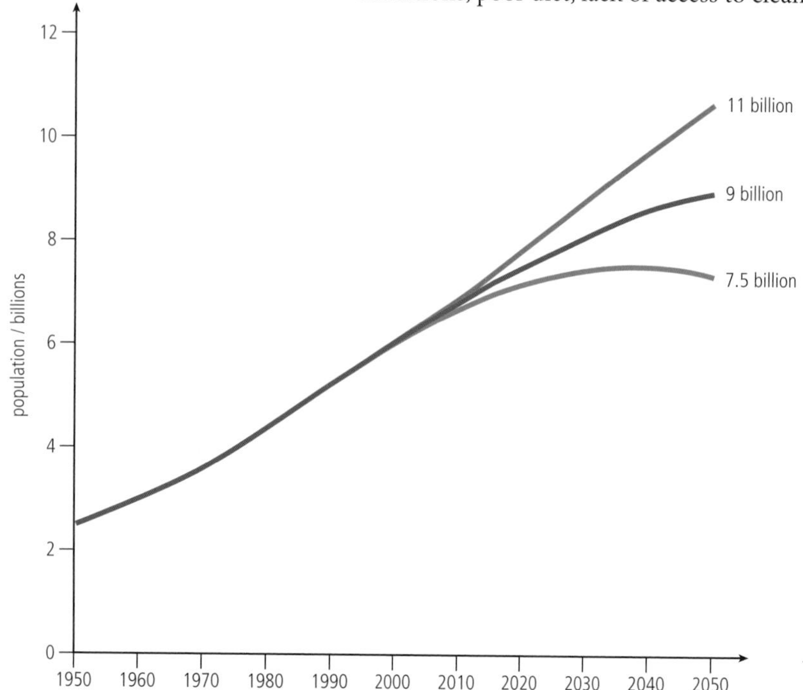

The range of factors affecting birth and death rates makes predicting the growth of the human population difficult (Figure 8.8).

National population policies

Population policies refer to official government actions to control the population in some way. Pro-natalist policies are in favour of increasing the birth rate. Anti-natalist policies attempt to limit the birth rate.

National and international development policies may also have an impact on human population dynamics.

Figure 8.8 Range of UN population predictions 1950–2050

Case study

National level anti-natalist policy: China

After considering options for family size and total population projections, the Chinese government felt the need to control its population (Figure 8.9).

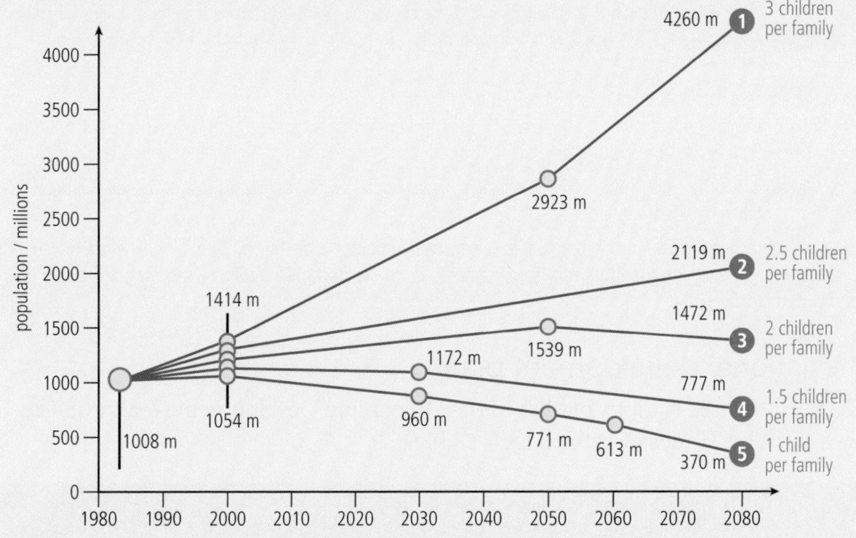

Figure 8.9 Options for family size and projected total population for China

The government introduced its one-child policy in 1979. The policy rewarded families that had only one child and penalized those that had more than one. For example, families that had two or more children paid higher taxes, and the parents were prevented from reaching high-level positions in their jobs. Other measures included forced sterilizations and abortions so that families were limited to one child.

The policy has been relaxed somewhat since October 1999. In most rural areas, couples can now have two children without penalties. Increasingly, rich farmers are able and willing to pay fines or bribes in order to get permission to have more children; poor families simply take the view that they have nothing much to lose, so they may have more than one child.

One of the results of the policy has been gender imbalance. In 2003, in China, 117 boys were born for every 100 girls, whereas the global average is 105 boys to every 100 girls. The disparity is greater in rural areas: 130 to 100. There has been an increase of female infanticide. Girls are hidden from the authorities, or die at a young age through neglect. China is offering to pay couples a premium for producing baby girls to counter this imbalance.

Most Chinese families in urban areas have only one child, and the growing urban middle classes do not discriminate against daughters as much. However, the rural populations remain traditionally focused on male heirs. But even in urban areas, boys are generally preferred because they are regarded as more able than girls to provide for their families, care for elderly relatives, and continue the family line. The preference for a son makes simple economic sense as sons are less likely to leave the family home after marrying and, as higher earners than women, are more able to provide for the extended family.

China now offers welfare incentives to couples with two daughters and has tightened the prohibition on sex-selective abortions. In some areas, couples with two daughters and no sons have been promised an annual payment of 600 renminbi (£38) once they reach 60 years of age. The money will also be given to families with only one child to discourage couples with a daughter from trying again for a boy.

Statistically, the one-child policy has had success. The government says it has prevented well over 400 million births since it was introduced and is fulfilling its initial aim of ensuring that China can combat rural poverty and improve standards of living across the board. China's population stood at 1.3 billion in 2014. As the country's economy continues to grow and transform at an unprecedented rate, pressure to relax the policy looks likely to intensify.

It is forecast that there will be a shortage of potential marriage mates which will lead to some social instability. By the end of the decade, demographers say China will have 24 million men who, because of the gender imbalance, will not be able to find a wife.

continued

CHALLENGE YOURSELF

Have a class debate on the motion that: 'The positive impacts of China's one-child policy outweigh the disadvantages.'

2013 Reforms

The loosening of the one-child policy, announced in November 2013, allows couples to have a second child if either parent is an only child. This is meant to signal the beginning of a more family-friendly bureaucracy. Provinces can set their own timetable for implementing the reform. The fertility rates in Beijing, Shanghai, and Guangzhou are among the lowest in the world.

Party conservatives still fear two things about loosening population controls. The first is that without proper controls, the population may grow beyond the country's planned capacity to feed itself (1.5 billion people by the year 2033). The second is that loosening too quickly may spur a baby boom that would strain public services.

But there are reasons to be optimistic. The declining fertility rate, now about 1.5 births per women, has eased food-security fears. However, the downsides of maintaining a one-child quota include a shrinking labour force and an ageing population.

Factors other than the one-child policy, such as a lack of social security support, have also encouraged couples to limit their offspring.

However, the policy's relaxation is unlikely to lead to a population boom. Many urbanites, already burdened by increasing education and housing costs, consider that having two or more children can only be afforded by the rich.

International development plans

In 2000, the United Nations published the Millennium Development Goals (MDGs). These included 21 measureable targets (Table 8.2).

Table 8.2 Millennium Development Goals and Targets

Millennium Development Goal	Targets
1 Eradicate extreme poverty and hunger	• reduce by 50 per cent the proportion of people living on less than US$1 a day • reduce by 50 per cent the proportion of people suffering from hunger
2 Achieve universal primary education	• ensure all children complete a full course of primary schooling
3 Promote gender equality and empower women	• eliminate gender disparity in primary and secondary education by 2005 (all levels by 2025) • ensure literacy parity between young men and women • women's equal representation in national parliaments
4 Reduce child mortality	• reduce by two-thirds the under-five mortality rate • universal child immunization against measles
5 Improve maternal health	• reduce the maternal mortality ratio by 75 per cent • increase access to reproductive health
6 Combat HIV/AIDS, malaria, and other diseases	• halt and begin to reverse the spread of HIV/AIDS • halt and begin to reverse the incidence of malaria • halt and begin to reverse the incidence of tuberculosis
7 Ensure environmental sustainability	• reverse loss of forests • halve proportion without improved drinking water in urban areas • halve proportion without improved drinking water in rural areas • halve proportion without sanitation in urban areas • halve proportion without sanitation in rural areas • improve the lives of at least 100 million slum dwellers by 2020
8 Develop global partnership for development	• reduce youth unemployment • increase internet use

In 2015, the MDGs were replaced by the sustainable developments goals (SDGs). Both the MDGs and the SDGs have direct and indirect implications for women's health and fertility levels.

Exercises

1. In 2014, the UK had a total population of 63 742 977. There were 813 200 births and 558 800 deaths that year. Calculate the crude birth rate and crude death rate for the UK in 2014.
2. The table below shows data for number of births, deaths and total population for the world, MEDCs and LEDCs. Calculate the crude birth rate, crude death rate and natural increase for the world, MEDCs and LEDCs.

Key population data for the world, MEDCs and LEDCs

	World	MEDCs	LEDCs
total population	7 238 184 000	1 248 950 000	5 989 225 000
births	143 341 000	13 794 000	129 547 000
deaths	56 759 000	12 328 000	44 547 000

3. Study the table below, then fill in the gaps.

Selected population data

	World	USA	UK	China	India	Highest	Lowest
crude birth rate/1000	20	13	12	12	22	Niger 46.12	Monaco 6.72
crude death rate/1000	8	8	9	7	7	South Africa 17.49	Qatar 1.53
natural increase rate (%)							
total fertility rate	2.5	5.4	1.9	1.6	2.4	Niger 6.89	Singapore 0.8
growth rate (%)	1.06	0.77	0.54	0.44	1.25	Lebanon 9.37	Syria 9.73
doubling time (years)							

(Data *CIA World Factbook*, 2014)

4. Copy this grid then use it to plot the changes in China's birth rates, death rates, and total population between 1950 and 1990 using the information in the table below.

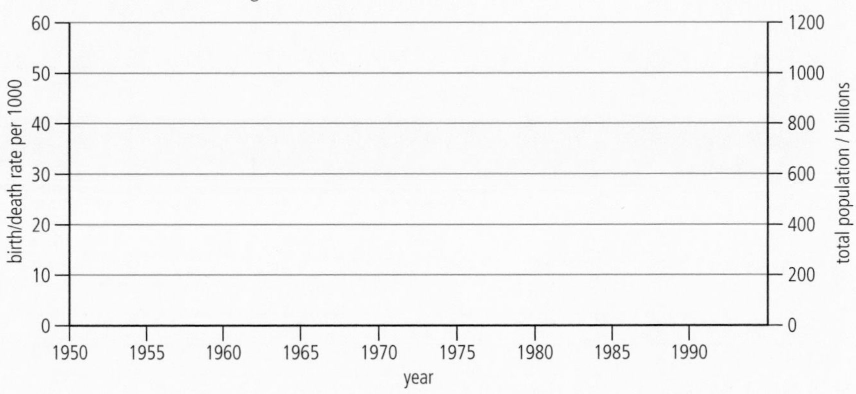

continued

Key population data for China

Year	Birth rate/1000	Death rate/1000	Total population/ millions
1950	44	38	570
1955	44	25	600
1960	29	22	670
1965	40	12	750
1970	38	10	825
1975	25	9	950
1980	20	8	1000
1985	18	8	1050
1990	16	9	1160
1995	17	7	1237
2000	14	6	1280
2005	12	7	1318
2010	12	7	1359
2015	12	7	1401

Note: In 1961 the death rate suddenly soared to 45 per thousand due to famine and starvation.

5. The figure below shows China's population pyramid for 1990.

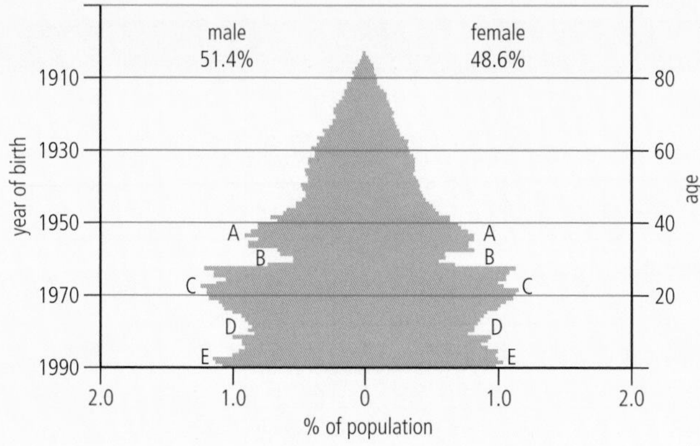

Match each of the letters A to E with one of the following explanations:

a. Famine and starvation cause millions to die (population declines steeply)

b. Effect of the one-child policy (population declines)

c. Relaxation of the one-child policy (population expands)

d. Baby boom following the creation of the People's Republic of China in 1949 (increase in population)

e. Large-scale population growth prompts a need for population control.

6. The table below shows demographic data for China in the years 1980, 2015, and 2025. The values represent the percentage of the total population for each year in each cohort by gender.

Demographic data for China: 1980, 2015, 2025

Age	1980		2015		2050	
	Male	Female	Male	Female	Male	Female
>80			0.7	1.0	2.8	3.7
75–79	0.6	0.9	0.8	0.9	2.6	2.9
70–74	0.7	0.8	1.2	1.2	2.6	2.7
65–69	1.0	1.1	1.8	1.8	3.3	3.3
60–64	1.4	1.4	2.8	2.7	4.6	4.3
55–59	1.6	1.6	2.9	2.7	3.9	3.5
50–54	2.0	1.8	3.7	3.5	3.1	2.7
45–49	2.4	2.1	4.4	4.1	2.9	2.5

Age	1980		2015		2050	
	Male	Female	Male	Female	Male	Female
40–44	2.6	2.2	4.3	4.1	3.2	2.8
35–39	2.7	2.4	3.4	3.3	3.5	3.0
30–34	3.4	3.1	3.8	3.6	3.2	2.8
25–29	4.7	4.3	5.0	4.6	2.9	2.5
20–24	4.6	4.2	4.1	3.6	2.7	2.4
15–19	5.7	5.3	3.2	2.7	2.6	2.4
10–14	6.7	6.2	3.0	2.6	2.6	2.4
5–9	6.4	6.1	3.3	2.8	2.6	2.4
0–4	5.2	4.9	3.5	3.0	2.5	2.3

   a. Draw the population pyramids for China in 1980, 2015, and 2050.
   b. Compare the two pyramids in terms of:
      i. percentage population under the age of 19 years
      ii. percentage population over the age of 60 years
      iii. gender imbalances.
   c. Suggest reasons for the differences that you have noted.

7. Distinguish between anti-natalist and pro-natalist policies.

8. Refer back to Figure 8.9.
   a. Briefly explain why an anti-natalist policy was considered necessary in China.
   b. What are the predicted impacts of the one-child policy over the next 50 years?

9. Outline ways in which the Millennium Development Goals may affect population growth.

Big questions

Having read this section, you can now discuss the following big questions:

- What strengths and weaknesses of the systems approach and the use of models have been revealed through this topic?
- To what extent have the solutions emerging from this topic been directed at *preventing* environmental impacts, *limiting* the extent of the environmental impacts, or *restoring* systems in which environmental impacts have already occurred?
- What value systems can you identify at play in the causes and approaches to resolving the issues addressed in this topic?
- How does your own value system compare with others you have encountered in the context of issues raised in this topic?
- How are the issues addressed in this topic of relevance to sustainability or sustainable development?
- In what ways might the solutions explored in this topic alter your predictions for the state of human societies and the biosphere some decades from now?

Points you may want to consider in your discussions:

- How do models help our understanding of human dynamics?
- To what extent have population policies been effective in their aims?
- How do environmental value systems affect population dynamics? Give examples to support your answer.
- What are your views on these?
- Examine the relationship between population dynamics related to sustainable development.

8.2 Resource use in society

Significant ideas

The renewability of natural capital has implications for its sustainable use.

The status and economic value of natural capital is dynamic.

Big questions

As you read this section, consider the following big questions:

- What strengths and weaknesses of the systems approach and the use of models have been revealed through this topic?
- To what extent have the solutions emerging from this topic been directed at *preventing* environmental impacts, *limiting* the extent of the environmental impacts, or *restoring* systems in which environmental impacts have already occurred?
- What value systems can you identify at play in the causes and approaches to resolving the issues addressed in this topic?
- How does your own value system compare with others you have encountered in the context of issues raised in this topic?
- How are the issues addressed in this topic of relevance to sustainability or sustainable development?
- In what ways might the solutions explored in this topic alter your predictions for the state of human societies and the biosphere some decades from now?

Knowledge and understanding

- Renewable natural capital can be generated and/or replaced as fast as it is being used. It includes living species and ecosystems that use solar energy and photosynthesis. It also includes non-living items, such as groundwater and the ozone layer.
- Non-renewable natural capital is either irreplaceable or can only be replaced over geological timescales (e.g. fossil fuels, soil, and minerals).
- Renewable natural capital can be used sustainably or unsustainably. If renewable natural capital is used beyond its natural income, this use becomes unsustainable.
- The impacts of extraction, transport, and processing of a renewable natural capital may cause damage making this natural capital unsustainable.
- Natural capital provides goods (e.g. tangible products) and services (e.g. climate regulation) that have value. This value may be aesthetic, cultural, economic, environmental, ethical, intrinsic, social, spiritual, or technological.
- The concept of natural capital is dynamic. Whether or not something has the status of 'natural capital', and the marketable value of that capital, varies regionally and over time. This is influenced by cultural, social, economic, environmental, technological, and political factors (examples include cork, uranium, lithium).

Renewable natural capital can be generated and/or replaced as fast as it is being used. It includes living species and ecosystems that use solar energy and photosynthesis. It also includes non-living items, such as groundwater and the ozone layer.

Renewable natural capital

A resource is only a resource when it becomes useful to people – uranium only became a resource in the 20th century despite existing for millions of years. The Earth contains many resources that support its natural systems: the core and crust of the planet; the biosphere (the living part) containing forests, grassland, deserts, tundra and other biomes, and the upper layers of the atmosphere. These resources are all extensively used by humans to provide food, water, shelter, and life-support systems. Humans tend to have an anthropocentric (human-centred) view of these resources and their use. In this section, you will examine the way in which resources are discussed in terms of their use by and relationship to human populations.

Ecologically minded economists describe resources as **natural capital**. This is equivalent to the store of the planet (or stock) – the present accumulated quantity of natural capital. If properly managed, **renewable natural capital** can produce **natural income** indefinitely in the form of valuable **goods** and **services**.

- Goods are marketable commodities such as timber and grain.
- Ecological services might be flood and erosion protection, climate stabilization, maintenance of soil fertility.

Renewable resources can be used over and over again. In order to provide income indefinitely, the products and services used should not reduce the original resource (or capital). For example, if a forest is to provide ongoing income in the form of timber, the amount of original capital (the forest) must remain the same while income is generated from new growth. This is the same idea as living on the interest from a bank account – the original money is not used and only the interest is removed and spent.

Non-renewable natural capital is either irreplaceable or can only be replaced over geological timescales (e.g. fossil fuels, soil, and minerals).

These **non-renewable resources** will eventually run out if they are not replaced. Using economic terms, these resources can be considered as parallel to those forms of economic capital that cannot generate wealth (i.e. income) without liquidation of the estate. In other words, the capital in the bank account is spent.

Predictions about how long many of Earth's minerals and metals will last before they run out are usually guesstimates. They may not take into account any increase in demand due to new technologies, and they may assume that production equals consumption. Accurate estimates of global reserves and precise figures for consumption are needed for more exact predictions. However, it is clear that key non-renewable natural resources are limited and that there is a need to conserve, minimize waste, recycle, reuse, and, where possible, replace rare elements with more abundant ones.

To view an audit of Earth's natural wealth, go to www.pearsonhotlinks.co.uk, enter the book title or ISBN, and click on weblink 8.5.

Sustainable and unsustainable use of renewable natural capital

Renewable natural capital can be used sustainably or unsustainably. If renewable natural capital is used beyond its natural income, this use becomes unsustainable.

Sustainability means using global resources at a rate that allows natural regeneration and assimilation of pollution (Chapter 1, pages 48–56). When human well-being is dependent on the goods and services provided by certain forms of natural capital, then long-term harvest and pollution rates should not exceed rates of capital renewal. For example, a system harvesting renewable resources at a rate that enables replacement by natural growth shows sustainability.

Sustainability is living within the means of nature (i.e. on the 'interest' or sustainable income generated by natural capital) and ensuring resources are not degraded (i.e. natural capital is not depleted and/or polluted) so that future generations can continue to use the resource. The concept can be applied in our everyday lives.

An example of irresponsible use of a resource concerns groundwater. Pollutants from agricultural products and run-off from storage tanks, landfills, and septic tanks are reducing the water quality. Unsustainable extraction from groundwater sources (aquifers) means that water tables are lowered, which can lead to the intrusion of saltwater in coastal areas and further contamination of the supply (e.g. the Gaza Strip). Excessive use of surface water, often for agriculture, means that groundwater supplies are not replenished. The effect of groundwater pollution and reduction is decreased availability of water resources. This has a knock-on impact on agriculture – less water is available for irrigation, so yields decline. At the same time, the cost of water for industry and agriculture increases, which has serious implications for the economy. Water shortages can lead to tensions and conflict over the limited resource (e.g. the Israeli–Palestinian conflict).

CONCEPTS: Environmental value systems

People with a technocentric worldview see humanity as being ultimately able to solve shortages of natural capital by finding alternative technological solutions. Such people tend to be from MEDCs, where continuation of the lifestyle enjoyed in these countries depends on such solutions being found. People from cultures in closer contact with the environment would seek to find solutions by limiting our use of non-renewable natural capital and replacing them with renewable and sustainable sources (an ecocentric approach).

CONCEPTS: Sustainability

The impacts of extraction, transport, and processing of a renewable natural capital may cause damage making this natural capital unsustainable. For example, the removal of forests may have many impacts. Decreased tree cover can lead to:

- increased soil erosion
- a reduction in habitat
- climate change
- increased risk of flooding (Figure 8.10).

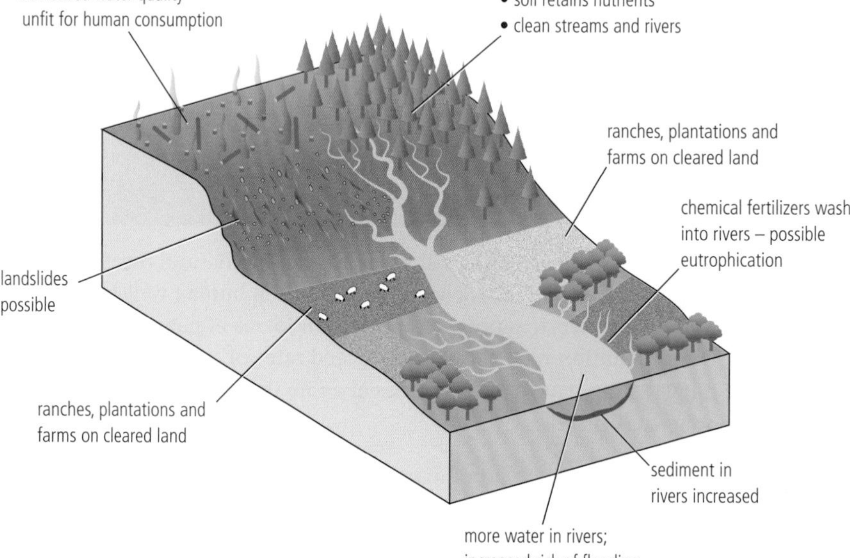

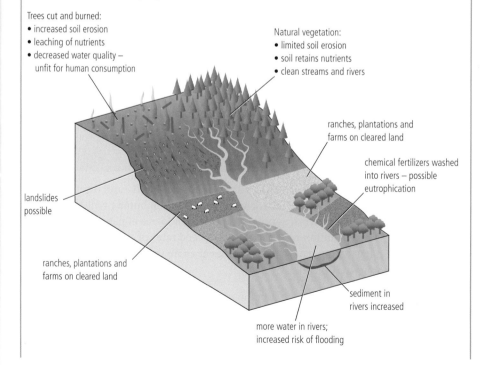

Figure 8.10 The effects of unsustainable forest removal

The transport of sawn logs releases greenhouse gases. Removal of vegetation to make access roads into the forest reduces habitat and threatens biodiversity. The processing of timber to make paper or for industrial use also uses greenhouse gases in the running of machines. Deforestation has been widespread in parts of Malaysia and Indonesia although there the rate of deforestation in the Amazon has slowed.

Similarly, fishing can be seen as an example of unsustainable use of a renewable resource. For example, the Pacific bluefin tuna is a meaty fish prized in Japan. Its numbers have declined by over 30 per cent over the last 20 years. The IUCN has reclassified it as 'vulnerable' on the Red List (pages 176–184). The main threat to the species is its value as sashimi - one fish can cost over £60 000.

Catching fish has environmental impacts including use of greenhouse gases for transport, use of dynamite on coral, and over-fishing. Likewise, the processing of fish uses fossil fuels to run machines. Fossil fuels are also used to transport the processed food to markets and shops.

You should be able to outline an example of how renewable and non-renewable natural capital has been mismanaged.

Natural capital provides goods (e.g. tangible products) and services (e.g. climate regulation) that have value. This value may be aesthetic, cultural, economic, environmental, ethical, intrinsic, social, spiritual, or technological.

Types of ecosystem service

Supporting services

These are the essentials for life and include primary productivity, soil formation, and the cycling of nutrients. All other ecosystem services depend on these.

Regulating services

These are a diverse set of services and include pollination, regulation of pests and diseases, and production of goods, such as food, fibre, and wood. Other services include climate and hazard regulation and water quality regulation.

Provisioning services

These are the services people obtain from ecosystems and from which they obtain goods such as food, fibre, fuel (peat, wood and non-woody biomass), and water from aquifers, rivers and lakes. The production of such goods can be within heavily managed ecosystems (intensive farms and fish farms) or from semi-natural ones (hunting and fishing).

Cultural services

These are derived from places where people interact with nature, enjoying cultural goods and benefits. Open spaces – such as gardens, parks, rivers, forests, lakes, the sea-shore, and wilderness – provide opportunities for outdoor recreation, learning, spiritual well-being, and improvements to human health.

Table 8.3 summarizes ecosystem services.

Table 8.3 Ecosystem services

Mountains, moorlands and heaths	Woodlands
Provisioning services	
food* fibre* fuel* fresh water*	timber* species diversity* fuelwood* fresh water*
Regulating services	
climate regulation† flood regulation† wildfire regulation† water quality regulation† erosion control†	climate regulation† flood regulation† erosion control† disease and pest control† wildfire regulation†

Mountains, moorlands and heaths	Woodlands
	air and water quality regulation† soil quality regulation† noise regulation†
Cultural services	
recreation and tourism* aesthetic values* cultural heritage* spiritual values* education* sense of place* health benefits*	recreation and tourism* aesthetic values* cultural heritage* employment* education* sense of place* health benefits*

* goods † services

(The supporting services, including primary production and nutrient cycling, are not listed for individual habitats as they are necessary for the production of all other ecosystem services.)

Natural capital has various values. We usually, rightly or wrongly, assess worth in monetary terms. *Economic value* can be determined from the market price of the goods and services a resource produces. *Ecological values*, however, have no formal market price: soil erosion control, nitrogen fixation, and photosynthesis are all essential for human existence but have no direct monetary value. Similarly, *aesthetic values* (e.g. the appreciation of a landscape for its visual attraction) have no market price. Ecological and aesthetic values do not provide easily identifiable commodities, so it is difficult to assess the economic contributions of these values using traditional methods of accounting. They are usually undervalued from an economic viewpoint (Figure 8.11).

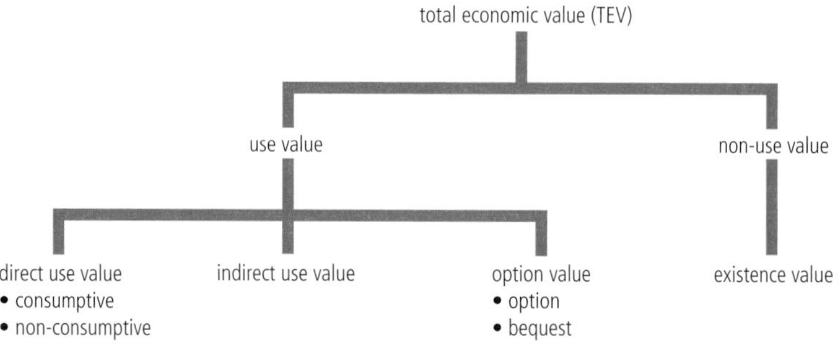

Figure 8.11 Ways of assessing the value of natural capital

Ethical, spiritual, and philosophical perspectives tend to give organisms and ecosystems *intrinsic value* (i.e. value in their own right, irrespective of economic value). They are valued regardless of their potential use to humans. The evaluation of natural capital therefore requires many diverse perspectives that lie outside the remit of conventional economics.

Direct use values are ecosystem goods and service that are directly used by humans, most often by people visiting or residing in the ecosystem. **Consumptive use** includes harvesting food products, timber for fuel or housing, medicinal products and hunting animals for food and clothing. **Non-consumptive use** includes recreational and cultural activities that do not require harvesting of products. **Indirect use** values are derived from ecosystem services that provide benefits outside the ecosystem itself (e.g. natural water filtration which may benefit people downstream). **Optional values** are derived from potential future use of ecosystem goods and services not currently used

– either by yourself (*option value*) or your future offspring (*bequest value*). *Non-use values* include aesthetic and intrinsic values, and are sometimes called **existence values**.

Other ways of measuring the value of a resource (besides calculating the direct price of its products) include calculating or estimating:

- the cost of replacing it with something else
- the cost of mitigating its loss
- the cost of averting the cost of its degradation
- its contribution to other income or production
- how much people are prepared to pay for it.

There are attempts to acknowledge these diverse valuations of nature. For example, scientists are examining the importance of biodiversity for ecosystem functioning by looking at connections between species diversity and the integrity of ecosystem processes (e.g. the role of pollinators such as wasps and bees in maintaining flowering and fruiting in rainforest). Biodiversity also has value in its contribution to ongoing evolution and speciation, and as a genetic resource. We need to find ways to value nature more rigorously against common economic values (e.g. GNP). However, some argue that these valuations are impossible to quantify and price realistically (Figure 8.12). Not surprisingly, much of the sustainability debate centres on the problem of how to weigh conflicting values in our treatment of natural capital.

To learn more about the economic value of biodiversity, go to www.pearsonhotlinks.co.uk, enter the book title or ISBN, and click on weblink 8.6.

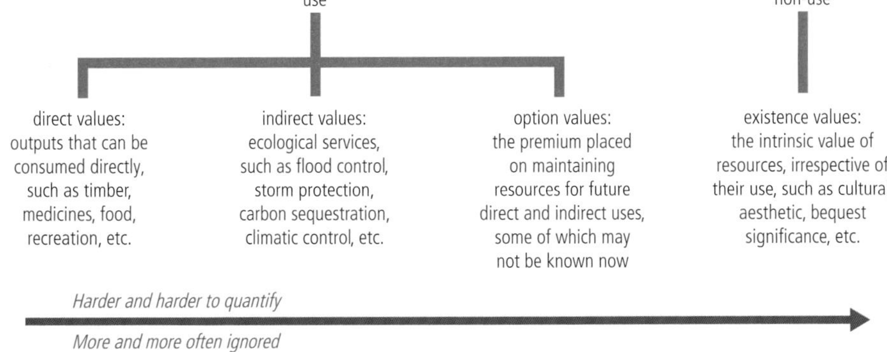

Figure 8.12 Levels of difficulty in assessing the economic value of natural capital

Natural resources have **recreational value**, as holiday destinations and places for people to relax. Ecotourism is a growing source of revenue for countries with natural resources that are attractive to tourists, and can provide an alternative income that is sustainable and does not deplete the source of natural capital. For example, rainforests are under threat from logging which is non-sustainable and damages the original stock; however, ecotourism provides income at the same time as requiring the resource to remain intact so as to attract tourists. The income is therefore sustainable.

Case study

Valuing oceans: the US$2 trillion question

Economists at the Stockholm Environment Institute have estimated that the cost of climate change on the oceans will amount to nearly US$2 trillion annually by 2100, or about 0.4 per cent of global GDP. The estimate was based on five measures:

- loss of fisheries
- reduced tourism revenues
- the economic cost of rising sea levels
- the cost of increased storm activity
- ocean acidification.

If temperatures rise by 4 °C by 2100, they estimate the total cost will come to US$1.98 trillion whereas if temperatures rise by only 2.2 °C, it will be US$612 billion.

Estimates of the world's GDP a century from now depend on too many variables to calculate with any precision. The same is true for the rise in temperature by 2100. the time-scale is too great to suggest an accurate prediction.

How can we quantify values such as aesthetic value, which are inherently qualitative?

CONCEPTS: Environmental value systems

The value that different cultures place on natural capital ultimately depends on the environmental value system through which those cultures see the world.

The concept of natural capital is dynamic. Whether or not something has the status of 'natural capital', and the marketable value of that capital, varies regionally and over time. This is influenced by cultural, social, economic, environmental, technological, and political factors. Examples include cork, uranium, lithium.

Dynamic nature and concept of a resource

The value of a resource should be seen as dynamic, with the possibility that its status may change over time. As humans advance culturally and technologically, and our resource base changes, the importance of a resource may be transformed. Resources become more valuable as new technologies need them. For example, flint – once an important resource as a hand tool – is now redundant; it was superseded by the development of metal extraction from ores (i.e. technological progress).

Uranium, in contrast, was of little value before the advent of the nuclear age. Nuclear fission involves the bombardment of uranium atoms with neutrons. A neutron splits a uranium atom, releasing a great amount of energy as heat and radiation. A different process, nuclear fusion, powers the Sun. In nuclear fusion, the nuclei of atoms (e.g. deuterium and tritium, both isotopes of hydrogen) fuse together, causing a much greater release of energy than in nuclear fission. If we ever learn to generate power by harnessing the energy from nuclear fusion, uranium, like flint before it, will lose its value.

Stone Age tools were made from a variety of stones such as flint. Stones were shaped by chipping to provide a sharp edge. The tools could be attached to arrows or hand-held to skin animals and cut meat.

Exercises

1. Using examples, distinguish between renewable and non-renewable forms of natural capital.
2. Outline the range of goods and services that are provided by the natural environment.
3. Using an example, explain the dynamic nature of natural capital.

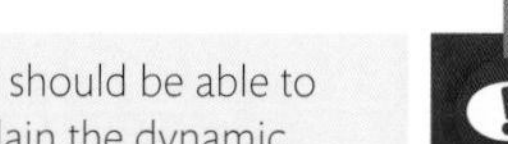

You should be able to explain the dynamic nature of the concept of natural capital.

Big questions

Having read this section, you can now discuss the following big questions:

- What strengths and weaknesses of the systems approach and the use of models have been revealed through this topic?
- To what extent have the solutions emerging from this topic been directed at *preventing* environmental impacts, *limiting* the extent of the environmental impacts, or *restoring* systems in which environmental impacts have already occurred?
- What value systems can you identify at play in the causes and approaches to resolving the issues addressed in this topic?
- How does your own value system compare with others you have encountered in the context of issues raised in this topic?
- How are the issues addressed in this topic of relevance to sustainability or sustainable development?
- In what ways might the solutions explored in this topic alter your predictions for the state of human societies and the biosphere some decades from now?

Points you may want to consider in your discussions:

- How do models and/or a systems approach help our understanding of resource use in society?
- Why do people use non-renewable resources rather than renewable resources?
- How do environmental value systems influence the use of renewable and non-renewable resources?
- What are your views on this?
- Outline the relationship between renewable and non-renewable resources and sustainability.
- How do you think society is likely to change in the coming years? Give reasons for your answer.

8.3 Solid domestic waste

Significant ideas

Solid domestic waste (SDW) is increasing as a result of growing human population and standards of living (consumption).

Both the production and management of SDW can have a significant influence on sustainability.

Big questions

As you read this section, consider the following big questions:

- What strengths and weaknesses of the systems approach and the use of models have been revealed through this topic?
- To what extent have the solutions emerging from this topic been directed at *preventing* environmental impacts, *limiting* the extent of the environmental impacts, or *restoring* systems in which environmental impacts have already occurred?
- What value systems can you identify at play in the causes and approaches to resolving the issues addressed in this topic?
- How does your own value system compare with others you have encountered in the context of issues raised in this topic?
- How are the issues addressed in this topic of relevance to sustainability or sustainable development?
- In what ways might the solutions explored in this topic alter your predictions for the state of human societies and the biosphere some decades from now?

Knowledge and understanding

- There are different types of SDW of which the volume and composition changes over time.
- The abundance and prevalence of non-biodegradable (e.g. plastic, batteries, e-waste) pollution in particular has become a major environmental issue.
- Waste disposal options include landfill, incineration, recycling, and composting.
- There are a variety of strategies that can be used to manage SDW (influenced by cultural, economic, technological, and political barriers. These strategies include:
  - altering human activity: includes reduction of consumption and composting of food waste

- controlling release of pollutant: governments create legislation to encourage recycling and reuse initiatives and impose tax for SDW collection, impose taxes on disposable items
- reclaiming land-fills, use of SDW for *trash-to-energy* programmes, implementing initiatives to remove plastics from the Great Pacific Garbage Patch (clean-up and restoration).

There are different types of SDW of which the volume and composition changes over time.

Types of solid domestic waste

The amount of waste produced by the global population is steadily increasing. The world faces an ongoing problem in how and where to dispose of this waste. Household waste is composed of a wide variety of materials. There is limited compositional data available, but the best overall estimates currently available are shown in Figure 8.13.

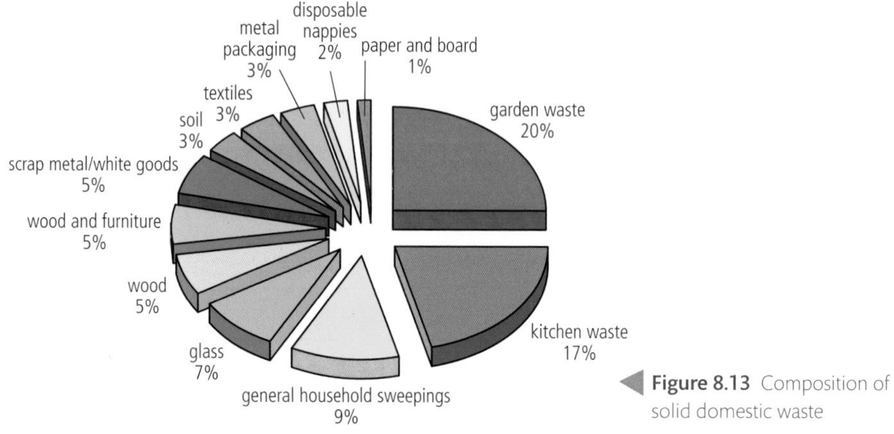

Figure 8.13 Composition of solid domestic waste

The WEEE man: a robotic figure 7 m tall, weighing 3.3 tonnes and made from waste electrical and electronic equipment

General domestic waste

The volume of waste varies by society and over time. MEDCs generate more waste than LEDCs (Figure 8.14). It also changes over time – there is now more non-biodegradable waste (e-waste and plastics). It increases as countries become more developed. It also increases when there are festivities such as Christmas, Easter, Ramadan, Diwali, birthdays, and so on.

To learn about the Story of Stuff - an organization that aims to promote awareness about waste and other issues - go to www.pearsonhotlinks.co.uk, enter the book title or ISBN, and click on weblink 8.7. To watch various short movies that have been produced by the Story of Stuff (including their original movie on waste), click on weblink 8.8.

Some facts about solid domestic waste in the UK.

- The UK population produces approximately 28 million tonnes of SDW per year.
- This is almost 500 kg per person per year.
- The figure is growing by about 3 per cent per year.
- The disposal of this waste has local, natural, and global consequences for the environment.

Up to half of the world's population lack access to the most basic of waste collection and safe disposal. Almost 40 per cent of the world's waste ends up in huge rubbish tips, mostly found near urban populations in LEDCs, posing a serious threat to human health and the environment. The Mbeumbeus waste dump in Senegal covers some 175 ha. The dump used to receive a few thousand tonnes of rubbish a year in the 1960s. Now it takes in 475 000 tonnes of rubbish a year, increasingly e-waste from computers, televisions and mobile phones.

See inset

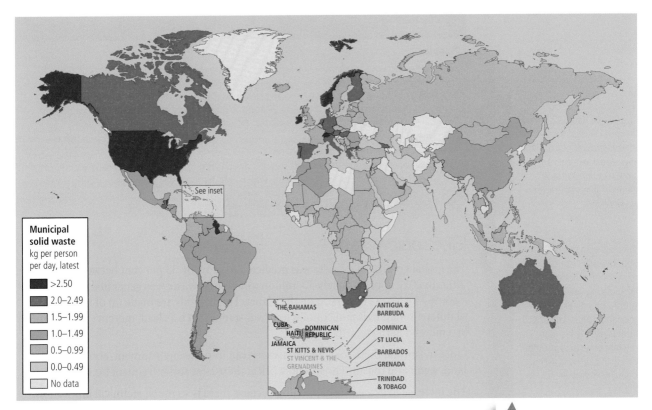

Municipal solid waste
kg per person per day, latest

- >2.50
- 2.0–2.49
- 1.5–1.99
- 1.0–1.49
- 0.5–0.99
- 0.0–0.49
- No data

THE BAHAMAS
CUBA
HAITI
DOMINICAN REPUBLIC
JAMAICA
ST KITTS & NEVIS
ST VINCENT & THE GRENADINES
ANTIGUA & BARBUDA
DOMINICA
ST LUCIA
BARBADOS
GRENADA
TRINIDAD & TOBAGO

Figure 8.14 The world's rubbish

Much of the world's rubbish is generated by city-dwellers. According to the World Bank, the potential costs of dealing with increasing rubbish are high. The world's cities currently generate around 1.3 billion tonnes of SDW a year, or 1.2 kg per city-dweller per day. Almost half of this comes from rich countries. With increasing urbanization, this is expected to rise to 2.2 billion tonnes by 2025, or 1.4 kg per person. China's urban population will throw away 1.4 billion tonnes in 2025, up from 520 million tonnes today. By contrast, the USA's urban population will throw away 700 million tonnes compared with 620 million tonnes today.

The world's largest dumps are increasingly polluting rivers, groundwater, air and soil, and are having an adverse impact of those who live and work nearby or on the dump. Waste pickers often have no protective clothing. The most common human health problems are gastro-intestinal (stomach) disorders, skin disorders, respiratory disorders, and genetic disorders.

Case study

Plastic waste in the Thames

Scientists from London collected rubbish over a 3-month period at the end of 2012 from seven locations along the Thames estuary. They collected over 8400 items including plastic cups, food wrapping, and cigarette packaging.

The two most contaminated sites were close to sewage treatment works, and could suggest that plants were not filtering out larger waste, or were letting sewage overflow when heavy rains created extra waste. However, the scientists were unable to estimate the volume of litter that was actually entering the North Sea.

The potential impacts this could have for wildlife are far-reaching: not only are the species that live in and around rivers affected, but also those in seas that rivers feed into.

Plastic waste in the River Thames

Larger pieces of plastic are being continuously rolled backward and forward by tidal movements and broken down into smaller and smaller fragments that are easily ingested by birds, fish, and smaller species such as crabs. The toxic chemicals they contain, in high doses, could harm the health of wildlife.

Case study

Nappies

Some waste material is more problematic than others. Nappies are a particular problem. Nappy waste is harmful and expensive. It costs £40 million a year to dispose of an estimated 1 million tonnes of nappy waste, of which 75 per cent is urine and faeces. Most nappy waste is taken to landfill sites, where nappies can take an estimated 500 years to break down, and add to the build-up of methane gas. Environmentalists say that using washable nappies would represent a saving of £500 per baby.

About 1 million tonnes of absorbent hygiene products are generated in the UK each year. In 2011, Knowaste set up the UK's first recycling facility for absorbent hygiene products such as disposable nappies. In its first 2 years of operation, the facility recycled over 77 million nappies. The company isolates plastics and fibres from the products, separating them from human waste so that the nappy or incontinence pad can be recycled. The recycled material can be used as an additive for concrete, plastic sheets, flood defence systems, and containers used for disposable nappies.

To learn more about the World Bank Global Review of Solid Waste Management, go to www.pearsonhotlinks.co.uk, enter the book title or ISBN, and click on weblink 8.9.

Electronic waste

According to the UN Step Initiative (set up to tackle the world's growing e-waste problem), millions of mobile phones, laptops, tablets, electronic toys, digital cameras, and other electronic devices are being dumped illegally in LEDCs.

The global volume of e-waste was predicted to grow by 33 per cent between 2013 and 2017. In 2012 approximately 50 million tonnes of e-waste was generated worldwide – some 7 kg for every person on the planet. These goods are made up of hundreds of different materials and contain toxic substances such as lead, mercury, cadmium, arsenic, and flame retardants.

Once in landfill, these toxic materials seep out into the environment, contaminating land, water and the air. Those who work at these sites suffer frequent bouts of illness.

The increase in e-waste is happening because there is so much technical innovation. TVs, mobile phones and computers are all being replaced more and more quickly. The lifetime of products is shortening.

Recycling e-waste in China

In 2012, China generated 11.1 million tonnes of e-waste, followed by the USA with 10 million tonnes. However, *per capita* figures were reversed. On average, each American generated 29.5 kg of e-waste, compared to less than 5 kg per person in China. In Europe, Germany discards the most e-waste in total, but Norway and Liechtenstein throw away more per person.

The European Environment Agency estimates that between 250 000 tonnes and 1.3 million tonnes of used electrical products are shipped out of the EU every year, mostly to west Africa and Asia. Research by the Massachusetts Institute of Technology (MIT) suggests that, in 2010, the USA discarded 258.2 million computers, monitors, TVs, and mobile phones, of which only 66 per cent was recycled. The life of a mobile phone is now less than 2 years. In 2011, in the USA, only 12 million mobile phones were collected for recycling even though 120 million were bought.

Most phones contain precious metals. The circuit board can contain copper, gold, zinc, beryllium, and tantalum. The coatings are typically made of lead and phone-makers are now increasingly using lithium batteries. Yet fewer than 10 per cent of mobile phones are dismantled and reused. Part of the problem is that computers, phones and other devices are becoming increasingly complex and made of smaller and smaller components. The failure to recycle is also leading to shortages of rare-earth minerals to make future generations of electronic equipment.

To learn more about recycling of e-waste in China, go to www.pearsonhotlinks.co.uk, enter the book title or ISBN, and click on weblink 8.10.

Guiyu in China has been described as the e-waste capital of the world. Most of the recycling takes place in people's homes. The industry is worth US$75 million to the town each year. Guiyu's population has elevated rates of lead poisoning, cancer caused by dioxins, and miscarriages.

Solid domestic waste in Europe

The percentage make-up of waste in Europe other than the UK is shown in Table 8.4.

Country	Organics	Paper/ board	Glass	Metals	Plastics	Textiles	Other
Belgium	43	28	9	4	7	9	none
Denmark	37	30	6	3	7	18	none
Germany	32	24	8	5	9	none	22
France	21	27	7	4	11	2	28
Greece	49	20	5	4	9	13	none
Ireland	42	15	6	4	11	8	14
Italy	32	27	8	4	7	3	19
Luxembourg	41	16	4	3	8	3	25
Netherlands	39	25	8	5	8	15	none
Portugal	39	20	4	2	9	5	21
Spain	44	21	7	4	11	5	8

Table 8.4 The composition of Europe's waste (per cent of total by country)

TOK: SDW is one of the world's major development issues. Over 25 per cent of the world's population do not have proper sanitation. The link between contaminated water and disease is very strong. Can we afford to ignore solid domestic waste?

Disposal options for SDW

Waste disposal options include landfill, incineration, recycling, and composting.

There are a number of methods of dealing with solid domestic waste. The most common are recycling, composting, landfill, and incineration. In addition, it is possible to reduce the amount of waste generated, and reuse goods to extend their lifespan.

You should be able to evaluate SDW disposal options.

Recycling

The UK has a poor record of recycling, which is much worse than in many other countries of the EU (Figure 8.15). The UK long lagged behind with recycling mainly because there were many more landfill sites which are cheaper to use, such as old quarries or mines. In 2005, 410 000 tonnes of plastic were collected, 43 per cent of

Up to 60 per cent of household waste in the USA is recyclable or compostable. But Americans compost only 8 per cent of their waste. Surveys suggest that the main reason Americans don't compost is because they think it is a complicated process. In contrast, the Zabbaleen, who are responsible for much of the waste collection in Cairo, Egypt, recycle as much as 80 per cent of the waste collected.

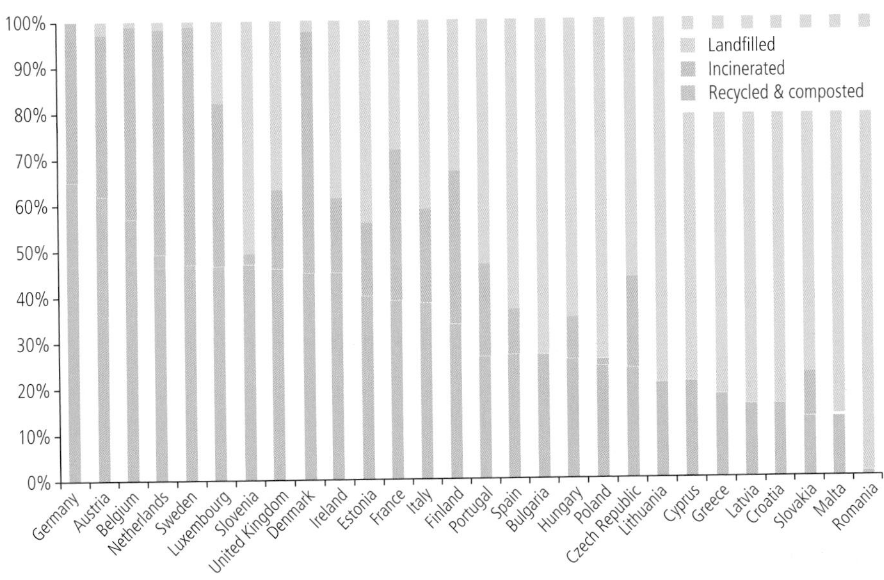

Figure 8.15 Waste management in Europe

which was recycled in the UK and almost 57 per cent was exported for recycling, mainly to Asia. However, there is growing concern that exported plastic never reaches recycling plants.

Composting

Composting is the decomposition of biodegradable material. It recycles organic household waste into a humus-like soil. It returns valuable nutrients to the soil.

Landfill

Landfill may be cheap but it is not always healthy (an indirect cost) – and will eventually run out. Recent research and surveys show that living near to landfill sites increases the risk of health problems including heart problems and birth defects. Landfills need to be located relatively close to the source of waste to be economic, so they tend be found near areas of high population density. Landfill can give off gases such as methane and may contaminate water supplies. However, landfills are generally designed to prevent leaching by lining them with impermeable clay. Many landfills are sited in old quarries which could be turned into lakes or nature reserves instead.

Currently, much domestic waste ends up in landfill sites. However, the reliance on landfill is unsustainable for a number of reasons.

- There are already areas struggling to find suitable new landfill sites. This shortage of space will become more acute as the amount of waste continues to grow.
- When biodegradable waste, such as food, decomposes anaerobically it releases methane which, as a greenhouse gas, contributes to global warming. It is also explosive.
- Chemicals and heavy metals can pollute the soil and groundwater. Leachate, produced from organic waste, breaks down causing the same problem.
- Communities are often violently opposed to the creation of any new sites (the not-in-my-back-yard (NIMBY) principle).

Incineration

Incineration means burning. It converts the waste into ash and gas particulates. The heat can be used to generate electricity. Incinerators can reduce the volume of the original waste by as much as 80–90 per cent. Thus, the technology could significantly reduce the volume of waste for which for landfill disposal is necessary. Incineration has particularly strong benefits for the treatment of certain types of waste, notably clinical waste and hazardous waste, as it can dispose of these products safely.

Problems with landfill and incineration

Landfill is considered the cheapest approach only because the economics of waste disposal do not yet fully take into account environmental costs. There are local problems with noise, smells and vermin.

The major environmental problems associated with landfill are as follows.

- When biodegradable matter (paper, food waste, and garden waste) decomposes in anaerobic conditions (without oxygen), as is the case in a landfill site, methane (a greenhouse gas) is given off.
- Some parts of any landfill site are not anaerobic, so carbon dioxide is also given off – another greenhouse gas.
- Leachates from the landfill can contaminate aquifers (underground water stores).

Unfortunately, the measures used to try to reduce the escape of gases and leachates (liner systems, waste compaction, and capping) also stop oxygen entering and this increases the generation of methane. Methane is 21 times more powerful as a greenhouse gas than carbon dioxide. Methane can seep into buildings where it represents a fire and explosive risk, or into sewers where it may suffocate workers.

The major environmental problems associated with incineration are as follows:

- Air pollution – carbon dioxide, sulfur dioxide, nitrogen dioxide, nitrous oxide, chlorine, dioxins, and particulates all result from incineration. In turn, these lead to other environmental problems such as acid rain, smogs, lung disease.
- The volume of traffic generated is increased due to the need to get the waste to the incinerators – again leading to greater air pollution, noise, vibration, and accidents.
- The ash which results (usually equal to 10–20 per cent of the mass of the original waste) is often toxic and still needs to be disposed of in landfill.

Building incinerators represents a high initial capital cost.

In 2005, incineration produced 4.8 per cent of the electricity consumed and 13.7 per cent of the total domestic heat consumption in Denmark. A number of other European countries including Germany, the Netherlands, France, and Luxemburg rely heavily on incineration for disposing of municipal waste.

CONCEPTS: Sustainability

Many municipal incinerators are trying to produce energy from waste with energy from waste (EfW) or waste to energy (WTE) programmes. However, in many places, communities prefer these facilities to be located 'somewhere else'.

The USA has some 1900 municipal landfill sites. The largest is Puente Hills landfill in Los Angeles. The dump is over 160 m high. With 60 years' worth of decomposing rubbish, Puente Hills produces enough methane to generate electricity to 70 000 homes.

Strategies for managing SDW

There are a variety of strategies that can be used to manage SDW, influenced by cultural, economic, technological, and political barriers.

These strategies for managing SDW include:

- altering human activity: includes reduction of consumption and composting of food waste
- controlling release of pollutant: governments create legislation to encourage recycling and reuse initiatives and impose tax for SDW collection, impose taxes on disposable items
- reclaiming landfills, use of SDW for trash-to-energy programmes, implementing initiatives to remove plastics from the Great Pacific Garbage Patch (clean-up and restoration).

These may be influenced by cultural factors (e.g. is it acceptable?), economic factors (e.g. is it affordable?), technological factors (e.g. can it be achieved?), and political factors (e.g. is there support for the strategy?).

Table 8.5 summarizes waste management options for SDW.

Waste management options	How it works
reduce the amount of waste	• producers think more about the lifespan of goods and reduce packaging • consumers consider packaging and lifespan when buying goods
reuse goods to extend their lifespan	• bring-back schemes where containers are refilled (e.g. milk bottles) • refurbish/recondition goods to extend their useful life (e.g. use of old car tyres to stabilize slopes/reduce soil erosion) • used goods put to another use rather than thrown out (e.g. plastic bags used as bin liners; old clothes used as cleaning cloths) • charity shops pass on goods to new owners
recover value	• recycle goods such as glass bottles and paper • compost biodegradable waste for use as fertilizer • incinerate (burn) waste – collect electricity and heat from it
dispose of waste in landfill sites	• put waste into a hole (natural or the result of quarrying) or use to make artificial hills

Table 8.5 Waste management options for SDW

You should be able to compare and contrast pollution management strategies for SDW, and evaluate them, with reference to Figure 1.35, by considering recycling, incineration, composting, and landfills.

Development organizations frequently provide aid for conservation, clean water and housing developments. Rarely do they provide funding for solid domestic waste projects. Why should sanitation be the 'Cinderella option' in the development process?

The IB ESS Guide summarizes pollution management strategies as shown in Figure 1.35.

The Great Pacific Garbage Patch

The **Great Pacific Garbage Patch** is an area of marine debris, approximately 135° to 155° W and 35° to 42° N (Figure 8.16). It shifts its exact position every year; it remains within the North Pacific Gyre (a system of circular ocean currents) as it is confined by ocean currents.

Estimates of its size vary from 700 000 km^2 to more than 15 000 000 km^2 – between 0.41 per cent and 8.1 per cent of the size of the Pacific Ocean. Figure 8.16 suggests that it is at least three times the size of Spain and Portugal combined. The GPGP is also estimated to contain up to 100 million tons of rubbish.

Plastics never biodegrade. They do not break down into natural substances. Instead they go through a photodegradation process, splitting into smaller and smaller particles that are still plastic.

Problems caused by plastic

- Plastic fouls beaches throughout the world and reduces potential income from tourism and recreation.
- Plastic entangles marine animals and drowns them, strangles them, and makes them immobile.
- Plastic garbage when washed ashore destroys habitats.
- Plastic gets inside ships propellers and keels making ship maintenance more expensive.
- Plastic does not degrade; it also makes an ideal medium for the transfer of invasive species.

Hence, on account of its sheer size and the small nature of some of its content, it is nearly impossible to clean up the Garbage Patch. Plastic garbage can be collected when washed up on beaches – but there are relatively few beaches in the region of the GPGP. Some have suggested putting booms around the Patch and hauling in the plastic. The size makes this impractical. Others have suggested 'hoovering' up the

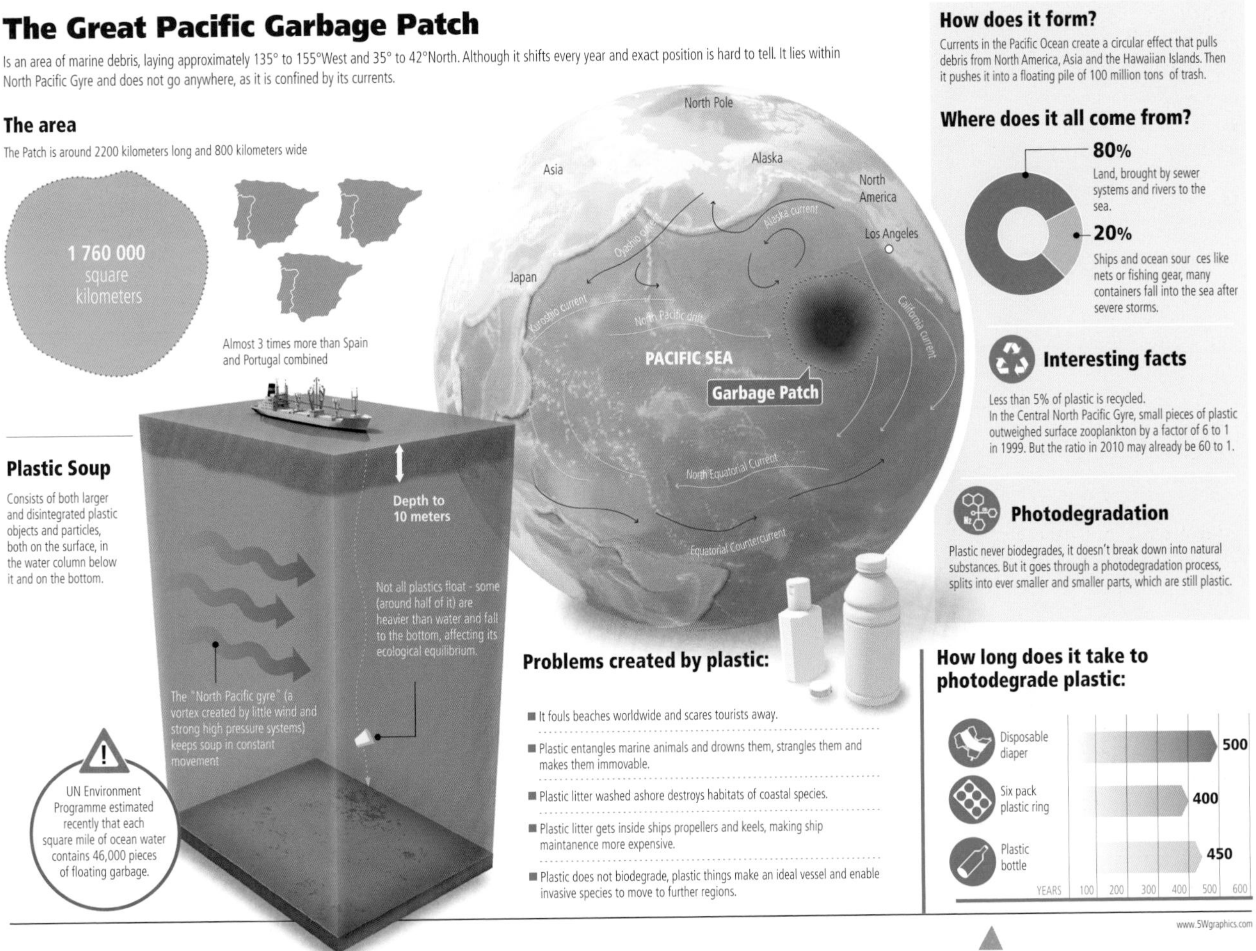

The Great Pacific Garbage Patch

Is an area of marine debris, laying approximately 135° to 155°West and 35° to 42°North. Although it shifts every year and exact position is hard to tell. It lies within North Pacific Gyre and does not go anywhere, as it is confined by its currents.

The area

The Patch is around 2200 kilometers long and 800 kilometers wide

1 760 000 square kilometers

Almost 3 times more than Spain and Portugal combined

North Pole

Asia

Alaska

North America

Los Angeles

Japan

Oyashio current

Alaska current

Kuroshio current

North Pacific drift

California current

PACIFIC SEA

Garbage Patch

North Equatorial Current

Equatorial Countercurrent

Plastic Soup

Consists of both larger and disintegrated plastic objects and particles, both on the surface, in the water column below it and on the bottom.

Depth to 10 meters

Not all plastics float - some (around half of it) are heavier than water and fall to the bottom, affecting its ecological equilibrium.

The "North Pacific gyre" (a vortex created by little wind and strong high pressure systems) keeps soup in constant movement

UN Environment Programme estimated recently that each square mile of ocean water contains 46,000 pieces of floating garbage.

How does it form?

Currents in the Pacific Ocean create a circular effect that pulls debris from North America, Asia and the Hawaiian Islands. Then it pushes it into a floating pile of 100 million tons of trash.

Where does it all come from?

80% Land, brought by sewer systems and rivers to the sea.

20% Ships and ocean sour ces like nets or fishing gear, many containers fall into the sea after severe storms.

Interesting facts

Less than 5% of plastic is recycled.
In the Central North Pacific Gyre, small pieces of plastic outweighed surface zooplankton by a factor of 6 to 1 in 1999. But the ratio in 2010 may already be 60 to 1.

Photodegradation

Plastic never biodegrades, it doesn't break down into natural substances. But it goes through a photodegradation process, splits into ever smaller and smaller parts, which are still plastic.

Problems created by plastic:

- It fouls beaches worldwide and scares tourists away.
- Plastic entangles marine animals and drowns them, strangles them and makes them immovable.
- Plastic litter washed ashore destroys habitats of coastal species.
- Plastic litter gets inside ships propellers and keels, making ship maintanence more expensive.
- Plastic does not biodegrade, plastic things make an ideal vessel and enable invasive species to move to further regions.

How long does it take to photodegrade plastic:

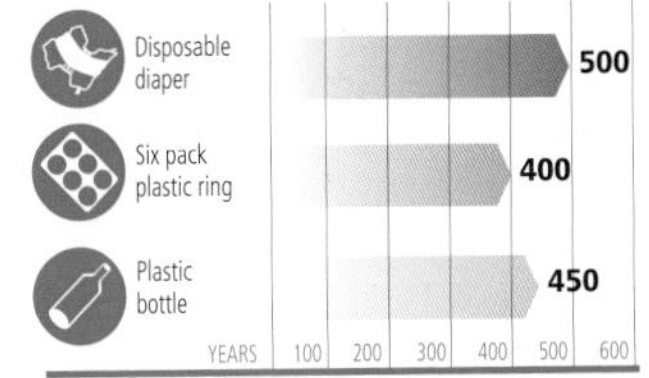

www.5Wgraphics.com

Figure 8.16 The Great Pacific Garbage Patch

plastic – again the sheer scale of this makes is unlikely. Ultimately, it would be far easier to control the pollution at source – less waste, more recycling, reuse etc.

Exercises

1. Outline the types of solid domestic waste.
2. Describe and evaluate pollution management strategies for solid domestic (municipal) waste.
3. Describe how landfill works as a type of waste disposal.
4. What are the advantages of landfill?
5. What are the disadvantages of landfill?
6. In what ways is incineration a better option than landfill?
7. What are the disadvantages of incineration?

CHALLENGE YOURSELF

1 Monitor your waste for a week and keep a waste diary. After the week, copy then complete the table below.

Waste diary

Activity	Is waste produced?	Type of waste produced	What happens to the waste?
At home			
cooking			
eating			
drinking			
washing/cleaning			
watching TV			
homework			
play in garden			
In the community			
meeting friends			
going to films, etc.			
sport			
part-time work			
At school			
lessons			
break time			
other			

Big questions

Having read this section, you can now discuss the following big questions:

- What strengths and weaknesses of the systems approach and the use of models have been revealed through this topic?
- To what extent have the solutions emerging from this topic been directed at *preventing* environmental impacts, *limiting* the extent of the environmental impacts, or *restoring* systems in which environmental impacts have already occurred?
- What value systems can you identify at play in the causes and approaches to resolving the issues addressed in this topic?
- How does your own value system compare with others you have encountered in the context of issues raised in this topic?
- How are the issues addressed in this topic of relevance to sustainability or sustainable development?
- In what ways might the solutions explored in this topic alter your predictions for the state of human societies and the biosphere some decades from now?

Points you may want to consider in your discussions:

- How do models and/or a systems approach help our understanding of solid domestic waste?

- Evaluate the alternatives for the disposal of solid domestic waste.
- How do environmental value systems influence the disposal of solid domestic waste?
- What are your views on this?
- Examine the relationship between solid domestic waste and sustainability.
- How is solid domestic waste likely to change over the next few decades? Give reasons for your answer.

8.4 Carrying capacity and ecological footprints

Significant ideas

Human carrying capacity is difficult to quantify.

An ecological footprint is a model that makes it possible to determine whether human populations are living within the carrying capacity of their environment.

Big questions

As you read this section, consider the following big questions:

- What strengths and weaknesses of the systems approach and the use of models have been revealed through this topic?
- To what extent have the solutions emerging from this topic been directed at *preventing* environmental impacts, *limiting* the extent of the environmental impacts, or *restoring* systems in which environmental impacts have already occurred?
- What value systems can you identify at play in the causes and approaches to resolving the issues addressed in this topic?
- How does your own value system compare with others you have encountered in the context of issues raised in this topic?
- How are the issues addressed in this topic of relevance to sustainability or sustainable development?
- In what ways might the solutions explored in this topic alter your predictions for the state of human societies and the biosphere some decades from now?

Knowledge and understanding

- Carrying capacity is the maximum number of a species or 'load' that can be sustainably supported by a given area.
- It is possible to estimate the carrying capacity of an environment for a given species but this is problematic in the case of human populations for a number of reasons.
- An ecological footprint is the area of land and water required to support a defined human population at a given standard of living. The measure takes into account of the area required to provide all the resources needed by the population, and the assimilation of all wastes.
- Ecological footprints are models used to estimate the demands that human populations place on the environment.

- Ecological footprints may vary significantly from country to country and person to person. They include aspects such as lifestyle choices (environmental value systems), productivity of food production systems, land use, and industry. If the ecological footprint of a human population is greater than the land area available to the population, this indicates that the population is unsustainable and exceeds the carrying capacity of that area.
- Degradation of the environment together with the use of finite resources is expected to limit human population growth.
- If human populations do not live sustainably, they will exceed carrying capacity and risk collapse.

Carrying capacity is the maximum number of a species or load that can be sustainably supported by a given area.

Carrying capacity

The concept of a population ceiling is one of a saturation level where population equals the **carrying capacity** of the local environment. Figure 8.17 shows three models of a population growing exponentially and approaching carrying capacity.

Model 1

There is no reduction in the rate of increase until the ceiling is reached at which point the increase drops to zero. This highly unlikely situation is unsupported by evidence from either human or animal populations.

Model 2

The population increase begins to slow down as the carrying capacity is approached and levels off when the ceiling is reached. It is claimed that populations which are large in size, have long lives, and low fertility rates conform to this S-curve pattern.

Model 3

The rapid rise in population overshoots the carrying capacity resulting in a sudden check (e.g. by famine, birth control). The population then recovers and fluctuates

Figure 8.17 Population growth and carrying capacity

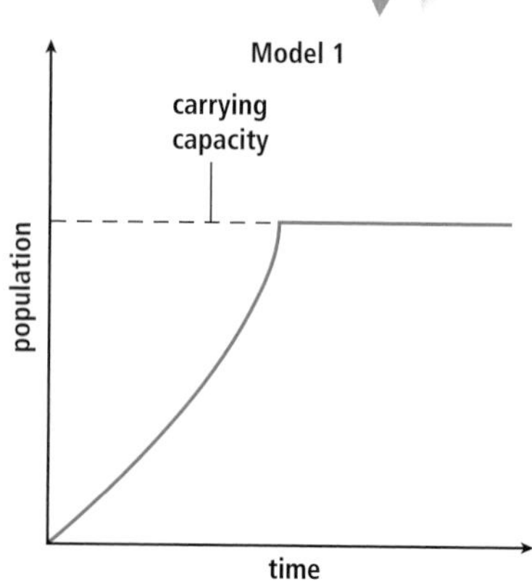

There is no reduction in the rate of increase until the ceiling is reached at which point the increase drops to zero. This highly unlikely situation is unsupported by evidence from either human or animal populations.

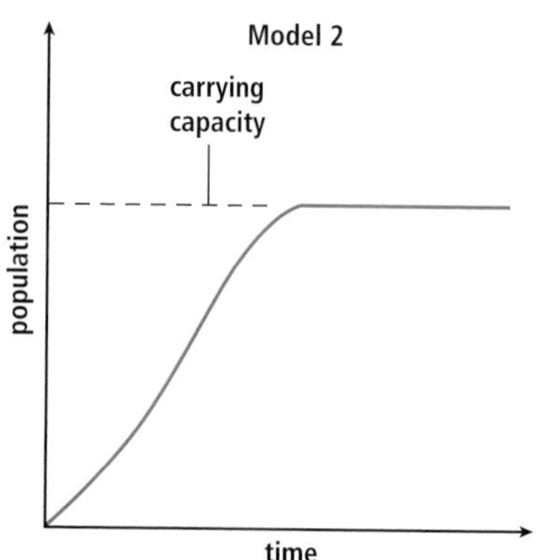

The population increase begins to taper off as the carrying capacity is approached and levels off when the ceiling is reached. It is claimed that populations which are large in size, have long lives, and low fertility rates conform to this S-curve pattern.

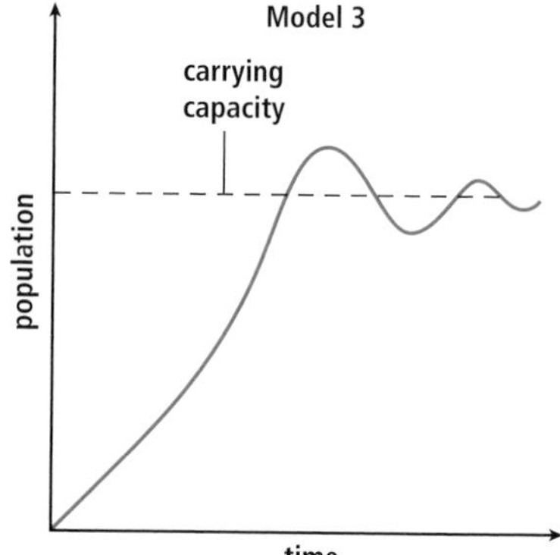

The rapid rise in population overshoots the carrying capacity resulting in a sudden check (e.g. by famine, birth control, etc.). The population then recovers and fluctuates eventually settling at the carrying capacity. This J-shaped curve appears more applicable to populations which are small in number, have short lives and high fertility rates.

eventually settling at the carrying capacity. This J-shaped curve appears more applicable to populations which are small in number, have short lives and high fertility rates.

Optimum-, over-, and under-population

Optimum population is the number of people who, when using all the available resources, will produce the highest *per capita* economic return (Figure 8.18). It is the point at which the population has the highest standard of living and quality of life. If the size of the population increases or decreases from the optimum, the standard of living will fall. This concept is dynamic and changes with time as techniques improve, as population totals and structures change, and as new materials are discovered.

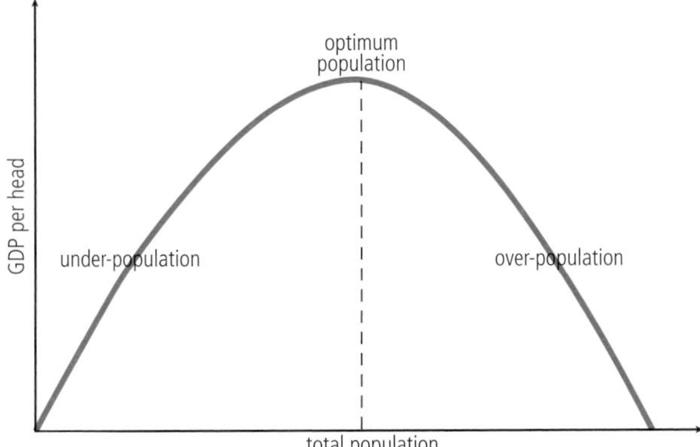

Figure 8.18 Under-, optimum-, and over-population

Standard of living is the result of the interaction between physical and human resources and can be expressed as:

$$\text{Standard of living} = \frac{\text{natural resources} \times \text{technology}}{\text{population}}$$

Over-population occurs when there are too many people relative to the resources and technology locally available to attain the optimum standard of living. Bangladesh (population density of 132 per km^2), Somalia and parts of Brazil and India are over-populated as they have insufficient food and materials. They suffer from natural disasters such as drought and famine and are characterized by low incomes, poverty, poor living conditions, and a high level of emigration.

Under-population occurs when there are far more resources in an area (e.g. food production, energy and minerals) than can be used by the people living there in order to reach the optimum population. Canada could theoretically double its population and still maintain its standard of living. Countries like Canada and Australia can export their surplus food, energy, and mineral resources. It is possible that the standard of living would increase through increased production if population were to increase.

To learn more about human carrying capacity, go to www.pearsonhotlinks.co.uk, enter the book title or ISBN, and click on weblink 8.11.

Population growth and food resources

Thomas Malthus

In 1798, the Reverend Thomas Malthus (1766–1834) produced his *Essays on the Principle of Population Growth.* He believed that there was finite optimum population size in relation to food supply and that any increase in population beyond this point

It is possible to estimate the carrying capacity of an environment for a given species but this is problematic in the case of human populations for a number of reasons.

would lead to a decline in the standard of living and to war, famine, and disease. His theory was based on two principles:

- in the absence of checks, population would grow at a geometric or exponential rate and could double every 25 years
- food supply at best increases at an arithmetic rate (Figure 8.19).

It is difficult to reliably estimate the environmental carrying capacity for human populations because:

- the variety of resources used is great
- humans can substitute one resource for another when the first becomes depleted
- lifestyle affects resource requirement
- technological developments change resources required and available (e.g. developments in renewable energy could reduce demand for fossil fuels)
- resources can be imported.

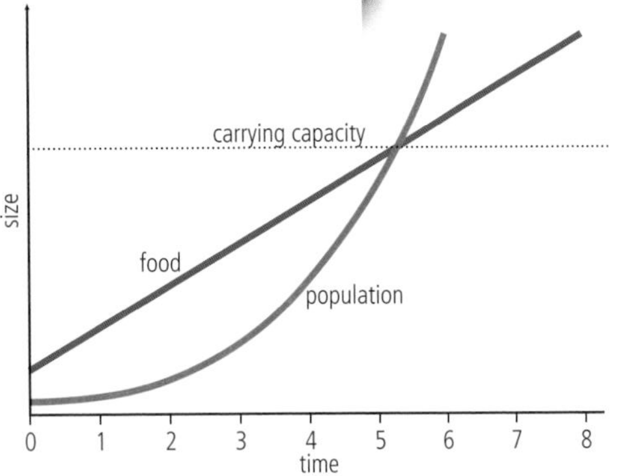

Figure 8.19 Malthus's views on population growth and growth of food resources

These principles state *potential* and not the *actual* growth of population and food production. Thus, the limit of food production creates a block or ceiling to the population growth in a given country. Malthus suggested preventive and positive checks as two main ways by which population growth could be reduced once this ceiling had been reached.

- Preventive checks include abstinence from marriage, a delay in the timing of marriage, and abstinence from sex within marriage. All these would reduce fertility rate.
- Positive checks such as lack of food, disease, and war directly affect mortality rates.

Malthus suggested that optimum population is related to resources and the level of technology. We now consider the concepts of over-population and under-population rather than the optimum population. Optimum population is difficult to identify and may vary as technology improves and attitudes change.

Unsustainable fishing – tuna landed at the Tokyo fish market

During the industrial revolution, the rate of growth in production was very rapid – greater than arithmetic rate and exceeding the rate of population growth. Malthus's ceiling was always ahead of and moving away from the population. Industrial development also had a positive effect on agricultural production through both intensification (labour and capital) and extension (more land). Nevertheless, as population increased and standard of living rose, so consumption of resources increased and diet changed from grain-and-vegetable based to dairy-and-fish based. A meat diet cannot sustain as many people as a plant-based diet.

Since Malthus's time, people have increased food production in many ways:

- draining marshlands
- reclaiming land from the sea
- cross-breeding cattle
- developing high-yield varieties of plants
- terracing steep slopes
- growing crops in greenhouses
- growing crops in water (hydroponics)
- using more sophisticated irrigation techniques

- using new foods such as soya
- making artificial fertilizers
- farming native species of crops and animals
- fish farming
- increased use of hormones/antibiotics.

Esther Boserup

Esther Boserup (1910–99) had a different view. She believed that people have the resources to increase food production. The greatest resource is knowledge and technology. When a need arises, someone will find a solution.

Whereas Malthus thought that food supply limited population size, Boserup suggested that in a pre-industrial society, an increase in population stimulated a change in agricultural techniques so that more food could be produced. Population growth has thus enabled agricultural development to occur.

She examined different land-use systems according to their intensity of production (measured by the frequency of cropping). At one extreme was shifting cultivation: at its least intensive, any one plot would be used less than once every 100 years. At the other extreme was multicropping with more than one harvest per year. Boserup suggested that there was a close connection between agricultural techniques and the type of land-use system. The most primitive was shifting cultivation, and the most advanced was ploughing with multiple cropping. She considered that any increase in the intensity of productivity by the adoption of new techniques would be unlikely unless population increased. Thus, population growth leads to agricultural development and the growth of the food supply.

Boserup's theory was based on the idea that people knew of the techniques required by more intensive systems, but adopted them only when the population grew. If knowledge was not available, the local agricultural system would regulate the population size in a given area.

Changing carrying capacities

Human carrying capacity of the environment is determined by:

- rate of energy and material consumption
- level of pollution
- interference with environmental life- support systems.

Reuse and recycling can increase human carrying capacity of the environment.

Reuse of plastic bottles

CONCEPTS: Sustainability

Sustainable tourism has a role to play in increasing the carrying capacity of the environment. The Casuarina Beach Club was one of the best examples of sustainable tourism in the Caribbean. The hotel showed considerable environmental awareness and responsibility by:

- meeting the internationally recognized Green Globe 21 criteria for sound environmental practices
- making partnerships with national and local governments, NGOs, and the local community
- conserving natural resources
- environmental awareness training for the staff, other hoteliers, learning institutes, and schools
- making massive reductions in waste by composting and other reuse and recycling initiatives
- promoting local culture, history, music, and furniture

- protecting turtle-nesting habitats
- undertaking revegetation projects
- conserving coastal forest strip to act as a hurricane defence.

This metal drum was used to compost vegetable waste.

The biggest reuse initiative was the collection and modification of 320 plastic containers in which the cooking oil was delivered. After modification, the containers were used for the garbage collection in the guest rooms. Composting of food was facilitated by means of four compostumblers.

The Casuarina adopted the policy of 'reduce, reuse, and recycle'. To reduce, the hotel contacted suppliers and requested less packaging. Individual portions of ketchup and butter, for example, were not provided in the restaurants. Instead hard, reusable bottles and containers were used. Paper towels were replaced by hand driers. Large shampoo and conditioner dispensers replaced individual sachets.

Water loss was reduced by installing low-flow devices on showers and taps. Beach and poolside showers were fitted with 'pull chain flush valve' systems. Waste-water from the beach showers was used to irrigate the gardens. The well water on the property was used to irrigate the grounds. There were signs in the hotel's rooms regarding the choice of frequency with which towels may be changed, which also reduces the consumption of water.

Recycling initiatives include the manufacture of pot pourri from cut flowers, and recycled paper from waste generated on the property. Separation of garbage for recycling is apparent throughout the property.

Recycling, Casuarina

Other initiatives included:

- purchasing local produce as much as possible
- using only degradable plastic bags in the hotel
- employing local handicapped people in the hotel.

Sadly, in 2005, the Casuarina Beach Club was bought by the Almond Beach hotel and converted into an all-inclusive hotel, where the focus on recycling and reuse was not continued.

However, sustainable tourism is still advocated in many parts of the Caribbean. The Caribbean Alliance for Sustainable Tourism aims to enhance the practices of the region's hotel and tourism operators by providing high-quality education and training related to sustainable tourism; promoting the industry's efforts and successes to the travelling public and interested parties.

Ecological footprints

Whereas carrying capacity is the number of individuals or species that an environment can sustain (providing resources and absorbing waste), an **ecological footprint** is area of land (and water) required to support an individual or population (providing all resources and absorbing waste). The ecological footprint is a theoretical area, whereas carrying capacity refers to a real area. These concepts are therefore the inverse of each other. Carrying capacity involves sustainable support of a population, whereas ecological footprints are not necessarily sustainable.

An ecological footprint is the hypothetical area of land and water required to support a defined human population at a given standard of living. The measure takes into account of the area required to provide all the resources needed by the population, and the assimilation of all wastes.

The term *ecological footprint* was coined by William Rees in 1992, and further developed with Mathis Wackernagel in *Our Ecological Footprint: Reducing Human Impact on the Earth*. A country described as having an ecological footprint of 2.4 times its own geographical area is consuming resources and assimilating its wastes on a scale that would require a land area 2.4 times larger than the actual size of the country in order to be sustainable. Ecological footprint can act as a model for monitoring environmental impact. It can also allow for direct comparisons between groups and individuals, such as comparing LEDCs and MEDCs. It can highlight sustainable and unsustainable lifestyles: populations with a larger ecological footprint than actual land area they are in, are living beyond sustainable limits. Wackernagel and Rees originally estimated that the available biological capacity for the population of the Earth (around 6 billion people at that time but over 7 billion now) was about 1.3 global hectares (gha) per person (or 1.8 gha if marine areas are included as a source of productivity).

Ecological footprint is a model used to estimate the demands that human populations place on the environment.

Ecological footprint is increased by:

- greater reliance on fossil fuels
- increased use of technology and, therefore, energy (but technology can also reduce ecological footprint)
- high levels of imported resources (which have high transport costs)
- large *per capita* production of carbon waste (i.e. high energy use, high fossil fuel use)
- large *per capita* consumption of food
- a meat-rich diet.

Local biomes with high productivity produce a lower footprint as they absorb carbon dioxide (net emission of carbon dioxide is used in the calculation of footprint size).

Ecological footprint can be reduced by:

- reducing amounts of resources used
- recycling resources
- reusing resources
- improving efficiency of resource use
- reducing amount of pollution produced
- transporting waste to other countries
- improving technology to increase carrying capacity
- importing more resources from other countries
- reducing population to reduce resource use
- using technology to increase carrying capacity (e.g. use GM crops to increase yield on the same amount of land)
- using technology to intensify land use.

Many innovations are still in the early stages (e.g. renewable technologies) but these could have a huge impact on ecological footprints in the future. The funding to

Ecological footprints vary significantly from country to country and person to person, and includes aspects such as lifestyle choices (environmental value systems), productivity of food production systems, land use, and industry.

Recycling

support technological change exists in MEDCs, which currently face the biggest problem with their ecological footprints (Figure 8.20). There is a real incentive to address the issue.

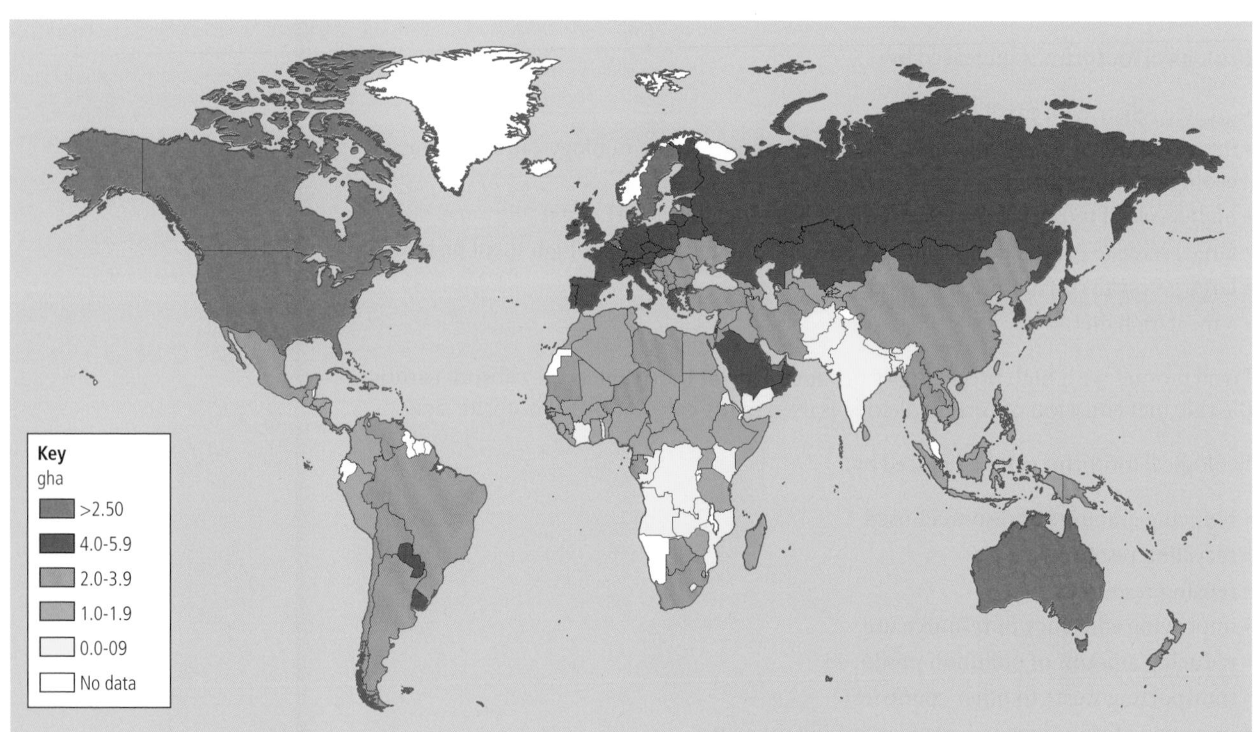

Figure 8.20 The ecological footprints of countries around the world, 2014

Calculating ecological footprint

Ideally, all resource consumption and land uses are included in an ecological footprint calculation (full calculation). But this would make the calculation very complex. Ecological footprints are usually simplified and an approximation achieved, by using only net carbon dioxide emissions (Figure 8.21).

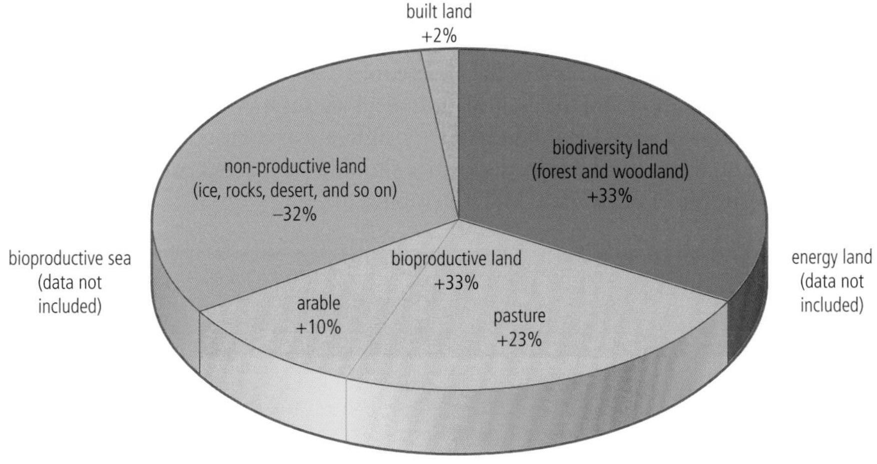

Figure 8.21 Some of the factors used to calculate a full ecological footprint. Pie chart represents averages for the planet but does not include figures for energy land or bioproductive sea.

Factors used in a full ecological footprint calculation would include the following list:

- Bioproductive (currently used) land: land used for food and materials such as farmland, gardens, pasture, and managed forest.
- Bioproductive sea: sea area used for human consumption (often limited to coastal areas).
- Energy land: an equivalent amount of land that would be required to support renewable energy instead of non-renewable energy. The amount of energy land depends on the method of energy generation (large in the case of fossil fuel use) and is difficult to estimate for the planet.
- Built (consumed) land: land that is used for development such as roads and buildings.
- Biodiversity land: land required to support all of the non-human species.
- Non-productive land: land such as deserts is subtracted from the total land available.

Thus, the simplified calculation of ecological footprint clearly ignores the following factors that influence the amount of land a population needs to support itself:

- the land or water required to provide any aquatic and atmospheric resources
- land or water needed to assimilate wastes other than carbon dioxide
- land used to produce materials imported into the country to subsidize arable land and increase yields
- replacement of productive land lost through urbanization.

In North America, a high *per capita* grain consumption relates to the meat-rich diet of MEDCs, where high grain production supports intensive cattle farming. Grain productivity is high in Africa reflecting the warmer conditions and concentrated solar radiation, which increase net primary productivity (pages 90–94). The same conditions are also reflected by the high net carbon dioxide fixation by local vegetation. Carbon dioxide emissions are much higher in North America, reflecting an industrialized society reliant on fossil fuels.

Since carrying capacity for human populations is difficult to calculate, it is also difficult to estimate the extent to which those populations are approaching or exceeding carrying capacity. The use of ecological footprint provides a model that estimates the area of environment necessary to sustainably support that particular population. How great this area is, compared with the area available to the population then gives an indication of whether the population is living sustainably and within the carrying capacity of the environment.

Degradation of the environment together with the use of finite resources is expected to limit human population growth.

Ecological footprints of MEDCs and LEDCs

Ecological footprint takes account of the area required to provide all the resources needed by the population and the assimilation of all wastes. Given the different standards of living between LEDCs and MEDCs, differences in resource consumption, energy usage and waste production, disparities should be expected between the ecological footprints of LEDCs and MEDCs (Table 8.6).

Rank	Country	Ecological footprint / gha per person	Rank	Country	Ecological footprint / gha per person
1	United Arab Emirates	15.99	109	India	1.06
2	USA	12.22	114	Nepal	1.01
3	Kuwait	10.31	128	Ethiopia	0.85
4	Denmark	9.88	135	Haiti	0.78
5	New Zealand	9.54	141	Bangladesh	0.89

Table 8.6 The world's largest and smallest ecological footprints

To calculate your ecological footprint, go to www.pearsonhotlinks.co.uk, enter the book title or ISBN, and click on weblink 8.12.

LEDCs tend to have smaller ecological footprints than MEDCs. MEDCs generally have much greater rates of resource consumption than LEDCs. This is partly because people have more disposable income, and demand for energy resources is high. Consumption is also high because resource use is often wasteful. MEDCs produce far more waste and pollution as by-products of production. LEDCs are often characterized by lower consumption because people have less to spend. The informal economy in LEDCs is responsible for recycling many resources. However, as LEDCs develop, their ecological footprint size increases.

A meat-eating diet, prevalent in MEDCs where 30 per cent of diet may be based on animal protein, requires the use of much more land than a vegetarian diet. More of the energy from the crop goes to humans if the crop is eaten directly (as in LEDCs where meat is about 12 per cent of the diet). Data for food consumption are often given in **grain equivalents**. So a population with a meat-rich diet consumes a higher grain equivalent than a population feeding directly on grain.

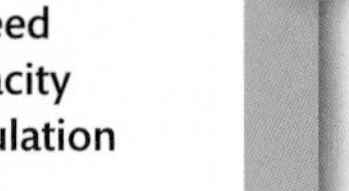

If human populations do not live sustainably they will exceed carrying capacity and risk population collapse.

In MEDCs, about twice as much energy in the diet is provided by animal products than in LEDCs. Grain production is therefore higher, using high-yield farming strategies. Greenhouse gas emissions from agriculture also affect footprint totals. According to the IPCC, the agricultural sector emits between 5.1 and 6.1 billion tonnes of greenhouse gases annually, about 10–12 per cent of the total greenhouse gas emissions. The main sources of these gases are nitrogen oxides from fertilizer, methane emissions from cows, and biomass burning.

Populations more dependent on fossil fuels have higher carbon dioxide emissions. Higher rates of emissions, contribute to the higher ecological footprints in MEDCs.

To learn more about national ecological footprints, go to www.pearsonhotlinks.co.uk, enter the book title or ISBN, and click on weblink 8.13.

CONCEPTS: Environmental value systems

People in MEDCs generally have a technocentric worldview, which encourages continued high consumption of resources, in the expectation that technology will provide solutions to minimize the environmental impact. LEDCs have not only had an historically low consumption of non-renewable resources, but have also adopted environmental value systems that have encouraged working in balance with nature, particularly where failure to do so would result in direct negative impact on the community (e.g. cutting down a forest on which you directly depend for food and shelter).

Case study

National level: Peru versus Canada

Table 8.7 shows the breakdown of the ecological footprint for an average Canadian.

Part of footprint	Energy	Agricultural land	Forest	Built environment	Total
housing	0.5	0.0	1.0	0.1	1.6
food	0.4	0.9	0.0	0.1	1.1
transport	1.0	0.0	0.0	0.1	1.1
consumer goods	0.6	0.2	0.2	0.0	0.4
resources in services	0.4	0.0	0.0	0.0	0.4
total	2.9	1.1	1.2	0.2	5.4

Table 8.7 Ecological footprint (in hectares) for an average Canadian

In 2001, the *per capita* ecological footprint of Canada was 5.4 gha whereas for Peru it was 0.9 gha. Peru is an LEDC with an energy component of 16.0 per cent in its ecological footprint, whereas Canada has an energy component of 53.7 per cent. Canada has a larger consumer-driven economy, a greater car culture, uses more energy for heating, and has higher consumer spending *per capita* than Peru, all of which contribute to the high percentage of energy within the Canadian ecological footprint. Non-renewable energy generation in Canada, using fossil fuels, adds to the carbon dioxide emission component of the footprint. The higher rates of photosynthesis and net primary productivity in Peruvian vegetation, due to its location nearer the equator, contribute to Peru's lower net contributions to atmospheric carbon dioxide levels.

CHALLENGE YOURSELF

Research skills (ATL)

How can you reduce your ecological footprint? What are the main obstacles to reducing your ecological footprint?

Case study

Reducing the ecological footprint in Calgary, Canada

Calgary is an urban area in Canada. In 2005, it had the dubious distinction of having the largest ecological footprint for a large urban area in Canada (9.9 gha). By 2012, it had transformed its footprint. Calgary now buys all its electricity from renewable sources, having built a windpark to produce wind energy. In addition, Calgary has:

- improved the efficiency of energy-intensive systems such as water-treatment plants
- required Leadership in Energy and Environmental Design (LEED) for all new buildings and older buildings that were renovated
- used smart, energy-efficient street lighting
- upgraded and modernized public transport
- generated electricity from methane recovered from rubbish tips.

Although Calgary's ecological footprint has been reduced to 9.5 gha, it is still much higher than the Canadian average footprint of 7.1 gha.

To use an ecological footprint interactive graph, go to www.pearsonhotlinks.co.uk, enter the book title or ISBN, and click on weblink 8.14.

Exercises

1. Explain the difficulties in applying the concept of carrying capacity to local human populations.
2. Explain how absolute reductions in energy and material use, reuse and recycling can affect human carrying capacity.
3. Study the figure below which shows the ecological footprint for selected countries in 2014.

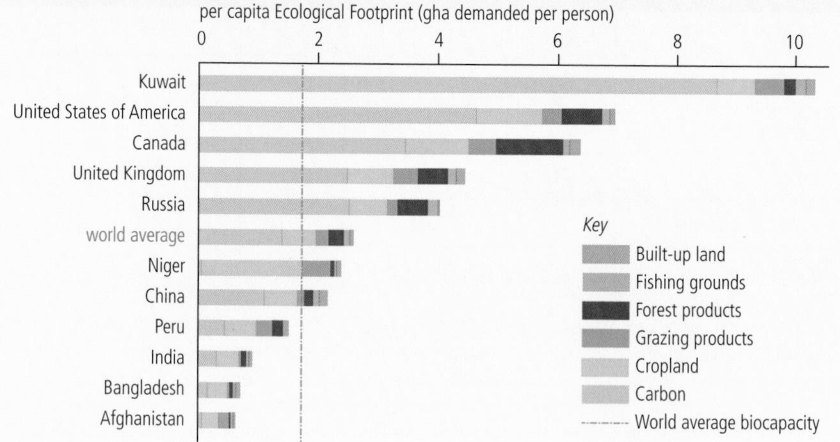

   a. Describe the composition of Kuwait's ecological footprint.
   b. Compare the ecological footprints of Canada and Peru.
   c. Compare the ecological footprint of India with that of Bangladesh.
   d. Suggest how China's ecological footprint is likely to change as the country develops.

Big questions

Having read this section, you can now discuss the following big questions:

- How useful are the systems approach and the use of models in the study of carrying capacity and ecological footprints?
- To what extent are solutions directed at *preventing* environmental impacts, *limiting* the extent of the environmental impacts, or *restoring* systems likely to be most successful in the management of pollution?
- Outline contrasting value systems in the development of population policies.
- How does your own value system compare with others you have encountered with regard to resource use?
- Can human use of resources ever lead to sustainable development?
- How far is it possible for human society to live in balance with the biosphere?

Points you may want to consider in your discussions:

- How do models and/or a systems approach help our understanding of carrying capacity?
- Why are some carrying capacities larger than others?
- What can be done to reduce carrying capacities?
- How do environmental value systems influence carrying capacities?
- What are your views on how best to reduce carrying capacities?
- Examine the relationship between carrying capacity and sustainability.
- How might carrying capacities change in the decades to come. Justify your answer.

actice questions

1 Study the population pyramids below.

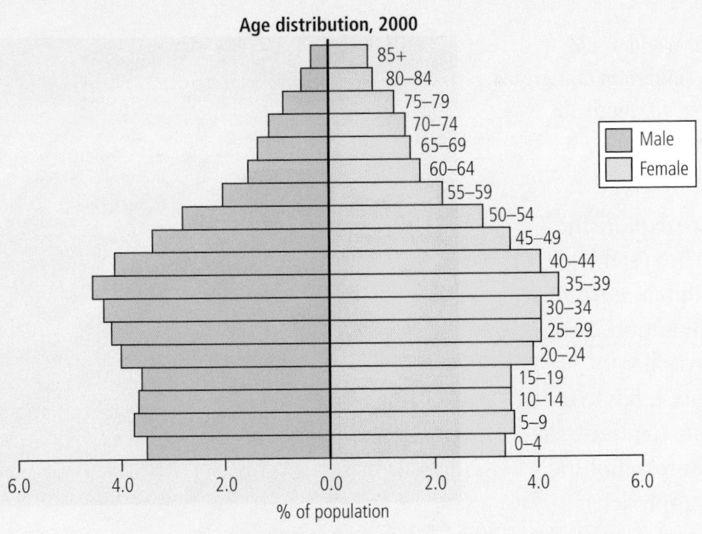

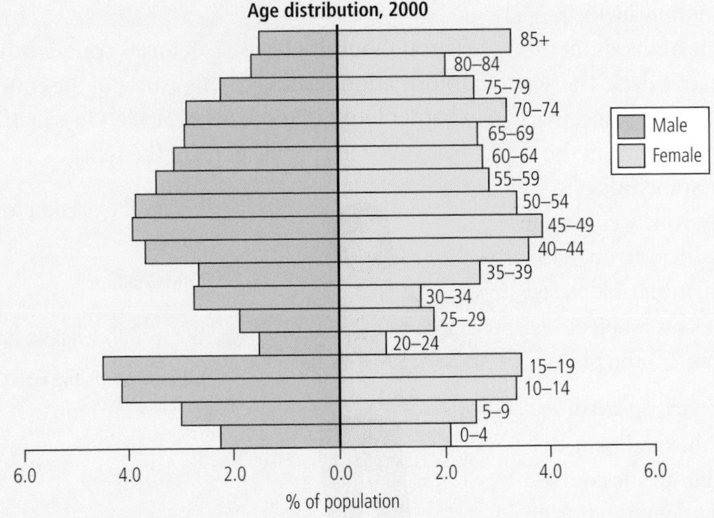

- **a** Label the two population pyramids to describe the main characteristics of the population. [6]
- **b** Identify which of the pyramids is likely to be drawn from an urban area and which is likely to be from a rural location. [2]
- **c** Give reasons for your answer to (b). [4]
- **d** What are the demographic issues likely to be faced in the two contrasting areas? [8]

2 **a** Discuss the potential ecological services and goods provided by a named ecosystem. [6]

b Outline how culture, economics and technology have influenced the value of a named resource in different regions or in different historical periods. [6]

3 Evaluate landfill and incineration as strategies for the disposal of solid domestic waste. [6]

4 Compare and contrast the ecological footprint of the MEDCs with that of LEDCs. [6]

Theory of Knowledge

> *If all mankind were to disappear, the world would regenerate back to the rich state of equilibrium that existed ten thousand years ago. If insects were to vanish, the environment would collapse into chaos.*
> **Edward O Wilson**

There are plenty of opportunities to explore the theory of knowledge within ESS. The systems approach used throughout ESS is different from traditional models of scientific exploration. This allows us to compare the two approaches to understanding. Conventional science tends to use a reductionist approach to looking at scientific issues, whereas the systems approach requires a holistic understanding. While the systems approach is frequently quantitative in its representation of data, it also addresses the challenge of handling a wide range of qualitative data. This leads to questions about the value of qualitative versus quantitative data. There are many checks and guidelines to ensure objectivity in quantitative data collection and handling in the purely physical sciences, but these standards of objectivity are more difficult to rigorously control in ecological and biological sciences. In addition, ESS is a transdisciplinary subject, the material addressed often crosses what may seem to be clear subject boundaries (e.g. geography, economics, and politics).

The systems approach allows comparisons to be made across disciplines, and the value and issues regarding this are discussed throughout the course and in this book. In exploring and understanding an environmental issue, you must be able to integrate the hard, scientific, quantitative facts with the qualitative value-judgements of politics, sociology and ethics. All this makes particularly fertile ground for discussions related to theory of knowledge.

> *Whoever undertakes to set himself up as a judge in the field of Truth and Knowledge is shipwrecked by the laughter of the gods.*
> **Albert Einstein**

> *Genius is 1 per cent inspiration and 99 per cent perspiration.*
> **Thomas Edison**

Care for the planet – it's the only one we have

Throughout this book, ToK boxes contain advice and information relating to this aspect of the course. This chapter looks in more detail at ways in which ToK can be applied in specific parts of the syllabus.

> *Chance favours the prepared mind.*
> **Louis Pasteur**

> *We see only what we know.*
> **Johann Wolfgang von Goethe**

JUST SAY KNOW

Is knowledge all you need?

Can too much knowledge be a bad thing?

1 Foundations of environmental systems and societies

This topic covers the concepts that are central to the ESS course, namely the systems approach, models, equilibrium, sustainability and pollution. Throughout Chapter 1, relevant ToK ideas are raised at appropriate parts of the text.

Environmental value systems (EVSs) are influenced by education, family, friends, culture and other inputs from the society we live in. These EVSs influence how we see the world and respond to it. This topic offers many opportunities to discuss the interaction between EVSs and societies' responses to the environmental issues covered in the course.

Many of the strategies proposed during the course to tackle environmental concerns have alternative options. Experts sometimes disagree about pollution management strategies, for example — how do we decide which strategy is best, and on what basis might we decide between the judgements of the experts if they disagree? How do we decide between alternative perspectives, and can any one management strategy be considered as final?

Case study

Holism versus reductionism

The emphasis in this course is on understanding the sum of the parts of a system (i.e. a holistic approach) rather than considering the components separately. This contrasts with the reductionist approach of conventional science. Data collection is involved with measuring the inputs and outputs of a system, and processing the data reveals understanding of the processes within the system. The main difference between the systems approach and conventional science is that the former describes patterns and models of the whole system, whereas the latter aims at explaining cause-and-effect relationships within it. Is one approach better than the other, or is it a matter of perspective as to which approach brings real benefits in understanding?

> *All models are wrong, but some are useful.*
>
> **George Box (innovator in statistical analysis)**

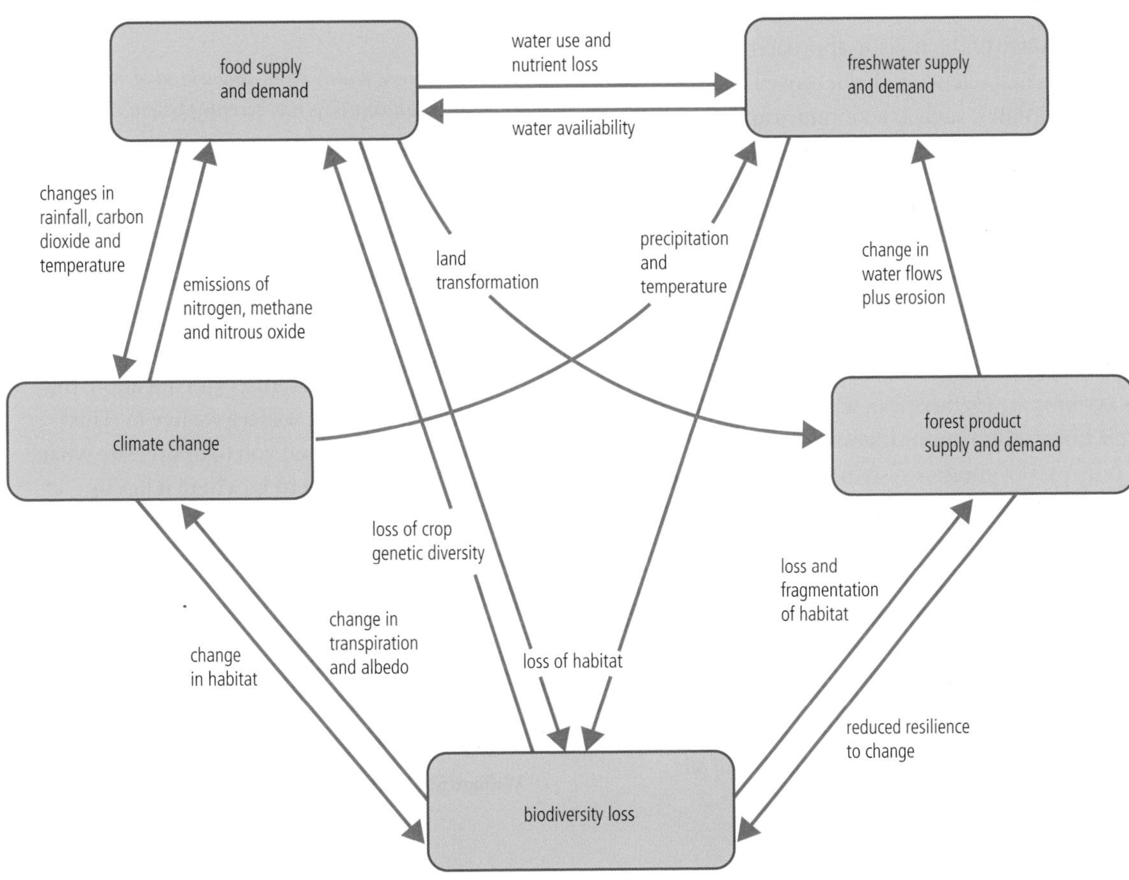

Holism, such as the Gaia hypothesis, looks at the whole system.

A reductionist approach

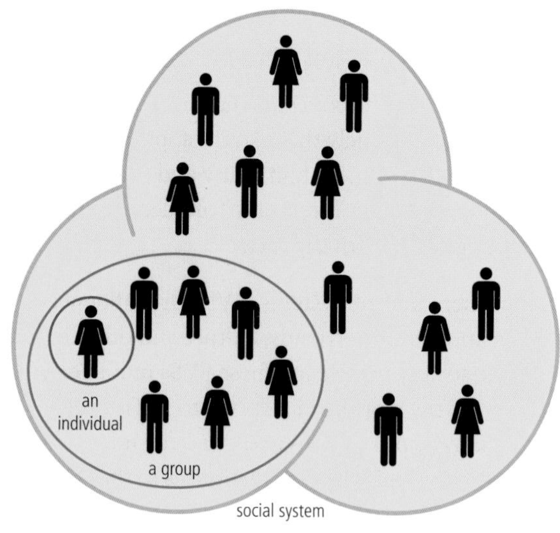

Nested systems within social systems

Holism and science

Does the holistic approach really differ from conventional science, or is it just a matter of using different terminology?

Advantage of holism

The advantage of the holistic approach in environmental science is that it is used extensively in other disciplines, such as economics and sociology, and so allows integration of these different subjects in a way that would not be possible (or at least not so easy) in conventional science.

Shall I refuse my dinner because I do not fully understand the process of digestion?

Oliver Heaviside

As systems are hierarchical, what may be seen as the whole system in one investigation may be seen as only part of another system in a different study (e.g. a human can be seen as a whole system with inputs of food and water and outputs of waste, or as part of a larger system such as an ecosystem or a social system). Difficulties may arise at where the boundaries are placed, and how this choice is made.

The strongest arguments prove nothing so long as the conclusions are not verified by experience. Experimental science is the queen of sciences and the goal of all speculation.

Roger Bacon

Reductionism is a dirty word, and a kind of 'holistier than thou' self-righteousness has become fashionable.

Richard Dawkins

Environmental value systems

There are assumptions, values and beliefs, and worldviews that affect the way in which we view the world. These are influenced by the way we are brought up by our parents, our education, the friends we have, and the society we live in. This course should have helped you to appreciate what your personal value system is, where it lies in a spectrum of other worldviews, and have enabled you to justify and evaluate your position on a range of environmental issues.

The great end of life is not knowledge but action.

Thomas Huxley

We know what we are, but know not what we may be.

William Shakespeare

2 Ecosystems and ecology

This topic gives scope for discussion on how ecological relationships can be represented by different models, such as food chains, food webs, and ecological pyramids. How can we decide when one model is better than another? Ecosystems and ecology also allow exploration of the benefits of the systems approach, and how ecological research compares to investigations in the physical sciences (e.g. physics and chemistry). For example, a reductionist approach looks at the individual parts of a system: this approach is usually used in traditional scientific investigations; a holistic approach looks at how the parts of a system work together as a whole: this approach is usually used in modern ecological investigations.

▲ How would you evaluate your personal standpoint about the environment and the issues raised throughout the course?

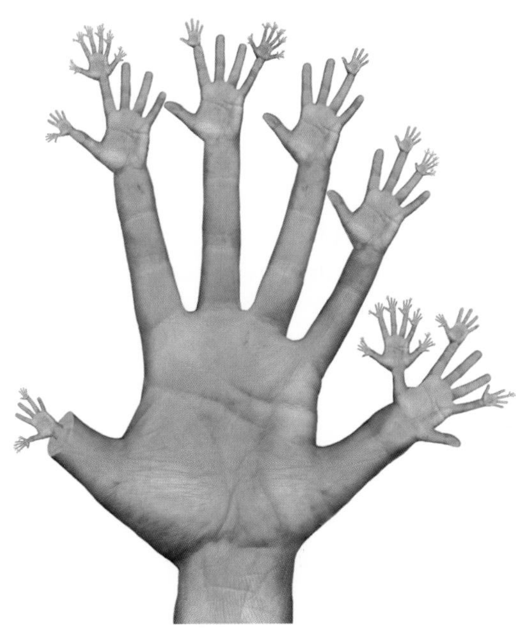

▲ Global cooperation is needed to tackle many environmental problems, and to ensure a sustainable future for all. We hold the future in our hands.

> *If facts are the seeds that later produce knowledge and wisdom, then the emotions and the impressions of the senses are the fertile soil in which the seeds must grow.*
>
> **Rachel Carson**

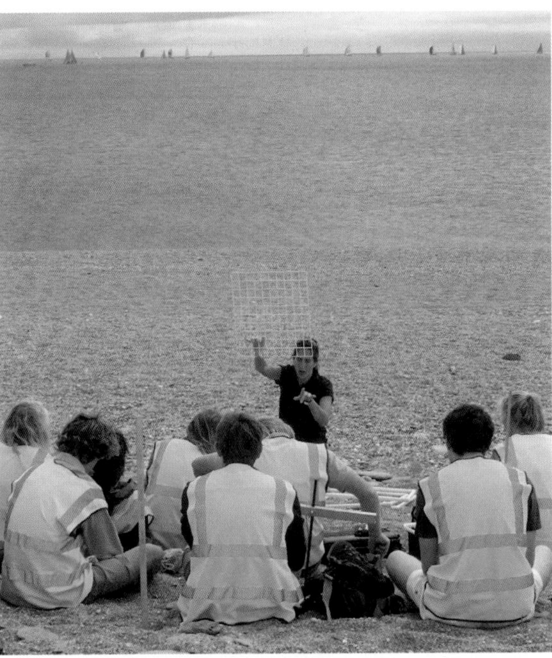

▲ Students on an ecology field trip are instructed on sampling techniques.

Ecological research applies techniques that aim to remove subjectivity (e.g. the randomized location of quadrats when studying the distribution of species in a habitat rules out site-selection by the researcher). The interpretation of data, however, is open to interpretation and personal judgement. It has been said that historians cannot be unbiased: could the same be said of environmental scientists when making knowledge claims?

Sampling freshwater invertebrates in a lake

> *I often say that when you can measure what you are speaking about, and express it in numbers, you know something about it; but when you cannot measure it, when you cannot express it in numbers, your knowledge is of a meagre and unsatisfactory kind.*
>
> **Lord Kelvin**

Succession

TOK

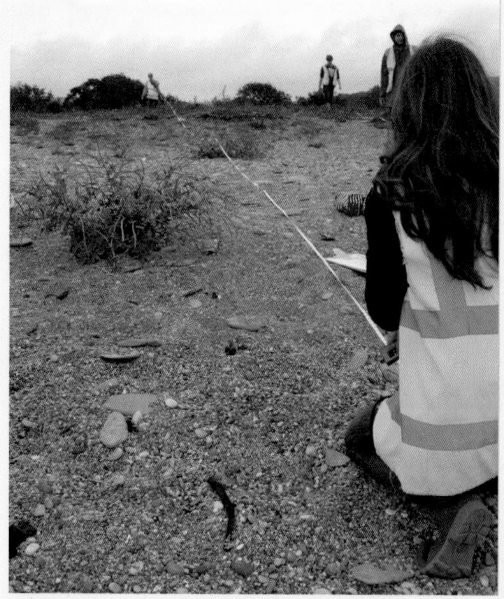

Setting up a transect to study succession on a shingle ridge

Can we substitute space for time? Is space (and spatial change) a surrogate for time (and temporal change)? In studying succession, we use spatial changes to make inferences about temporal changes. Should this be allowed? Succession may be affected by external factors, such as global warming.

> *No effect that requires more than 10 per cent accuracy in measurement is worth investigating.*
>
> **Walther Nernst**

Ecology provides many opportunities to explore issues regarding the reliability and validity of data and how it is collected. It also addresses the pros and cons of subjective as opposed to objective data. Nevertheless, the interpretation of data – even of objective data – is open to widely different viewpoints (see the case study 'A matter of interpretation').

Ecology relies on the collection of both biotic and abiotic data. Abiotic data can be collected using instruments that avoid issues of objectivity as they directly record quantitative data. The measurement of the biotic (or living) component is often more subjective, relying on your interpretation of different measuring techniques to provide data. It is rare in environmental investigations to be able to provide ways of measuring variables that are as precise and reliable as those in the conventional (i.e. physical) sciences. Working in the field means that variables cannot be controlled, only measured, and fluctuations in environmental conditions can cause problems when recording data. Standards of acceptable margins of error are therefore different. Will this affect the value of the data collected and the validity of the knowledge? Applying the rigorous standards used in a physics investigation, for example, would render most environmental studies unworkable, and we would miss out on gaining a useful understanding of the environment. A

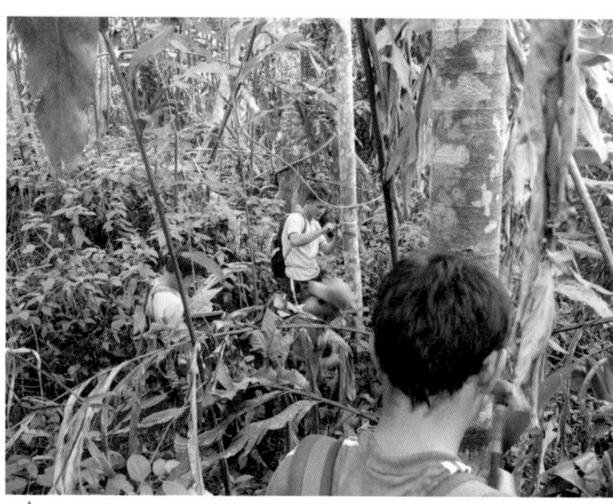

Field workers measuring the diameter of trees at breast height (DBH) to assess the recovery of tropical rainforest following logging in Borneo.

Case study

A matter of interpretation

Ecosystem	Mean NPP / kg m^{-2} yr^{-1}	Mean biomass / kg m^{-2}	NPP / biomass
tropical rainforest	2.2	45	
tropical deciduous forest	1.6	35	
tropical scrub	0.37	3	
savannah	0.9	4	
Mediterranean sclerophyll	0.5	6	
desert	0.003	0.002	
temperate grassland	0.6	1.6	
temperate forest	1.2	32.5	
boreal forest	0.8	20	
tundra and mountain	0.14	0.6	
open ocean	0.12	0.003	
continental shelf	0.36	0.001	
estuaries	1.5	1	

Questions

1. Which are the three most productive ecosystems in terms of NPP?
   - tropical rainforest 2.2
   - tropical deciduous forest 1.6
   - estuaries 1.5
2. Which are the three most productive ecosystems in terms of NPP per unit of biomass?
   - continental shelf 360 kg m^{-2} yr^{-1} per kg m^{-2}
   - open oceans 40 kg m^{-2} yr^{-1} per kg m^{-2}
   - estuaries 1.5 kg m^{-2} yr^{-1} per kg m^{-2}
   - deserts 1.5 kg m^{-2} yr^{-1} per kg m^{-2}
3. How do you explain the differences between your two answers?

Much of the biomass of a forest is woody, non-photosynthesizing material. This means it is non-productive, thus the NPP per unit of biomass is low.

pragmatic approach is called for in ecological studies, but this leaves the subject open to criticism from physical scientists regarding the rigour with which studies are done. Is some understanding better than no understanding at all?

Controlled laboratory experiments are often seen as the hallmark of the scientific method, but are not possible in fieldwork–to what extent is the knowledge obtained by observational natural experiment less scientific than the manipulated laboratory experiment?

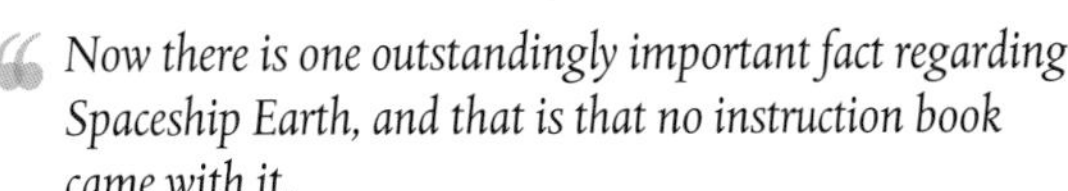

Now there is one outstandingly important fact regarding Spaceship Earth, and that is that no instruction book came with it.

Buckminster Fuller

3 Biodiversity and conservation

This topic offers the opportunity to discuss what is meant by the term 'biodiversity'. Diversity indices are not absolute measures in the same way that temperature is, for example. Diversity indices involve a subjective judgement on the combination of two measures – proportion and richness. Diversity measures are sometimes misread, or confused with species richness (pages 138–140). This can have implications for the way in which the impacts of human disturbance are interpreted.

This topic also offers different perspectives on species and habitat conservation. Which strategy is best for conservation, and how do societies decide the best approach? How do we know when critical points are reached, beyond which damage to ecosystems and biodiversity may become irreversible (e.g. leading to species extinction)? Should the people who cause environmental damage be held morally responsible for the long-term consequences of their actions?

One further topic for discussion is the different views people have on the origin of life on Earth. The established scientific explanation is that all species have evolved through the process of natural selection, although other people take alternative views.

Mark on this grid where your opinions lie in terms of the truth of creationism and evolution.

Evolution versus creationism

What constitutes 'good science' and what makes a 'good theory'? Can we have confidence in scientific theories that rely on indirect evidence and that happen over such long periods of time as to make testability a problem?

In 1859, Charles Darwin's book *On the Origin of Species* revolutionized biology and the way it is studied. Despite this, some people still refute its claims: one such group are the creationists, who believe in the literal truth of the biblical Genesis story. Can their views be reconciled with the scientific evidence? What do their views say about the scientific method and what constitutes good science?

TOK

- Science concerns testable ideas.
- Science therefore focuses on recurrent, repeatable events.

Is this true for all science?

Is evolution by natural selection a testable theory? Are there organisms that are suitable for experimentation to see natural selection in operation?

Creationist claims

1 **Is evolution scientific?**
- Evolution within a species can be tested and is well established, but doesn't explain the creation of species.
- If evolution occurs over millions of years, it is untestable and therefore unscientific.

2 **Evolution contradicts physics**
- Physics, second law of thermodynamics: The entropy (disorder) in a system will always increase over time.
- Evolution: Life appears from disorder, becoming increasingly ordered and complex over time.

3 **Counter-evidence**
- Fossilized allegedly human tracks with a trail of typical dinosaur tracks, in the same rock layer.
- Radio-isotope dating is the basis for almost all estimates of evolutionary time. It was applied in 1986 to lava from Mt St Helens, which erupted in 1980, and produced dates of millions of years ago. Since the dating is almost a million times too old, dating of fossils must likewise be a million times too old.

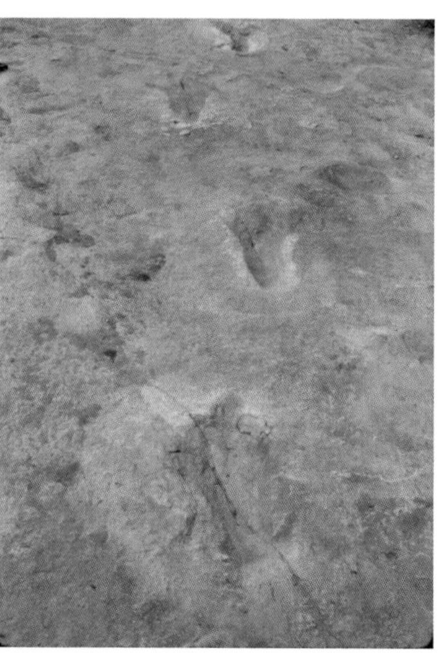

▲ These metatarsal (heel-impressed) dinosaur tracks in the Cretaceous limestone of the Paluxy River, near Glen Rose, Texas, were once considered by many creationists to be human tracks together with a trail of typical dinosaur tracks. It was thought that this was evidence for humans and dinosaurs living together before Noah's flood.

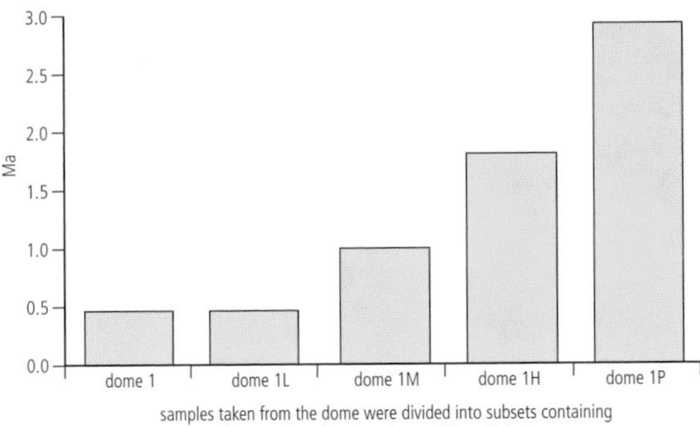

Potassium/argon dating of the volcanic dome formed by lava at Mt St Helens

- The key 'missing link' fossils – intermediates between major groups – are still missing.
- *Archaeopteryx* – the famously feathered reptile – has been debunked as a fake, along with 'Piltdown man' and many others.

4 **Key creationists in science**

- Physics – Newton, Maxwell, Kelvin
- Chemistry – Boyle, Dalton, Ramsay
- Biology – Linnaeus, Mendel, Pasteur
- Astronomy – Copernicus, Galileo, Kepler, Herschel
- Mathematics – Pascal, Leibnitz.

5 **Common sense**

- The idea that we were created via a purely random process of mutation ...
  ... is statistically absurd
  ... contradicts the obvious signs of design all around us
  ... denies humanity: it allows no meaning for creativity, love, or purpose
  ... just gives selfish people the justification to act without regard for morality.

> *Ignorance more frequently begets confidence than does knowledge: it is those who know little, not those who know much, who so positively assert that this or that problem will never be solved by science.*
>
> **Charles Darwin**

If you were to re-assess your view of which was true, creationism or evolution, based on belief in the arguments expressed above, where would you mark the grid?

Creation  Evolution

> *Neither science nor maths can ever be complete.*
>
> **Kurt Gödel**

Re-assessing the creationist argument

1 **Distortion**

- The second law of thermodynamics is true as quoted but it only applies to isolated systems. Life is part of a system in which entropy does increase overall.
- Mutation is indeed random, but natural selection is not, so it can work cumulatively to bring about apparent design.
- Evolution by natural selection can be demonstrated in organisms with short generations (MRSA bacteria is an example).

2 **Highly selective use of data**

- For every creationist scientist mentioned, hundreds, even thousands of others are not. Moreover, scientists who pre-date evolutionary theory cannot be called creationists since there was no creationist/evolutionist argument at that time.
- Huge areas of evolutionary data are completely ignored:
  - homologous structures
  - biogeographical evidence
  - molecular evidence (e.g. DNA)
  - embryological evidence.

3 **Disinformation and misinterpretation**

- *Archaeopteryx* is widely accepted as authentic by the scientific community; there is no basis for its 'debunking'.
- We now understand, from dedicated work in the 1980s, how the tracks at Paluxy River were formed. Dinosaur footprints normally recognized as such were created by dinosaurs walking or running on their toes (these are the deep three-toed tracks, called digitigrade tracks). Dinosaurs walking on their soles or heels (metatarsal bones) create different impressions called metatarsal tracks which are longer than the digitigrade ones. Unlike digitigrade tracks, metatarsal tracks may look superficially like human tracks after erosion or if mud-movements followed formation of the track.

Metatarsal impressions are not the only reasons for creationists misinterpreting the tracks at Paluxy River. Several other phenomena have also been mistaken for human tracks including erosional features and some carvings on loose stone blocks. To learn more about the Paluxy River site, go to www.pearsonhotlinks.co.uk, enter the book title or ISBN, and click on weblink 9.1.

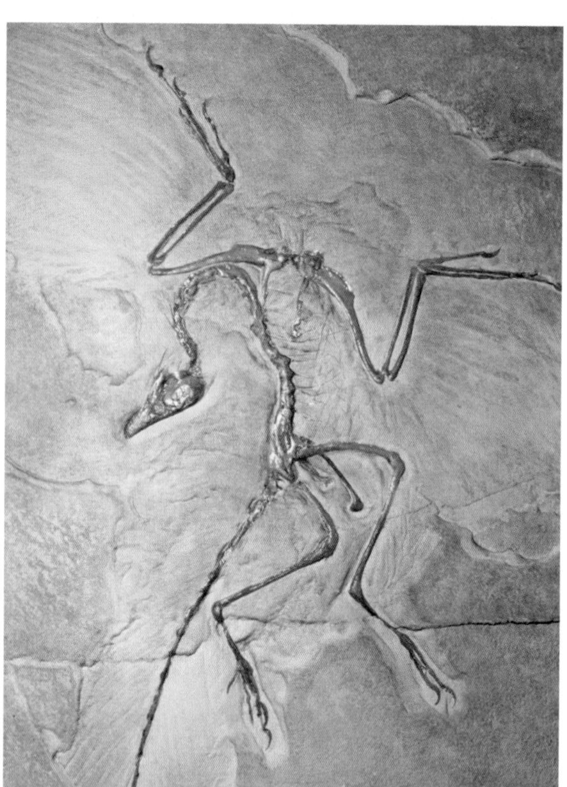

Archaeopteryx – a link between reptiles and birds

Many species living today are intermediates between other groups (see Richard Dawkins' book *The Greatest Show on Earth* for extensive evidence). It is not true that evidence for missing links is absent (see photos).

A 47-million-year-old fossil (named Ida, top) seems to be a link in primate evolution, bridging the evolutionary split between higher primates such as monkeys, apes, and humans and their more distant relatives such as lemurs (lower photo).

4 **Hype and spin**

- Some real concerns about isotope dating voiced by scientists are hyped up to make the whole process appear void.
- Sensible explanations of anomalous results (e.g. for Mt St Helens) are spun as 'desperate evolutionists patching up a defunct theory'.

5 **Learn to live with uncertainty**

- There is much uncertainty in both science and faith.
- Doubt and questioning are creative.
- In science, uncertainty leads to new ideas.
- In faith, too, doubt can lead us to ask new questions and find new meaning.
- At the same time, you need to know what your core values are.

6 **Learn to spot 'pseudo-science'**

Pseudo-science	Good science
• shows fixed ideas (dogma)	• shows willingness to change
• selects favourable findings	• accepts and attempts to explain all findings
• does not have peer-review	• has ruthless peer-review
• is unable to predict	• has predictive power
• has unverifiable claims	• is experimentally verifiable
• has a hidden agenda	• makes few assumptions
• lacks consistency	• is usually consistent

Archbishop James Ussher (1581–1656), Church of Ireland (protestant) Bishop of Armagh, claimed to have established the date the Earth was formed as Sunday 23 October, 4004 BC whereas the evolutionist approach now puts the Earth as at least 4.5 billion years old.

TOK

Extraordinary claims require extraordinary evidence.

Carl Sagan

Case study

How should we decide what to protect?

Humans make judgements about the natural world, and the ways in which it can be protected. Do species have an intrinsic right to exist even if they are of no economic value at the moment? Should as much as possible of the environment be protected, or do we need more pragmatic approaches based on realistic expectations?

How do we justify the species we choose to protect?

Is there a focus on animals we find attractive? Is there a natural bias within the system? Sometimes the choices we make are based on emotion rather than reason: does this affect their validity?

People are not going to care about animal conservation unless they think that animals are worthwhile.

David Attenborough

Cheetahs have a very small gene pool with little genetic variation. They are especially prone to changes in their environment or the outbreak of disease. Should we focus conservation on species which are more resilient and more likely to survive into the future?

Do tigers have a greater right to exist than endangered and endemic species of rat?

> *The nation behaves well if it treats the natural resources as assets which it must turn over to the next generation increased, and not impaired, in value.*
>
> **Theodore Roosevelt**

Describing species

Historically, taxonomists (scientists who describe new species) focused on groups that interested them. These tend to be the larger more attractive groups (e.g. mammals, birds, flowering plants). Is there a bias in the way in which species are described? What about small and more obscure groups (e.g. nematodes) or smaller organisms that are difficult to collect and identify, or which have not attracted scientific attention? What impact does this have on estimations of the total number of species on the Earth? Can we reliably comment on species' extinction rates?

Most of the species of animals on the planet are beetles. Do you think the number of described species reflect this? What type of organisms have scientists historically focused attention on?

> *Like the resource it seeks to protect, wildlife conservation must be dynamic, changing as conditions change, seeking always to become more effective.*
>
> **Rachel Carson**

> *A little knowledge that acts is worth infinitely more than much knowledge that is idle.*
>
> **Kahlil Gibran**

4 Water, aquatic food production systems, and societies

Topic 4 provides the opportunity to discuss the value and limitations of models. The hydrological cycle is represented as a systems model: to what extent can such diagrams effectively model reality, given that they are based on limited observable features?

Water scarcity around the globe raises the issue of how aid agencies often use emotive advertisements to promote their cause. To what extent can emotion be used to manipulate EVSs and the actions that follow on from them? Do the ends justify the means?

Many societies have traditions of food production that may go against our own EVS. The Inuit people, for example, have an historical tradition of whaling – something that would go against the EVS of many, if not most, societies. To what extent does our culture determine or influence our ethical judgements?

Inuit whale hunting off the coast of Alaska

Topic 4 also looks at how water quality can be tested. A wide range of parameters are used to test the quality of water, but to what extent can scientists be sure that they have correctly identified cause-and-effect relationships (e.g. that pollution directly affects species diversity in a stream), given that they can only ever observe correlation?

Case study

The Tragedy of the Commons

Fishing net catch, North Sea

Renewable resources, such as fish, need not be depleted provided that the rate of use does not exceed maximum sustainable yield. In other words, if the rate of use is within the limit of natural replacement and regeneration. If resources become over-exploited, then depletion and degradation will lead to scarcity. If more than one nation is exploiting a resource, which is clearly the case in the fishing industry, resource degradation is often the result. Garrett Hardin (1968) has suggested a metaphor, the 'Tragedy of the Commons', to explain the tendency: this refers to the way that there may be little control over the way common resources are used, and the selfish acts of a few individuals can destroy the resource for others.

In any given ocean, a number of nations may be fishing. Apart from the seas close to land, where there is an Economic Exclusive Zone, no country owns the oceans, or the resources that they contain. But countries may use the resources. If one country takes more fish from the oceans, their profit increases. However, other countries do not benefit from this. To maintain the same relative profitability, other countries may increase their catch, so that they are not losing out relative to their competitors. The 'tragedy' is that other countries feel compelled to increase their catch, to match the catch of the one that initially increased its catch. Thus, the rate of use may exceed maximum sustainable yield and the resources become depleted.

Although simplistic, the Tragedy of the Commons does explain the tendency to over-exploit shared resources and the need for agreements over common management.

Questions

1. Who should set the limits for fish yields? Justify your answer.
2. Who should decide on common management? Justify your choice. Who would suffer as a result of this choice?

5 Soil systems, terrestrial food production systems, and societies

Topic 5 raises issues concerning different methods of food production. Are the intensive methods of food production carried out in many MEDCs detrimental to the environment, or are they in reality the best way to provide food for ever-growing populations? Do intensive farming methods, in fact, have environmental benefits? Intensive chicken farming (broiler production systems), for example, has been shown to produce a lower carbon footprint than free-range/organic methods. What ethical issues do different types of animal feed production raise?

This topic also provides points for discussion concerning our perception of time compared to the time scales that environmental systems operate under. Fertile soil can be considered as a non-renewable resource because once depleted, it can take significant time to restore the fertility: how does our perception of time influence our understanding of change?

Case study

Food deserts

Food deserts are areas without ready access to fresh, healthy, and affordable food. These communities may have only fast food restaurants and shops that offer few healthy, affordable food options. Such factors contribute to a poor diet and can lead to high levels of obesity, diabetes and cancer.

A food desert is a geographic area where affordable and nutritious food is difficult to obtain, especially for those without a car. The term 'food desert' can be defined as any census area where at least 20 per cent of inhabitants are below the poverty line and 33 per cent live more than a mile from a supermarket.

According to the US Department of Agriculture (USDA), 10 per cent of the USA is a 'food desert'. It claims that there are thousands of areas where low-income families have limited or no access to healthy fresh food. The concept of 'food deserts' was originally identified in Scotland in the 1990s. They are associated with urban decay and are characterized by numerous fast-food restaurants and convenience stores serving fatty, sugary junk food to overweight consumers.

The USDA links food deserts to a growing weight problem. In the USA, childhood obesity has tripled since 1980. The annual cost of treating obesity is nearly $150 billion.

Critics note that only about 15 per cent of customers shop within their own census area and that the focus on supermarkets means that the USDA ignores tens of thousands of larger and smaller retailers, farmers' markets and roadside greengrocers.

An example of a food desert is found in the South Side district of Chicago. Though crisps, sweets and doughnuts are easy to come by, fresh fruit is a rare commodity. Nevertheless, between 2006 and 2011, due to the arrival of some new grocery stores, Chicago's food desert decreased by 40 per cent. It sometimes takes only one shop to make a big difference – the Food-4-Less store in Englewood improved access to fresh food for over 40 000 people. Moreover, the opening of a decent grocery store can have a multiplier effect and lead to the arrival of other better-class shops in the area, which in turn fuels a local economic revival.

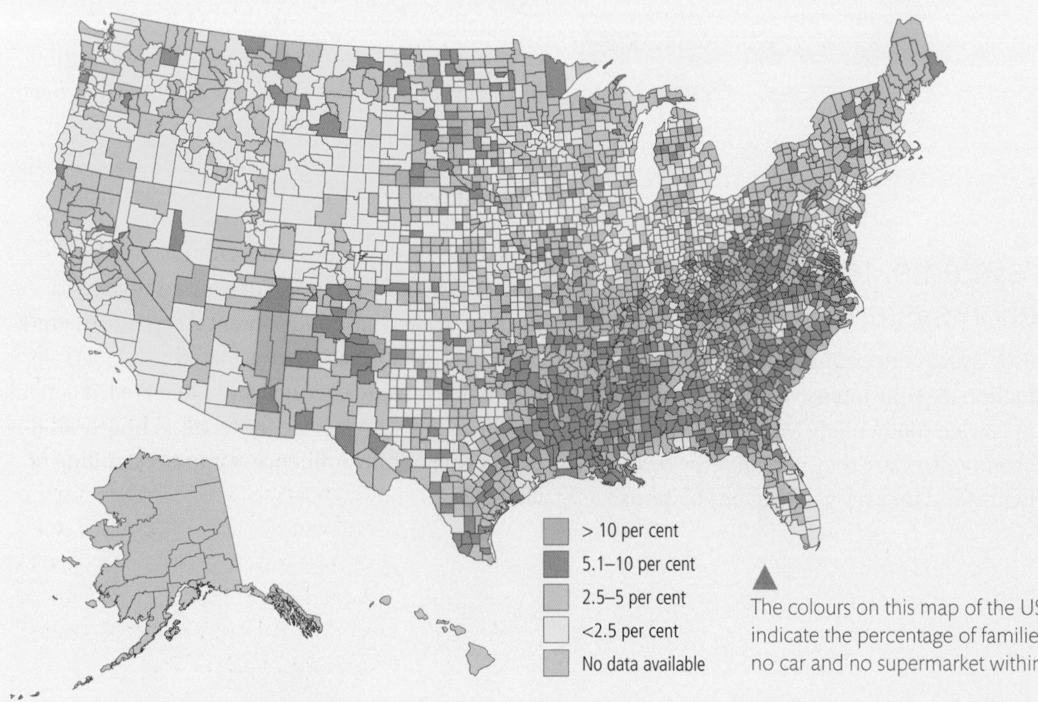

The colours on this map of the USA indicate the percentage of families with no car and no supermarket within a mile.

To learn more about food deserts, go to www.pearsonhotlinks.co.uk, enter the book title or ISBN, and click on weblink 9.2.

Vegetation destroyed by a combination of acid rain, sulfur emissions and lahars (mudflows) from the Soufrière volcano, Montserrat.

Questions

1. Evaluate the definition of 'food desert'.
2. Suggest why the term 'food desert' is used rather than 'food poverty' or 'low-quality food area'.
3. Comment on the implications of food deserts.

6 Atmospheric systems and societies

This topic examines the atmospheric systems that control and regulate climate systems. It also explores the effects that human societies have had on these systems (e.g. through pollution). International meetings have endeavoured to limit the emissions of pollutants that have an adverse effect on the atmosphere. To what extent have international agreements such as the UN-organized Montreal Protocol (pages 327–328) been successful? Unless treaties are legally binding, is there any point in countries signing up to them? Can one group or organization decide what is best for the rest of the world?

Natural pollution?

Are all forms of pollution necessarily human in origin? No. Many are natural. Some, like acidification, may be completely natural in some areas and anthropogenic in others. However, it is the case that acidification is largely related to human activity. It is an 'industrial form of ruination, which pays little heed to international boundaries'. Many countries produce acid pollutants and some export them. Nevertheless, there are natural causes of acidification – bog moss secretes acid, heather increases acidity, and conifer plantations acidify soils. Volcanoes are important sources of atmospheric pollution – especially sulfur dioxide and hydrogen dioxide. For example, before the eruption of the Soufrière volcano, Montserrat had some of the finest cloud forest in the Caribbean, but by 1996 vegetation loss from acid rain, gases, heat, and dust was severe. In 1996, the pH of the lake at the top of Chances Peak was recorded at 2.0 (i.e. 1000 times more acidic than a pH of 5.0).

Acid rain caused by volcanic eruption

Case study

Atmospheric pollution and the 'prisoners' dilemma'

Someone else's problem – atmospheric pollution in one country leads to harmful effects on the environment elsewhere.

The 'prisoners' dilemma' model (aka the Nash equilibrium, after the mathematician who developed it) can be used to explore the ethical issues involved in atmospheric pollution, and the ways in which these problems can be resolved. Non-point source pollution (page 50), such as that produced by the combustion of fossil fuels leading to acid rain, may mean that an individual polluting a common resource suffers little. Indeed, such an individual may even benefit from their disposal of pollutants, but the non-polluting users of the resource are affected by the pollution and they do not benefit in any way. For example, atmospheric pollution produced in the UK and carried to Scandinavia has led to the destruction of fresh water ecosystems (fjords) there. There is, therefore, a benefit for those who continue to pollute. This conundrum underlies many issues regarding the management of non-point pollution, from local (e.g. a lake) to global (e.g. the atmosphere). Pollution leading to global warming and acid rain is affected by the problem of making the polluter pay for the damage caused. The course explores ways in which solutions to the prisoners' dilemma can be found to solve both local and international issues of pollution. The ways in which legislation and public opinion can be used to address these problems is rich territory for ToK – is a system of rules better than a programme that educates and informs the public?

7 Climate change and energy production

> *If you don't read the newspaper, you are uninformed; if you do read the newspaper, you are misinformed.*
>
> **Mark Twain**

Power plant emissions in America

This topic addresses the way in which science can sometimes not be 100 per cent certain about particular issues, especially ones as complex as climate change. Is there a correlation between temperature rise and carbon dioxide emissions, or is the evidence effectively circumstantial? To what extent do politicians and environmentalists take advantage of the lack of consensus on the issue and use this to their own advantage? Are there parallels between the influence of religious communities in the past and the influence of the political community today on the science of climate change and how it is interpreted? The choice of energy sources is controversial and complex: how can we distinguish between scientific and pseudoscientific claims when making choices? Regardless of the lack of hard scientific certainty regarding the evidence, should we nevertheless take preventive measures to avoid potential future catastrophe? With the degree of uncertainty relating to the extent and effect of climate change, how can we best decide what to do, given our understanding is based on often provisional or incomplete knowledge?

> *Some of the scientists, I believe, haven't they been changing their opinion a little bit on global warming? There's a lot of differing opinions and before we react I think it's best to have the full accounting, full understanding of what's taking place.*
>
> **George W Bush**

Case study

Bias and spin

Global warming challenges views of certainty within the sciences. In the popular perception, global warming is having a negative impact on the world. There is, moreover, some confusion in the public mind between global warming and the greenhouse effect. The greenhouse effect is a natural process, without which there would be no life on Earth. There is, however, an *enhanced* or accelerated greenhouse effect which is implicated in global warming. The enhanced greenhouse effect is largely due to human (anthropogenic) forces, although feedback mechanisms may trigger some natural forces, too. Lobby groups and politicians may take views which suit their own economic and political ends – it is possible to hide other agendas behind the uncertainties around global warming (causes, consequences and potential solutions). In the USA, the strength of the oil companies during the Bush administration was seen by many as an example of an economically powerful group, and the politicians it supported, choosing a stance which was not in the long-term environmental, social or economic interest of the world. However, there were short-term benefits for both the oil companies and the politicians they supported.

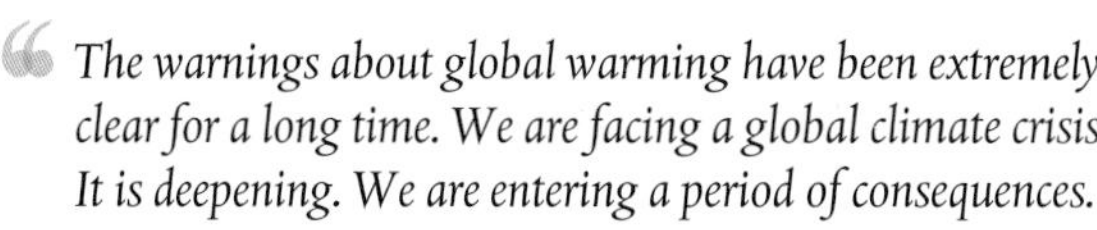

> *The warnings about global warming have been extremely clear for a long time. We are facing a global climate crisis. It is deepening. We are entering a period of consequences.*
>
> **Al Gore**

8 Human systems and resource use

This topic examines the intrinsic values (e.g. aesthetic and indirect) of nature as opposed to the values that are measured on economic grounds. This exemplifies the problem of trying to give a value to (i.e. quantifying) factors that are qualitative in nature. The value of the systems approach is especially highlighted in this part of the course. The concepts of resource and carrying capacity are given a fresh look by using the systems approach, and models of ecological footprint and natural capital/income bring a new moral and political perspective to these subjects.

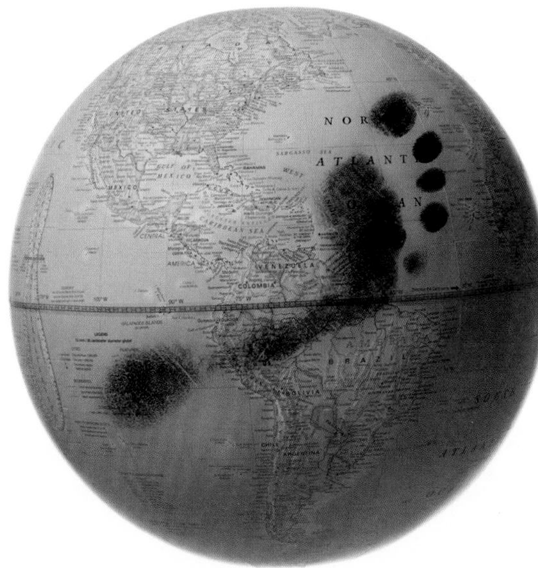

Humans are leaving a massive footprint on the planet. Is this damage reversible?

The term *natural capital* came from ecologically minded economists, and brings with it a value system which implies that resources must have an economic value. The baggage such terms come with encourages a particular view of the world. Terms can therefore influence the way we see the world. The term *ecological footprint* considers the environmental threat of a growing population, whereas *carrying capacity* makes us see the same issues in terms of the maximum number a population can reach sustainably. Does such use of language affect our understanding of concepts and environmental issues? It has been claimed that historians cannot be unbiased – could the same be said of environmental scientists when making knowledge claims? Human carrying capacity of the environment is difficult to quantify and contains elements of subjective judgement.

This topic also offers the opportunity to discuss the models and indicators that are employed to quantify human population dynamics: to what extent are the methods of the human sciences 'scientific'? Do they offer a quantitative assessment or a qualitative one?

> *Your descendants shall gather your fruits.*
>
> **Virgil**

Population and resources

- Population growth is going to use up the world's resources.
- Population growth will stimulate the development of new resources.

Both of these views are valid. Which one do you believe? It may depend on the time scale and spatial scale that you use. On a small time scale, there is evidence that population growth can lead to famine (e.g. in Ethiopia in 1984). However, during the 1984 famine, Ethiopia was exporting crops – not everyone had access to food and that is why there was famine. In addition, there was long-term drought. Human populations have so far managed to survive on Earth, despite massive increases in the size of the human population.

However, population decline in Easter Island suggests that environmental mismanagement could lead to population crashes. Maybe we just haven't been there on a global scale yet.

Case study

Population crash on Easter Island

Easter Island is famous for its statues of heads.

Easter Island was discovered by Europeans in 1722. The island is about 117 km^2 and situated about 3700 km west of the Chilean coast. It is one of the most remote inhabited islands in the world. It was colonized by Polynesian people in AD 700 and the population peaked at 12 000 around 1600. The population is now about 4000.

Pre-1600, the islanders had a diet of birds and fish. But after about 1600, palm forests disappeared and also the supply of birds and fish ran out. There was social disintegration, starvation, hardship, and conflict (Malthusian crisis) in the post-1600 period. The cause of the crisis appears to be total deforestation related to the cult of statue building. Trees were used to move the statues. Removal of the trees led to soil erosion, landslides, crop failures and famine. Thus, it appears to have been a human-made ecological disaster - namely overuse of resources.

However, by 1722 when the island was discovered, there was no sign of such a crisis. The islanders had reorganized their society to regulate their use of resources and control their distribution (Boserup: 'necessity is the mother of invention'). But between 1722 and 1822, the arrival of Europeans led to the spread of disease and the death rate increased. In 1862, slave traders from Peru took 1500 slaves (a third of the population). Only 15 returned home, and these brought smallpox with them. By 1877, the population was down to just 111. The population has now risen to over 4000 largely as a result of migration. Easter Island is now struggling to cope with a new distinction: it was recently named by UNESCO as a World Heritage Site and the pressure caused by tourism is having a negative impact on resource availability for some of the islanders.

Questions

- Is Easter Island an example of a Malthusian population–resource disaster?
- Is it an example of human ingenuity coping with a crisis?
- What do you think?

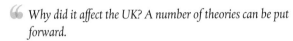

There is a sufficiency in the world for man's need but not for man's greed.

Mohandas K Gandhi

Moral issues

In 1997, an article on agricultural issues in the UK appeared. In a section entitled 'BSE (blame someone else)', the author wrote:

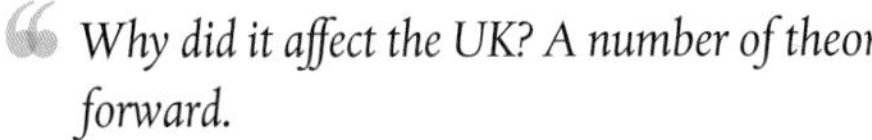

Why did it affect the UK? A number of theories can be put forward.

1. *It could be just bad luck.*
2. *Few places outside the UK suffer from scrapie and also raise large numbers of cattle.*
3. *Cattle carcases in the UK are burned at relatively low temperatures.*
4. *Cattle in the UK derive up to 5 per cent of their cattle ration from meat and bone meal.*

Consumers are worried about whether eating beef is safe. Steak almost certainly is. Muscle (meat) does not seem to carry BSE. The danger lies in eating pies, burgers and sausages which might have bits of infectious brain or spinal cord in them.

The publishers of this article received an irate letter from the National Farmers Union (NFU) in which it was claimed that there was no scientific evidence to suggest that BSE was due to farming practices in the UK, that the author was scaremongering and just trying to be sensationalist in order to get noticed.

As a result, the publishers sent out a letter highlighting the views of the NFU. Should they have done so? Should the author offer an apology? Should the author and publisher publish what they believe to be right? One of the issues here is, where do you draw the line? If an oil company were to say there is no link between burning fossil fuels and global warming, would you believe them? What interests does the NFU have? What interests do the author and the publisher have?

There are four different assessment objectives examined in ESS.

1. **Demonstrate knowledge and understanding**
   - This objective relates to the factual understanding of the course (e.g. concepts, methodologies and techniques, values and attitudes).
2. **Apply knowledge and understanding**
   - In this objective, you are expected to be able to apply the knowledge and understanding gained in Objective 1 in the analysis of different aspects of the course (e.g. explanations, concepts and theories; data and models; case studies in unfamiliar contexts; arguments and value systems).
3. **Evaluate, justify and synthesize material covered by the syllabus**
   - In this objective, you are expected to be able to make an appraisal by weighing up the strengths and limitations of specific aspects of the syllabus (i.e. to evaluate). You are also expected to be able to provide evidence to support or defend a choice, decision, strategy or course of action (i.e. to justify), and to synthesize ideas (i.e. make links and draw conclusions).
   - You may be asked to fulfil this objective with regard to: explanations, theories and models; arguments and proposed solutions; methods of fieldwork and investigation; cultural viewpoints and value systems.
4. **Engage with investigations of environmental and societal issues at the local and global level**
   - This objective is fulfilled by selecting and applying the appropriate research and practical skills necessary to carry out investigations (i.e. carrying out Internal Investigation work); evaluating the political, economic and social contexts of issues; suggesting collaborative and innovative solutions that demonstrate awareness and respect for the cultural differences and value systems of others.

Objectives 1 and 2 address simpler skills; objectives 3 and 4 relate to higher-order skills.

The following table shows how assessment objectives are applied in practice.

Assessment objectives	Which components test these objectives?	How is the assessment objective tested?
1–3	Paper 1 (1 hour, 40 marks) 25% of the total marks	Case study
1–3	Paper 2 (2 hours, 65 marks) 50% of the total marks	Section A: short-answer and data-response questions Section B: two structured essay questions (from a choice of four)
1–4	Internal assessment (10 hours, 30 marks) 25% of the total marks	Individual investigation assessed using mark-bands

The objectives will be tested in the examinations through the use of the command terms (pages 481–482).

Further details about the exams are contained in the chapter Examination strategies (pages 478–480).

The purpose of the Internal Assessment (IA) is to enable you to demonstrate skills and knowledge you have gained during the course, and to pursue a topic that is of personal interest to you. The IA focuses on a particular aspect of an ESS issue and applies the results to a broader environmental and/or social context.

The IA involves the completion of an individual investigation of an ESS research question that you have designed and implemented. The investigation is submitted as a written report. The IA needs to address specific assessment criteria (see below). If you are undertaking an ESS extended essay, it must not be based on the research question of the ESS Internal Assessment.

You are allowed a total of 10 hours to complete the IA. The 10 hours include:

- time for initial explanation of the IA requirements by your teacher
- time to ask questions
- time for consultation with your teacher to discuss the research question before the investigation is carried out, and throughout the execution of the IA
- time to develop the method and collect data
- time to review and monitor progress
- time for final report writing.

You are allowed 10 hours to carry out the IA. This includes time for initial discussions, planning, gathering data, and writing the report.

Successful IAs have research questions that are based on a topic within the syllabus that is of particular interest to you. Each person in your class needs to have a different research question. You need to make sure you carefully follow the IA criteria and hit the marking points that the IB are looking for. You should be aiming for the highest marks possible in your IA as good marks will give you confidence as you approach the exams, and will help support your overall mark. The IA is worth 25% of you final ESS mark, and a good performance can raise you to the next grade if you are borderline between two grades.

Successful IAs have research questions that are based on a topic within the syllabus that is of particular interest to you.

The main problems students encounter with IAs revolve around design (especially suitable sample sizes and sampling techniques), proper treatment of data (this is closely linked to lack of data stemming from poor design), vigorous discussions of the data in a broader context and an analysis of strengths and weaknesses of design. Specific issues concerning each criterion are discussed below.

The report should be 1500 to 2250 words long. External moderators (who check the marks given by your teacher) will not read beyond 2250 words.

The IA report should be 1500 to 2250 words long.

The Internal Assessment investigation consists of:

- identifying an ESS issue and focusing on one of its specific aspects
- developing methodologies to generate data that are analysed to produce knowledge and understanding of this focused aspect
- applying the outcomes of the focused investigation to provide understanding or solutions in the broader ESS context.

The focused research question should arise from a broader area of environmental interest (the context), so that in conjunction with evaluating the research process and findings of your study, you can discuss the extent to which your study applies to the environmental issue that interests you at a local, regional or global level (the application). The discussion should lead you to develop creative thinking and novel solutions, or to inform current political and management decisions relating to the issue. For example, if you carry out a study on the impact of wind turbines that had been erected in the vicinity of your school, you may suggest solutions for the erection of wind turbines in other areas based on your findings.

The following methodologies may be applied:

- values and attitude surveys or questionnaires
- interviews
- issues-based inquiries to inform decision-making
- observational fieldwork (natural experiments)
- field manipulation experiments
- ecosystem modelling (including mesocosms or bottle experiments)
- laboratory work
- models of sustainability
- use of systems diagrams or other valid holistic modelling approaches
- elements of environmental impact assessments
- secondary demographic, development and environmental data
- collection of both qualitative and quantitative data.

The following analytical techniques may be applied to data:

- estimations of NPP/GPP or NSP/GSP (Chapter 2, pages 93–94)
- application of descriptive statistics (measures of spread and average) and inferential statistics (testing of null hypotheses) (Appendix)
- other complex calculations
- cartographic analysis
- use of spreadsheets or databases
- detailed calculations of footprints (including ecological, carbon, water footprints).

Investigations may consist of appropriate qualitative work or quantitative work. In some cases, these are descriptive approaches and may involve the collection of considerable qualitative data. In others, establishing cause and effect through inferential statistical analysis (a scientific approach) may be used.

Internal Assessment criteria

For Internal Assessment, the following assessment criteria will be used. The number of marks allocated to each criterion is indicated in the table.

Internal Assessment criteria

Identifying the context	Planning	Results, analysis, and conclusion	Discussion and evaluation	Applications	Communication	Total
6 (20%)	6 (20%)	6 (20%)	6 (20%)	3 (10%)	3 (10%)	30 (100%)

There are 6 different assessment criteria for the IA:

- **identifying the context**
- **planning**
- **results, analysis and conclusion**
- **discussion and evaluation**
- **applications**
- **communication.**

The maximum number of marks for the IA is 30. It makes up 25% of your final total mark in ESS.

Identifying the context (6 marks)

This criterion assesses the extent to which you establish and explore an environmental issue (either local or global) for an investigation and develop this to state a relevant and focused research question.

Achievement levels for identifying the context

Achievement level	Descriptor
0	The student's report does not reach a standard described by any of the descriptors given below.
1–2	The student's report: • **states** a research question, but there is a lack of focus • **outlines** an environmental issue (either local or global) that is linked to the research question • **lists** connections between the environmental issue (either local or global) and the research question but there are significant omissions.
3–4	The student's report: • **states** a relevant research question • **outlines** an environmental issue (either local or global) that provides the context to the research question • **describes** connections between the environmental issue (either local or global) and the research question, but there are omissions.
5–6	The student's report: • **states** a relevant, coherent and focused research question • **discusses** a relevant environmental issue (either local or global) that provides the context for the research question • **explains** the connections between the environmental issue (either local or global) and the research question.

Planning (6 marks)

This criterion assesses the extent to which you have developed appropriate methods to gather data relevant to the research question. This data could be primary or secondary, qualitative or quantitative, and may use techniques associated with both experimental or social science methods of inquiry. There is an assessment of safety, environmental and ethical considerations where applicable.

Achievement levels for planning

Achievement level	Descriptor
0	The student's report does not reach a standard described by any of the descriptors given below.
1–2	The student's report: • **designs** a method that is inappropriate because it will not allow for the collection of relevant data • **outlines** the choice of sampling strategy but with some errors and omissions • **lists** some risks and ethical considerations where applicable.
3–4	The student's report: • **designs** a repeatable* method appropriate to the research question but the method does not allow for the collection of sufficient relevant data • **describes** the choice of sampling strategy • **outlines** the risk assessment and ethical considerations where applicable.

Achievement level	Descriptor
5–6	The student's report: • **designs** a repeatable* method appropriate to the research question that allows for the collection of sufficient relevant data • **justifies** the choice of sampling strategy used • **describes** the risk assessment and ethical considerations where applicable.

*Repeatable, in this context, means that sufficient detail is provided for the reader to be able to replicate the data collection for another environment or society. It does not necessarily mean repeatable in the sense of replicating it under laboratory conditions to obtain a number of runs or repeats in which all the control variables are exactly the same.

Results, analysis and conclusion (6 marks)

This criterion assesses the extent to which you have collected, recorded, processed and interpreted the data in ways that are relevant to the research question. The patterns in the data must be correctly interpreted to reach a valid conclusion.

Achievement levels for results, analysis and conclusion

Achievement level	Descriptor
0	The student's report does not reach a standard described by any of the descriptors given below.
1–2	The student's report: • **constructs** some diagrams, charts or graphs of quantitative and/or qualitative data, but there are significant errors or omissions • **analyses** some of the data but there are significant errors and/or omissions • **states** a conclusion that is not supported by the data.
3–4	The student's report: • **constructs** diagrams, charts or graphs of quantitative and/or qualitative data which are appropriate but there are some omissions • **analyses** the data correctly but the analysis is incomplete • **interprets** some trends, patterns or relationships in the data so that a conclusion with some validity is deduced.
5–6	The student's report: • **constructs** diagrams, charts or graphs of all relevant quantitative and/or qualitative data appropriately • **analyses** the data correctly and completely so that all relevant patterns are displayed • **interprets** trends, patterns or relationships in the data, so that a valid conclusion to the research question is deduced.

Discussion and evaluation (6 marks)

This criterion assesses the extent to which you discuss the conclusion in the context of the environmental issue, and carry out an evaluation of the investigation.

Achievement level	Descriptor
0	The student's report does not reach a standard described by any of the descriptors given below.
1–2	The student's report: • **describes** how some aspects of the conclusion are related to the environmental issue • **identifies** some strengths, weaknesses and limitations of the method • **suggests** superficial modifications and/or further areas of research.
3–4	The student's report: • **evaluates** the conclusion in the context of the environmental issue but there are omissions • **describes** some strengths, weaknesses and limitations within the method used • **suggests** modifications and further areas of research.
5–6	The student's report: • **evaluates** the conclusion in the context of the environmental issue • **discusses** strengths, weaknesses and limitations within the method used • **suggests** modifications addressing one or more significant weaknesses with large effect and further areas of research.

Achievement levels for discussion and evaluation

Applications (3 marks)

This criterion assesses the extent to which you identify and evaluate one way to apply the outcomes of the investigation in relation to the broader environmental issue that was identified at the start of the project.

Achievement level	Descriptor
0	The student's report does not reach a standard described by any of the descriptors given below.
1	The student's report: • **states** one potential application and/or solution to the environmental issue that has been discussed in the context • **describes** some strengths, weaknesses and limitations of this solution.
2	The student's report: • **describes** one potential application and/or solution to the environmental issue that has been discussed in the context, based on the findings of the study, but the justification is weak or missing • **evaluates** some relevant strengths, weaknesses and limitations of this solution.
3	The student's report: • **justifies** one potential application and/or solution to the environmental issue that has been discussed in the context, based on the findings of the study • **evaluates** relevant strengths, weaknesses and limitations of this solution.

Achievement levels for applications

Communication (3 marks)

This criterion assesses whether the report has been presented in a way that supports effective communication in terms of structure, coherence and clarity. The focus, process and outcomes of the report are all well presented.

Achievement levels for communication

Achievement level	Descriptor
0	The student's report does not reach a standard described by any of the descriptors given below.
1	The investigation has limited structure and organization: • the report makes limited use of appropriate terminology and it is not concise • the presentation of the report limits the reader's understanding.
2	The report has structure and organization but this is not sustained throughout the report: • the report either makes use of appropriate terminology or is concise • the report is mainly logical and coherent, but is difficult to follow in parts.
3	The report is well-structured and well-organized: • the report makes consistent use of appropriate terminology and is concise • the report is logical and coherent.

Read these pages carefully – they provide advice for carrying out the IA successfully and obtaining good marks.

Advice for your IA

Identifying the context

You need to state explicitly why you have selected the research question chosen, and your personal interest in the topic.

Planning

- Many students lose marks for not knowing the difference between independent, dependent and controlled variables. You need to clearly identify these variables in your report: the terminology is not compulsory, and some students refer to the variable that they will manipulate and that which will respond, and those that will be held constant. The concepts of 'control' and 'control variables' are often confused: control variables are required for a fair test, where only one variable is changed (the independent variable) and the rest kept the same (the control variables). The dependent variable is the one you are measuring. A 'control' refers to an experiment where the independent variable is removed, so that the scientist can see what happens when the factor that they think is having the effect is taken away – this proves that the independent variable is the one having the effect rather than other factors. In ecological IAs, it may not be possible to control other variables as you will be working out-of-doors where conditions vary: in these investigations, you need to say that you will monitor other variables that may affect your dependent variable.
- When explaining your sampling method, you must outline how those samples are to be selected, ensuring that there is no significant bias. You need to be able to develop a method that results in a 'fair' test or one in which reasonable attempts have been made to remove bias. For example, a practical that includes sampling of quadrats (pages 135–136) should include some description of how these are to be selected. It is

not sufficient simply to indicate that quadrats were selected randomly – the method to ensure randomness should be outlined. If you are comparing germination of plants under different salinity conditions, for example, the method should indicate how temperature, moisture, and other variables are being controlled in order to ensure that the results are comparable.

- Most students succeed in obtaining data that are relevant to the question or topic that is being studied, but lose marks by collecting data insufficient in quantity. Normally, *five* is the minimum number of samples required per site, treatment, repetition, and so on. For example, if you are measuring changes in rate of oxygen release with respect to light intensity in *Elodea*, it would be expected that you would take at least five readings for each light intensity. Lack of sufficient data can have knock-on effects for other marking criteria; for example, if only a single measurement is collected per treatment, the data does not lend itself for processing and by extension is not suitable for the presentation of processed data. When carrying out ecological fieldwork, time constraints can be problematic, and it is appreciated that in these cases you may need to collect fewer than five samples or transects. In such cases, three may be acceptable, but you need to explain in your report about the time constraints you encountered.

Results, analysis and conclusion

Students often *lose* marks for the following errors.

- Tables and graphs are not labelled correctly. Tables should have an adequate title and appropriate headings are needed. Axes of graphs should be labelled and units included.
- Putting units in the cells of a table and not in the column or row headings where they should be.
- Reporting data to a varying number of decimal places within the same column or row. In a table, for example, the temperature data and dissolved oxygen data may have different numbers of decimal places, but all the temperature readings must have the same number of decimals.
- It is not possible to carry out a good analysis when there is insufficient data. If the design calls for one pH sample from each of five locations in a stream, then there is no significant analysis that can be carried out with these data and therefore you are likely to perform poorly. Five repeats at each site would have been necessary for good data analysis.
- Data are often unprocessed. It is expected that you do something with your data (e.g. calculate indices, averages, standard deviations, and so on). Statistical techniques such as Chi squared, regressions, *t*-test can also be done: although these are not specifically required, they do provide a way to achieve full marks in this aspect, although to achieve full marks they must be done well (Appendix).
- When processing data, accuracy is sometimes increased through mathematical means. Processed data should be to the same level of accuracy as raw data. If a mean is calculated from numbers with two decimal places, for example, this should not be reported to four decimal places.

Presentation of processed data usually takes the form of a scatter plot, pie chart, histogram, bar chart, or some other method of visually portraying the analysed data. Do not just plot unprocessed/raw data (e.g. if you take temperature readings at 10 different sites on a river, do not just draw a graph of these – plot mean values).

The data you collect must be recorded at the level of accuracy made possible by the precision of the equipment you are using. For example, if plant lengths are measured with a ruler that reports millimetres, the average of these data should not be reported to 8 decimal places in the data tables, but rather to the nearest millimetre. Similarly, if a light meter records to two decimal places, then this is the level of accuracy that should be used when calculating means.

In the conclusion, marks are often lost by not being specific enough. You should cite your data in your conclusions (e.g. if you conclude that in a study of soil moisture along a slope, there is a trend towards increasing moisture down slope, this should be illustrated with the actual data. There should be a brief explanation as well: for example, 'The increase in soil moisture down the slope may be due to run-off and infiltration.'

Discussion and evaluation

- The best reports cite literature, indicate how close data is to what might be expected, contain discussion about why data did not support theory, and include comments about the relative reliability of the data. Calculation of standard deviations allows discussion about the reliability of the data. Although it is not intended that the discussion should turn into a dissertation of several pages, there does need to be a critical look at the quality of the data and how it relates to what is known.
- A good discussion should identify patterns in the data (or comment on their absence), place the research in a context that relates it to theory and/or research, and assess the quality of the data generated. This is much easier to do if the planning and results sections have been done well. If the research question is tightly focused, and there is sufficient data to address the question, then a discussion is more likely to produce interesting insight. For example, if you have carried out a study of the relationship between temperature and dissolved oxygen at sites above and below a pollution source, you should address the quality of the data. Is it reliable? Why, or why not? This is where having means and standard deviations can be useful. Standard deviation (which can easily be worked out on a scientific calculator) shows the variation in the data: if there is a very large standard deviation, you would be expected to comment on this fact and interpret it (i.e. large variation means that the data are less reliable).
- The discussion should be thought provoking and will almost certainly be the most challenging (and perhaps lengthiest) part of the report. Are there important differences among the data? Are there trends? Do these trends support/refute accepted theory? Are the standard deviations in the data so huge as to make differences meaningless? Are there anomalies in the data? These should be discussed, and if they are to be ignored or excluded from the analysis, a case for this decision should be made. Were the samples collected without significant bias? Are there literature values that can be used for comparison? If there are, these should be mentioned. If these are non-existent or unavailable, a note to this effect should be included.
- You need to look at your method critically and offer improvements. Many students, however, miss the most obvious improvement (i.e. collection of more data, repeating the experiment, and calculation of averages). Potential marks are generally lost by making suggestions that are either too simple or unrealistic. In the evaluation, data quality issues that may have been noted in the discussion should be addressed. Was the standard deviation very high? How can it be reduced? Is the data representative? If not, how can that be addressed? What improvements will address the issues that have been identified? All these questions should be answered in this section of the report.

Applications

Make sure you identify and evaluate one way to apply the outcomes of the investigation in relation to the broader environmental issue. You need to justify one potential application and/or solution to the environmental issue based on the findings of the study, and evaluate the relevant strengths, weaknesses and limitations of this application/solution.

Communication

Make sure that your report is well structured and well organized. The report should make consistent use of appropriate terminology and be concise, follow a logical order, and be clearly written.

The Extended Essay is an in-depth study of a focused topic that promotes intellectual discovery, creativity and writing skills. It provides you with an opportunity to explore and engage with an academic idea or problem in your favourite International Baccalaureate diploma subject. It will develop your research skills (something needed at universities and tertiary education in general), and provide you the opportunity to produce an individualized and personal piece of work. The essay is a major piece of structured writing that is formally presented. Many students find the Extended Essay a valuable stimulus for discussion in interviews for university or employment. You are expected to spend approximately 40 hours on the project, and the finished piece of work should be no more than 4000 words.

The essay should be no more than 4000 words. You are expected to spend approximately 40 hours on the essay.

You will have a supervisor for your essay, who will help you to decide on a suitable topic, and check that you are keeping to the timing and regulations. You can meet with your supervisor for 3 to 5 hours over the course of your essay: several short meetings are recommended (e.g. once a fortnight for 20 minutes) rather than a few long ones, as this will help you and your supervisor to exchange ideas and feedback on a regular basis. Your supervisor will:

- give you a copy of the assessment criteria and subject specific details (available from the IB)
- give you advice on the skills of undertaking research
- help you with shaping your research question and the subsequent structure and content of your essay
- give you examples of excellent Extended Essays
- read and comment on your work (but cannot edit it)
- give you advice on the format of the bibliography, the abstract and referencing
- conduct a short concluding interview (*viva voce*) with you once you have finished the essay.

To make the most of your Extended Essay, you need to make sure you:

- undertake the work agreed by you and your supervisor
- keep appointments and deadlines
- are honest about your progress and any problems you may be facing
- pace yourself so you do not have a lot of work at the last minute.

The maximum number of marks for the Extended Essay is 36. It is marked using 11 different assessment criteria.

The Extended Essay is marked according to certain assessment criteria which you will receive from your supervisor. The maximum total number of marks you can receive is 36, and you can see from the criteria how many marks are allotted to each aspect of your Essay (there are 11 different marking criteria, pages 475–477).

Bibliography (references)

To learn more about how to cite sources in the *Chicago Manual of Style* format, go to www.pearsonhotlinks.co.uk, enter the book title or ISBN, and click on weblink 10.1.

A bibliography should be in alphabetical order according to author's surname. It should be on a separate sheet of paper with a title at the top. You must make sure you consider whether any source you use is likely to be reliable; this is especially true for internet resources, where there are relatively few quality controls. If at all possible, you should include the name of the author of any article from the web. You should also include the date a web page was accessed. The references in your Extended Essay should follow the *Chicago Manual of Style* (Footnotes).

Detail specific to Environmental systems and societies

An Extended Essay in Environmental systems and societies (ESS) will provide you with the opportunity to explore an environmental topic or issue of particular interest to you and your locality. As this is an interdisciplinary subject, you will need to integrate theory from the course with practical methodologies which are relevant to your topic. A systems approach is particularly effective and, as this is something emphasized throughout the course, this should be familiar to you. You will be expected to show appreciation and use of this approach in the analysis and interpretation of the data gathered.

The course focuses on the interaction and integration of natural environmental systems and human societies (i.e. includes aspects of both group 3 and group 4), and your essay needs to achieve this as well. It should not deal exclusively with ecological processes or with societal activities, but instead should give significant (though not necessarily equal) weight to both these dimensions. For example, the ESS syllabus includes the study of pure ecological principles, but an Extended Essay would have to explore ecological principles within the context of some human interaction with an environmental system. A specific natural system needs to be studied, rather than general systems that have been covered in the course.

An ESS Extended Essay must integrate aspects of both group 3 and group 4.

If you aim to obtain largely descriptive or narrative data, of the type produced in the human sciences, a group 3 essay may be more appropriate. If you want to collect quantitative data typical of the experimental sciences, then a group 4 essay may be more appropriate. An ESS essay *must* cover both group 3 and group 4 criteria and be fully transdisciplinary in nature.

A crucial feature of any suitable topic is that it must be open to analytical argument. For example, rather than simply describing a given nature reserve, you would need to evaluate its relationship with a local community, or compare its achievement with original objectives, or with a similar initiative elsewhere. The topic must leave room for you to be able to form an argument that you both construct and support, using analysis of your own data, rather than simply reporting analysed data obtained from other sources. Certain topics should be avoided for ethical or safety reasons (e.g. experiments likely to inflict pain on living organisms, cause unwarranted environmental damage, or put pressure on others to behave unethically). Experiments that pose a threat to health (e.g. using toxic or dangerous chemicals, or putting oneself at physical risk during fieldwork) should also be avoided unless adequate safety apparatus and qualified supervision are available.

Your research question must be open to analytical argument.

Focus

Essential to a successful Extended Essay is the focus of the topic chosen. If a topic is too broad, it can lead you into superficial treatment and it is unlikely you will be able to produce any fresh analysis, or novel and interesting conclusions of your own. So, for example, topics on the left of Table 1 are better than topics on the right.

An essay with a sharply focused research question will be more successful than one that has a topic that is too broad.

Table 1 Focusing the topic of your essay

Focused	Unfocused
The ecological recovery of worked-out bauxite quarries in Jarrahdale, Western Australia	Environmental effects of mining
A comparison of the energy efficiency of grain production in the Netherlands and Swaziland	Efficiency of world food production
The comparative significance of different sources of carbon dioxide pollution in New York and Sacramento	Impacts of global warming
Managing the environmental impact of paper use at a Welsh college	Paper recycling

Topics with a sharper focus enable you to channel your research to produce interesting and original conclusions and discussions. A short and precise statement outlining the overall approach of your investigation is also helpful in determining the focus of your essay, and making sure you stick to it. For example, if your topic is an examination of the ecological footprint of your school canteen, the research question could be: From the major inputs and outputs of the school canteen, what overall estimate of its environmental impact can be made in terms of an ecological footprint? The approach would include an analysis of the records and practical measurements that assess the inputs and outputs of the canteen, and an analysis of data into a holistic environmental footprint model that indicates environmental impact. For some investigations, particularly those that are experimental, a clearly stated hypothesis may be just as acceptable as, and possibly better than, a research question.

An Extended Essay in ESS may be investigated either through primary data collection (i.e. from fieldwork, laboratory experimentation, surveys or interviews) or through secondary data collection (i.e. from literature or other media).

An Extended Essay in ESS may be investigated either through primary data collection (i.e. from fieldwork, laboratory experimentation, surveys or interviews) or through secondary data collection (i.e. from literature or other media). It may even involve a combination of the two. However, given the limited time available and the word limit for the essay, the emphasis should be clearly with one or the other to avoid the danger of both becoming superficial. Experience shows that data based on questionnaires and interviews are to be avoided, as such data is difficult to analyse and conclusions difficult to arrive at. Fieldwork and lab experiments are a more reliable way to gather data for your essay.

If the essay is focused largely on the collection of primary data, you must check carefully with the literature to make sure you select the most appropriate method for obtaining valid quantitative data. You must ensure you reference these secondary sources of information in your bibliography. If the essay is focused on secondary data, you need to take great care in selecting sources, ensuring that there is a sufficient quantity and range, and that they are all reliable. The internet and other media contain a great many unfounded and unsupported claims that you need to be wary of – checking information from several different sources will help you evaluate its value and accuracy. You must sort through your sources and use only those that have some academic credibility. For an essay of this type, you are expected to produce a substantial bibliography and not be limited to just a few sources.

Once you have assembled your data, you must produce your own analysis and argue your own conclusions. This will happen more naturally if the essay is based on primary data since such data has not been previously analysed. A source of secondary data may come with its own analysis and conclusions. If you use secondary data, it is essential that you further manipulate it, or possibly combine it with other sources, so that there is clear evidence in the essay of your personal involvement in analysis and drawing of conclusions. You are expected to be academically honest in your essay – plagiarism (direct copying) is a very serious matter which the IBO deals with severely.

A central theme in the syllabus for ESS is the systems approach. This should be reflected to some degree in your essay which should include an attempt to model, at least partially, the system or systems in question. The term *model* can be applied in its broadest sense to include, for example, mathematical formulae, maps, graphs and flow diagrams. Systems terminology (e.g. input, output, processes) should be used where appropriate.

Assessment criteria

Read the advice on these pages carefully. Check your essay to see that you have fulfilled all criteria.

Your essay will be marked according to specific assessment criteria. There are 11 different criteria. The maximum total number of marks you can receive is 36. For each criterion, the number of marks allocated and the details that should be covered in relation to ESS are outlined below. In each case, details are presented as a series of questions to help you check you have fully realized each one.

Criterion A: Research question (2 marks)

- Do you have a sharply focused research question clearly defining the purpose of the essay?
- Have you considered formulating the research question as a clearly stated hypothesis? This is especially appropriate in experimental investigations.
- Does the hypothesis lead to clear critical arguments concerning the extent to which your results will support or argue against it?

Criterion B: Introduction (2 marks)

- Does the introduction set the research question in context?
- Does the introduction give the reader a sense of why the question is worth asking?
- Does the introduction outline theoretical principles underlying the research question (e.g. what topics of theory does the research question explore)?
- Have you outlined the history or geography of any location you will be studying that is central to the issue under discussion?

Criterion C: Investigation (4 marks)

- If the essay involves experimentation or practical fieldwork, have you included a detailed description of your methods, ideally including diagrams and photos?
- Have you included details of the experimental design, including quantification, and a description of all variables being considered, including control variables, replication, and random sampling, where appropriate?
- Have you explained the selection of techniques, and justified them? Have you clearly stated any assumptions on which they depend?
- If secondary data has been used, have you ensured that the sources are reliable? Have you referenced sources? Have you indicated how the secondary data was generated?

Criterion D: Knowledge and understanding of the topic studied (4 marks)

- Have you demonstrated sound understanding of the ESS course, and shown that you have read beyond the syllabus and carried out your own independent study?
- Have you shown sufficient knowledge of the topic, and shown that you have handled the issues and arguments effectively?

- Have you shown links between your study and previous work from references you have found? Have you demonstrated use of theoretical knowledge to underpin your essay?

Criterion E: Reasoned argument (4 marks)

- Does your essay show a clear, step-by-step, logical argument linking the raw data to the final conclusions?
- Is each step defended and supported with evidence?

Criterion F: Application of analytical and evaluative skills appropriate to the subject (4 marks)

- Have you used analytical skills to manipulate and present your data?
- Have you evaluated your data and commented on its reliability and validity?
- Have you included a model of the system studied and used the correct terminology when discussing it?

Criterion G: Use of language appropriate to the subject (4 marks)

- Have you used terminology appropriate to ESS throughout your essay? Both scientific and systems terminology should be used. A systems approach is especially important (Chapter 1).

Criterion H: Conclusion (2 marks)

- Have you separated the conclusion with its own heading within your essay?
- Does your conclusion contain a brief summary of the direct conclusions of your research question or hypothesis, supported by evidence and arguments already presented? You conclusion should not contain new evidence or discussion.
- Have you identified any outstanding gaps in your research or new questions that have arisen which you think would deserve further attention?

Criterion I: Formal presentation (4 marks)

- Have you checked the assessment criteria to make sure that you have met all the formal requirements for the Extended Essay?
- Have you labelled all your graphs, diagrams, illustrations, and tables of data with a figure or table number, a title, and a citation where appropriate? Have you placed all your figures and tables in the body of the essay, as close as possible to their first reference?

The abstract does not count towards the 4000-word limit for the essay. The abstract can be up to 300 words long. You are allowed up to 4000 words for the essay plus up to 300 for the abstract.

Criterion J: Abstract (2 marks)

- Have you included a brief summary of the essay? The abstract is judged on the clarity of the overview it presents, not on the quality of the research, arguments, or conclusions.

Criterion K: Holistic judgement (4 marks)

- Have you demonstrated personal engagement, initiative and insight in your topic?

- Have you stressed the inter-relatedness of systems and their components? An essay that recognizes these underlying principles and the inter-relatedness of components will most clearly demonstrate an element of the 'insight and depth of understanding' referred to in this criterion.

At the end of your course, you will sit two exam papers: Paper 1 and Paper 2.

Paper 1 is worth 25 per cent of the final marks and Paper 2 is worth 50 per cent of the final marks. The other assessed part of the course (25 per cent) is the Internal Assessment (practical work – see pages 463–471), which is marked by your teacher.

There are no options in ESS and therefore all topics need to be thoroughly revised for both papers. Here is some general advice for the exams.

- The applications and skills listed in your syllabus use specific command terms that let you know the approach needed in exams (e.g. evaluate, explain, outline, discuss). Make sure you have learned the command terms. Questions which use the command term 'compare and contrast' require you, for example, to relate the *similarities* as well as *differences* between two sets of data. If you are asked to 'discuss' you should identify and present at least two alternative views; if you are asked to 'list three factors' you will not gain any extra credit for listing more than three factors. There is a tendency to focus on the content in an exam question rather than the command term, but it is essential that your answer addresses what the command term is asking of you (command terms are listed on pages 481–482).
- Answer all questions and do not leave gaps.
- Do not write outside the answer boxes provided to answer a question – if you do so, this work will not be marked. Your answers will be scanned and only the material within the boxes is sent to examiners. If you run out of room on the page, use continuation sheets and indicate clearly that you have done this at the end of the question box and on the cover sheet. (The fact that the answer continues on another sheet of paper needs to be clearly indicated in the text box provided.)
- Plan your time carefully before the exams – make sure you have time to revise all topics and to practise past papers.

Paper 1

Paper 1 (1 hour) focuses on a case study. The case study will be unfamiliar to you, and you are expected to apply the knowledge and understanding gained during the course to this new scenario. You will be given a range of data in various forms (e.g. maps, photos, diagrams, graphs and tables). You are required to answer a series of questions, which can involve a variety of command terms, by analysing these data. The total number of marks for this paper is 40.

There are plenty of case studies available from the ESS course (up to 2016 these were part of Paper 2, with data presented in a Resource booklet) – practise as many of them as you can. Remember that because the case studies test knowledge from the whole course, you will not be able to properly tackle a full Paper 1 until you have completed the course.

In the exam, make sure you read the case study carefully. You will be given 5 minutes reading time before the exam begins – use this time to look through each figure and to read the questions.

Paper 2

Paper 2 (2 hours) contains short answers and structured essays. The total number of marks for this paper is 65, with questions covering the whole breadth of the syllabus.

- The short answer section (Part A) carries only 25 marks: the questions cannot therefore cover all aspects of the syllabus. However, it is essential that you thoroughly revise the whole syllabus so that you can tackle any question that comes up.

- Part A also asks you to answer data-based questions. Data-based questions present you with information in some form and ask you questions about it. Some questions will ask you to read the data displayed and some will ask you to draw conclusions from it. You are expected to use the data provided in the question. Try to get into the habit of using the data when you practise data-based questions. This will make it natural to do the same when sitting the exam. Become familiar with unit expressions such as kJ m^{-2} yr^{-1} (read as kilojoules per metre squared per year). If you are not comfortable with the unit expressions you see in data-based questions, see your teacher for help.
- The size of the answer boxes, and number of marks available, give you an indication of the length of answer expected – make sure your answers are concise. If 3 marks are awarded, the examiner will be looking for three different points. Make sure you do not contradict yourself.
- Part B requires you to answer two structured essay questions from a choice of four. Each question is worth 20 marks.
- The final part of each essay in section B (9 marks) will be marked using mark-bands. Here are the descriptions for these mark-bands.

Marks	Description of details included in the answer
0	The response does not reach a standard described by the descriptors below and is not relevant to the question.
1–3	The response contains: • minimal evidence of knowledge and understanding of ESS issues or concepts • fragmented knowledge statements poorly linked to the context of the question • some appropriate use of ESS terminology • no examples where required, or examples with insufficient explanation/ relevance • superficial analysis that amounts to no more than a list of facts/ideas • judgements/conclusions that are vague or not supported by evidence/ argument.
4–6	The response contains: • some evidence of sound knowledge and understanding of ESS issues and concepts • knowledge statements effectively linked to the context of the question • largely appropriate use of ESS terminology • some use of relevant examples where required, but with limited explanation • clear analysis that shows a degree of balance • some clear judgements/conclusions, supported by limited evidence/ arguments.
7–9	The response contains: • substantial evidence of sound knowledge and understanding of ESS issues and concepts • a wide breadth of knowledge statements effectively linked with each other, and to the context of the question • consistently appropriate and precise use of ESS terminology • effective use of pertinent, well-explained examples, where required, showing some originality • thorough, well-balanced, insightful analysis • explicit judgements/conclusions that are well-supported by evidence/ arguments and that include some critical reflection.

- The examiner will decide which band your answer fits best, and whether it is at the top, middle, or bottom or the mark-band. Marks will be awarded accordingly.

Make sure you plan your strategy for the paper before you sit it: how much time will you spend on section A, and how much time on the essays (section B)? Practising past papers will help you work out how much time you need to take on each of the sections. This will vary from student to student, but here is some general advice.

- Some students will want to answer the essays first, and some the short answer questions. Tackle the paper in whichever order suits you best.
- The essays need to be thought about carefully and planned – aim to spend a *minimum* of 35 minutes per essay but to move on if you are still on the first one after 40 minutes.
- Choose your essays carefully. Look at all sections of an essay before making your choices. There are usually several sections in an essay question – make sure you answer all parts.
- Some students write pages on sections in essays worth only a few marks, and then run out of time later on. Look carefully at the number of marks available for each part of the question and adjust the amount of time you spend on that question accordingly. Writing a plan for your essays will help you.
- Using case studies you have learned during the course will help you answer Paper 2 essay questions (examples are given in this book). You should use your own case studies to answer the essay questions rather than taking ideas from the case study used in Paper 1.
- Essays should be subdivided into sections, not written as one long paragraph – examiners like this because it makes the paper easier to read and mark.
- Leave at least one line between sections of an essay for clarity, and note on your scripts if a continuation sheet has been used.

Tips for exams

Remember these tips.

Do not write outside the box provided for your answer. If you require extra room, use an extension booklet and indicate this in your answer.

- The examiner does not know you. You must communicate fully what you know and not expect the examiner to do your thinking for you.
- State the obvious in your answers. Many of the items in a mark-scheme will be information that is very basic in relation to the question.
- Do not use abbreviations that may be unfamiliar to someone else. Always use the full words first and put the abbreviations in brackets. Be clear and concise with your choice of words.
- If your handwriting is very small or unclear, print your response. If the examiner cannot read your writing, you will not get any marks.
- Make sure to use extra paper if you need it.

Remember that the written papers form only part of the overall assessment. The Internal Assessment is graded by your teacher and moderated by an examiner.

During the exam use your time appropriately:

- read each question twice before beginning to write
- plan your time – allocate time according to the number of marks per question.
- If you have time at the end, re-read your answers and make sure you have said exactly what you want to say.

Command terms

Make sure you learn, and can apply, command terms. Command terms indicate the depth of treatment required for a given assessment statement. It is essential that you are familiar with these terms, for both Papers 1 and 2, so that you are able to recognise the type and depth of response you are expected to provide. Command terms are grouped according to the different assessment objectives: objectives 1 and 2 address simpler skills; objectives 3 and 4 relate to higher-order skills.

Objective 1

Define: Give the precise meaning of a word, phrase, concept or physical quantity.

Draw: Represent by means of a labelled, accurate diagram or graph, using a pencil. A ruler (straight edge) should be used for straight lines. Diagrams should be drawn to scale. Graphs should have points correctly plotted (if appropriate) and joined in a straight line or smooth curve.

Label: Add labels to a diagram.

List: Give a sequence of brief answers with no explanation.

Measure: Obtain a value for a quantity.

State: Give a specific name, value or other brief answer without explanation or calculation.

Objective 2

Annotate: Add brief notes to a diagram or graph.

Apply: Use an idea, equation, principle, theory or law in relation to a given problem or issue.

Calculate: Obtain a numerical answer showing the relevant stages of working.

Describe: Give a detailed account.

Distinguish: Make clear the differences between two or more concepts or items.

Estimate: Obtain an approximate value.

Identify: Provide an answer from a number of possibilities.

Interpret: Use knowledge and understanding to recognize trends and draw conclusions from given information.

Outline: Give a brief account or summary.

Objectives 3 and 4

Analyse: Break down in order to bring out the essential elements or structure.

Comment: Give a judgement based on a given statement or result of a calculation.

Compare and contrast: Give an account of similarities and differences between two (or more) items or situations, referring to both (all) of them throughout.

Construct: Display information in a diagrammatic or logical form.

Deduce: Reach a conclusion from the information given.

Demonstrate: Make clear by reasoning or evidence, illustrating with examples or practical application.

Derive: Manipulate a mathematical relationship to give a new equation or relationship.

Design: Produce a plan, simulation or model.

Determine: Obtain the only possible answer.

Discuss: Offer a considered and balanced review that includes a range of arguments, factors or hypotheses. Opinions or conclusions should be presented clearly and supported by appropriate evidence.

Evaluate: Make an appraisal by weighing up the strengths and limitations.

Explain: Give a detailed account, including reasons or causes.

Examine: Consider an argument or concept in a way that uncovers the assumptions and interrelationships of the issue.

Justify: Provide evidence to support or defend a choice, decision, strategy or course of action.

Predict: Give an expected result.

Sketch: Represent by means of a diagram or graph (labelled as appropriate). The sketch should give a general idea of the required shape or relationship, and should include relevant features.

Suggest: Propose a solution, hypothesis or other possible answer.

To what extent: Consider the merits or otherwise of an argument or concept. Opinions and conclusions should be presented clearly and supported with appropriate evidence and sound argument.

Mathematical requirements

You will be expected to be able to carry out the following mathematical analysis, in both Paper 1 and Paper 2.

- Perform the basic arithmetic functions: addition, subtraction, multiplication and division.
- Carry out calculations involving means, decimals, fractions, percentages, ratios, approximations and reciprocals.
- Use standard notation (for example, 3.6×10^6).
- Use direct and inverse proportion.
- Solve simple algebraic equations.
- Plot graphs (with suitable scales and axes) including two variables that show linear and non-linear relationships.
- Interpret graphs, including the significance of gradients, changes in gradients, intercepts and areas.
- Interpret data presented in various forms (for example, bar charts, histograms and pie charts).

There are usually several marks for mathematical analysis in both Paper 1 and Paper 2 (section A). Make sure you bring an appropriate calculator to each exam (your teacher will tell you which type you are allowed to use).

T